T0183541

Springer Undergraduate Texts in Mathematics and Technology

For other titles published in this series, go to
http://www.springer.com/series/7438

Vladimir Rovenski

Modeling of Curves and Surfaces
with MATLAB®

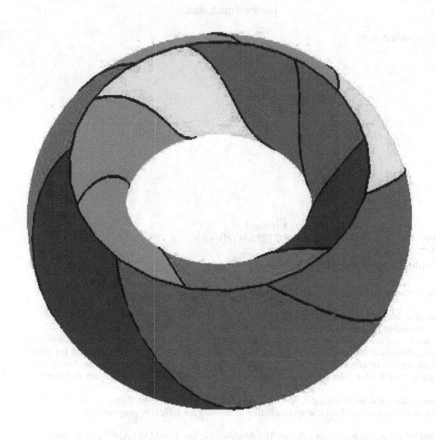

Springer

Vladimir Rovenski
Department of Mathematics and Computer Science
University of Haifa
Mount Carmel, Haifa 31905
Israel
rovenski@math.haifa.ac.il

Series Editors
Jonathan M. Borwein, FRSC
Professor Laureate
Director Centre for Computer Assisted Research
  Mathematics and its Applications, CARMA
School of Mathematical & Physical Sciences
University of Newcastle
Callaghan NSW 2308
Australia
Jonathan.Borwein@newcastle.edu.au

Helge Holden
Department of Mathematical Sciences
Norwegian University of Science and Technology
Alfred Getz vei 1
NO-7491 Trondheim
Norway
holden@math.ntnu.no

ISSN 1867-5506   ISSN 1867-5514 (eBook)
ISBN 978-1-4939-3896-4   ISBN 978-0-387-71278-9 (eBook)
DOI 10.1007/978-0-387-71278-9
Springer New York Dordrecht Heidelberg London

Mathematics Subject Classification (2010): 51Nxx, 65Dxx, 68Uxx

MATLAB® and Simulink® are registered trademarks of The MathWorks, Inc. For MATLAB® and Simulink®
product information, please contact: The MathWorks, Inc., 3 Apple Hill Drive, Natick, MA, 01760-2098 USA,
E-mail: info@mathworks.com, Web: www.mathworks.com.

Printed on acid-free paper

Springer is part of Springer Science+Business Media (www.springer.com)

Dedicated to Irina, my wife, and to my daughter Nadya. I am greatly indebted to them for their encouragement and support.

Foreword

The textbook *Modeling of Curves and Surfaces with MATLAB®* by Professor Vladimir Rovenski contains interesting geometrical topics with examples and exercises which illustrate solution techniques in numeric and symbolic computing and visualizing the objects. Covering many aspects of geometry and algebra, this book exposes readers to geometrical concepts through the use of a modern computing tool — MATLAB®.

This work is based on the author's previous book *Geometry of Curves and Surfaces with MAPLE*, but it is a greatly expanded version with new chapters and excellent chosen themes.

The book is organized in two parts. The first part covers basic topics: graphs of functions and transformations of space, classical polyhedra and non-Euclidean geometries. The second part treats curves and surfaces from the discrete flavor of the first part to the area of classical analysis including approximation and fitting problems.

This new edition is aimed at advanced undergraduate and postgraduate students, and can be recommended to engineers and applied mathematicians who are interested in mathematical modeling and visualization.

Irina Albinsky, Ph.D.
Computer Sciences and Modeling Specialist

Preface

This text on geometry modeling is devoted to a number of central geometrical topics — graphs of functions, transformations, (non-)Euclidean geometries, curves and surfaces — and presents some elementary methods for analytical modeling and visualization of them.

In 1872 F. Klein proposed his Erlangen Programme in which he suggested that different geometries can be studied by the properties of the groups of transformations acting on the geometry. The following geometries are represented in this way: Euclidean, affine, projective, inversive, spherical, and hyperbolic.

B. Riemann's (1868) idea was to represent geometries by a metric (differential) form in a curvilinear coordinate system. The distance between points is measured as the minimum length of curves (calculated using the metric form!) joining them. The intrinsic geometry of a surface in space and Euclidean, spherical, and hyperbolic geometries are represented in this way.

We systematically examine such powerful tools of MATLAB® as transformations and projections of geometric images using matrices, and then study more complex geometrical modeling problems related to curves and surfaces. MATLAB® is becoming the most popular software package in engineering and mathematical education. The symbolic capability of the latest version of MATLAB® uses a part of the MuPAD engine (MAPLE prior to 2008). The aim of the book is to help the reader to develop experience in visualization of objects and computations with MATLAB®. Throughout this book the latest version of MATLAB® at the time of writing is assumed — Version 7.8 Release 2009a.

This book with MATLAB is based on the author's textbook *Geometry of Curves and Surfaces with Maple* (Birkhäuser, 2000). The text has been revised throughout and complemented with new chapters and topics on using complex numbers, quaternions, matrices and transformations, hyperbolic geometry in the upper half-space, spline surfaces, etc. It contains basic theory and more practical material, examples and exercises.

The interested reader can find further theory and additional reading on geometry and using MATLAB in the SUMS series textbooks and other books in the bibliography. The book is illustrated with a number of figures producible using MATLAB.

Key features:

- Aimed at a broad audience of advanced undergraduate and postgraduate students, engineers and computer scientists, instructors of applied mathematics, this book will be an excellent classroom resource or self-study reference.
- The book demonstrates the potential of symbolic/numeric computational tools to support the development of solutions using MATLAB.
- With a number of stimulating exercises, problems and solutions, *Modeling of Curves and Surfaces with* MATLAB integrates traditional and non-Euclidean geometries, differential and computational geometry with more current computer systems in a practical and user-friendly format.

Outline:

Part 1 (four chapters) studies functions and their graphs, curves and surfaces in classical and modern geometries.

Chapter 1 is devoted to functions and their graphs and transformations, and it can serve as the introduction to MATLAB symbolic/numeric calculations, programming, and basic graphing as needed in later chapters. *Section 1.1* discusses the geometry of (integer, real, complex, and quaternion) numbers. *Sections 1.2–1.4* investigate elementary, special, and piecewise functions and how to plot their graphs in Cartesian coordinates and (for some curves) in polar coordinates. *Sections 1.5–1.7* study basic fitting of curves and surface graphs. MATLAB commands are used to carry out *polynomial* and *spline interpolation* and to present concrete applications of splines for generating curves and surfaces. Some optimization problems including the classical method of least squares are briefly considered.

Chapter 2 is devoted to methods and visualizing of Euclidean geometry based on coordinate presentation of objects. The following topics are covered: vectors and matrices and their use in analytic geometry *(Section 2.1)*, rigid motions or isometries *(Section 2.2)*, applications of complex/quaternion numbers, geometry of a sphere in space *(Section 2.3)*, and polyhedra *(Section 2.4)*. The study of polyhedra is organized in the sequence of themes: Platonic and Archimedean solids, star-shaped and compound polyhedra, symmetries, etc.

Chapter 3 takes a look at affine and projective transformations and their presentation using matrices and homogeneous coordinates. Such transformations provide the reader with basic mathematical means of picture manipulation. The inspiration for projective geometry is human experience and the artist's perspective; it has gained added significance in recent years with the advance of computer graphics.

Chapter 4 investigates non-Euclidean geometry based on inversive and Möbius (i.e., fractional linear) transformations. Non-Euclidean geometry fundamentally changes our views about geometry and mathematics in general, and has a great importance in developing our imagination. Complex numbers play an essential role in the two-dimensional case, while the quaternion presentation of transformations is used to operate in three-dimensional space. We look at reflection (inversion) in a sphere (or a circle) as a tool for proving results in Euclidean and hyperbolic geometry. We do some calculations in the upper half-plane with the Poincaré metric and display the resulting curves (lines, transversals, circles, equidistants, horocycles, fifth lines, etc.). Next we provide some calculations in hyperbolic space (modeled in the upper half-space) and display the resulting figures.

Part 2 (five chapters) studies differential-geometric models of curves and surfaces and approximation/interpolation of curves and surfaces. The methods (matrices and transformations, complex numbers and quaternions, interpolating and approximating) of Part 1 are also applicable to these specific constructions.

Chapter 5 discusses examples and constructions with curves. Key concepts are illustrated by named curves of historical and scientific significance. We define a regular curve and study cycloidal and other remarkable parameterized and implicitly given plane curves. We employ Euler's equations (calculus of variations) to study elastica. Next we discuss the basic MATLAB capabilities for plotting space curves using various coordinate systems and (parallel and perspective) projections, and present the curves with shadows on planar, cylindrical or spherical displays. We also study fractal curves (Peano, Koch, Sierpiński, Menger) defined recursively.

Chapter 6 is devoted to the differential geometry of curves. It studies tangents and normals at regular points, arc length, singular points, mathematical embroidery (curves appear as envelopes of families of lines or circles), the curvature and the torsion of a curve, evolutes, caustics and parallel curves (equidistants), and other constructions with curves. Several programs for detecting singular points on curves, deriving characteristics of space curves, plotting the moving Frenet frame field and the osculating circle, and reconstructing a plane curve from its curvature function are developed.

In *Chapter 7* we study surfaces in space: we define a parametric surface, examine a regular surface, and produce examples of surfaces using MATLAB commands. We calculate and plot tangent planes and normal vectors of a surface, the osculating paraboloid, and the Gaussian and mean curvatures, solve conditional extremum problems, plot geodesics, study surfaces with singularities, and provide a short program for investigating the differential geometry of a surface.

Chapter 8 studies three very important and commonly occurring classes of surfaces (algebraic surfaces, surfaces of revolution, and ruled surfaces) and considers several interesting constructions with surfaces: translation, parallel, pedal, and podoid surfaces, tubes, the envelope of a family of surfaces, etc.

Chapter 9 studies approximation and interpolation of curves and surfaces. *Sections 9.1 and 9.2* study Bézier curves (an interesting application of them is to font design) and discuss spline curves and surfaces. *Section 9.3* applies the Hermite interpolation technique to construct piecewise curves. The (geometrical) β-spline curves (studied in *Section 9.4*) are based on the geometric C^k-continuity of piecewise curves. The (classical) B-splines appear as a particular case when the shape parameters are $\beta_1 = 1, \beta_2 = 0$. *Sections 9.5 and 9.6* extend these ideas to surfaces.

Although many of the chapters can be studied independently, the concepts introduced in earlier chapters are often the underlying tools used in subsequent chapters.

The MATLAB examples and exercises often contain short bits of code that the reader can apply or modify. In addition, we developed several MATLAB M-files to implement some solutions and the graphical displays that may be useful in various cases. In some cases, readers can write larger codes using these M-files. The *Appendix* collects these M-files for modeling curves, surfaces and polyhedra in Euclidean and hyperbolic worlds. Some of them can be downloaded from the Web at http://www.mathworks.com/matlabcentral/fileexchange/.

Additional programs and remarks will be posted as they are developed on the author's Web page http://math.haifa.ac.il/ROVENSKI/rovenski.html.

Readers are encouraged to send comments and suggestions to the author at the e-mail address rovenski@math.haifa.ac.il.

The materials of the book were used in courses given by the author for the students of Haifa University.

Acknowledgments. The author would like to thank Professor David Blanc (University of Haifa) and Professor Robert Wolak (Mathematical Institute of Krakov) for their helpful corrections concerning the manuscript, and Ann Kostant (Executive Editor) for support in the publishing process. Finally, I wish to thank the entire staff at Springer, Brian Treadway at Outer Sum, and the anonymous reviewers whose input significantly improved the text.

Mathematics Department   *Vladimir Rovenski*
University of Haifa
June 2009

Contents

Part I
Functions and Transformations

1

Functions and Graphs

Section 1.1 discusses the geometry of (integer, real, complex, and quaternion) numbers. In Section 1.2 we plot graphs of some elementary, special, and piecewise functions, and investigate functions using derivatives. In Section 1.3 we study piecewise functions of one variable that are defined by several formulae for different values (intervals) of that variable. Section 1.4 is an excursion into remarkable curves (graphs) in polar coordinates. In Sections 1.5 and 1.6 we use several MATLAB® functions that implement various interpolation and approximation algorithms. Chapter 1 can be considered as the introduction to MATLAB® symbolic/numeric calculations, programming, and basic graphing as needed in later chapters.

1.1 Numbers

When MATLAB starts, its desktop opens with the *Menu* window, the *Command* window (where the special >> prompt appears), the *Workspace* window, the *Command History* window, and the *Current Directory* window.

Throughout this book we usually type/edit the code in the *Editor* window, then place it (copy–paste) in the Command window, and execute. To save space, we present code and MATLAB answers in compact form (not identical to the view on the display).

MATLAB can be used in two distinct modes (see Example, p. 4):

- it offers immediate execution of statements in the Command window, or
- it also offers programming by means of script and function M-files.

A script M-file collects a sequence of commands that constitute a program. It is executed when one enters the name of the script M-file.

V. Rovenski, *Modeling of Curves and Surfaces with MATLAB®*,
Springer Undergraduate Texts in Mathematics and Technology 7, DOI 10.1007/978-0-387-71278-9_1,
© Springer Science+Business Media, LLC 2010

We type % to designate a group of words (in a line) as a comment. The M-files created by ourselves we type with a bold font. The reader should place them in the MATLAB Current Directory, say D : /work; one may choose it manually or type in the Command window

```
cd  d:/work
```

1.1.1 Integers, rationals, and reals

Let $\mathbb{Z} = \{0, \pm 1, \pm 2, \dots\}$ be *integers*, and $\mathbb{N} = \{1, 2, 3, \dots\} \subset \mathbb{Z}$ be *natural numbers*. The following table contains some scalar and array arithmetic in MATLAB:

N	Symbol	Operation	MATLAB
1	$+$ and $-$	addition/subtraction	a + b - c
2	$*$ and .$*$	multiplication	a * b
3	/ and ./	right division	a / b
4	\ and .\	left division	b \ a
5	^ and .^	exponentiation	a^b

Example. Type 2+3 after the $>>$ prompt, followed by **Enter** (i.e., press the **Enter** key). Next try the following:

```
2*3,  4^2,  1/2,  2\1
```

Try the commands

```
x = [1, 2, 3]
x.^2                   % a dot in front of *, /, ^ is used when we work with arrays.
ans = 1   4   9        % next we present MATLAB answers in a compact form.
```

Compound numbers of the form n^2 are called *squares* for obvious reasons. The *triangular numbers* are $t_n = \frac{1}{2}n(n+1)$, etc.

The MATLAB command for allows a statement or a group of statements to be repeated. One may type two lines in the Command window:

```
for n = 1 : 10;  x(n) = n^2;  end;                       % squares
x                        % Answer: x = 1  4  9  16  25  36  49  64  81  100
```

Alternatively, we write a program **my_squares.m** with one line above in the file edit window. For practice, create an M-file **my_triangles.m**:

```
for n = 1 : 10;  t(n) = n*(n+1)/2;  end;                 % triangle numbers
```

To run this script (after saving **my_triangles.m** in the Current Directory), go to the MATLAB Command window and enter two commands:

my_triangles;

t   % Answer: t = 1 3 6 10 15 21 28 36 45 55

(MATLAB executes the instructions in the order in which the commands are stored in the **my_triangles.m** file).

If $i, j \in \mathbb{N}$, $j \neq 0$, and $i = kj$ for some integer k, then we say that j *divides* i and that j is a *divisor* of i. We write $j \mid i$. From the Euclidean algorithm for integers there follows: if i, j are integers that have a *greatest common divisor* d, then there exist integers s, t such that $si + tj = d$.

Two integers i, j are said to be *congruent modulo m* if m divides $i - j$. In that case we shall write $i \equiv j \pmod{m}$. For example,

mod(8, 3)   % $8 \equiv 2 \pmod 3$. Answer: 2

Modular arithmetic. A set of all "remainders" of any integer divided by n is denoted by $\mathbb{Z}_n = \{0, \ldots, n-1\}$, where $n \in \mathbb{Z}$. The operations of *addition* and *multiplication modulo n*, $\oplus_n, \otimes_n \colon \mathbb{Z}_n \times \mathbb{Z}_n \to \mathbb{Z}_n$, are defined naturally: $a \oplus_n b = a + b \pmod n$, $a \otimes_n b = a \cdot b \pmod n$.

For example, $5 \oplus_7 6 = 4$, $0 \oplus_n (-1) = n - 1$, and $3 \otimes_6 7 = 4$, $3 \otimes_6 4 = 0$.

mod(5 + 6, 7), mod(3 * 7, 6), mod(3 * 4, 6)   % Answer: 4, 3 and 0

Note that $(\mathbb{Z}_n, \oplus)$ is a cyclic group C_n (see the definition in Section 2.4).

Definition 1.1. A *skew field* is a triple $(K, +, \cdot)$, where
(1) $(K, +)$ is an abelian group (with identity denoted as "0"),
(2) $(K \setminus \{0\}, \cdot)$ is a group,
(3) multiplication $\cdot$ is left and right distributive over $+$.
A skew field in which multiplication is commutative is called a *field*.

Rational numbers $\mathbb{Q}$ form a field. If p is a prime, then $\mathbb{Z}_p$ is a field.

The **fundamental theorem of arithmetic** reads that

Every integer $n \geq 2$ can be factorized into a product of primes in only one way apart from the order of the factors.

For example, $2009 = 7^2 \cdot 41$, $2010 = 2 \cdot 3 \cdot 5 \cdot 67$,

factor(2009), factor(2010)   % Answer: 7 7 41 and 2 3 5 67

If a, b are members of a set A then (instead of $a = b$) we write $a \sim b$ and say that this is an *equivalence relation* if it satisfies the following three properties: (1) *reflexivity*: always $a \sim a$; (2) *symmetry*: if $a \sim b$ then $b \sim a$; (3) *transitivity*: if $a \sim b$ and $b \sim c$ then $a \sim c$.

The notation $\frac{n}{m}$ $(n, m \in \mathbb{Z})$ is a representation of an equivalence class because $\frac{kn}{km}$ is also the same number.

A *continued fraction* is an expression of the form $a_0 + \cfrac{1}{a_1 + \cfrac{1}{a_2 + \cdots}}$.

To write an ordinary fraction as a continued fraction we need to use the *Euclidean algorithm*. For example, $43/5 = 8 + \cfrac{1}{1 + \frac{2}{3}} = 8 + \cfrac{1}{1 + \frac{1}{1 + 1/2}}$.

We will use If-Else-End structures. The simplest form of this is:

> if *expression*
>   *commands evaluated if expression is True*
> else   % when there are three or more alternatives, elseif is used
>   *commands evaluated if expression is False*
> end

Example. Real numbers can be presented in MATLAB in decimal form, and also in scientific notation, e.g.,

```
1.2345, -.001                              % decimal form
1.2345e3                % scientific notation:  1.2345 × 10³ = 1234.5
```

Define the function M-file for future use:

```
function f = fract(n)
    syms a;
    if n > 0;   f = a + 1/fract(n - 1);   else   f = a;          % recursion
end;
```

The number $\sqrt{2}$ can be written as the infinite continued fraction $\sqrt{2} = 1 + \cfrac{1}{2 + \cfrac{1}{2 + \cdots}}$.
To show this, let the right-hand side be x. Then $1 + x = 2 + \frac{1}{1+x}$ and hence $x = \sqrt{2}$. Let us call the program **fract.m**:

```
q = fract(5)                              % symbolic expression
q = a + 1/(a + 1/(a + 1/(a + 1/a))))       % Answer.
vpa(subs(q, 'a', 2), 5)       % a = 2. Answer: 2.4143 (compare with √2)
```

The function vpa (variable precision arithmetic) converts the expression into a decimal number of arbitrary precision.

Real numbers $\mathbb{R}$ form a field (with respect to addition and multiplication). They include not only the rational numbers but also the *irrational numbers*, say $\sqrt{5}$ or $\sqrt[3]{2}$, which cannot be expressed as a ratio of integers. The irrational numbers include the *algebraic numbers* (i.e., roots of nonzero polynomials with integer coefficients) and the *transcendental numbers* (such as $\pi \approx 3.14$ and $e \approx 2.7$, *Euler's constant*), which by definition are not roots of any polynomials with integer coefficients. However, actually verifying that any particular number is transcendental seems very hard.

Predefined MATLAB constant values are

pi — the number $\pi = 3.14159\ldots$,   Inf — infinity,   etc.

Examples. π and the golden section τ.

1. The number π is the ratio of the length of the circumference of a circle to its diameter. We inscribe a polygon $ABC\ldots$ whose vertices lie on a circle of unit diameter and circumscribe a polygon $PQR\ldots$ whose edges touch the circle. We would expect the perimeter of the circle to lie between those of the two polygons. Now if we let the number of vertices of the polygons increase to infinity and the length of the edges decrease to zero, their perimeters run (increase or decrease, respectively) to π. The length of an edge of a regular inscribed *octagon* is $\frac{1}{2}\sqrt{2 - \sqrt{2}}$ and the perimeter is $4\sqrt{2 - \sqrt{2}}$. Similarly the lengths of an edge of a regular inscribed 16-gon, 32-gon,

etc., are $8\sqrt{2 - \sqrt{2 + \sqrt{2}}}$, $16\sqrt{2 - \sqrt{2 + \sqrt{2 + \sqrt{2}}}}$, $\ldots$. Repeating ad infinitum we

get the following limit: $2^n \underbrace{\sqrt{2 - \sqrt{2 + \sqrt{2 + \cdots + \sqrt{2}}}}}_{n \text{ roots}} \to \pi$.

We prepare the M-file

```
function g = g1(n)
   if n > 0;  g = sqrt(2 + g1(n - 1));   else  g = 0;        % recursion
   end;
```

Then we calculate 10 terms (that approximate π):

```
for n = 1 : 10
  f = vpa(2^n*sqrt(2 - g1(n - 1)), 9);
end
disp ([n f])                          % Answer: [ 10, 3.14159142]
```

2. The *golden section* $\tau = \frac{1}{2} + \frac{\sqrt{5}}{2}$ satisfies the equation $\tau = 1 + \frac{1}{\tau}$ (verify!). Show that τ can be written as the infinite continued fraction $\tau = 1 + \cfrac{1}{1 + \cfrac{1}{1 + \cdots}}$. The successive approximations of τ are

$$1 + \frac{1}{1} = \frac{2}{1}, \; 1 + \frac{1}{1 + 1/1} = \frac{3}{2}, \; 1 + 1/(1 + \frac{1}{1 + 1/1}) = \frac{5}{3}, \; \ldots, \; \frac{f_{n+1}}{f_n},$$

where $f_n = f_{n-1} + f_{n-2}$, $f_0 = f_1 = 1$, are the *Fibonacci numbers*.

```
q = fract(10);  vpa(subs(q, 'a', 1), 6)              % compare with τ ≈ 1.6180.
```

The following table lists some commonly used (elementary) functions, where variables x and y can be numbers, vectors, or matrices:

Function	MATLAB	Function	MATLAB
cosine	cos(x)	sine	sin(x)
arccosine	acos(x)	arcsine	asin(x)
tangent	tan(x)	arctangent	atan(x)
signum function	sign(x)	absolute value	abs(x)
maximum value	max(x)	minimum value	min(x)
natural logarithm	log(x)	common logarithm	log10(x)
exponential	exp(x)	square root	sqrt(x)
remainder after division	rem(x)	round towards ∞	ceil(x)
round to nearest integer	round(x)	round towards $-\infty$	floor(x)

In addition, many other specialized functions are available. There are also functions of linear algebra that find information about matrices, such as eig(A), which finds the eigenvalues of a matrix A.

Typing help elfun or help specfun calls up full lists of elementary and special functions, respectively.

MATLAB basically has two kinds of functions: numerical functions and symbolic expressions of functions. A *numerical function* is really a short program that operates on numbers to produce numbers. A *symbolic expression* of a function operates on symbolic variables to produce symbolic results. These symbolic expressions can be manipulated with operations like differentiation and integration.

For each simplification procedure we look in turn (see the table in what follows) at its effect on trigonometric functions, the exponential function and logarithms, powers, radical expressions, and other functions.

command	trig	exp and logs	powers
expand	$\cos 2x \rightarrow 2\cos^2 x - 1$	$\lg(2x) \rightarrow \lg 2 + \lg x$	$(2x)^y \rightarrow 2^y x^y$
combine	$2\cos^2 x \rightarrow \cos(2x) + 1$	$2\lg 3 \rightarrow \lg 9$	$\sqrt{x-1}\sqrt{x} \rightarrow \sqrt{x^2-x}$
simplify	$\tan x \rightarrow \sin x / \cos x$	$e^x e^y \rightarrow e^{x+y}$	$x^y x^z \rightarrow x^{y+z}$
convert	$\cos x \rightarrow \frac{1}{2}(e^{ix} + e^{-ix})$	$e^{ix} \leftrightarrow \cos x + i\sin x$	

Example. Symbolic tools. Declare x, y as symbolic variables:

    syms x y;

In expand, the factors are first multiplied out, and secondly similar terms are collected.

    expand((x + y)^3)   % Answer: $x^3 + 3x^2 y + 3xy^2 + y^3$

The procedure factor is "inverse" to expand. It computes the factorization over different domains in irreducible factors.

factor(2009)   % Answer: 7, 7, 41. (So, $2009 = 7^2 \cdot 41$).

The procedure sort is used to rearrange terms, to sort terms in some suitable ordering.

sort(['a' 'c' 'e' 'b' 'd'])   % Answer: [a b c d e].

The procedure collect is used to group coefficients of like terms in a polynomial.

collect((exp(x) + x)*(x + 2))   % Answer: $2\exp(x) + (\exp(x) + 2)x + x^2$.

The simple function returns the shortest result by independently applying simplify, collect, factor, and other simplification functions to an expression and keeping track of the lengths of the results.

1.1.2 Complex numbers

The *complex numbers* $\mathbb{C}$ arise by denoting the point (or pair of reals) (x, y) by a new symbol $x + yi$, and then introducing simple algebraic rules for the numbers $z = x + yi$ where $i^2 = -1$. For example,

z = complex(2, 3)   % Answer: $z = 2 + 3i$

Addition and *multiplication* in terms of pairs are given as

$$(a,b) + (c,d) = (a+c, b+d), \quad (a,b) \cdot (c,d) = (ac - bd, ad + bc). \tag{1.1}$$

Note that from definition (1.1) follows $i^2 = (0 + 1i)(0 + 1i) = -1 + 0 = -1$.

Complex numbers $\mathbb{C}$ form a field with respect to the addition and multiplication defined above (verify!).

Every nonzero complex number has a multiplicative inverse, namely, $z^{-1} = \frac{x - yi}{x^2 + y^2}$. Hence we define division by $z/w = zw^{-1}$. The *complex conjugate* $\bar{z}$ of z is given by $\bar{z} = x - yi$, where $z = x + iy$, and geometrically, z and $\bar{z}$ are mirror images of each other in the real axis. We call $\operatorname{Re} z = x$ the *real part*, and $\operatorname{Im} z = y$ the *imaginary part* of $z = x + yi$. Hence $\operatorname{Re} z = (z + \bar{z})/2$, $\operatorname{Im} z = (z - \bar{z})/2i$. One may show that

$$\overline{z + w} = \bar{z} + \bar{w}, \qquad \overline{zw} = \bar{z}\bar{w}, \qquad z\bar{z} = x^2 + y^2.$$

The *modulus* of z is defined by $|z| = \sqrt{x^2 + y^2}$. Given $z \neq 0$, the modulus $r = |z|$ is the length of the line segment from 0 to z.

The *distance* between z_1 and z_2 is defined as $|z_2 - z_1|$. Multiplication is distance-preserving: $|z_1 z_2| = |z_1| \cdot |z_2|$, and addition satisfies the *triangle inequality*: $|z_1 + z_2| \le |z_1| + |z_2|$. Repeating the last inequality we obtain

$$|z_1 + \cdots + z_n| \le |z_1| + \cdots + |z_n| \qquad \text{for all } n > 2.$$

The *argument* $\arg z$ of z is the angle θ between the positive real axis and the segment from 0 to z measured in the counterclockwise direction. Then (r, θ) are the *polar coordinates* of z, and

$$x = |z| \cos \theta, \quad y = |z| \sin \theta \quad \text{where} \quad \theta = \arg z.$$

It is convenient to write $\cos \theta + (\sin \theta) i = e^{\theta i}$. The *polar form* of the product zw in terms of the polar forms of $z = r_1 e^{\theta_1 i}$ and $w = r_2 e^{\theta_2 i}$ is $zw = (r_1 r_2) e^{(\theta_1 + \theta_2) i}$. From this follows *de Moivre's formula*:

$$(\cos \theta + \sin \theta\, i)^n = \cos(n \theta) + \sin(n \theta)\, i \qquad \forall n \in \mathbb{N}.$$

The following table lists some operations with complex numbers:

N	Object	Equation	MATLAB
1	modulus of $z = a + ib$	$|z| = \sqrt{a^2 + b^2}$	abs(z)
2	real part of $z = a + ib$	$\mathrm{Re}(z) = a$	real(z)
3	imaginary part of $z = a + ib$	$\mathrm{Im}(z) = b$	imag(z)
4	complex conjugate of $z = a + ib$	$\mathrm{conj}(z) = a - ib$	conj(z)
5	exp function of $z_1 = ib$	$\exp(z) = \cos(b) + i \sin(b)$	exp(i*b)

The fundamental theorem of algebra. *Let $p(z) = a_0 + a_1 z + \cdots + a_n z^n$ be a given polynomial, where $n \ge 1$ and $a_n \ne 0$. Then there are $z_1, \ldots, z_n \in \mathbb{C}$ such that, for all z, $p(z) = a_n(z - z_1) \cdots (z - z_n)$.*

The problem of finding (exactly) the zeros of a given polynomial of degree $n > 4$ is extremely difficult.

From the fundamental theorem of algebra there follows the claim:
Let p and q be polynomials of degree at most n. If $p(z) = q(z)$ at $n + 1$ distinct points then $p(z) = q(z)$ for all z.

Complex numbers provide an easy way to describe lines and circles. A *(straight) line l* in the plane is the set of points that are equidistant from two distinct points u and v. Hence, the equation of a line l is

$$|z - u|^2 = |z - v|^2 \iff \bar{a} z + a \bar{z} + b = 0, \quad \text{where } b \in \mathbb{R}. \tag{1.2}$$

```
syms u v z                                        % Find a,b
```

```
collect(expand((z - u)*conj(z - u) - (z - v)*conj(z - v)), [z, conj(z)])
(conj(v) - conj(u))*z + (v - u)*conj(z) + u*conj(u) - v*conj(v)          % Answer
```

Hence, $a = v - u, b = u\bar{u} - v\bar{v}$.

The *circle* $C(a, r)$ with center a and radius r has equation $|z - a| = r$, or

$$(z - a)\overline{(z - a)} = r^2 \iff z\bar{z} - \bar{a}z - a\bar{z} + |a|^2 - r^2 = 0. \qquad (1.3)$$

A *regular n-gon* is a polygon whose n vertices are evenly spaced around a circle, and the angle at a vertex of a regular n-gon is $(n - 2)\pi/n$.

Examples.

1. The *angle function* can be expressed as $\arg(z) = \arctan(\text{Im}(z), \text{Re}(z))$.

```
z = 1 + i*3;
t = atan2(imag(z), real(z))          % Answer: θ = 1.25, or use imag(log(z))
```

2. Find the roots of polynomials.

```
syms b c z;
solve(z^2 + b*z + c)                          % 2nd degree polynomial
solve(z^3 + b*z + c),  eval(solve(z^3+3*z+1))     % 3rd degree polynomial
          % Answer: −0.3222, 0.1611 − 1.7544i, 0.1611 + 1.7544i
solve(z^4 + b*z + c),  eval(solve(z^4+z+1))       % 4th degree polynomial
     % Answer: −.7271 + .43i, −.7271 − .43i, .7271 + .9341i, .7271 − .9341i
```

3. The function $f(z) = az + b$ with $a \neq 0$ preserves the ratio of 3 points, $[z_1, z_2, z_3] = \frac{z_1 - z_3}{z_1 - z_2}$. Namely, $[z_1, z_2, z_3] = [f(z_1), f(z_2), f(z_3)]$ for all $z_i \in \mathbb{C} \cup \{\infty\}$. To show this, define the functions **rat3** and $f(z)$

```
rat3 = @(z1, z2, z3) (z1 - z3)/(z1 - z2)          % using @ to create a function
     syms a b;    f = @(z) a*z + b;
```

Then verify the property

```
syms z1 z2  z3;
simplify(rat3(f(z1), f(z2), f(z3)))          % Answer: (z1 − z3)/(z1 − z2)
```

4. Given $z_1 \neq z_2$ and $w_1 \neq w_2$ in $\mathbb{C}$, there is a function $f(z) = az + b$ with $a \neq 0$ such that $w_i = f(z_i)$ ($i = 1, 2$). To prove this, we solve the equation $[z_1, z_2, z] = [w_1, w_2, f(z)]$.

```
syms z1 z2 z3 w1 w2 w3 z w;
f = solve(rat3(z1, z2, z) - rat3(w1, w2, w), w)
```
$$f = (zw_1 - zw_2 - w_1z_2 + z_1w_2)/(z_1 - z_2) \qquad\qquad \text{% Answer}$$

5. The set of *unimodular* complex numbers $S^1 = \{z \in \mathbb{C} : |z| = 1\}$ (the unit circle) is a group with respect to multiplication.

6. The nth roots of 1 are the distinct complex numbers $1, \omega, \omega^2, \ldots, \omega^{n-1}$, where $\omega = e^{2\pi i/n}$. These points are equally spaced around the circle $|z| = 1$ starting at 1, and they form a group with respect to multiplication.

```
n = 7;                                              % define n
k = 1 : n;  Z = exp(2*pi*k*i/n)     % n points cos(2πi k/n) + i sin(2πi k/n).
a = 2;  b = 4;  Z(a)*Z(b) - Z(a + b)      % Answer: 0 (define a,b ∈ ℤ).
```

7. Let $w = re^{\varphi i}$, $z' = r^{\frac{1}{n}} e^{i\frac{\varphi}{n}}$. The solutions of $z^n = w$ are $z', z'\omega, \ldots, z'\omega^{n-1}$.

```
n = 7;  w = 1 + i;                          % define n and w
syms z;  solve(z^n - w)                  % Answer: n roots of z^n − w.
```

1.1.3 *Quaternions*

Let $D = \{1, i, j, k\}$ be the canonical basis for $\mathbb{R}^4$ over $f = \mathbb{R}$. Define an operation $\cdot$ (multiplication) on D by the table

$\cdot$	i	j	k
i	-1	k	$-j$
j	$-k$	-1	i
k	j	$-i$	-1

and extend this operation to $\mathbb{R}^4$ by linearity. The elements of $\mathbb{R}^4$, together with this operation, are called the *real quaternions*. They were first defined by Hamilton in 1844 as a generalization of the field $\mathbb{C}$, and earlier studied by Gauss, who did not publish his results. Quaternions were generalized by Clifford and used in the study of non-Euclidean spaces. Lately, they have also been used in computer graphics and in signal analysis. Define the *quaternionic conjugate* $\bar{q}$ of $q = q_0 + q_1 i + q_2 j + q_3 k$ to be $\bar{q} = q_0 - q_1 i - q_2 j - q_3 k$. The map $q \to \bar{q}$ can be thought of as reflection in the real line $\mathbb{R}$. For any quaternions q, p, the quaternionic conjugate satisfies

$$\overline{q + p} = \bar{q} + \bar{p}, \qquad\qquad \overline{q \cdot p} = \bar{p} \cdot \bar{q}.$$

As usual let $|q| = \sqrt{q_0^2 + q_1^2 + q_2^2 + q_3^2}$ be the *norm* of q. Since quaternionic multiplication is associative and real numbers commute with quaternions, we can prove Hamilton's law of the moduli:

$$|q \cdot p|^2 = (q \cdot p)(\overline{q \cdot p}) = (q \cdot p)(\bar{p} \cdot \bar{q}) = q(p \cdot \bar{p})\bar{q} = |p|^2 q \cdot \bar{q} = |q|^2 |p|^2.$$

If $q \neq 0$ then q has an *inverse* $q^{-1} = \bar{q}/|q|^2$. We call q a *unit quaternion* if $|q| = 1$. The unit quaternions lie on the unit 3-sphere, $S^3 \subset \mathbb{R}^4$, and form a group under quaternion multiplication.

If $q = q_0 + q_1 i + q_2 j + q_3 k$ then q_0 is the *real* (or temporal) part of q and $q_1 i + q_2 j + q_3 k$ is the *pure* (or spatial) part. The decomposition $q = (q + \bar{q})/2 + (q - \bar{q})/2$ exhibits a quaternion as the sum of its two parts.

Just as we think of a complex number as a pair of reals, we can think of a quaternion $q = q_0 + q_1 i + q_2 j + q_3 k$ as a pair of complex numbers. So, $q = z + wj$, where $z = q_0 + q_1 i$, $w = q_2 + q_3 i$.

Multiplication of quaternions in terms of pairs is given as

$$(z, w) \cdot (z_1, w_1) = (z z_1 - \overline{w_1} w, \, w_1 z + w \overline{z_1}).$$

(Verify using $j w_1 = \overline{w_1} j$ and $j z_1 = \overline{z_1} j$.)

A pure quaternion can be thought of as an element (vector) of $\mathbb{R}^3$ and has the usual scalar and vector products; see Section 2.1. Two pure quaternions p, q have an interesting formula for their product:

$$p \cdot q = -(\mathbf{p}, \mathbf{q}) + \mathbf{p} \times \mathbf{q}.$$

The following table lists some operations with quaternions:

N	Object	Equation	MATLAB
1	modulus of q	$|q| = \sqrt{\sum_{i=0}^{3} q_i^2}$	quatmod(q)
2	norm of q	$\mathrm{norm}(q) = \sum_{i=0}^{3} q_i^2$	quatnorm(q)
3	normalized q	$\mathrm{normal}(q) = q/|q|$	quatnormalize(q)
4	inverse of q	q^{-1}	quatinv(q)
5	q divide by p	q/p	quatdivide(q, p)
6	q multiply by p	qp	quatmultiply(q, p)
7	conjugate of q	$\mathrm{conj}(q) = q_0 - i q_1 - j q_2 - k q_3$	quatconj(q)

The following commands are also implemented in MATLAB:

n = quatrotate(q, v) rotates a vector **v** by a quaternion q (see Section 2.2),
q = euler2quat(ea) calculates the quaternion q for Euler's angles *ea*,
ea = quat2euler(q) is the inverse to the above.

Two quaternions p, q are said to be *similar* if there is a nonzero quaternion u such that $u^{-1} p u = q$; this is written as $p \sim q$.

One may verify that $\sim$ is an equivalence relation on the quaternions. Two quaternions $p = p_0 + p_1 i + p_2 j + p_3 k$, $q = q_0 + q_1 i + q_2 j + q_3 k$ are similar if and only if $p_0 = q_0$ and $p_1^2 + p_2^2 + p_3^2 = q_1^2 + q_2^2 + q_3^2$; see Exercises 11, 12, p. 69.

The fundamental theorem of algebra. Let a_i $(0 \le i \le n)$ be quaternions, $n \ge 1$ and $a_n \ne 0$. Then the polynomial equation $a_0 + a_1 x + \cdots + a_n x^n = 0$ has at least one solution in $\mathbb{H}$.

We shall pay special attention to the space $\mathbb{H}_0$ of pure quaternions which we identify with $\mathbb{R}^3$. If $H \in S^2$ (i.e., a pure unit quaternion), then $(\mathbf{H}, \mathbf{H}) = 1$ and so $H \cdot H = -1$. Just as for complex numbers, a quaternion q can be written in a *polar form*:

$$q = r(\cos\theta + \sin\theta\, H) \quad \text{where } r = |q|, \ H \cdot H = -1.$$

The angle θ is a generalization of the argument of a complex number. A quaternion is real if and only if θ is 0 or π (verify!).

A non-real quaternion determines a *copy of the complex numbers* $\mathbb{C}_q$ (the two-dimensional subspace of $\mathbb{R}^4$ spanned by $1, H$ where $H \cdot H = -1$). If $q = r(\cos\theta + H \sin\theta)$ then $\mathbb{C}_q$ is isomorphic to $\mathbb{C}$ by the correspondence $x + yi \leftrightarrow x + Hy$.

Examples.

1. Let us find the conjugate of q = [q0 q1 q2 q3].

```
syms q0 q1 q2 q3 real;                      % q0, q1, q2, q3 are real
cq = quatconj([q0 q1 q2 q3])          % Answer: cq = [q0, −q1, −q2, −q3]
```

Note that conj([q0 q1 q2 q3]) gives a different result $[q_0, q_1, q_2, q_3]$.

2. The *commutator* of two quaternions is $[q, p] = q \cdot p - p \cdot q$.

Show that $\overline{[q, p]} = -[q, p]$ and that $[q, p]$ is a pure quaternion. Deduce that q and p commute if and only if either one is real or if $p \in \mathbb{C}_q$.

Solution. Use the program

```
syms q0 q1 q2 q3 p0 p1 p2 p3 real;
q = [q0 q1 q2 q3];   p = [p0 p1 p2 p3];              % quaternions q, p
F1 = quatconj( quatmultiply(q, p) - quatmultiply(p, q) )      % [q, p]
F2 = - ( quatmultiply(q, p) - quatmultiply(p, q) )          % −[q, p]
F1 - F2                                          % obtain [0, 0, 0, 0].
```

1.2 Elementary and Special Functions

MATLAB provides three ways to define functions: (a) using @ to create an anonymous function, (b) using inline, (c) or using an M-file.

Example. Define a function $f(x)$ by a symbolic expression.

```
syms x a                                          % symbolic variables
f = x^2 + 3*x -a
```

Evaluate the function f at $x = 2$.

```
subs(f, x, 2)                                     % Answer: 10 − a.
```

Another way to represent a function at the command line is by creating an anonymous function from a string expression.

```
g = @(x)x^2 + 3*x - a;
```

One may then evaluate g at 2.0 as

```
g(2.0)                                            % Answer: 10 − a.
```

To solve the equation $f(x) = 0$, we use the solve command.

```
solve(f)
```

One can differentiate and integrate a function $f(x)$ using the commands

```
diff(f),  diff(f, 2),  diff(f, a),  int(f),  int(f, a),  int(f, 0, 2),   etc.
```

The command taylortool computes Taylor's polynomial of arbitrary degree for a given function f.

```
taylortool('x*cos(x)')                    % A Taylor series (and a graph) calculator.
```

1.2.1   *Plotting in two dimensions*

This section is a short introduction to MATLAB two-dimensional plots. One can type

help graphics for general graphics commands,
help graph2d for two-dimensional graphics commands,
demo matlab graphics for more examples.

The simplest geometric object is a *point* in the plane. The location of a point in Cartesian coordinates is specified by a pair of numbers called the x- and y-coordinates of the point and is written as (x, y). Of course, in the space $\mathbb{R}^3$, the vectors and points (of attachment) have three components. The following commands are used to plot the points $P = (1, 2)$, $Q = (3, -4)$:

```
plot(1, 2, '*');    plot(3, -4, 'o');
```

The "∗" is used to mark the point that we are plotting. Other authorized symbols for point displays are presented in the following table:

Line		Color		Symbol	
solid	-	red	r	diamond	d ◇
dashed	- -	green	g	circle	o
dotted	:	blue	b	x-mark	×
dashdot	-.	white	w	star	*
no line	none	yellow	y	plus	+
		cyan	c	square	s □
		magenta	m	point	.

Vectors in space are directed line segments, used to represent velocity, acceleration, forces, etc. A vector in the plane is given in terms of 2 components, as $\mathbf{v} = [a, b]$. The norm of the vector is $|\mathbf{v}| = \sqrt{a^2 + b^2}$. The magnitude of a vector $\mathbf{v}$ attached to a point $P_0 = (x_0, y_0)$ can be computed by the command norm:

```
v = [1 3];  norm(v)                  % Answer: 3.1623  (the norm of a vector v).
P = [2 5];  arrow(P, v)          % vector v attached at P (M-file arrow.m, p. 70)
```

Once you have created a figure, there are two basic ways to manipulate it:

(a) by typing MATLAB commands in the Command window, such as the commands title and axis equal, etc., or

(b) with the mouse, using the menus and icons in the figure window itself.

Examples.

1. The *dot product* can be obtained only by the multiplication of a row on the left and a column of the same length on the right. The row vector $\mathbf{u}$ is turned into a column vector with the command $\mathbf{u}'$.

```
u = [1 5 3 7];  v = [2 4 6 8]
u*v';  v'*u;  u'*v
u*v                                  % we cannot multiply two rows
u*u';  norm(u)^2                              % compare!
```

Observe that the length of a vector is equal to its dot product with itself.

2. View the right angles of a rectangle using the option axis equal.

```
x1 = [0 12 16 4 0];  y1 = [0 16 13 -3 0];
plot(x1, y1);  axis equal;                                  % rectangle
```

The same option sometimes does not allow one to see the graph.

```
x = linspace(0, 10*pi, 2000);
plot(x, sin(x));  axis equal;
```

There are many ways to use plot. For example, we can change the style and color of the line. Using plot(t, y, ':') gives a dotted line. To get a green dashdot line, we type plot(t, x, 'g-.'). The plot command also accepts matrix arguments. If X and Y are both m by n, plot(X,Y) superimposes the plots created by corresponding columns of X and Y. One can use the command help plot for detailed information.

The following table lists some elementary geometric objects:

N	Object	Equation	MATLAB Commands
1a	point	$P(a,b)$	P = [2 6]; plot(P)
1b	complex number	$z = a + ib$	z = 2 + 3*i; plot(z)
2a	line	$Ax + By + C = 0$	plot([x0, x1], [y0, y1])
2b			plot([x0, x1], [y0, y1], [z0, z1])
3	segment	$[P,Q]$	x=[Px Qx]; y=[Py Qy]; plot(x,y)
4	vector		P = [2 5]; v = [1 3]; **arrow**(P, v)
5	polygonal line	$[P_1, \ldots P_n]$	x =[x1, ... xn]; y = [y1, ... yn] z = [z1, ... zn]; plot(x, y, z)
6	triangle	$[P_1, P_2, P_3]$	x = [x1,x2,x3, x1]; y = [y1,y2,y3, y1] plot(x, y)

Example. A *line* in the plane is represented as a linear equation of the form $Ax + By + C = 0$, where $A^2 + B^2 \neq 0$; A, B, and C are called the coefficients of the line and determine its position and direction. One can plot a *line* $l(P; \mathbf{a})$ in the plane through the point $P = (x_0, y_0)$ and parallel to the vector $\mathbf{a} = [a_1, a_2]$ using the command

plot([x0, x0 + a1], [y0, y0 + a2])   % define x_0, y_0, a_1, a_2.

A part of a line that has definite starting and ending points is called a *line segment* and is represented in MATLAB as

x = [x0, x1];   y = [y0, y1];
plot(x, y);

where $P = (x_0, y_0)$ and $Q = (x_1, y_1)$ are the coordinates of the starting and ending points, respectively, of the line segment.

A very powerful feature of MATLAB is its ability to render an *animation*. For example, suppose that we want to visualize the rotation of a point. The necessary steps to implement this objective are:

1. Generate a two-dimensional plot of the point at arbitrary time.

2. Use the for... and pause commands to display consecutively the different frames obtained in step 1.

Example. The command comet(x) produces an *animated*, comet-like, x–y plot of a graph. For example,

```
x = 0 : 0.001 : 1;  comet(x, sqrt(1 - x.^2), 0);
```

For more complicated animations, a typical program to use is

```
X = 0 : 0.01 : 1;
for n = 0 : 50;
 plot(X, sin(n*pi/5)*sin(pi*X)), axis([0, 1, -2, 2]);
 pause(0.1);                                    % define the value of pause
end;
```

The axis command here ensures that each frame of the movie is drawn with the same coordinate axes.

The following table lists some graphing commands:

Command	Purpose
plot	draw two-dimensional graphs
xlabel, ylabel, zlabel	define axis labels
title	define graph title axis set various axis parameters (min, max, etc.)
legend	show labels for plot lines
shg	bring graphics window to foreground
text	place text at selected locations
grid	turns grid lines on or off
mesh	draw surface using colored lines
surf	draw surface using colored patches
hold	fix the graph limits between successive plots
view	change surface viewing position
drawnow	empty graphics buffer immediately
zoom	magnify graph or surface plot
clf	clear graphics window
contour	draw contour plot
ginput	read coordinates interactively

1.2.2 *Plotting graphs*

The *graph of a real function* $y = f(x)$ $(x \in I)$ is the set in the plane $\mathbb{R}^2$ defined by $G_f = \{(x,y) \in \mathbb{R}^2 : y = f(x), x \in I\}$.

The graph of $f(x)$ of class C^1 (i.e., the derivative $f'(x)$ exists and is continuous) is smooth: the tangent line $y = f(x_0) + f'(x_0)(x - x_0)$ (see Section 6.1) depends continuously on x_0.

Graphs are visualized using the plot command. The command ezplot is used primarily to graph functions that are defined symbolically. *Labels* and a *title* can be attached to the graph with additional commands. Moreover, we can use plot$(x, y, string)$, where *string* combines up to three elements that control the *color*, *marker*, and *line style*. The reader can try funtool — a visual function calculator that manipulates functions of one variable.

In stem types of graphs, vertical lines terminating with a symbol such as a circle are drawn from the zero value point on the y-axis to the values of the elements in the vector **y**. The function stairs(x, y) creates a step graph of your data at the level specified by the elements of **y**.

Multiple (x, y) pair arguments create multiple graphs with a single call to plot. The command hold on is used to plot several graphs on the same figure; the hold off turns off the hold on feature. One can place text over a figure (and names of graphs) using the command text.

Examples.

1. One can graph a numerical function in this way:

```
x = linspace(0, pi, 51);    f = sin(x);
plot(x, f, 'r')                                        % f = sin x in red
```

Alternatively, we may use the symbolic method:

```
ezplot('sin(x)', [0, 2*pi]);
ezplot('tan(x)');  axis ([-pi, pi, -10, 10]);
```

To graph quickly the second derivative of f, we add the line

```
f = 'sin(x)';
g = diff(f, 2);  ezplot(g, [0, 2*pi])
```

2. Plot the polygon (stem and stairs) passing through 20 points that lie on the parabola $y = x^2$: $\{x_i = -2 + \frac{4}{19}i, \ 0 \le i \le 19\}$.

```
x = -2 : 0.4 : 2;
stem(x, x.^2, '.');    stairs(x, x.^2)              % x.^2, not x^2;  Figure 1.1(c)
```

3. MATLAB's plot command can be used for simple "join-the-dots" x–y plots. Let the function $y = f(x)$ be given by the table of values (x_i, y_i), where $i = 1, 2, \ldots, n$. Suppose that we have measured the air temperature (*Celsius*) in the morning during 10 days in April:

Date:	12	13	14	15	16	17	18	19	20	21
Temp.:	15	17	17.5	19	20	19.5	18	17	17	19

We form arrays with x- and y-coordinates of the given points:

Date = [12 : 21]; TempC = [15 17 17.5 19 20 19.5 18 17 17 19];

We can plot the points, using the command plot with the option 'o':

```
plot(Date, TempC, 'o');
```

We plot the polygon through these points using the command plot, labeling the coordinate axes, Figure 1.1(a):

```
plot(Date, TempC); title('Graph'); xlabel('Date'); ylabel('TempC');
```

Now we plot the diagrams of vertical or horizontal segments, Figure 1.1(b).

```
stem(Date, TempC, '.');    stairs(Date, TempC, '-');    bar(Date, TempC)
```

The temperature is transformed from Celsius to Fahrenheit by the formula $F = F(C)$. Preliminarily we fix the number of significant digits to 4.

```
format bank; TempF = 9/5*TempC + 32
[59.00, 62.60, 63.50, 66.20, 68.00, 67.10, 64.40, 62.60, 62.60, 64.40]
```

Now we calculate TempF by a different method:

```
syms t;    F = 9/5*t + 32;    TempF = subs(F, t, TempC);
```

Using loglog instead of plot causes the axes to be scaled logarithmically. Related functions are semilogx and semilogy, for which only the x- or y-axis, respectively, is logarithmically scaled.

```
semilogy(Date, TempC)
```

4. We plot several curves on the same graph with the same color (but different types of lines):

```
x = -2 : 0.4 : 2;
hold on;   plot(sin(x));   plot(cos(x), '--');   plot(tan(x), ':');
```

or

```
syms x;  f = sin(x);
hold on; ezplot(f, [0,2*pi]); ezplot(diff(f), [0,2*pi]); ezplot(diff(f,2), [0,2*pi]);
```

A second way makes the graphs (of the three curves) in different colors:

```
x = -3*pi : 0.001 : pi;
plot(x, sin(x), x, tan(x), x, exp(x));  axis ([-3*pi pi -5 5]);
legend('sin', 'tan', 'exp')
```

5. Using **text** we plot several lines, see Figure 1.3(a):

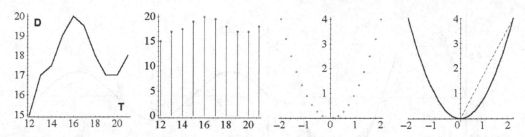

Fig. 1.1 Temperature during 10 days. Points on a parabola. Graphs of the function x^2 and its derivative.

```
x = -2 : 2/500 : 2;  hold on;  axis([-2 2 -2 2]);  axis equal;
plot(x,-2*x, x,-x, x,-x/2, x,x/2, x,x, x,2*x);
text(-0.8, 2, 'y = -2x');  text(-2, 2, 'y = -x');  text(-2, 1, 'y = -x/2');
text(2, 1, 'y = x/2');  text(2, 2, 'y = x');  text(0.8, 2, 'y = 2x');
```

6. Plot the table of graphs.

```
subplot(1, 3, 1);  ezplot('sin(1*x)');    subplot(1, 3, 2);  ezplot('sin(3*x)')
subplot(1, 3, 3);  ezplot('sin(6*x)');                    % line of three plots
y = 0 : 0.01 : 2*pi;
for i = 1 : 4;
 subplot(2, 2, i);  plot(y, sin(i*y));                    % table of plots
end;
```

7. The command compose(f, g) returns $f(g(y))$ where $f = f(x)$, $g = g(y)$. Using the *composition* of functions, visualize iterations of a function.

```
syms x;
f = sin(x);  F = f;
hold on;
for i = 1 : 9;
 ezplot(F);  axis([0 2*pi -1 1]);  F = compose(F, f);
end;                                                      % Figure 1.2(a)
```

8. Taylor decomposition is very useful in studying functions. Compare Taylor polynomials of a function with the given function, Figure 1.2(b).

```
syms x;  f = sin(x);
hold on;  axis equal;  ezplot(f, 'r');
for i = 1 : 9;
 axis([0 3*pi -3 3]);  ezplot(taylor(f, x, 2*i - 1));
end;
```

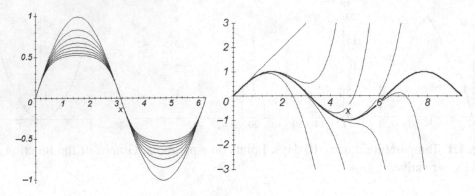

Fig. 1.2 Iterations of $y = \sin(x)$ and Taylor polynomials.

When a vector (or matrix) is involved in a logical expression, the comparison is carried out element by element, the resulting vector, which is called the *logical vector* (or 0–1 vector), consisting of units and zeros.

Let a and b be vectors of the same length. The logical expression $a = b$ returns a 0–1 vector:

```
a = 1:4;   b = [0, 2, 3, 5];
a == b                          % no semicolon !   Answer: 0 1 1 0
```

The reader can also try another logical expression, say $a \leq 2$, etc.

```
a <= 2
```

A useful application of *logical vectors* is in plotting graphs of functions:

   (i) piecewise continuous functions,
  (ii) avoiding division by zero,
 (iii) avoiding infinity.

For example,

```
x = linspace(-4*pi, 4*pi, 501);                        % for 3 examples below

y = sin(x);    plot(x,  y.*( y > 0))                    % a discontinuous graph
x = x + ( x == 0)*eps;   y = sin(x)./x;
plot(x,  y)                            % avoiding "Divide by zero" warning
y = 100*tan(x);  z = y.* (abs(y) < 1e3);
plot(x, y,  x, z, 'g', 'LineWidth', 2)      % avoiding infinity (compare 2 graphs)
```

1.2.3 *Elementary functions*

Let us consider some further examples.

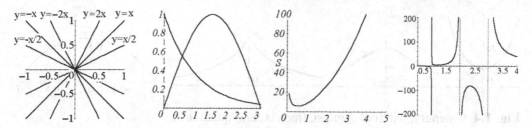

Fig. 1.3 Placing text. Solving equations. Rational function.

Examples.

1. Plot graphs of the functions (set $a = 1$).

    (a) *Catenary* $y = a\cosh(x/a)$, Figure 1.4(a), is "similar" to a parabola.

```
x = -5 : 0.01 : 5;
plot(x, cosh(x)), grid on
```

    (b) *Witch of Agnesi (versiera)* $y = \dfrac{a^3}{a^2 + x^2}$, Figure 1.4(b).

```
x = -5 : 0.001 : 5;
plot(x, 1./(1 + x.^2)), grid on; axis equal
```

    (c) *Cycloid* $x = a\arccos\dfrac{a-y}{a} - \sqrt{2ay - y^2}$, Figure 5.1.

```
a = 1; y = 0 : 0.01: 2;
plot(y, acos((a - y)/a) - sqrt(2*a*y - y.^2)); axis equal
```

    (d) *Tractrix* $x = a\ln\dfrac{a - \sqrt{a^2 - y^2}}{y} + \sqrt{a^2 - y^2}$ $(0 < y \le a)$, Figure 1.4(c).

```
y = 0 : 0.001 : 1;  f = log((1 - sqrt(1 - y.^2))/y) + sqrt(1 - y.^2);
plot(y, f,  y, -f); axis equal
```

    (e) *Dinostratus' quadratrix* $y = x\cot\left(\dfrac{\pi x}{2a}\right)$ $(|x| \le a)$, Figure 1.4(d).

```
x = -.5 : 0.001 : .5;
f = x.*cot(pi*x)/2;
plot(x, f),  grid on;  axis equal
```

    (f) *Beats*. A complicated case arises in summing oscillations with different frequencies: $y = A(\cos(\omega_1) + \cos(\omega_2)) = 2A\cos\left(\frac{\omega_1 - \omega_2}{2}\right)\cos\left(\frac{\omega_1 + \omega_2}{2}\right)$.

If ω_1 and ω_2 have a small difference, then the factor $\cos\left(\frac{\omega_1 - \omega_2}{2}\right)$ changes slowly, and the factor $\cos\left(\frac{\omega_1 + \omega_2}{2}\right)$ has almost the same frequency as each of the given oscillations. Beats are sometimes presented (approximately but visually) as harmonic oscillations with slowly changing amplitude.

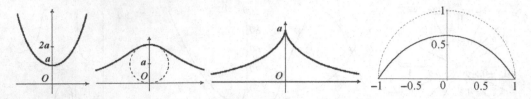

Fig. 1.4 Catenary, witch of Agnesi, tractrix, and quadratrix.

```
x = -pi : 0.001 : 9*pi;
f = 6*sin(x/4);  g = f.*sin(4*x);
plot(x, g, x, f, '--', x, -f, '--'),  grid on;  axis equal          % Figure 1.5(a)
```

(g) *Curve of damped oscillations* $y = A\exp(-kx)\sin(\omega x + a)$. For $k = 0$ the oscillations are periodic (without resistance of the medium). Also plot this graph with the
option x=0..infinity.

```
x = -10 : 0.001 : 10;
f = exp(1).^(-x/2);  g = f.*sin(4*x);
plot(x, g, x, f, '--', x, -f, '--'),  grid on;  axis equal          % Figure 1.5(b)
```

(h) A *remarkable limit* $y = \sin(x)/x$, Figure 1.5(c) (or $y = \tan(x)/x$).

```
syms x;  limit(sin(x)/x, x, 0)                                      % Answer: 1
x = -10 : 0.001 : 10;  plot(x, sin(x)./x), grid on
```

This can also be done in one step with the fplot command. Consider the plotting of the
following function, known as the sinc or *sampling function* $y = \tan(\pi x)/(\pi x)$:

```
fplot('sinc', [-10 10]),  xlabel('x'),  ylabel('sinc(x)')
```

A *sequence* f(n) is a function on $\mathbb{N}$. One may calculate the sequence $n\sin(1/n)$ for
$12 \le n \le 16$ and compare with the previous results.

```
n = 12 : 16;  sin(1./n).*n        % Answer: 0.998   0.998   0.998   1.00   1.00
```

2. *Curves of growth* are curves that describe the laws of development of events in
time. The process of analytical smoothing of a dynamical line using some functions,
i.e., adjusting it to these functions, is in most cases the best way to represent a set of
empirical data. Plot the following curves of growth (a)–(c) for some values of their
parameters.

(a) *Logarithmic parabola* $y = b^x c^{x^2}$ $(b, c > 0)$. One can write the equation in the
form $y = e^{x\ln b + x^2 \ln c}$. For $c > 1$ the branches of the parabola $x\ln b + x^2 \ln c$ are directed
upward, and for $0 < c < 1$ they are directed downward. For $c > 1$ and $b > 1$ the curve
is displaced to the left, and for $c < b < 1$ it is displaced to the right.

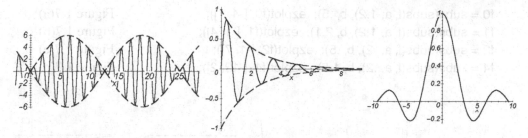

Fig. 1.5 Beats, damped oscillations, remarkable limit.

```
syms b c x;
f = b^x.*c^(x.^2);
f0 = subs(subs(f, b, 2), c, 2),  f1 = subs(subs(f, b, 2), c, .5);
ezplot(f0, [-2, 1]);                                    Figure 1.6(a)
ezplot(f1, [-2, 3])                                     Figure 1.6(b)
```

(b) *Logarithmic curve* $y = \dfrac{k}{1 + be^{-ax}}$ $(k, a, b > 0)$, Figure 1.6(c).

```
syms k a b x;
f = k/(1 + b*exp(-a*x));  solve(diff(f, x, 2), x)        % Answer: log(b)/a
f0 = subs(subs(subs(f, k, 2), b, .2), a, 1);  ezplot(f0, [-7, 3]),  grid on
```

The horizontal asymptote is $y = k$; $x = \ln(b)/a$ is the point of inflection.

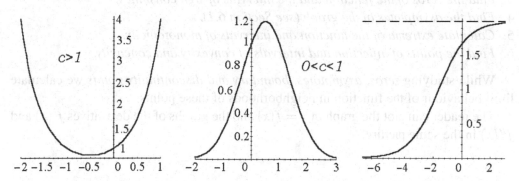

Fig. 1.6 Logarithmic growth curves.

(c) *Growth curves of Gompertz* $y = a^{b^x}$ $(a, b > 0)$. One can write the equation in the form $y = e^{b^x \ln a}$ and consider four cases, Figure 1.7. It happens that $y = 1$ is the horizontal asymptote.

```
syms a b x;
f = a^(b^x);
```

f0 = subs(subs(f, a, 1.2), b, .5); ezplot(f0, [-4, 1]);   Figure 1.7(a)
f1 = subs(subs(f, a, 1.2), b, 2.1); ezplot(f1, [-1, 4]);   Figure 1.7(b)
f2 = subs(subs(f, a, .2), b, .5); ezplot(f2, [-1, 7]);   Figure 1.7(c)
f4 = subs(subs(f, a, .2), b, 1.5); ezplot(f4, [-12, 2]);   Figure 1.7(d)

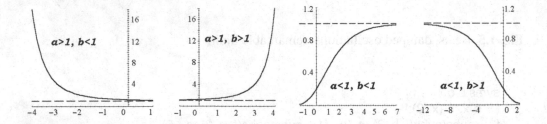

Fig. 1.7 Growth curves of Gompertz.

One can investigate functions using derivatives by the following plan (the graphs that MATLAB plots help us to realize this plan), which is sufficient for most functions in practice.

1. *Find the domain of definition and points of discontinuity; derive the function (or its limits) at boundary points of the domain of definition.*
2. *Check whether the function is odd or even, periodic, or not.*
3. *Find the zeros of the function and the intervals of sign constancy.*
4. *Find the asymptotes of the graph (see Section 6.1).*
5. *Calculate extrema of the function and intervals of monotonicity.*
6. *Find the points of inflection and intervals of convexity and concavity.*

While studying *zeros, asymptotes, boundary and discontinuity points* we calculate limit behaviour of the function in neighborhoods of these points.

The reader can plot the graph of $y = f(x)$ and the graphs of the derivatives $f'(x)$ and $f''(x)$ in the same picture.

1.2.4 *Special functions*

The name *special* is usually used for *functions* that generally cannot be represented in terms of elementary functions. Most of these arise as the solutions of ordinary differential equations (ODEs) of special forms and can be expressed using integrals. Over 50 of the special functions of classical applied mathematics (among the most important of which are the *beta*, *gamma*, and *zeta* functions, *elliptic* functions, *Bessel* functions, *theta* function, and *hypergeometric* functions) are available in MATLAB; see the commands mfunlist, mfun.

Examples. Plot graphs of the following six special functions.

1. The *incomplete gamma function* $y = \Gamma(x,z) = \int_z^{\infty} e^{-t}t^{x-1}\,dt$, one of the most popular special functions.

   The *gamma function* $y = \Gamma(x) = \int_0^{\infty} e^{-t}t^{x-1}\,dt$ has vertical asymptotes at the points of discontinuity $x = -n$ $(n = 0,1,2,\ldots)$.

```
axis ([-4 5 -10 10]);  ezplot('gamma(x)'), grid on          % Figure 1.8(a)
```

2. *Integral of probabilities* $y = \mathrm{erf}(x) = \dfrac{2}{\sqrt{\pi}}\int_0^x e^{-t^2}\,dt$ (*error function*):

```
syms x;  ezplot(erf(x), [-2, 2]),  grid on;                 % Figure 1.8(b)
```

3. The *exponential integral function* $y = \mathrm{Ei}(n,x) = \int_1^{\infty} \dfrac{e^{-xt}}{t^n}\,dt$ $(n \in \mathbb{N},\ x > 0)$ is related to the gamma function by $\mathrm{Ei}(n,x) = x^{n-1}\Gamma(1-n,x)$, Figure 1.8(c).

```
x = .1 : 0.01 : 1;  hold on;
for n = 0 : 5;  plot(mfun('Ei',n, x));  end
```

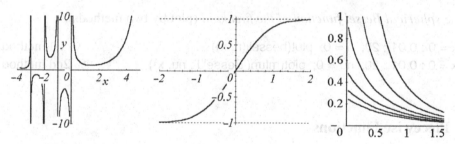

Fig. 1.8 Graphs of special functions $\Gamma(x)$, $\mathrm{erf}(x)$, $\mathrm{Ei}(n,x)$.

4. The *sine integral function* $y = \mathrm{Si}(x) = \int_0^x \dfrac{\sin(t)}{t}\,dt$ is derived using Taylor series; its horizontal asymptote is $y = \frac{\pi}{2}$, Figure 1.9(a). The *cosine integral function* is $y = \mathrm{Ci}(x) = \gamma + \ln(x) + \int_0^x \dfrac{\cos(t)-1}{t}\,dt$. Here $\gamma \approx 0.57721$ is the *Euler constant*; the horizontal asymptote is $y = 0$, Figure 1.9(b).

```
x = 0 : 0.05 : 21;  plot(mfun('Si', x)),  grid on
x = 0 : 0.05 : 21;  plot(mfun('Ci', x)),  grid on
```

5. *Elliptic integrals of the first* or *second kinds* are given by the formulae

Elliptic F(z,x)= $\int_{t=0}^{z} 1/(\sqrt{(1-t^2)}\sqrt{(1-x^2t^2)})\,dt$,   Figure 1.9(c)

Elliptic E(z,x)= $\int_{t=0}^{z}\sqrt{(1-x^2t^2)}/\sqrt{(1-t^2)}\,dt$,   Figure 1.9(d)

Elliptic K(x)=Elliptic F(1,x), Elliptic E(x)=Elliptic E(1,x).

```
x = 0 : 0.01 : 1;   k = 1;
plot(mfun('EllipticE', k, x)),  grid on
plot(mfun('EllipticF', k, x));      plot(mfun('EllipticK', x));
```

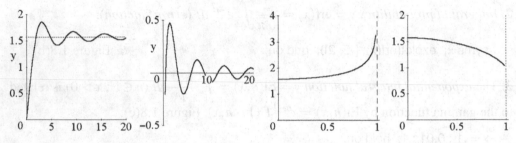

Fig. 1.9 Graphs of Si(x), Ci(x), and elliptic integrals.

6. The *spherical Bessel function of the first kind* (plot by two methods).

```
x = 0 : 0.01 : 20;  n = 0;  plot(besselj(n, x))          % 1st method
x = 0 : 0.01 : 20;  nu = 0;  plot(mfun('BesselJ', nu, x))  % 2nd method
```

1.3 Piecewise Functions

The *continuity class* of a function is the number of derivatives that are continuous over the whole domain. It determines how smoothly the pieces of a piecewise-defined function join together. If a function f has continuity class n this is denoted $f \in C^n$. Differentiating a function decreases its continuity class by 1, and integrating it may increase it by 1.

A function $f(x)$ given on the interval $I = (a, b)$ is called *piecewise continuous (staircase)* if one can break this interval into a finite number of *intervals of continuity* (resp., *intervals of constancy*) on each of which the function $f(x)$ is continuous (resp., constant). Values of a function at the points of discontinuity are not always well-defined.

A continuous function $f(x)$ given on $I = (a,b)$ is called *piecewise differentiable* if I can be broken into a finite number of segments on each of which $f(x)$ is of class C^1. (In other words, $f'(x)$ is piecewise continuous).

We will define and plot the graphs of piecewise-defined functions,

$$F(\text{cond}_1, f_1, \ldots, \text{cond}_n, f_n, f) = \{f_i \text{ if cond}_i, \text{ where } i \leq n; \text{ otherwise } f\}.$$

Examples.

1. The function $F(x) = \{\sin(x) \text{ if } -2\pi \leq x \leq -\pi; \; \pi - |x| \text{ if } -\pi \leq x \leq \pi; \; -\sin^3(x) \text{ if } \pi \leq x \leq 2\pi\}$ is composed of 3 parts.

```
x1 = -2*pi : pi/50 : -pi;   y1 = sin(x1);
x2 = -pi : pi/50 : pi;   y2 = pi - abs(x2);
x3 = pi : pi/50 : 2*pi;   y3 = -sin(x3).^3;
x = [x1 x2 x3];   y = [y1 y2 y3];
```

We plot it:

```
plot(x, y),  grid on;                          % Method 1: graph with one color
plot(x1, y1, 'r+', x2, y2, 'g:', x3, y3, 'b--')     % Method 2: 3 parts with colors
```

2. A function is *recursive* (i.e., is based on recursion) if it contains one or more calls to itself or to another function that calls the given function.

Plot the graph of the *recursively defined function*

$$f(x+1) = 2f(x) \qquad \text{and} \qquad f(x) = x(1-x) \quad (0 \leq x \leq 1).$$

```
function f = f1(x)                             % Create M-file!
    if x < 0;  f = f1(x + 1)/2;
    elseif x < 1;  f = x*(x - 1);
    else f = 2*f1(x - 1);
    end
end
```

```
fplot('f1(x)', [-2, 3])                         % graph, Figure 1.10(a)
```

1.3.1 *Piecewise constant and linear functions*

An important discontinuous example is the *unit step* (*Heaviside*) function

$$H(x) = \{0 \text{ for } x < 0, \; 1 \text{ for } x > 0, \text{ and NaN for } x = 0\}.$$

It is convenient for expressing other piecewise (composed) functions. The derivative of $H(x)$ is the Dirac (*delta*) function, which is equal to 0 everywhere except the singular

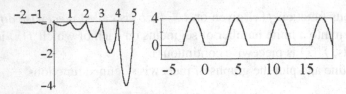

Fig. 1.10 Recursively defined functions.

point $x = 0$. Another standard function sign is

$$y = \text{sign}(x) = \{-1 \text{ for } x < 0,\ 1 \text{ for } x > 0,\ 0 \text{ for } x = 0\}.$$

Examples. 1. Let us plot the graph of $H(x)$, Figure 1.11(a).

```
ezplot('heaviside(x)', [-2, 2]);
```

Then plot the graph of its derivative

```
diff('heaviside(x)', 'x'), ezplot(dirac(x), [-2, 2])
for i = -2 : 2; dirac(i), end          % Answer: dirac(x)    0 0 ∞ 0 0
```

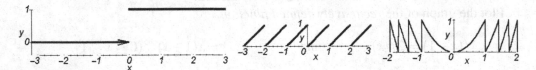

Fig. 1.11 Graphs of the functions $heaviside(x)$, $\{x\}$, $\{x^2\}$.

2. "Cut out" a real function $f(x)$ $(x \in \mathbb{R})$ on a segment $[a, b]$.

```
syms x;
a = 1;  b = 3;  f = x^2;                              % define a function
fab = f*(heaviside(x - a) - heaviside(x - b));
ezplot(fab, [-4, 6])                                  % Figure 1.13(c)
```

3. Finally, we plot the graph of $\text{sign}(x)$

```
ezplot('sign(x)', [-2, 2])
fplot('sign(x)', [-2, 2])                             % obtain a polygonal curve.
```

4. We define and plot the *saw-shaped function*, Figure 1.12(a):

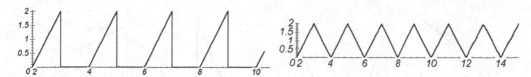

Fig. 1.12 Saw and triangular waves.

fplot('2*mod(x - 1, 2)*heaviside(mod(x, 2) - 1)', [-4, 14]), axis equal

Then we show that *"triangular waves"*, Figure 1.12(b), are obtained in a similar way
(the *abs* function can be presented using heaviside)

fplot('2*abs(mod(x, 2) - 1)', [0, 14]), axis equal, grid on

or using one of the formulae $y = \arccos(\cos(x))$, $y = \arcsin(\sin(x))$:

fplot('2/pi*acos(cos(pi*x))', [0, 14]), axis equal, grid on   %Figure 1.12(b)

Using heaviside (or sign) we define (and plot the graphs of) the characteristic functions of the segment $[a, b]$, the interval (a, b), and the half-interval.

$$\chi(x, [a,b]) = H(x-a) - H(x-b) = (\mathrm{sign}(x-a) + \mathrm{sign}(x-b))/2,$$
$$\chi(x, (a,b)) = \chi(x, [a,b]) - \chi(x, [a,a]) - \chi(x, [b,b]),$$
$$\chi(x, [a,b)) = \chi(x, [a,b]) - \chi(x, [b,b]), \quad \chi(x, (a,b]) = \chi(x, [a,b]) - \chi(x, [a,a]).$$

1.3.2 *The functions* max *and* min

The functions max (*maximum*) and min (*minimum*) of several differentiable functions
are piecewise differentiable.

Examples.

1. Consider the function $g = \min(x^2, x + 3)$.

   g = 'min(x^2, x + 3))'; solve('x^2 = x + 3')   % Answer: $(1 \pm \sqrt{13})/2$.

   Hence $g = \begin{cases} x^2, & (1 - \sqrt{13})/2 < x < (1 + \sqrt{13})/2, \\ x+3, & \text{otherwise.} \end{cases}$

   fplot(g, [-2, 5]), axis equal   % Figure 1.13(a)

2. The graph of max (min) of several linear functions is a convex (concave) polygon.
One may illustrate this using the program

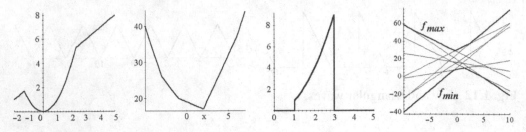

Fig. 1.13 Graphs of piecewise differentiable functions.

```
n = 9;
k = round(5*rand(1, n));  b = round(30*rand(1, n));
syms x;    F = k.*x + b;
x = -10 : 0.01 : 10;
m = max(eval(F(1)), eval(F(2)));    M = min(eval(F(1)), eval(F(2)));
for i = 2 : n;
 M = min(M, eval(F(i)));  m = max(m, eval(F(i)));
end
hold on;  plot(x, M, x, m, 'b', 'LineWidth', 2);              % Figure 1.13(d),
for i = 1 : n;  plot(x, eval(F(i)), 'g');  end         % upper/lower green polygons.
```

1.3.3 *Functions containing the operation* abs

Example. Given sequences of points $\{a_i\}$ and of weights $\{b_i\}$, where $1 \le i \le n$, find the minimum of the function $f_n(x) = \sum_{i=1}^{n} b_n |x - a_i|$. First we plot the graphs:

```
n = 6;  a = @(j)(2 + sqrt(j));  b = @(j)1;
syms x;  for i = 1 : n;  F(i) = b(i)*abs(x - a(i));  end
f = sum(F);  ezplot(f, [a(1) - 1, a(n) + 1]);
```

An application is the *problem of machine tools* with weight coefficients: place the machine tool at the point x such that the sum of the distances from it to machine tools at the given points $-3, -1, 2$, and 6 with given weights 2, 1, 3, 1 is minimal.

```
ezplot('2*abs(x+3)+abs(x+1)+3*abs(x-2)+abs(x-6)', [-5, 6]);  % Figure 1.13(b)
```

For unit weights $b_i = 1$ explain the following rule of calculation of the *optimum* point x (where the function $f_n(x)$ takes its minimum).

Hint. For odd n write down the numbers $a(1), \ldots, a(n)$ in increasing order, then take the optimum point x equal to the middle number of this sequence. What does the graph of $f_n(x)$ look like, and what is the optimum point x for even n?

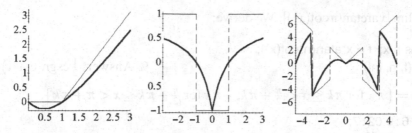

Fig. 1.14 Animation of the limits of functions.

1.3.4 *Functions that are defined using* limit

One may diversify the plotting of graphs of functions defined in terms of limits by applying the commands for ... and pause.

Examples. Plot the graphs of the functions (using animation) and find their formulae.

(1) $y = \lim\limits_{n \to \infty} \dfrac{x^n}{1 + x^n}$ $(x \geq 0)$. We have

$$y = \{\, 0 \text{ for } 0 \leq x < 1,\ 1 \text{ for } x > 1,\ \text{and } 1/2 \text{ for } x = 1 \,\}.$$

```
x = 0:0.02:3;   for n = 0:50;
  plot(x, (x.^n)./(1 + x.^n)), axis([0, 3, -1, 2]);    pause(0.1)
end;
```

(2) $y = \lim\limits_{n \to \infty} [(x - 1)\arctan x^n]$ $(x > -1)$. We have

$$y = \{\, 0 \text{ for } -1 \leq x \leq 1 \text{ and } \tfrac{\pi}{2}(x - 1) \text{ for } x > 1 \,\}.$$

```
x = 0:0.02:3;  for n = 0:50;
  plot(x, (x - 1).*atan(x.^n)), axis([-1, 3, -1, 2]);    pause(0.2)
end;                                                          % Figure 1.14(a)
```

(3) $y = \lim\limits_{n \to \infty} \dfrac{|x|^n - 1}{|x|^n + 1}$ $(x \neq 0)$. We have

$$y = \{\, 1 \text{ for } |x| > 1,\ -1 \text{ for } |x| < 1 \,\}.$$

```
x = -3:0.02:3;  for n = 0:50;
  plot(x, (abs(x).^n -1)./(abs(x).^n +1)), axis([-3, 3, -2, 2]);    pause(0.1)
end;                                                          %Figure 1.14(b)
```

(4) $y = \lim\limits_{n \to \infty} [x \arctan(n \cot(x))]$. We derive:

```
syms n x;  f = x*atan(n*cot(x));
limit(f, n, inf)
```
                                                    % Answer: $\frac{1}{2} \operatorname{csgn}(\cot(x)) \pi x$

Hence $y = \{ \pi x \text{ for } \pi k \leq x < \frac{\pi}{2} + \pi k, \ -\pi x \text{ for } \frac{\pi}{2} + \pi k \leq x < \pi + \pi k \}$.

```
x = -6 : 0.02 : 7;
for n = 0 : 50;
 plot(x, x.*atan(n*cot(x))),  axis([-4, 4, -6, 7]);  pause(0.1)
end;
```
                                                    %Figure 1.14(c)

1.3.5 *Statistical functions*

The main *statistical functions* of MATLAB include the density of a probability function $F(x) = P\{\xi < x\}$ for various distributions. Using examples of Section 1.3.1, one can plot the graphs $F(x)$ of some *statistical functions*.

Example.

1. The *Bernoulli distribution* with the parameter $p \in (0,1)$.

```
f = @(x, p)(1 - p)*heaviside(x) + p*heaviside(x - 1);
x = -1 : 0.01 : 4;  plot(x, f(x, 0.2)),  axis equal
```

2. The *binomial distribution* $P\{\xi = k\} = C_n^k p^k (1-p)^{n-k}$ $(0 \leq k \leq n)$ (density function) with parameters n, p $(0 < p < 1, n \geq 1)$.

```
n = 10;  p = 0.4;
x = 0 : n;  y = binopdf(x, n, p);  stem(x, y, '.')          % Figure  1.15(a,b)
x = 0 : n;  hold on;
for k = 1 : n;  plot([x(k), x(k + 1)], [y(k), y(k)]);  end
```

The *binomial distribution function* is

$$F(x) = \begin{cases} \sum_{k=1}^{m} C_n^k p^k (1-p)^{n-k}, & m < x \leq m+1, \quad m \in (0,n], \\ 1, & x > n, \\ 0, & x \leq 0. \end{cases}$$

```
yy(1) = y(1);
for k = 1 : n;  yy(k+1) = yy(k) + y(k+1);  end;                % array F(x)
hold on;  for k = 1 : n;  plot([x(k), x(k + 1)], [yy(k), yy(k)]);  end
```

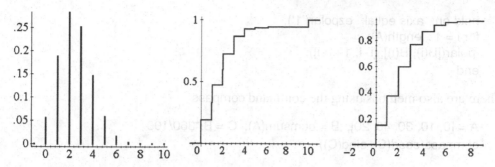

Fig. 1.15 Binomial and Poisson distributions.

1.4 Graphs in Polar Coordinates

First, we introduce the reader to using polar coordinates with MATLAB. Then we plot remarkable curves (spirals, roses and crosses, etc.) as polar graphs and use the transformation of inversion.

1.4.1 *Basic plots in polar coordinates*

The location of a point M in the plane with origin O is uniquely defined by the *distance* $|OM| = \rho$ and the *angle* $\varphi \in [-\pi, \pi]$ between the segment OM and the *polar axis*. The pair (ρ, φ) comprises the *polar coordinates* of the point M. The relation between polar coordinates (ρ, φ) and rectangular coordinates (x, y), when the axis OX is the polar axis, is the following (given by Newton in 1670): $x = \rho \cos \varphi$, $y = \rho \sin \varphi$. If $\rho = \rho(\varphi)$ is the polar equation of a curve, then its equations in rectangular coordinates are

$$x = \rho(t) \cos(t), \quad y = \rho(t) \sin(t).$$

Hence, the complex parametric equation is $z = \rho(t)e^{it}$.

The graph of a function $\rho = f(t)$ in polar coordinates is plotted using the command polar. For plotting the parameterized curve $\rho = f(t)$, $\varphi = g(t)$ $(a \leq t \leq b)$ in polar coordinates one uses the commands polar and ezpolar. Polygons and graphs of functions, see Figure 1.16(a,b), in polar coordinates can be plotted similarly to the case of rectangular coordinates.

Examples.

1. Plot the *circular diagram* of Figure 1.16(c).

    A = [0, 10, 30, 40, 20]; B = cumsum(A)   % enter data A

```
hold on;  axis equal;  ezpolar('1');
for i = 1 : length(A);
 polar([B(i), B(i)], [i - i, 1 + i - i]);
end
```

There are also methods using the command compass

```
A = [0, 10, 30, 40, 20];  B = cumsum(A);  C = B.*360/100
compass(cosd(C), sind(C))
```

and the commands pie or pie3.

```
data = [10 30 40 20];  parts = [1 0 0 0];
pie(data, parts)
pie3(data, parts)
```

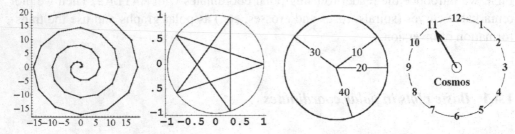

Fig. 1.16 Polar coordinates: polygon, star n-gon, circular diagram, a stopwatch.

2. The *spiral of Archimedes* $\rho = a\varphi$ was studied by Archimedes in the third century BC in relation to the problem of the trisection of an angle.

```
t = 0 : .01 : 4*pi;   polar(t, t)          % two coils of Archimedes' spiral
plot(t.*exp(i*t))                          % Continue: using complex equation
```

3. Aside from the regular simple polygons there also exist *regular star* (or simply *star*) *polygons*: one for $n = 5$ (Figure 1.18(b), pentagram), none for $n = 6$ (other cases decompose into two triangles), two for $n = 7$ (Figure 1.17), one for $n = 8$ (Figure 1.18(d), other cases decompose into two squares), etc.

A *star polygon* $\{n/m\}$, with n, m positive integers, is a figure formed by connecting with straight lines every mth point out of n regularly spaced points lying on a circumference. Here n is the number of sides of the polygon, and m is called the *density* of the star polygon. Without loss of generality, take $m < n/2$. If m is not relatively prime to n, then the star polygon has more than one component. For each natural number n there exist as many distinct (not similar to each other) *connected* regular star n-gons as there are integers between 1 and $\frac{n}{2}$ relatively prime to n, i.e., $\frac{1}{2}\varphi(n)$, where $\varphi(n)$ is *Euler's* *φ-function*.

We plot a regular *star* (m,n)-*gon* with relatively prime m and n.

```
n = 5; m = 2;
i = 0 : n; polar(m*i.*2*pi/n, 1 + i - i, '-')          % Figure 1.16(b)
```

Using a similar program, we plot the *disconnected star* (m,n)-*gon*.

```
n = 12; m = 3; i = 0 : n;              % m copies (components) of n/m-gon
hold on; for j = 0 : m - 1;
  polar((i.*m+j)*2*pi/n, 1 + i - i, '-'); axis equal;
end
```

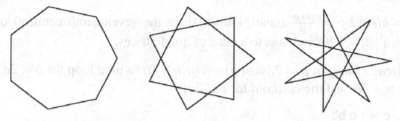

Fig. 1.17 Three regular heptagons.

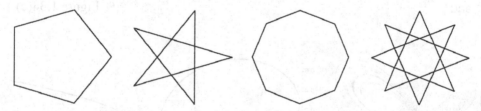

Fig. 1.18 Pairs of regular pentagons   and   octagons.

4. If for an arbitrary *plane conic section* we fix the point O at a focus and if the axis of the section through the foci plays the role of the polar axis, then the equation of the conic section in polar coordinates will take the form $\rho = \frac{p}{1-e\cos\varphi}$, where $0 \le e < 1$ for an *ellipse* (a circle when $e = 0$), $e = 1$ for a *parabola*, and $e > 1$ for a *hyperbola*. To obtain the second branch of the curve one replaces 1 by -1 in the denominator.

We plot conic sections using the following program, where f(i/15) are ellipses, f(1) is a parabola, and f(1 + (i - 8)/5) are hyperbolas; Figure 1.19(c).

```
syms t;
p = 1;  f = @(e)p/(1 - e*cos(t));
hold on;  axis equal;
ezpolar(f(1), [pi/2 - 1.1, 3*pi/2 + 1.1]);
for i = 9 : 13;
 ezpolar(f(1 + (i - 8)/5),  [pi/2 - .4, 3*pi/2 + .4]);
 ezpolar(f(i/15), [-pi, pi]);
end                                          % all types of conic sections
```

5. Plot the *butterfly* and the cochleoid, Figure 1.19(a,b).

```
t = linspace(-12*pi,12*pi,1000);
a = .8;
polar(t, exp(1)^cos(t) - 2*cos(4*t) + a*sin(t/12).^5)         % butterfly
t = linspace(-6*pi, 6*pi, 1001);  polar(t, sin(t)./t)        % cochleoid
```

6. The *cochleoid* $\rho = \frac{\sin\varphi}{\varphi}$ (snail, known since the seventeenth century) of Figure 1.19(b) with $|\varphi| < 720°$ belongs to a class of quadratrices.

7. Plot *Pascal's limaçon* $\rho = 2a\cos(\varphi) - b$ $(a, b > 0)$ (with a loop for $b < 2a$, without a loop for $b > 2a$, and the *cardioid* for $b = 2a$).

```
a = 1;  c = 'r g b';
t = linspace(-pi, pi, 101);
hold on;  axis equal;
for b = 1 : 3;
 polar(t, 2*a.*cos(t) + b, c(b));
end                                          % Figure 1.34(c)
```

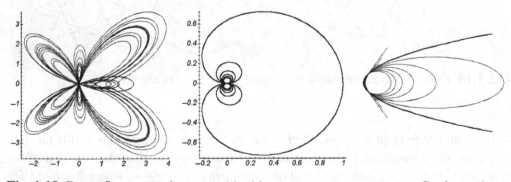

Fig. 1.19 Butterfly   and   cochleoid.   Conic sections.

1.4.2 *Remarkable curves*

Implicit equations of (algebraic) curves are often simplified when we use polar coordinates instead of rectangular ones.

Examples. Several curves are given by their geometric or kinematic properties. Determining their equations is more convenient using polar coordinates.

1. Let us fix the circle ω with diameter $|OA| = 2a$ and the tangent line l at the point A. On the ray r emanating from O and rotating around the point O one places the line segment OM of length equal to the segment of the ray between the circle and the line l. The set of all points M (on the ray r as it rotates) is called the *cissoid* (ivy-shaped). It was studied by Diocles in the third century BC in relation to the problem of *doubling a cube*. Its equation is $\rho = 2a \frac{\sin^2 \varphi}{\cos \varphi}$, Figure 1.20(b).

```
ezpolar('2*sin(t).^2/cos(t)', [-pi/2+.2, pi/2-.2])
```

Solution. Let $K = r \cap l$ and $L = r \cap \omega$ (different from O). Then $|\rho_M| = \rho_K - \rho_L$, where $\rho_K = 2a/\cos \varphi$, $\rho_L = 2a \cos \varphi$.

The construction of the cissoid can be applied to two arbitrary curves, given in polar coordinates by equations $\rho = \rho_1(\varphi)$ and $\rho = \rho_2(\varphi)$, with the aim of plotting the curve $\rho = \rho_1(\varphi) - \rho_2(\varphi)$, which is called the *cissoidal transformation*. The analogous transformation in rectangular coordinates is known as the *difference of graphs* $f(x) = f_1(x) - f_2(x)$ of two functions.

2. Fix the line g parallel to the polar axis at distance a from it. On the ray emanating from O and rotating around the point O place on both sides of the point B of intersection of the ray with g the segments BM_1 and BM_2 of length l. The set of all points M_1, M_2 (called the *conchoid*; shell-shaped) was studied by Nicomedes in the third century BC in relation to the *problem of the trisection of an angle*. Its equation is $\rho = \frac{a}{\sin \varphi} \pm l$. For $l > a$ the conchoid has a loop; for $0 < l < a$ it has a cuspidal point of the first kind. In Figure 1.20(c) the curve is rotated by 90°.

```
syms t;
a = 5; l = 4*a;
hold on; ezpolar(a./sin(t) + l, [.15, pi - .15]);
ezpolar(a/sin(t) - l, [.07, pi - .07]); axis equal
```

An obvious generalization of this construction is the *conchoid of a plane curve*, which can be obtained by increasing or decreasing the position vector at each point on the given curve by the constant segment l. If $\rho = f(\varphi)$ is the equation of the given curve in polar coordinates, then the equation of its conchoid is $\rho = f(\varphi) \pm l$. In other words, Nicomedes studied the conchoid of a line. One can show that the conchoid of a circle is Pascal's limaçon, Figure 1.34(c).

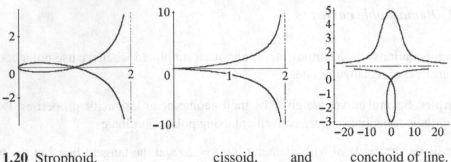

Fig. 1.20 Strophoid,   cissoid,   and   conchoid of line.

3. Fix a point A and a line l at distance a from this point; AC is the perpendicular from A onto the line l. On a ray rotating around the point A place the segments BM_1 and BM_2 from the point B of its intersection with the line l. Moreover, $BM_1 = BM_2 = CB$ is assumed. The set of all points M_1, M_2 is called the *strophoid* (twisted strip). It was studied by Torricelli in 1645. Its equation is $\rho = a\frac{1\pm\sin\varphi}{\cos\varphi}$, Figure 1.20(a).

```
hold on;  ezpolar('(1 + sin(t))/cos(t)', [-pi/2, pi/2-.5])
ezpolar('(1 + sin(t))/cos(t)', [pi/2+.5, 3*pi/2]);  axis equal
```

4. The *ovals of Cassini* are defined to be the sets of points in the plane for which the product of the distances to two fixed points is constant. It is easy to deduce the following polynomial equation of the ovals of Cassini: $(x^2 + y^2)^2 - 2c^2(x^2 - y^2) = a^4 - c^4$, or in polar coordinates, $\rho^2 = c^2\{\cos(2\varphi) \pm \sqrt{\cos(2\varphi)^2 + (a^4/c^4 - 1)}\}$. For $0 < a < c$ the curve consists of two simple closed components (for $a = 0$ it degenerates into two points), for $c < a < c\sqrt{2}$ the curve is closed with a waist that disappears for $a \geq c\sqrt{2}$. All points on the ovals with the tangent line parallel to the axis OX lie on the circle $x^2 + y^2 = c^2$. The problem of plotting these curves on the computer display is fascinating.

```
syms t;  n = 24;
f1 = @(a) 3*sqrt(abs(cos(2*t) + sqrt(cos(2*t).^2 - cos(2*a).^2)));
f2 = @(a) 3*sqrt(abs(cos(2*t) - sqrt(cos(2*t).^2 - cos(2*a).^2)));
f3 = @(b) 3*sqrt(cos(2*t) + sqrt((cos(2*t).^2 + b)));
hold on;  axis equal
for i = 1 : 6;  ti = i.*pi/n;
 ezpolar( f1(ti), [-ti, ti]);    ezpolar(-f1(ti), [-ti, ti]);
 ezpolar( f2(ti), [-ti, ti]);    ezpolar(-f2(ti), [-ti, ti]);
 ezpolar( f3(ti), [-pi, pi]);    ezpolar(-f3(ti), [-pi, pi]);
end;                                            % Figure 1.21(c)
```

For $a = c$ (see the program above) we obtain the *lemniscate of Bernoulli*

$$\rho = c\sqrt{2\cos(2\varphi)}$$

(in Latin *lemniscate* means "adorned with ribbons"), which has a point of self-intersection.

```
f = sqrt(abs(cos(2*t)));  e = pi/4;
hold on;  axis equal;  ezpolar( f, [-e, e]);  ezpolar(-f, [-e, e]);
```

5. *Kappa* is the curve consisting of all points of tangency on tangent lines from the origin to a circle of fixed radius a with moving center along the axis OX. The shape of *kappa* is similar to the Greek letter $\varkappa$ of that name. Its equation is $\rho = a\cot\varphi$.

```
syms t;  a = 1;  e = .4;  f = a*cot(t)
hold on;  axis equal;
ezpolar(f, [e, pi - e]);  ezpolar(-f, [e, pi - e]);          % Figure 1.21(a)
```

The lines $y = \pm a$ are horizontal asymptotes, the point O is the node. *Kappa* is a member of the family of curves $\rho = a\cot(k\varphi)$, called *nodal curves*. This family includes the strophoid ($k = 1/2$) and the *windmill*, having $k = 2$.

```
syms t;  a = 1;  e = 0.2;  f = a*cot(2*t);
hold on;  axis equal
ezpolar( f, [e, pi/2 - e]);  ezpolar(-f, [e, pi/2 - e]);
ezpolar( f, [-pi/2 + e, -e]);  ezpolar(-f, [-pi/2 + e, -e]);          % Figure 1.21(b)
```

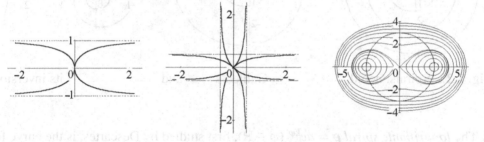

Fig. 1.21 Kappa,   windmill,   ovals of Cassini (and the lemniscate).

1.4.3 *Spirals*

The *spirals* $\rho = f(\varphi)$ (where f is a monotone function) have especially simple equations in polar coordinates. In contrast, their equations in rectangular coordinates are complicated. The spiral of Archimedes (considered in Example 2, p. 36) is a member of the family of *algebraic spirals*, which are given by polynomial equations $F(\rho, \varphi) = 0$.

Examples.

1. *Galileo's spiral* $\rho = a\varphi^2 - l$ has been known since the seventeenth century from the problem of the trajectory of a point falling near the earth's equator, starting with initial velocity from the usual rotation of the earth. The *inverse Galileo's spiral* for $l = 0$ has the equation $\rho = \frac{a}{\varphi^2}$.

    ezpolar('0.01*t.^2 - 0.02', [0, 6*pi])   % Figure 1.22(b)
    ezpolar('100/t^2', [6, 10*pi])   % Figure 1.22(c)

2. *Fermat's spiral* $\rho = a\sqrt{\varphi}$ was discovered only in 1636.

    ezpolar('sqrt(t)', [0, 6*pi])   % Figure 1.23(a)

Its conchoid is the *parabolic spiral* $\rho = a\sqrt{\varphi} + l$, $l > 0$, Figure 1.23(b).

    hold on; ezpolar('sqrt(t) + .5', [0, 4*pi]);
    ezpolar('-sqrt(t) + .5', [0, 4*pi]), axis equal

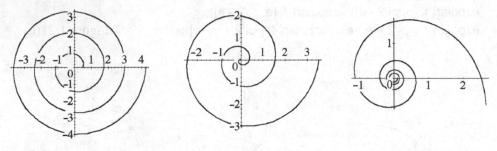

Fig. 1.22 Spirals: neoid,   Galileo's   and   its inversion.

3. The *logarithmic spiral* $\rho = ae^{k\varphi}$ ($\varphi \in \mathbb{R}$), first studied by Descartes, is the curve for which the angle between the polar radius and a tangent line at its endpoint is constant. In view of this property, this curve is intensively used in applications. In the natural world, some shells look like a logarithmic spiral.

    ezpolar('1.1^t', [-6*pi, 4*pi])   % Figure 1.23(c)

4. The *neoid* $\rho = a\varphi + l$ (the conchoid of Archimedes' spiral) is used in the construction of a spinning machine.

    ezpolar('0.2*t + 0.5', [0, 6*pi])   % Figure 1.22(a)

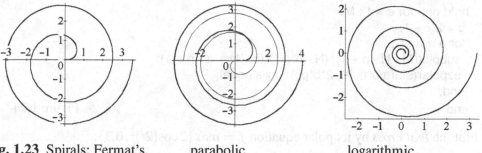

Fig. 1.23 Spirals: Fermat's,   parabolic,   logarithmic.

1.4.4 *Roses and crosses*

Since the function $\sin(k\varphi)$ is periodic, the *rose* $\rho = R\sin(k\varphi)$ consists of equal *leaves*; Figure 1.24. If the modulus $k \in \mathbb{Q}$ (a rational), then the rose is a closed algebraic curve of even order. For $k \in \mathbb{Z}$ the curve consists of k leaves for odd k and of $2k$ leaves for even k. If $k = p/q$ $(q > 0)$ with p, q relatively prime, then the rose consists of p leaves when both p and q are odd, and of $2p$ leaves when one of these integers is even; moreover, in contrast to the first case, each subsequent leaf partially covers the previous one. If $k \notin \mathbb{Q}$, then the rose consists of an infinite number of overlapping leaves.

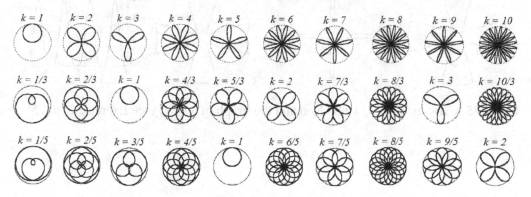

Fig. 1.24 Table with roses for $\rho = \sin(k\varphi)$.

Examples.

1. We plot the roses in Figure 1.24 using the program below:

```
syms t;
h = 1;  N = 5;  M = 3;  n = 1;              % second part: N = 10; n = 6;
```

```
hold on; for c = 1 : M;
b = 2*c - 1;
for a = n : N;
 subplot(M,  N - n + 1,  (N - n + 1)*(c - 1) + a - n + 1);
 ezpolar(sin(a/b*t), [0  2*b*pi]); axis equal;
end;
end                                              % Figure 1.24
```

2. Plot the *leaf cross* by its polar equation $f = \max\{2\cos(2t)^2, 0.3\}$.

```
f = ' max(2*cos(2*t)^2, 0.3)';
hold on;  ezpolar('0.25');  ezpolar('2.05');
ezpolar(f);  axis equal;                         % Figure 1.25(1)
```

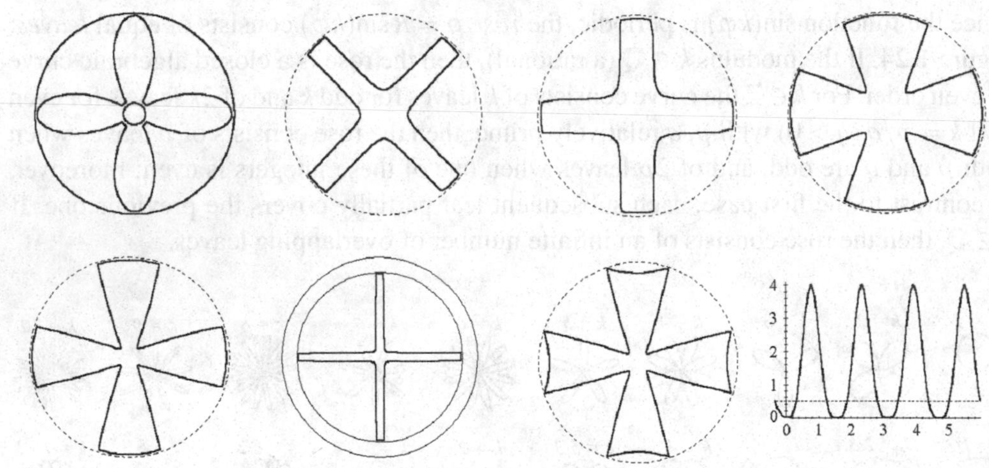

Fig. 1.25 Crosses: 1, 2, 3, 4 a,b, 5, 6 and the graph of $1/g(t)$ (see Exercise 2, p. 75).

1.5 Polynomial Interpolation of Functions

In mathematics and its applications one often needs to find a continuous function given at a finite set of points taken from its domain of definition.

The problem of *interpolation* is to find a continuous function $y = f(x)$ $(x \in I)$ such that $f(x_i) = y_i$ for all $i = 0, 1, \ldots, n$. All real, the points $x_0, x_1, \ldots, x_n$ (*nodes of a net*) are assumed different and given in increasing order. The choice of a function depends on additional boundary conditions. For example, by connecting the successive points

(x_i, y_i), one obtains a piecewise linear function. Several kinds of interpolants could be envisaged, such as:

- polynomial interpolant, $p(x) = a_0 + a_1 x + a_2 x^2 + \cdots + a_n x^n$;
- trigonometric interpolant, $p(x) = a_{-n} e^{-inx} + \cdots + a_0 + \cdots + a_n e^{inx}$;
- rational interpolant, $f(x) = \dfrac{a_0 + a_1 x + \cdots + a_n x^n}{b_0 + b_1 x + \cdots + b_x^n}$.

1.5.1 *Representation of polynomials*

MATLAB represents a polynomial $P_n(x) = p_1 x^n + p_2 x^{n-1} + \cdots + p_n x + p_{n+1}$ in two forms: a row vector p = [p(1) p(2) ... p(n + 1)] of the coefficients and a symbolic form. The two forms are equivalent from the algebraic point of view.

Here are three problems related to polynomials:

- given a polynomial's coefficients, evaluate the polynomial at one or more points;
- given a polynomial's coefficients, find the roots of the polynomial;
- given a set of data $\{(x_i, y_i)\}_{i=1}^{m}$, find a polynomial that "fits" the data.

The function polyval carries out Horner's method, which corresponds to the nested representation $p(x) = (\cdots ((p_1 x + p_2) x + p_3) x + \cdots + p_n) x + p_{n+1}$.

The roots (or zeros) of $p(x)$ are obtained with z = roots(p). The function poly carries out the converse operation: given a set of roots, it constructs a polynomial. Thus, if $\mathbf{z}$ is an n-vector, p = poly(z) gives the coefficients of the polynomial

$$p_1 x^n + p_2 x^{n-1} + \cdots + p_n x + p_{n+1} = (x - z_1)(x - z_2) \cdots (x - z_n).$$

When $p(x)$ is of degree n, then there are exactly n roots, which may be repeated or complex roots. The function poly2sym converts the result into a symbolic expression. The function sym2poly provides the inverse operation, i.e., converts the symbolic polynomial into a row vector of its coefficients.

The function conv(p, q) returns the coefficient vector for the polynomial resulting from the product of polynomials represented by p and q.

The commands polyint(p) and polyder(p) provide, respectively, the coefficients of the primitive (vanishing at $x = 0$) and those of the derivative of the polynomial (vector) p; polyder(p, q) returns the derivative of the product of the polynomials p and q.

Examples.

1. Evaluate and plot $P(x)$ over the interval $[-2, 2]$:

```
p = [1 4 -7 -10];                           % a row vector (polynomial)
x = linspace(-2, 2, 101);   P = polyval(p, x);        % values over a set x
plot(x, P)                                       % a graph
```

The symbolic alternative is the following:

```
syms x;   P = poly2sym(p, x)                    % Answer: P = x³ + 4x² − 7x − 10
ezplot(P, [-2, 2])                                             % a graph
```

2. Find primitive and derivative:

```
p = [1 3 5];
dp = polyder(p),    ip = polyint(p)
```

$dp = 3 \quad 3 \quad ip = 0.3333 \ 1.5000 \ 5.0000 \ 0$ % Answer

The symbolic alternative is the following:

```
syms x;   P = poly2sym(p, x)
dp = diff(P),   ip = int(P)
```

$dp = 2x + 3, \quad ip = (1/3)x^3 + (3/2)x^2 + 5x$ % Answer

3. MATLAB manipulates ratios of polynomials by considering the numerator and denominator polynomials separately. Verify the following:

```
n = [1 -5 5];                                        % a numerator
d = [1 5 25 0];                                      % a denominator
z = roots(n)                                         % the zeros of n(x)/d(x)
```
$2.5000 + 4.3301\,i, \quad 2.5000 - 4.3301\,i$ % Answer
```
p = roots(d)                                         % the poles of n(x)/d(x)
```
$-2.5000 + 4.3301\,i, \qquad -2.5000 - 4.3301\,i$ % Answer

Note that the symbolic polynomial fraction can be exhibited by this means:

```
P = poly2sym(n)/poly2sym(d)
```
$P = (x^2 - 5x + 25)/(x^3 + 5x^2 + 25x)$ % Answer

4. Find the partial-fraction expansion of a rational polynomial:

```
n = [1 -5 25];                                       % a numerator
d = [1 5 25 0];                                      % a denominator
[r, p, k] = residue(n, d)                            % partial-fraction expansion of n(x)/d(x)
```
$r = 0.0000 + 1.1547i \qquad 0.0000 - 1.1547i \qquad 1.0000$
$p = -2.5000 + 4.3301i \qquad -2.5000 - 4.3301i \qquad 0$
$k = [\ ]$ % there are no direct terms.

Here, r are the residues of partial-fraction coefficients, p are their poles, and k is the direct-term polynomial. For this example,

$$\frac{n(x)}{d(x)} = \frac{1.1547i}{x + 2.5000 - 4.3301i} + \frac{-1.1547i}{x + 2.5000 + 4.3301i} + \frac{1}{x}.$$

5. Given this information, the original rational polynomial can be reconstructed by using residue again:

    [nd, dd] = residue(r, p, k)   % numerator and denominator polynomials

MATLAB is based on the concept of a *matrix*. To enter the matrix $A = \begin{pmatrix} 1 & 2 & 3 \\ 4 & 5 & 6 \end{pmatrix}$ we use a semicolon to separate the rows:

    A = [1 2 3; 4 5 6]

The transpose of a real matrix is formed by the command A'. Elements of a vector or matrix are viewed by specifying their location:

    v(2); A(2, 3)

If there is a formula for the elements, say $x_j = 0.1\,j$ in $[0,2]$, we use either of these commands:

    x = 0 : .1 : 2;   x = linspace(0, 2, 20);

To create a matrix of *zeros* or *ones* of the same dimensions as a matrix A, there are these commands:

    B = zeros(size(A)), C = ones(size(A))

The same works for vectors. The $n \times n$ identity matrix is produced with the command eye(n).

There is an obvious way to add/subtract matrices or vectors (having the same dimensions). We can also multiply a matrix (or vector) by a scalar:

    A + C;   3*A;   A / 2

The matrix multiplication $A \cdot F$ is only defined for an $m \times n$ matrix A and an $n \times l$ matrix F. Note that $A \cdot A'$ is always defined.

    F = [-2 3; 4 5; 0 1];   A*F, A*A'

Some other commands of linear algebra are:

    inv   the inverse of a matrix;
    det   the determinant of a matrix;
    expm   the matrix exponential;
    [V, D] = eig(M)   the eigenvalues and eigenvectors of a matrix.

Let $f(x) = a_0 x^n + a_1 x^{n-1} + \cdots + a_n$ and $g(x) = b_0 x^m + b_1 x^{m-1} + \cdots + b_m$ (a_0, $b_0 \neq 0$) be two polynomials with coefficients in a field k. Their *resultant* is defined as $\mathrm{Res}(f,g) = \det M$, where $M = M(f,g)$ is the $(m+n)$-by-$(m+n)$ matrix

$$M = \begin{pmatrix} a_0 \ a_1 \ \ldots \ a_{n-1} \ & a_n \ \ldots \\ & a_0 \ a_1 \ \ \ldots \ a_{n-1} \ a_n \ \ldots \\ & \ldots \\ b_0 \ b_1 \ \ldots \ b_{m-1} \ & b_m \ \ldots \\ & b_0 \ b_1 \ \ \ldots \ b_{m-1} \ b_m \ \ldots \\ & \ldots \end{pmatrix}.$$

There are m rows for the coefficients a_i of $f(x)$ and n rows for the coefficients b_j of $g(x)$; all the entries other than those shown are assumed to be 0.

It can be proved that $\mathrm{Res}(f,g) = 0$ if and only if f and g have a nonconstant common divisor (or root). Hence, $\mathrm{Res}(f,f') = 0$ when f has a root of multiplicity greater than 1.

We prepare an M-file to compute the *resultant* of two polynomials. (Alternatively, the reader can use a similar procedure from MAPLE). Consider vectors of the coefficients $\mathbf{a} = [a_0, a_1, \ldots, a_n]$ and $\mathbf{b} = [b_0, b_1, \ldots, b_m]$.

```
function Rfg = Resfg(a, b)
    n = length(a) - 1;    m = length(b) - 1;
    A(1, :) = cat(2, a, zeros(1, m - 1));
    B(1, :) = cat(2, b, zeros(1, n - 1))
    for i = 2 : m;   A(i, :) = cat(2, 0, A(i - 1, 1 : n + m - 1));   end
    for i = 2 : n;   B(i, :) = cat(2, 0, B(i - 1, 1 : n + m - 1));   end
    Rfg = cat(1, A, B);                                    % matrix M
end
```

Example. Two polynomials $f = x^2 - 1$ and $g = x - 1$ have a common root $x = 1$. We detect this fact using **Resfg**.

```
a = [1, 0, -1];  b = [1, -1];
M1 = Resfg(a, b)          % obtain the matrix M1 = [1, 0, −1; 1, −1, 0; 0, 1, −1]
res1 = det(M1)                                          % obtain 0.
```

Another example: $f = x^2 + 1$ and $g = x^3 + x$ have a common divisor $x^2 + 1$.

```
a = [1, 0, 1];  b = [1, 0, 1, 0];   res2 = det(Resfg(a, b))          % obtain 0.
```

The polynomial $f = x^3 - 4x^2 + 4x$ has a root $x = 2$ of multiplicity 2. We detect this fact using **Resfg**.

```
a = [1, -4, 4 0];
b = polyder(a)                          % Answer: b = [3, −8, 4]
r3 = det(Resfg(a, b))                                   % obtain 0.
```

Further applications of **Resfg** are considered in Sections 6.2 and 8.2.

1.5.2   *Lagrange polynomials*

We all know that two points determine a first-degree polynomial, $P_1(x) = p_1 x + p_2$ (a line). Given $n+1$ points $\{(x_1, y_1), \ldots, (x_{n+1}, y_{n+1})\}$ where x_i are $n+1$ distinct real numbers, we wish to find an nth-degree *interpolation Lagrange polynomial* $P_n(x) = \sum_{j=1}^{n+1} p_j x^{n+1-j}$ whose coefficients p_i satisfy the following *linear system of $n+1$ equations*:

$$p_1 x_i^n + \cdots + p_n x_i + p_{n+1} = y_i, \quad i = 1, \ldots, n+1. \tag{1.4}$$

When $x \in (x_1, x_{n+1})$, we call $P_n(x)$ an *interpolated value*. If either $x < x_1$ or $x > x_{n+1}$, then $P_n(x)$ is called an *extrapolated value*. The matrix A of the linear system (1.4) is known as a (nonsingular!) *Vandermonde matrix*; its elements are $\{x_i^{n+1-j}\}_{i,j=1}^{n+1}$. Hence, we conclude:

For any set $\{x_i, y_i\}_{i=1}^{n+1}$ with distinct nodes x_i there is a unique Lagrange polynomial $L_n(x)$ *of degree* $\leq n$ *such that* $L_n(x_i) = y_i$ ($1 \leq i \leq n+1$).

If $\{y_i\}_{i=1}^{n+1}$ represent the values of a continuous function f, L_n is called an *interpolant* of f, and is denoted by $L_n(f)$.

Example. Using vander(x) we build the matrix A, then solve the linear system $A\,\mathbf{y} = \mathbf{a}$, see (1.4). For example,

```
x = (2 : 8);   y = [32, 34, 36, 34, 39, 40, 37];
A = vander(x);
p = A \ y';                              % the operation A⁻¹y
xi = 1.8 : .01 : 8.2;   Y = polyval(p, xi)    % Lagrange polynomial L6(f)
plot (xi, Y, x, y, 'ro')                 % graph of L6(f), Figure 1.26(b)
```

As n increases, it may not be so easy to solve the system $A\,\mathbf{y} = \mathbf{a}$. Thus we look for alternatives to recover the coefficients p_i. It happens that

$$L_n(x) = y_1 \frac{(x - x_2) \cdots (x - x_{n+1})}{(x_1 - x_2) \cdots (x_1 - x_{n+1})} + y_2 \frac{(x - x_1)(x - x_3) \cdots (x - x_{n+1})}{(x_2 - x_1)(x_2 - x_3) \cdots (x_2 - x_{n+1})}$$
$$+ \cdots + y_{n+1} \frac{(x - x_1) \cdots (x - x_{n-1})(x - x_n)}{(x_n - x_1) \cdots (x_n - x_{n-1})(x_{n+1} - x_n))},$$

or, in short, $L_n(x) = \sum_{i=1}^{n+1} y_i L_{n,i}(x)$, where $L_{n,i}(x) = \prod_{j \neq i} \frac{x - x_j}{x_i - x_j}$. The graph of $L_n(x)$ matches every data point, $L_n(x_i) = y_i$ (verify!). The error term when the Lagrange polynomial is used to approximate a continuous function $f(x)$ is similar to the error term for the Taylor polynomial. Namely, if $f(x) \in C^{n+1}$ then $f(x) = P_n(x) + \varepsilon_n(x)$, where $\varepsilon_n(x) = \varphi(x) \frac{f^{(n+1)}(\xi)}{(n+1)!}$ for some $\xi \in (x_1, x_{n+1})$ and $\varphi(x) = \prod_{i=1}^{n+1}(x - x_i)$ is a polynomial of degree $n+1$. For details see a numerical analysis course.

The coefficients of $L_{n,i}(x)$ (the *Lagrange base*) are convenient when one needs to calculate a large number of interpolation polynomials for different initially given $\{y_i\}$ at the same nodes $\{x_i\}$ of the net. In this case one needs to calculate the base only once. The interpolating Lagrange polynomial is calculated using the command p = polyfit(x, y, n). The first two input arguments, x and y, are vectors of the same length that define the interpolating points. The third argument, n, is the degree of polynomial to be calculated. The output, p, is the vector with the $n+1$ coefficients of the polynomial. Note that the same command can be used with a smaller third argument, say 1, 2, or 3, for the purpose of polynomial regression.

Despite the ease of deriving the interpolation Lagrange polynomial, its tendency towards unlimited growth of the net requires careful attention. It is known (K. Weierstrass, 1885) that a continuous function defined on a segment can be approximated as closely as desired on this segment by a polynomial. In particular, *given n and $f(x)$, the sequence of polynomials*

$$B(f,n) = \sum_{i=0}^{n} C_n^i x^i (1-x)^{n-i} f(i/n)$$

converges uniformly to $f(x)$ *on* $[0,1]$. The interpolation polynomial $L_n(x)$ is not necessarily close to the function $f(x)$ at each point of the segment $[a,b]$; it can vary from this function by an arbitrarily large amount.

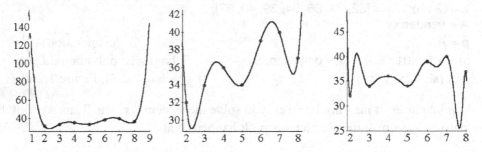

Fig. 1.26 Lagrange and Hermite polynomials.

One can conclude from examples and the results of exercises below something about weak *approximation properties* of interpolation polynomials.

Example. (K. Runge). Show that for an unbounded increase in the number of nodes on the segment $[a,b]$ for the function $f(x) = \frac{1}{1+25x^2}$, we have

$$\lim_{n\to\infty} \max_{-1\le x\le 1} |f(x) - L_n(x)| = \infty.$$

```
n = 6;   x = linspace(-1, 1, n);   y = 1./(1 + 25*x.^2);
f = polyfit(x, y, n - 1)          % f = 0.00, 1.20, −0.00, −1.73, 0.00, 0.57
```

$$f(x) \approx 1.202x^4 - 1.731x^2 + 0.567 \qquad \text{interpolation polynomial}$$

```
xi = linspace(-1, 1, 100);  p = polyval(f, xi);
plot(xi, p,  x, y, 'ro')                          % Figures 1.27(a,b)
```

By increasing the degree n, we do not necessarily obtain a better reconstruction of f. The problem occurs because the nodes are equally spaced!

Runge's phenomenon can be avoided if a suitable distribution of nodes is used. Namely, in an interval $[-1, 1]$ we can consider the *Chebyshev nodes* $x_i = -\cos(\pi i/n)$, where $i = 0, \dots, n$.

```
x = -cos(pi*[0 : n]/n);
```

For this special distribution of nodes and any C^1-regular function f in $[-1, 1]$, we have $L_n(x) \to f$ as $n \to \infty$ for all $x \in [-1, 1]$. Using

$$(n+1)!|\varepsilon_n(x)| \le |\varphi(x)| \max_{-1 \le x \le 1} |f^{(n+1)(x)}|,$$

our task is to follow Chebyshev's derivation to select the set of nodes x_i in a way that minimizes $\max_{-1 \le x \le 1} |\varphi(x)|$.

It so happens that, for the Lagrange interpolation $f(x) = P_n(x) + \varepsilon_n(x)$ on $[-1, 1]$, the minimum value of the error bound $\varepsilon_n(x)$ is achieved when the nodes $\{x_i\}$ are the *Chebyshev abscissas* (1.29) of $T_{n+1}(x)$. Moreover,

$$\max_{-1 \le x \le 1} |L_{n+1}(x)| \ge \max_{-1 \le x \le 1} |T_{n+1}(x)| = 1/2^{n+1}.$$

Example. (S. Bernstein). Check by calculation that the sequence of interpolation Lagrange polynomials $L_n(x)$ on uniform nets for the continuous function $f(x) = |x|$ on the segment $[-1, 1]$, with an increasing number of nodes n, does not converge to the function $f(x)$. For plotting using the previous program, we replace the specification of y with y = abs(x); see Figure 1.27(c,d).

1.5.3 *Hermite polynomial*

The *interpolation Hermite polynomials* arise when the derivatives at the given points are also known in addition to the given values of the function. Since the number of conditions is doubled, the interpolation polynomial has degree $2n + 1$. Given vectors $x = (x_1, x_2, \dots, x_{n+1})$, $y = (y_1, y_2, \dots, y_{n+1})$ and derivatives $y' = (y'_1, y'_2, \dots, y'_{n+1})$, the *interpolation Hermite polynomial* is

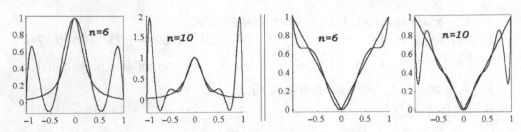

Fig. 1.27 Examples of Runge ($n = 6, 10$) and Bernstein ($n = 6, 10$).

$$H_n(x) = \sum_{i=1}^{n+1} y_i \left[1 - 2(x - x_i)\frac{d}{dx}\bar{L}_i(x_i)\right] \bar{L}_i(x)^2 + \sum_{i=1}^{n+1} y_i'(x - x_i)\bar{L}_i(x)^2,$$

where $\bar{L}_i(x)$ is the Lagrange polynomial with conditions $\bar{L}_i(x_j) = \delta_{i,j}$.

Examples.

1. Find $H_n(x)$ using polyfit(x, y, n+1). In the M-file **hermite_interp2d.m** the derivatives at nodes of the net are zero (for simplicity), i.e., tangent lines to the graph at the nodes are horizontal.

```
x = [2 : 8];  n = length(x) - 1;
y = [32, 34, 36, 34, 39, 40, 37];   z = [zeros(1, n + 1)];
H = hermite_interp2d(x, y, z)
```

$$H = 10^6[0, 0, 0, 0.0001, -0.0010, 0.0085, -0.0519, 0.2325,$$
$$-0.7681, 1.8456, -3.1327, 3.5553, -2.4165, 0.7425] \qquad \% \text{ Answer}$$

```
xi = linspace(1.9, 8.1, 201);   HERM = polyval(H, xi);
plot(xi, HERM, x, y, 'ro')                              % Figure 1.26(c)
```

2. Compare Lagrange and Hermite polynomials of the first half of the unit circle using only 7 nodes.

```
t = linspace (0, pi, 7);
x = cos(t);  y = sin(t);
xh = linspace (0, 1, 100);  yh = pchip(x, y, xh);       % 100 points on [0, 1]
pL = polyfit(x, y, 6);    yl = polyval(pL, xh);         % two interpolants
plot(x, y, 'o',  xh, yh,  xh, yl, '–r'),  axis equal
```

The command yh = pchip(x, y, xh) provides the piecewise cubic Hermite polynomial $p(x)$ (see Example 1, p. 52), which is locally monotone and interpolates the function as well as its first derivative at the nodes $\{x_i\}_{i=1}^{n+1}$. The first two input arguments, x and y, are vectors of the same length that define the interpolating points. The third input argument, xh, is a vector of points where the function is to be evaluated. The output, yh, is the same length as xh. The first derivative $p'(x)$ is continuous, $p''(x)$ may not be continuous; there may be jumps at the x_i.

The command interpft provides the *trigonometric* interpolant (obtained by a linear combination of sines and cosines) of a set of data. The Curve Fitting Toolbox provides a graphical interface for solving (non)linear fitting problems; see also the Optimization Toolbox.

Summary: Polynomial interpolation is successful only for a small number of points (not more than 15): if the number of points is increased, then the degree of the polynomial increases too, contributing little to reducing the approximation error (the Runge phenomenon). Moreover, large oscillations (the *polynomial wiggle*) take place on intervals between the points.

1.5.4 *Least-squares analysis*

Instead of exact matching at every data point, we can use an approximate function (not necessarily a polynomial) that describes the points as a whole with the smallest error in some sense. This is called *curve fitting*.

Given points $\{x_i, y_i\}_{i=1}^{n+1}$, the method of *least squares* (LS) consists in finding the functional form $p(x, a_1, \ldots, a_m)$ (containing several adjustable parameters) that minimizes the value $\delta^2 = \sum_{i=1}^{n+1}(p(x_i, \mathbf{a}) - y_i)^2$. The graph of obtained function lies near the given points (x_i, y_i).

Polynomial regression is a form of LS curve fitting in which the selected function is a polynomial and the parameters to be chosen are the polynomial coefficients. The *method of* LS is often used for smoothing purposes, and it is called by the command polyfit (the coefficients of a polynomial) and then polyval (the values of a polynomial on the net).

Given n, we find the optimum coefficients p_i of $p(x) = \sum_{j=0}^{n} p_j x^{n-j}$. This leads to the *Vandermonde matrix* $A = \{x_i^{n-j}\}_{i,j=1}^{n+1}$ using vander(x) (x_i are distinct real numbers), and to the linear system $\mathbf{a} = A\backslash\mathbf{y}$.

Examples.

1. *Linear regression* $y = c_1 x + c_2$. We find the best solution (c_1, c_2) of the over-determined linear system $c_1 x_i + c_2 = y_i$, $i = 1, 2, \ldots, n$. The linear approximation (by a straight line) is obtained by matrix calculation:

```
x = (2 : 8)';  y = [32, 34, 36, 34, 39, 40, 37]';          % enter data
A = [x  ones(size(x))];
c = A \ y          % Answer: c = [1.0714, 30.6429],  y = 1.0714x + 30.6429
Y = c(1)*x + c(2);
plot (x, Y,  x, y, '--r')                                   % graph
```

```
r = y - A*c;   d2 = r'*r                              % Answer: estimate the error
```

2. An nth-order polynomial approximation may be obtained by direct matrix calculation. We use the program below.

```
n = 3;   x = (2 : 8)';   y = [32, 34, 36, 34, 39, 40, 37]';              % enter data
N = length(x);   A = ones(N, 1);
xk = x;
for k = 1 : n,   A = [xk A];   xk = x.*xk;   end;                        % matrix A
c = A \ y;              % Answer: c = [−0.1111; 1.4762; −4.5794; 36.6429]
xi = linspace(x(1), x(N), 50);
Y = polyval(c, xi);            % y = −0.111x³ + 1.476x² − 4.579x + 36.643]
plot (xi, Y, x, y, '--r')                                               % graph
```

3. The simple form $f(x) = ax^b$ is a common solution to many physical problems
and is nonlinear in the parameter b. Taking the logarithm of this equation, $\log f = \log a + b \log x$, we apply linear regression. First, we should replace the data (x_i, y_i) by
$(\log x_i, \log y_i)$.

For example, to estimate the parameters a, b from the experimental values $f(6) = 15$, $f(7) = 12$, $f(8) = 9$, $f(9) = 6$, we write:

```
x = 6 : 9;   f = [15 12 9 6];
p = polyfit(log10(x), log10(f), 1), a = 10^p(2)     % p = −2.22, 2.93, a = 854.06.
```

The coefficients are $a \approx 854.06$, $b \approx -2.22$, and the derived relation is $f(x) \approx 854.06 x^{-2.22}$. The reader is invited to reformulate the previous codes into codes that
use these functions.

Other nonlinear relations that can be linearized to fit the least-squares algorithm are
listed in the following table, where the linearized function is always $y' = a'x' + b'$ for
suitable x', y', a', b':

N	Function to Fit	Variable Substitution/Parameter Restoration
1	$y = a/x + b$	$x' = 1/x$
2	$y = b/(x+a)$	$y' = 1/y,\ a = b'/a',\ b = 1/a'$
3	$y = ab^x$	$y' = \ln y,\ a = e^{b'},\ b = e^{a'}$
4	$y = be^{ax}$	$y' = \ln y,\ b = e^{b'}$
5	$y = c - be^{-ax}$	$y' = \ln(c - y)$
6	$y = ax^b$	$y' = \ln y,\ x' = \ln x$
7	$y = axe^{bx}$	$y' = \ln(y/x)$
8	$y = c/(1 + be^{ax})$	$y' = \ln(c/y - 1),\ b = e^{b'}$
9	$y = a \ln x + b$	$x' = \ln x$

The general form of the least-squares approximation consists of using, in $\mathbf{a} = A \backslash \mathbf{y}$,
forms of f and p_m that are no longer polynomials, but functions belonging to a space

V_n obtained by linearly combining $n + 1$ independent functions φ_j, $j = 0, \ldots, n$. The unknown coefficients of $\tilde{f}(x) = \sum_{j=0}^{n} a_j \varphi_j(x)$ can be obtained by solving the system $B'B\mathbf{a} = B'\mathbf{y}$ (verify!), where B is the rectangular $(n+1) \times (m+1)$ matrix of entries $b_{ij} = \varphi_j(x_i)$, $\mathbf{a}$ is the vector of the unknown coefficients, and y is the vector of the data.

In general, we find the optimum set of coefficients a_i in the linear combination $y(x) = \sum_{j=0}^{n} a_j \varphi_j(x)$ of a set of functions $\{\varphi_1(x), \ldots, \varphi_n(x)\}$. This leads to the *generalized Vandermonde matrix*

$$A = \begin{pmatrix} \varphi_1(x_1) & \varphi_2(x_1) & \cdots & \varphi_{n-1}(x_1) & \varphi_n(x_1) \\ \varphi_1(x_2) & \varphi_2(x_2) & \cdots & \varphi_{n-1}(x_2) & \varphi_n(x_2) \\ \vdots & \vdots & \ddots & \vdots & \vdots \\ \varphi_1(x_N) & \varphi_2(x_N) & \cdots & \varphi_{n-1}(x_N) & \varphi_n(x_N) \end{pmatrix}, \quad (1.5)$$

and we solve the equation $\mathbf{a} = A \backslash \mathbf{y}$.

1.6   Spline Interpolation of Functions

If the requirement on the smoothness of an interpolation function is weakened, then without greatly complicating the calculations one can obtain the nice approximation properties of the spline. The idea is to derive polynomials (or other standard functions) independently one from another on each segment $[x_i, x_{i+1}]$, then glue them at their endpoints.

1.6.1   Cubic spline

Before the first work on *splines* (Shenberg, 1946) there was earlier work of Leibniz and Euler by direct methods of the calculus of variations (for example, the Euler *polygon* for numerical solution of ODEs).

The notion of a *spline* ("flexible ruler") came from practice. Splines are used in various graphical packages and systems of design in industry.

A polynomial *cubic spline* defined by the vectors $\mathbf{x} = [x_1, \ldots, x_{n+1}]$ and $\mathbf{y} = [y_1, \ldots, y_{n+1}]$ is derived using the command $S = \text{spline}(\mathbf{x}, \mathbf{y})$. The result is the piecewise function $S(x) \in C^2$ composed of n polynomials $P_i = a_{i,0} + a_{i,1}x + a_{i,2}x^2 + a_{i,3}x^3$ of degree 3:

$$S = \begin{cases} P_1(x), & \text{if} \quad x \leq x_2, \\ P_i(x), & \text{if} \quad x_i \leq x \leq x_{i+1} \quad \text{and} \quad i = 2, 3, \ldots, n-1, \\ P_n(x), & \text{if} \quad x_n \leq x. \end{cases}$$

Example. Find a polynomial cubic spline.

```
pp = spline([0 : 3], [0, 1, 4, 3]);    pp.coefs
-1 4 -2 0;  -1 1 3 1;  -1 2 2 4;     % coefficients of p_1(x), p_2(x), p_3(x)
```

Strategy. We derive $4n$ coefficients of n polynomials $P_i(x)$. The equalities $S(x_i) = y_i$ for $1 \le i \le n+1$ give us $n+1$ equations. The condition $S(x) \in C^2$ means that the function $S(x)$ and its derivatives $S'(x)$ and $S''(x)$ at all $n-1$ inner nodes of the net are continuous. This gives an additional $3(n-1)$ equations for finding all the coefficients. Thus, we have obtained a total of $3(n-1) + (n+1) = 4n - 2$ equations for the coefficients a_{ij}. The two missing conditions come from restrictions on values of the spline and/or its derivatives at the endpoints x_1 and x_{n+1}. For example, $S'(x_1) = S'(x_{n+1}) = 0$ (*first kind*) or $S''(x_1) = S''(x_{n+1}) = 0$ (*second kind*) or $S'(x_1) = S'(x_{n+1})$, $S''(x_1) = S''(x_{n+1})$ (*periodic*).

In contrast to interpolation Lagrange polynomials, the sequence of interpolation cubic splines on a uniform net always converges to a continuous interpolation function. Moreover, the velocity of convergence increases when the differential properties of the function are strengthened.

The command spline constructs $S(t)$ in almost the same way that pchip constructs $p(t)$. However, spline chooses the slopes at the x_i differently, namely to make $S''(t)$ even continuous. The MATLAB toolbox splines allows one to explore several applications of spline functions.

The command interp1(x, y, xi) interpolates using alternative methods: 'linear'—linear interpolation (default); 'spline'— cubic spline interpolation; 'pchip'— piecewise cubic Hermite interpolation, etc.

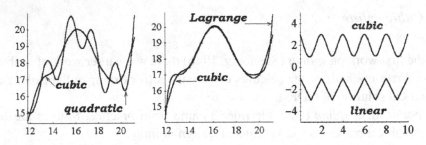

Fig. 1.28 Interpolation of data "$t°$ over 10 days".

Example. Compare the behavior of the Lagrange and Hermite polynomial interpolations with that of cubic and quadratic splines derived for the measurements of air temperature given in Example 3, p. 19.

```
Date = 12 : 21;  T = [15, 17, 17.5, 19, 20, 19.5, 18, 17, 17, 19];
x = linspace(11.8, 21.2, 41);                                    N = 41 points
f1 = interp1(Date, T, x, 'pchip');                        % Hermite cubic spline
f2 = polyfit(Date, T, length(T) - 1);  p2 = polyval(f2, x);            % Lagrange
f3 = interp1(Date, T, x, 'spline');                              % cubic spline
f4 = interp1(Date, T, x, 'square');                          % quadratic spline
plot(x, [f1; p2; f3; f4], Date, T, 'ro')        % (41 values) Figure 1.28(a,b)
```

1.6.2 *Construction of curves using splines*

In the treatment above, we have considered arrays of points whose abscissas form an increasing sequence of real numbers. With those methods it would not have been possible to compute a cubic spline for the spiral as a function of x, since there are multiple y values for each x near the origin. However, it is possible to compute a spline (or interpolating polynomial) for each axis with respect to the variable or parameter t, and to obtain interpolation or smoothing curves for arbitrary given points in the plane, in space, or in $\mathbb{R}^N$ ($N > 3$). The class of modeled curves is essentially enlarged: now they are defined by parametric equations $\mathbf{r}(t) = [x_1(t), \ldots, x_N(t)]$.

We graph the spline (or interpolating polynomial) using the following procedure:

1. *For a given segment $[a,b]$ of parameter t's domain, one defines an auxiliary net $a = t_1 < \cdots < t_n = b$, where n coincides with the number of control points.*
2. *Given an array $\mathbf{P} = \{P_i = (x_{i1}, \ldots, x_{iN})\}_{i=1}^n$ of n control points in $\mathbb{R}^N$, one defines N auxiliary arrays $X_j = \{(t_i, x_{ij})\}_{i=1}^n$, $j = 1, \ldots, N$.*
3. *For each array X_j ($j = 1, \ldots, N$) one derives the corresponding spline function (or interpolating polynomial) $x_j(t)$, where $t \in [a,b]$.*

Such a parametric curve may not be C^1-regular, because one cannot exclude the possibility of simultaneous vanishing (for some t_0) of the derivatives: $x'(t_0) = y'(t_0) = z'(t_0) = 0$. In Chapter 9 we study more powerful alternatives for plotting piecewise curves.

Examples.

1. Let us apply splines to interpolate spiral curves in $\mathbb{R}^2$.

```
t = linspace(0, 3*pi, 15);                                       % 15 points
x = sqrt(t).*cos(t); y = sqrt(t).*sin(t);   % data (Fermat's spiral, Figure 1.23(a))
ppxy = spline(t, [x; y]);
ti = linspace(0, 3*pi);                                  % total range, 100 points
xy = ppval(ppxy, ti);                                    % evaluate both splines
plot( xy(1, :), xy(2, :),  x, y, 'o- -');                % curve and polygonal line
```

2. As a concrete application of spline functions to space curves, we tie the following *knot*, well known to sailors, Figure 1.29(a).

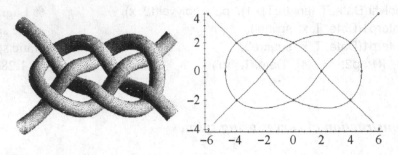

Fig. 1.29 Knot well known to sailors.

```
t = [0, 1/3, 2/3, 1, 4/3, 3/2, 5/3, 2, 7/3, 8/3, 3.1];
x = [-6, -4, -2, 0, 4, 4.8, 4, 0, -2, -4, -6];
y = [4, 2, 0, -2, -2, 0, 2, 2, 0, -2, -4];
z = [1, 1, -1, 1, -1, 0, 1, -1, 1, -1, -1];
T = linspace(0, 3.1, 500);
X = spline(t, x, T);   Y = spline(t, y, T);   Z = spline(t, z, T);        % splines
hold on;   view(3);
plot3( X, Y, Z, 'LineWidth', 14);      plot3(-X, Y, -Z, 'r', 'LineWidth', 14);
```

Alternatively, the MATLAB function spline can fit *any number n* of splines simultaneously and return a pp-form structure with all fits. In this approach, when $n = 3$, we replace the sixth line (in the program above) by these two lines:

```
S = spline(t, [x; y; z], T);          % vector function of three components
X = S(1, :);   Y = S(2, :);   Z = S(3, :);
```

To understand the idea of Figure 1.29(a), we plot the projection in the plane:

```
plot( X, Y, -X, Y, 'r', x, y, 'o--');                          % Figure 1.29(b)
```

1.6.3 *Extremal property of a cubic spline*

The interpolation cubic spline is the unique solution of the following extremal problem: *Derive a function $f(x)$ $(x \in [a,b])$ minimizing the functional $J(f) = \int_a^b (f''(x))^2 dx$ among all functions of class C^2 whose graphs contain the given points $\{(x_i, y_i)\}$ $(i = 1, 2, \ldots, n+1)$.*

Moreover, some additional conditions are required. The choice of the functional $J(f)$ is explained by the fact that, according to the Bernoulli–Euler physical law, the linearized equation of a bent ruler has the form $Ei \cdot f''(x) = -M(x)$, where $f(x)$ is a bend function, $M(x)$ varies linearly from bearing to bearing (the bending moment), and Ei is the rigidity of the ruler.

Denote by $W_2^2[a,b]$ the class of all real functions $f \in C^1[a,b]$ such that

$$\int_a^b (f''(x))^2 \, dx < \infty.$$

Note that $W_2^2[a,b] \subset C^2[a,b]$.

Theorem 1.1. *Assume that $f \in W_2^2[a,b]$ and that $S(x)$ is the unique cubic spline interpolant for $f(x)$ that passes through the points $\{(x_i, f(x_i))\}_{i=1}^{n+1}$ $(a = x_1 < \cdots < x_{n+1} = b)$ and satisfies either of the end conditions*

$$(a) \quad S''(a) = S''(b) = 0 \quad \text{if} \quad f''(a) = f''(b) = 0, \tag{1.6}$$

$$(b) \quad S'(a) = f'(a) \quad \text{and} \quad S'(b) = f'(b). \tag{1.7}$$

Then $\int_a^b (S''(x))^2 \, dx \le \int_a^b (f''(x))^2 \, dx.$

Proof. Integration by parts and either of the end conditions give us

$$\int_a^b S''(x)(f''(x) - S''(x)) \, dx = -\int_a^b S'''(x)(f'(x) - S'(x)) \, dx.$$

Since $S'''(x) = \text{const}$ on the subinterval $[x_i, x_{i+1}]$, it follows that

$$\int_{x_i}^{x_{i+1}} S'''(x)(f'(x) - S'(x)) \, dx = \text{const} \cdot (f(x) - S(x))|_{x_i}^{x_{i+1}} = 0.$$

Hence, $\int_a^b S''(x)(f''(x) - S''(x)) \, dx = 0$, and it follows that

$$\int_a^b S''(x) f''(x) \, dx = \int_a^b (S''(x))^2 \, dx. \tag{1.8}$$

Since $0 \le (f''(x) - S''(x))^2$, we obtain the integral relationship

$$0 \le \int_a^b [(f''(x))^2 - 2S''(x)f''(x) + (S''(x))^2] \, dx. \tag{1.9}$$

Now (1.8) is substituted into (1.9), which completes the proof. $\square$

The effectiveness of the spline in approximation can be explained to a considerable extent by its striking convergence properties.

Now we are ready to construct the cubic spline $S(x)$ directly. Given $\{(x_i, y_i)\}_{i=1}^{n+1}$, define $h_i = x_{i+1} - x_i$ and $m_i = S''(x_i)$. We will seek the segments $S_i(x)$ in the prescribed form involving only the unknowns $\{m_i\}$:

$$S_i(x) = \frac{m_i}{6h_i}(x_{i+1} - x)^3 + \frac{m_{i+1}}{6h_i}(x - x_i)^3$$

$$+ \left(\frac{y_i}{h_i} - \frac{m_i h_i}{6}\right)(x_{i+1} - x) + \left(\frac{y_{i+1}}{h_i} - \frac{m_{i+1} h_i}{6}\right)(x - x_i). \quad (1.10)$$

One may verify easily that $S_i(x_i) = y_i$, $S_i(x_{i+1}) = y_{i+1}$, $S_i''(x_i) = m_i$, $S_i''(x_{i+1}) = m_{i+1}$. To find the values $\{m_i\}$, use the continuity of the derivative of (1.10): $S_{i+1}'(x_i) = S_i'(x_i)$, where clearly

$$S_{i-1}'(x_i) = \frac{m_i}{3}h_{i-1} + \frac{m_{i-1}}{6}h_{i-1} + d_{i-1}, \qquad S_i'(x_i) = -\frac{m_i}{3}h_i - \frac{m_{i+1}}{6}h_i + d_i$$

with $d_i = (y_{i+1} - y_i)/h_i$. From this there follow the $n - 1$ linear equations

$$\frac{m_{i-1}}{6}h_{i-1} + \frac{m_i}{3}(h_i + h_{i-1}) + \frac{m_{i+1}}{6}h_i = d_i - d_{i-1}, \qquad i = 2, \ldots, n. \quad (1.11)$$

The endpoint constraints (see the table in what follows) complete (1.11) with two equations, and the resulting linear system has a unique solution. For the natural cubic spline,

$$\begin{pmatrix} 2(h_1 + h_2) & h_2 & & \\ h_2 & 2(h_2 + h_3) & h_3 & \\ & & \cdots & \\ & & h_{n-1} & 2(h_{n-1} + h_n) \end{pmatrix} \begin{pmatrix} m_2 \\ m_3 \\ \cdots \\ m_n \end{pmatrix} = 6 \begin{pmatrix} d_2 - d_1 \\ d_3 - d_2 \\ \cdots \\ d_n - d_{n-1} \end{pmatrix}. \quad (1.12)$$

N	Endpoint Constraints	Equations
1	Clamped cubic spline: Specify $S'(x_1)$, $S'(x_{n+1})$	$m_1 = \frac{3}{h_1}(d_1 - S'(x_1)) - \frac{1}{2}m_2$ $m_{n+1} = \frac{3}{h_n}(S'(x_{n+1}) - d_n) - \frac{1}{2}m_n$
2	Natural cubic spline	$m_1 = m_{n+1} = 0$
3	Extrapolate $S''(x)$ to the endpoints	$m_1 = m_2 - \frac{h_1}{h_2}(m_3 - m_2)$ $m_{n+1} = m_n + \frac{h_n}{h_{n-1}}(m_n - m_{n-1})$
4	$S''(x) = $ const near the endpoints	$m_1 = m_2$ $m_{n+1} = m_n$
5	Specify $S''(x)$ at each endpoint	$m_1 = S''(x_1)$ $m_{n+1} = S''(x_{n+1})$

The disadvantages of the cubic spline are the following:

1. There is no local control: Modifying the extreme tangent vectors changes (1.12) and results in a different set of n tangent vectors. The entire function (and its graph) is changed!

2. Equation (1.12) is a system of n equations that, for large values of n, may be too slow to solve.

Generalizing the interpolating splines, one comes to the notion of the smoothing polynomial spline.

A *smoothing cubic spline* is a function $f(x)$ $(x \in [a,b])$ of class C^2 which

- on each segment $[x_i, x_{i+1}]$ is a polynomial of degree $d = 3$, and which
- minimizes the functional

$$ J(f) = \alpha \int_a^b (f''(x))^2 \, dx + \sum_{i=1}^{n+1} \rho_i (f(x_i) - y_i)^2, \qquad (1.13) $$

where y_i and $\alpha, \rho_i > 0$ are given numbers.

- *End conditions* have, for example, either of the following forms:

$$ \begin{array}{ll} (a) & f''(a) = f''(b) = 0, \\ (b) & f'(a) = A \quad \text{and} \quad f'(b) = B \quad \text{(for some } A, B). \end{array} \qquad (1.14) $$

The choice of weight coefficients $\alpha, \rho_i > 0$ involved in the functional allows one in some sense to control the properties of smoothing splines. If all $\rho_i \to \infty$, then the spline is actually an interpolation spline, and for small values of ρ_i the spline is close to the straight line obtained by the least-squares method. Some generalizations of these smoothing splines are known for arbitrary degree $d > 3$.

Theorem 1.2. *Among all functions $f \in W_2^2[a,b]$, the cubic spline with either of the end conditions (1.14) is the unique function that minimizes the functional (1.13).*

Proof. Let the function $f \in W_2^2[a,b]$ minimize the functional (1.13), and assume f is not a cubic spline. Then there is a cubic spline $S(x)$ $(x \in [a,b])$ that interpolates $f(x)$ on the set $\{x_i\}$ and satisfies either of the conditions (1.14). The second term of (1.13) coincides for S and f, but, by Theorem 1.1, the first term is smaller for S. Hence $J(S) < J(f)$, a contradiction. The uniqueness of a spline also follows from Theorem 1.1. $\square$

1.7   Two-Dimensional Interpolation and Smoothing

Two-dimensional interpolation and smoothing deal with functions of two variables $f(x,y)$, and are based on the same ideas as the one-dimensional case.

1.7.1 *Two-dimensional interpolation*

MATLAB has two functions for two-dimensional interpolation (i.e., values on a surface): griddata and interp2. The syntax for the first is griddata(x, y, z, X1, Y1). The function interp2 has a similar argument list, but it requires x and y to be monotonic matrices in the form produced by meshgrid. The commands interp2(X, Y, Z, XI, YI, method) and griddata(x, y, z, XI, YI, method) specify an alternative interpolation method:

> 'nearest' – nearest-neighbor interpolation (i.e., piecewise constant),
> 'linear' – bilinear interpolation (default),
> 'spline' or 'cubic' – bicubic spline interpolation.

Example. Using the commands meshgrid, griddata.

```
x = rand(100, 1)*4 - 2;
y = rand(100, 1)*4 - 2;
z = x.*exp(-x.^2 - y.^2);
hi = -2 : .1 : 2;
[XI, YI] = meshgrid(hi);
ZI = griddata(x, y, z, XI, YI);
mesh(XI, YI, ZI),
hold on;   plot3(x, y, z, 'o')                          % Figure 1.31(a)
```

These concepts extend naturally to higher dimensions where other interpolation functions interp3, griddata3, interpn (for three-dimensional), and griddatan, ndgrid (for n-dimensional) interpolation apply.

The points of the arrays under consideration above are enumerated so that their abscissas and ordinates generate two strongly increased sequences. This determines the selection of the approximating class (graphs of functions) and the mode of their construction. Nevertheless, the method outlined above permits the construction of the interpolating surface with success in a more general case in which a spline is computed for each axis with respect to the variables or parameters (u,v), and interpolation or smoothing surfaces may be obtained for arbitrary given points in the space $\mathbb{R}^N$ ($N \geq 3$).

The class of modeled surfaces is significantly enlarged by the addition of self-intersecting, self-overlying surfaces, or surfaces which cannot be projected on any of the coordinate planes. Surfaces from the new class are described by parametric equations $\mathbf{r}(u,v) = [x_1(u,v), \ldots, x_N(u,v)]$.

The plan to visualize the spline surfaces is the following.

1. *Given a rectangle $D = [a,b] \times [c,d]$ of parameters (u,v), define an auxiliary net $\pi = \{a = u_1 < \cdots < u_n = b, \, c = v_1 < \cdots < v_m = d\}$ where nm coincides with the number of control vertices.*

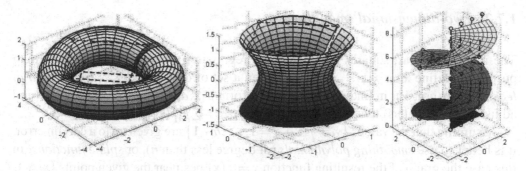

Fig. 1.30 Spline interpolation of surfaces: torus, hyperboloid and helicoid.

2. *Given an array of nm control vertices in* $\mathbb{R}^N$,

$$\mathbf{P} = \{P_{ij} = (x_{1;ij}, \ldots, x_{N;ij}), \; i = 0,1,\ldots,n; \; j = 0,1,\ldots,m\},$$

one defines auxiliary arrays $X_k = \{(u_{ij}, v_{ij}, x_{k;ij})\}_{j=1,\ldots,m}^{i=1,\ldots,n}$, $k = 1,\ldots,N$.

3. *For each array* X_k $(k = 1,\ldots,N)$ *one derives the corresponding spline function* $x_k(u,v)$, *where* $(u,v) \in D$.

Such a parameterized surface may not be C^1-regular, because one cannot exclude the possibility of collinearity (for some value (u_0, v_0)) of the velocity vectors $\mathbf{r}'_u(u_0, v_0)$, $\mathbf{r}'_v(u_0, v_0)$. In Section 9.6 we study further useful possibilities for approximating surfaces.

Example. Let us plot spline surfaces in $\mathbb{R}^3$, using interp2.

First, we prepare the data for a single-sheeted hyperboloid:

```
n = 4;   m = 5;                                        % enter n, m
[uu, vv] = meshgrid(-1 : 2/(n-1) : 1,  0 : 2*pi/(m-1) : 2*pi);
X = cosh(uu).*cos(vv);   Y = cosh(uu).*sin(vv);
Z = sinh(uu);                                          % enter X, Y, Z
[UI, VI] = meshgrid(-1 : .1 : 1,  0 : .2 : 2*pi);
```

Then we calculate splines

```
SX = interp2(uu, vv, X, UI, VI, 'spline');
SY = interp2(uu, vv, Y, UI, VI, 'spline');
SZ = interp2(uu, vv, Z, UI, VI, 'spline');
```

and plot the surface (with the control points) by the typical program:

```
hold on;   view(3);   grid on;   axis equal;
plot3(X, Y, Z, 'o- -k', 'LineWidth', 2);
surf(SX, SY, SZ);                                      % Figure 1.30(b)
```

1.7.2 *Two-dimensional smoothing*

The method of least squares (or polynomial regression) is often used for smoothing purposes. Define $\mathbf{x} = (x_1, \ldots, x_k) \in \mathbb{R}^k$. Given a set of data $\{\mathbf{x}_i; y_i\}_{i=1}^{n+1}$, the *method of least squares* consists in finding the functional form $p(\mathbf{x}, a_1, \ldots, a_m)$ (containing several adjustable parameters) that minimizes the value $\varepsilon^2 = \sum_{i=1}^{n+1} (p(\mathbf{x}_i, a_1, \ldots, a_m) - y_i)^2$.

Assuming that the values $\{y_i = f(\mathbf{x}_i),\ i = 1, \ldots, n+1\}$ are given up to a certain error, it is better to use *smoothing polynomials* (of degree less than n), or *spline functions*. In this case the graph of the resulting function $y = \tilde{f}(\mathbf{x})$ lies near the given points $(\mathbf{x}_i, y_i)$.

Note that the command polyfit does not work for $\mathbb{R}^k$, $k > 1$.

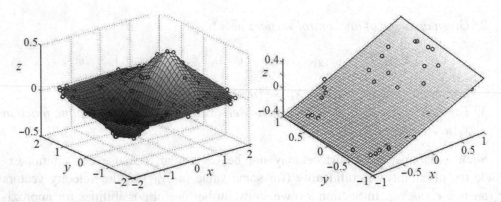

Fig. 1.31 (a) Bicubic spline interpolation. (b) Linear regression.

Example. *Linear regression* is a form of polynomial regression in which the polynomial is of first degree, i.e., $y = c_1 x_1 + \cdots + c_k x_k + c_0$. The linear regression method minimizes the functional $\delta^2 = \sum_{i=1}^{n} (Y_i - y_i)^2$, where $Y_i = c_1 x_{1i} + \cdots + c_k x_{ki} + c_0$. Hence we find the best solution $(c_0, c_1, \ldots, c_k)$ of the over-defined linear system

$$c_1 x_{1i} + \cdots + c_k x_{ki} + c_0 = y_i, \qquad i = 1, 2, \ldots, n,$$

or, in matrix form, $A\mathbf{c} = \mathbf{y}$, where $A = \begin{pmatrix} 1 & x_{11} & \ldots & x_{1k} \\ 1 & x_{21} & \ldots & x_{2k} \\ \ldots & \ldots & \ldots & \ldots \\ 1 & x_{n1} & \ldots & x_{nk} \end{pmatrix}$.

In MATLAB we use the method "\" (backslash)!

For instance, let $k = 2$. The graph of a solution is a plane in $\mathbb{R}^3$:

$$c_1 x_{1i} + c_2 x_{2i} + c_0 = y_i, \qquad i = 1, 2, \ldots, n.$$

The linear approximation may be obtained by direct matrix calculation. (The bilinear and bicubic approximations are considered in the exercises.)

```
x1 = rand(50,1)*2 - 1;
x2 = rand(50,1)*2 - 1;
y = x1.*exp(-x1.^2 - x2.^2);                               % data (k = 2)
A = [ones(size(x1)) x1 x2];                                % matrix (k = 2)
c = A \ y                                                  % solution coefficients c_i
```

Then we plot, see Figure 1.31(b):

```
hold on;  axis equal;  view(3);
syms u v;
ezsurf(u, v, c(1) + c(2)*u + c(3)*v, [-1, 1, -1, 1]);      % plane
plot3(x1, x2, y, 'o');                                     % points
```

The error δ^2 is estimated as:

```
Y = A*c;
r = y - Y;  d2 = r'*r
```

1.7.3 *Extremal property of a bicubic spline*

Let $D = [a,b] \times [c,d]$ be a given rectangle. The class $C^{2,2}(D)$ contains all functions on D such that the derivatives $f^{(i,j)} \stackrel{\text{def}}{=} \frac{\partial^{i+j}}{\partial x^i \partial y^j} f(x,y)$ $(i,j = 0,1,2)$ are continuous. Here $f^{(i,0)}$, $f^{(0,j)}$, and $f^{(0,0)}$ mean $\frac{\partial^i f}{\partial x^i}$, $\frac{\partial^j f}{\partial y^j}$, and f, respectively.

Denote by $W_2^{2,2}(D)$ the class of all real functions $f \in C^{1,1}(D)$ such that

$$\iint_D \left(f^{(2,2)}(x,y) \right)^2 dx\,dy < \infty.$$

Note that $C^{2,2}(D) \subset W_2^{2,2}(D)$. The class $C^{1,1}(D)$ consists of all functions on D such that $f^{(0,1)}$, $f^{(1,0)}$, and $f^{(1,1)}$ are also continuous on D.

We assume that $f(x,y)$ is in $W_2^{2,2}(D)$ and that the rectangle D is partitioned into subrectangles $D_{ij} = [x_i, x_{i+1}] \times [y_j, y_{j+1}]$ by a mesh

$$\pi = \{a = x_1 < \cdots < x_{n+1} = b, \ c = y_1 < \cdots < y_{n+1} = d\}.$$

For simplicity, we use a common n for both coordinates. By analogy with the case of curves (Section 1.6), consider the extremal problem below.

Problem. *Derive a function $f(x,y)$ on a rectangle D minimizing a functional $J(f) = \int_D (f^{(2,2)}(x,y))^2 dx\,dy$ among all functions of class $C^{2,2}(D)$ whose graphs contain the given points $\{(x_i, y_i, z_{ij})\}_{i,j=1}^{n+1}$.*

The interpolating bicubic splines are solutions of the problem in question in a far wider class of functions, namely, in the space $W_2^{2,2}(D)$.

In order that the solution of the problem be unique, additional requirements are necessary. Let us describe some of them. The type I and II representations of a spline $S(x,y)$ include

(I) $S^{(1,0)} = f^{(1,0)}$ at $P_{1j}, P_{n+1,j}$,

      $S^{(0,1)} = f^{(0,1)}$ at $P_{1j}, P_{n+1,j}$,

      $S^{(1,1)} = f^{(1,1)}$ at $P_{11}, P_{1,n+1}, P_{n+1,1}, P_{n+1,n+1}$,   (1.15)

(II) $S^{(2,0)} = 0$ if $f^{(2,0)} = 0$ at $P_{1j}, P_{n+1,j}$,

      $S^{(0,2)} = 0$ if $f^{(0,2)} = 0$ at $P_{i1}, P_{i,n+1}$,

      $S^{(2,2)} = 0$ if $f^{(2,2)} = 0$ at $P_{11}, P_{1,n+1}, P_{n+1,1}, P_{n+1,n+1}$.   (1.16)

The conditions for a *periodic* double cubic spline are not considered here.

Theorem 1.3. *Let $f \in W_2^{2,2}(D)$, and let $S(x,y)$ be the unique bicubic spline interpolant for f that passes through the points $\{(x_i, y_i, z_{ij})\}_{i,j=1}^{n+1}$ and satisfies either of the above boundary conditions (1.15), (1.16). Then*

$$\iint_D \left(S^{(2,2)}\right)^2 dx\,dy \leq \iint_D \left(f^{(2,2)}\right)^2 dx\,dy.$$ (1.17)

Proof. Denote by δ the difference $\delta = f - S$. We have the obvious identity

$$\iint_D (f^{(2,2)})^2 dx\,dy = \iint_D (S^{(2,2)})^2 dx\,dy + \iint_D (\delta^{(2,2)})^2 dx\,dy + 2\iint_D \delta^{(2,2)} S^{(2,2)} dx\,dy.$$
(1.18)

We modify the last term on the right-hand side of (1.18) by integrating twice by parts with respect to x and using $S^{(4,2)} = 0$:

$$\iint_D \delta^{(2,2)} S^{(2,2)} dx\,dy = \int_c^d \left[\sum_{i=1}^n \int_{x_i}^{x_{i+1}} \delta^{(2,2)} S^{(2,2)} dx\right] dy$$

$$= \int_c^d \sum_{i=1}^n \left[\delta^{(1,2)} S^{(2,2)} - \delta^{(0,2)} S^{(3,2)}\right]_{x_i}^{x_{i+1}} dy,$$

and after two more integrations by parts with respect to y,

$$\iint_D \delta^{(2,2)} S^{(2,2)} dx\,dy$$

$$= \sum_{j=1}^n \left\{\sum_{i=1}^n \left[\delta S^{(3,3)} - \delta^{(1,0)} S^{(2,3)} - \delta^{(0,1)} S^{(3,2)} + \delta^{(1,1)} S^{(2,2)}\right]_{x_i}^{x_{i+1}}\right\}_{y_j}^{y_{j+1}}.$$ (1.19)

Since $S^{(3,3)}$ is constant on each rectangle D_{ij}, it follows that

$$\sum_{j=1}^{n}\left[\sum_{i=1}^{n}\left(\delta \cdot S^{(3,3)}\right)_{x_i}^{x_{i+1}}\right]_{y_j}^{y_{j+1}} = 0. \tag{1.20}$$

If we substitute (1.20) and (1.19) into (1.18), we are led to the identity

$$\iint_D \left(f^{(2,2)}\right)^2 dx\,dy = \iint_D \left(S^{(2,2)}\right)^2 dx\,dy + \iint_D \left(\delta^{(2,2)}\right)^2 dx\,dy$$
$$+ 2\sum_{j=1}^{n}\left\{\sum_{i=1}^{n}\left[\delta^{(1,1)}S^{(2,2)} - \delta^{(1,0)}S^{(2,3)} - \delta^{(0,1)}S^{(3,2)}\right]_{x_i}^{x_{i+1}}\right\}_{y_j}^{y_{j+1}}, \tag{1.21}$$

in which $S(x,y)$ is a double cubic polynomial in each rectangle D_{ij}. Since $S(x,y)$ is a double cubic spline in D, the last identity is reduced to

$$\iint_D \left(f^{(2,2)}\right)^2 dx\,dy = \iint_D \left(S^{(2,2)}\right)^2 dx\,dy + \iint_D \left(\delta^{(2,2)}\right)^2 dx\,dy$$
$$+ 2\left[\left(\delta^{(1,1)}S^{(2,2)}\right)_a^b\right]_c^d - 2\left[\sum_{i=1}^{n}\left(\delta^{(1,0)}S^{(2,3)}\right)_{x_i}^{x_{i+1}}\right]_c^d - 2\left[\sum_{j=1}^{n}\left(\delta^{(0,1)}S^{(3,2)}\right)_{y_j}^{y_{j+1}}\right]_a^b. \tag{1.22}$$

Although we cannot expect $S^{(3,2)}$ to be continuous at a grid line $x = x_i$, both its left- and right-hand limits are continuous functions of y (and similarly for $S^{(2,3)}$ with the role of x and y interchanged). Under either of the boundary conditions (1.15), (1.16), the identity (1.22) reduces to the integral relation

$$\iint_D \left(f^{(2,2)}\right)^2 dx\,dy = \iint_D \left(S^{(2,2)}\right)^2 dx\,dy + \iint_D \left(\delta^{(2,2)}\right)^2 dx\,dy. \tag{1.23}$$

From (1.23) immediately follows the inequality (1.17). ☐

Generalizing the interpolating bicubic splines, one arrives at the notion of a smoothing bicubic polynomial spline.

Definition 1.2. A *smoothing bicubic spline* is a function $S(x,y)$ on $D = [a,b] \times [c,d]$ of class $C^{(2,2)}$ which

- is a bicubic polynomial on each subrectangle $D_{ij} = [x_i, x_{i+1}] \times [y_j, y_{j+1}]$, and which
- minimizes the functional

$$J(f) = \alpha \iint_D \left(f^{(2,2)} \right)^2 dx\,dy + \sum_{i,j=1}^{n+1} \rho_i \sigma_j \left(f(x_i, y_i) - z_{ij} \right)^2$$

$$+ \sum_{i=1}^{n+1} \rho_i \int_c^d \left(f^{(0,2)}(x_i, y) \right)^2 dy + \sum_{j=1}^{n+1} \sigma_j \int_a^b \left(f^{(2,0)}(x, y_j) \right)^2 dx, \quad (1.24)$$

where z_{ij} and $\alpha, \rho_i, \sigma_j > 0$ are given numbers.

• *Boundary conditions* may have either of the following forms:

(I) $f^{(1,0)} = A_j$ at $P_{1j}, P_{n+1,j},$

       $f^{(1,1)} = C_{ij}$ at $P_{11}, P_{1,n+1}, P_{n+1,1}, P_{n+1,n+1},$

       $f^{(0,1)} = B_i$ at $P_{1j}, P_{n+1,j},$ for some $A_i, B_j, C_{ij};$ (1.25)

(II) $f^{(2,0)} = 0$ at $P_{1j}, P_{n+1,j},$ $f^{(0,2)} = 0$ at $P_{i1}, P_{i,n+1},$

       $f^{(2,2)} = 0$ at $P_{11}, P_{1,n+1}, P_{n+1,1}, P_{n+1,n+1}.$ (1.26)

Theorem 1.4. *Among all functions $f \in W_2^{2,2}(D)$, the bicubic spline with either of the boundary conditions (1.25) or (1.26) is the unique function that minimizes the functional (1.24).*

Proof. Let the function $f(x,y) \in W_2^{2,2}(D)$ minimize the functional (1.24), but assume f is not a bicubic spline. Then there is a bicubic spline $S(x,y)$ on $\{D, \pi\}$ that interpolates $f(x,y)$ on the set π and satisfies either of the boundary conditions (1.15), (1.16). The second term of (1.24) coincides for S and f, but, by Theorem 1.3, the first term is smaller for S. Hence $J(S) < J(f)$, which is a contradiction. The uniqueness of the smoothing spline also follows from Theorem 1.3.   □

A comparison of interpolating and smoothing splines satisfying the same boundary conditions shows that the oscillation corresponding to the interpolating spline practically disappears when we use the smoothing spline.

1.8 Exercises

Section 1.1

1. Verify that a set with 3 elements $\mathbb{Z}_3 = \{0,1,2\}$ is a field if we define addition $\oplus_3$ and scalar multiplication $\otimes_3$ as follows:

$\oplus_3$	0	1	2
0	0	1	2
1	1	2	0
2	2	0	1

$\otimes_3$	0	1	2
0	0	0	0
1	0	1	2
2	0	2	1

(1.27)

2. Let $\varepsilon := \sqrt[3]{2}$. Then $\mathbb{Q} + \mathbb{Q}\varepsilon + \mathbb{Q}\varepsilon^2 := \{a + b\varepsilon + c\varepsilon^2 : a, b, c \in \mathbb{Q}\}$ is a field (a subfield of $\mathbb{R}$). Why is $\mathbb{Q} + \mathbb{Q}\varepsilon$ not a field?

3. Find
(a) all solutions of the equation $x^2 = x$ in $\mathbb{Z}_{12}$.
(b) integers u, v such that $31u + 17v = 1$, and hence find the multiplicative inverse of 17 in $\mathbb{Z}_{31}$.

4. Find all solutions in positive numbers $p, q \in \mathbb{Z}$
(a) of the equation $1/p + 1/q = 1/2$,
(b) of the inequality $1/p + 1/q > 1/2$.
(c) Repeat for the two cases given by $1/p + 1/q + 1/r > 1$.

5. Write $187/57$ and $-19/70$ as continued fractions.

6. Let f_n be Fibonacci numbers, τ the golden section. Show that
(a) $f_n = [\tau^n - (-\tau)^{-n}]/\sqrt{5}$ (Binet's formula).

(b) $\tau = \sqrt{1 + \sqrt{1 + \sqrt{1 + \cdots}}}$.

7. Prove that $|z_1 \pm z_2| \geq \big||z_1| - |z_2|\big|$ for any $z_1, z_2 \in \mathbb{C}$.

8. Plot a regular n-gon ($n = 5, 6, 7$) inscribed in the unit circle.

9. Verify the identities $\cos x = \frac{1}{2}(e^{ix} + e^{-ix})$, $\sin x = \frac{1}{2i}(e^{ix} - e^{-ix})$.

10. Find the equation of a circle through the points $z_1, z_2, z_3 \in \mathbb{C}$.

11. Show that $q = q_0 + q_1 i + q_2 j + q_3 k$ is *similar* to the complex number $q_0 \pm \sqrt{q_1^2 + q_2^2 + q_3^2}\, i$.

12. Show that
(a) all pure quaternions of norm one satisfy the equation $x^2 + 1 = 0$;
(b) the equation $x^2 i - ix^2 = 1$ has no solutions in $\mathbb{H}$ (*Hint:* Note that $i(x^2 i - ix^2) i = x^2 i - ix^2$ but $i1i = -1$);
(c) if $a, b, c \in \mathbb{H}$, and a, b are not similar, then the equation $xa + bx = x$ has a unique solution.

13. Show that (a) real or complex numbers form a field, (b) quaternions form a skew field.

14. Use MATLAB to solve the system of equations given by

$$x_1 + 3x_2 = 22, \quad 7x_1 + 11x_2 = -10.$$

Hint. Edit and execute the following:

```
M = [1 3;  7 11];   b = [22;  -10];
X = inv(M)*b
```

The vector X can also be obtained using the left slash notation:

```
X = M \ b
```
% see also Example, p. 49.

15. There is a multiplication rule for $\mathbb{R}^8$, called *octonion multiplication*,

	i	j	k	l	m	n	o
i	-1	k	$-j$	m	$-l$	$-o$	n
j	$-k$	-1	i	n	o	$-l$	$-m$
k	j	$-i$	-1	o	$-n$	m	$-l$
l	$-m$	$-n$	$-o$	-1	i	j	k
m	l	$-o$	n	$-i$	-1	$-k$	j
n	o	l	$-m$	$-j$	k	-1	$-i$
o	$-n$	m	l	$-k$	$-j$	i	-1

$$(1.28)$$

Show that it is not associative, so it makes $\mathbb{R}^8$ into something weaker than a skew field. Show that octonion multiplication preserves a norm. Find the inverse to $2 + i - m + 3o$.

Section 1.2   1. Plot a *triangle* with the vertices $P = (1,2)$, $Q = (3,-4)$, $R = (-1,0)$:

```
x = [1, 3, -1, 1];  y = [2, -4, 0, 2];   plot(x, y, 'LineWidth', 2)
```

2. Check that the M-file **arrow.m** plots an "arrow" with tail at the point (x_0, y_0) and tip at the point $(x_0 + a, y_0 + b)$. Writing $P = [x_0, y_0]$ and $V = [a, b]$, the complete call is **arrow**(P, V, color). The third argument, color, is optional. If the call is simply **arrow**(P, V), the arrow will be blue. To obtain a red arrow, use **arrow**(P, V, 'r').

```
function y = arrow(P, V, color)                          % see [Cooper]
   x0 = P(1);  y0 = P(2);
   a = V(1);  b = V(2);
   l = max(norm(V), eps);
   u = [x0 x0 + a];    v = [y0 y0 + b];
   hchek = ishold;
   plot(u, v, color);  hold on
   h = l - min(.2*l, .2);
   v = min(.2*l/sqrt(3), .2/sqrt(3) );
   a1 = (a*h - b*v)/l;    b1 = (b*h + a*v)/l;
   plot([x0 + a1,  x0 + a], [y0 + b1,  y0 + b], color)
   a2 = (a*h + b*v)/l;    b2 = (b*h - a*v)/l;
   plot([x0 + a2,  x0 + a], [y0+b2, y0+b],  color)
   if hchek == 0,  hold off,  end
end
```

---

3. Solve analytically and graph the equation $e^{-x} = \sin(x)$.

```
x = 0 : pi/1000 : pi;
plot(x, exp(-x), '--', x, sin(x));
```

Hint. The intersection of the two graphs consists of two points, Figure 1.3(b), corresponding to the roots $x_1 \approx 0.59$, $x_2 \approx 3.1$ of the equation $\exp(-x) = \sin(x)$. Find them numerically by clicking the mouse. The larger root, 3.10, is returned by the command fzero('exp(-x) - sin(x)', 3)   % Answer: 3.0964.

To obtain the root 0.589, write

fzero('exp(-x) - sin(x)', 0)

4. From a piece of cardboard 4 in. × 5 in. construct a rectangular box without a lid having maximal volume.
Hint. Let x be the height of the box. Obviously, $x \in [0,2]$. Calculate the volume by the formula $V(x) = x(4 - 2x)(5 - 2x)$ and plot the graph of the polynomial V.

syms x; V = x*(4 - 2*x)*(5 - 2*x); ezplot(V, [0, 2]);

The highest point of the graph is $\approx (0.74, 6.56)$. For an exact solution we solve the equation that sets the derivative of the polynomial equal to zero.

Vx = simplify(diff(V)),
x0 = double(solve(Vx)), subs(V, x, x0)
$V_x = 12x^2 - 36x + 20,$ 0.7362,   6.5642   % Answer

5. Find the minimal surface area of a cylinder with radius R if its volume V is 1.
Hint. Show that $V(R) = \frac{2}{R} + 2\pi R^2$, Figure 1.3(c). Use the fminbnd function to find a local minimizer of the function in a given interval.

W = @(R)(2/R + 2*pi*R^2); [R, fval] = fminbnd(W, 0, 5)
R = 0.5419   fval = 5.5358   % Answer

6. Plot the graph of the rational function $y = \frac{x^6 - x^2 + 1}{x^5 - 6x^4 + 12x^3 - 12x^2 + 11x - 6}$. Decompose this function into partial fractions.

syms x;
f = (x^6 - x^2 + 1)/(x^5 - 6*x^4 + 12*x^3 - 12*x^2 + 11*x - 6);
ezplot(f, [0.5, 4]);   %Figure 1.3(d)
partfrac(f)   % Answer: $x + 6 + \frac{1}{4}\frac{1}{x-1} - \frac{61}{5}\frac{1}{x-2} + \frac{721}{20}\frac{1}{x-3} - \frac{1}{10}\frac{x}{x^2+1}$

7. Investigate the functions using derivatives and plot their graphs:

(1) $y = \log x - \arctan x$ (2) $y = (x-3)/\sqrt{(4+x^2)}$ (3) $y = x^3 \log x$

(4) $y = |1+x|^{3/2}/\sqrt{x}$ (5) $y = x^2 e^{1/x}$ (6) $y = e^x/(4-x^2)$

(7) $y = \sqrt{1 - e^{-x^2}}$ (8) $y = (1+x)^{1/x}$ (9) $y = e^{1/(1-x^2)}/(1+x^2)$

(10) $y = 2^{\sqrt{x^2+1} - \sqrt{x^2-1}}$ (11) $y = \log_2(1-x^2)$

(12) $y = x + \log(x^2 - 1)$ (13) $y = \tan x + \sin x$

(14) $y = \arcsin((1-x^2)/(1+x^2))$ (15) $y = \arctan((x-3)/(x^2+4))$.

Section 1.3   1. Create a step function using the round function, which rounds numbers to the nearest integer, as follows:

```
H = sym(pi/3)*sym('round(x/pi)');
ezplot(H, [-2, 8])
```

Show that each step has width π and the jump between steps is $\pi/3$.

2. Verify the identity $|x| = x(H(x) - H(-x))$ using the function abs(x):

```
fplot('abs(x)', [-2, 2])
```

3. Plot the graph of the *integer part of* x: $y = [x]$ using the command floor(x) (see also the command ceil(x) = floor(x) + 1).

```
ezplot('floor(x)', [-5, 5]);
```

4. Verify that the graph $y = \{x\} = $ x - floor(x), Figure 1.11(b) (the *fractional part*), is similar in form to graphs of $y = \text{arccot}(\cot(x))$ and $y = \arctan(\tan(x))$.

5. A *periodic impulse* ("rectangular wave") $y = H(x) - 2H(x - T) + 2H(x - 2T) -$ $\cdots$ can be defined by various methods. Plot its graph for $T = 1$ using the command

```
fplot('heaviside(x)*sign(sin(pi*x))', [-2, 15]), axis equal        % Figure 1.32(a)
```

6. Plot *infinite stairs* $y = \sum_{i=0}^{\infty} H(x - i)$ using the command

```
fplot('heaviside(x)*floor(x)', [-2, 15])        % Figure 1.32(b)
```

Then plot *steep stairs* with the height of steps defined by the items of the sum $S = \sum_{n=0}^{\infty} a_n$. Create a function M-file and then plot:

```
function f = Hn(x)                              % M-file function
    an = @(n) n/4;  if x < 0;  f = 1;  else  f = Hn(x - 1) + an(floor(x));  end
end
```
```
fplot('Hn(x)', [-2, 8.2])                       %Figure 1.32(c)
```

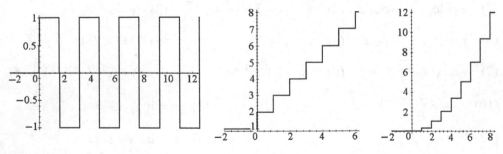

Fig. 1.32 Impulse, infinite stairs, steep stairs $\left\{\frac{n}{3}\right\}$.

7. Plot the graphs of functions containing the operations $\{\ \}, [\ ]$:
(1) $y = [f(x)]$, (2) $y = f([x])$, (3) $y = \{f(x)\}$, (4) $y = f(\{x\})$,
where $f = x^2, \sqrt{x}, 2^x, \cos x$; see Figure 1.11(b,c).

8. The reader is invited to consider some examples and formulate a plan for plotting the graphs of functions whose analytical expression contains the operation abs: $y = f(|x|), y = |f(x)|, y = |f(|x|)|$.

9. The primitive of a function of class C^k, where $k \geq 0$, is of class C^{k+1}. Direct symbolic integration of piecewise differentiable functions in MATLAB (version 7.8) does not always give the best result. For example, such symbolic integration of $|2 - |x||$ gives a discontinuous function:

```
syms x;
f = abs(2 - abs(x));   g1 = int(f, x)
```

$$g_1 = -(x - 2\,\text{sign}(x))^2 / (2\,\text{sign}(x)\,\text{sign}(2 - x/\text{sign}(x))) \qquad \% \text{ wrong answer}$$

The right answer is $g = \frac{x}{2}(|x-2| - |x| + |x+2| - 4) - |x-2| + |x+2| + 4$ or

$$g_1 = \begin{cases} -2x - x^2/2, & x \leq -2, \\ 2x + 4 + x^2/2, & x \leq 0, \\ 2x + 4 - x^2/2, & x \leq 2, \\ -2x + 8 + x^2/2, & 2 < x. \end{cases}$$

We verify that $\frac{dg}{dx} = f$ and plot the graphs.

```
g = (abs(x - 2) - abs(x) + abs(2 + x) - 4)*x/2 - abs(x - 2) + abs(2 + x) + 4;
ezplot(g, [-5, 5]);                                      % graph of g
f1 = simplify(diff(g, x))                                % f - f1 = 0
ezplot(f - f1, [-25, 5]);
```

10. Plot the graph of the *recursively defined function* $f(x + \pi) = f(x) + \sin(x)$ and $f(x) = 0$ ($0 \leq x \leq \pi$). *Hint:*

```
function f = f2(x)                                       % Create M-file!
    if x < 0;  f = f2(x + pi) + sin(x + pi);
    elseif x < pi;  f = 0;
    else f = f2(x - pi) + sin(x - pi);
    end;
end
```
```
fplot('f2(x)', [-2*pi, 4*pi])                            % Figure 1.10(b)
```

11. Plot the graphs of the functions (using animation) and find their formulae.
(1) $y = \lim_{n \to \infty} (1 - x)^{2n}$ ($|x| \leq 1$)   (2) $y = \lim_{n \to \infty} \sqrt[n]{1 + e^{n(x+1)}}$
(3) $y = \lim_{n \to \infty} \frac{\log(2^n + x^n)}{n}$ ($x \geq 0$)   (4) $y = \lim_{n \to \infty} \frac{x + x^2 e^{nx}}{1 + e^{nx}}$
(5) $y = \lim_{n \to \infty} \frac{x + e^{nx}}{1 + e^{nx}}$ (6) $y = \lim_{n \to \infty} \frac{x^{2n} \sin(\pi x/2) + x^2}{1 + x^{2n}}$.

12. Plot the graphs of the statistical functions:

(1) The *Poisson distribution* $P\{\xi = k\} = \frac{\lambda^k e^{-\lambda}}{k!}$ $(k = 0, 1, 2, \ldots)$ (density function) with parameter $\lambda > 0$.

```
lambda = 5;  x = 0 : 15;
y = poisspdf(x, lambda);
stem(x, y, '.');   stairs(x, y)
```

Plot the graph of the Poisson distribution function, see Figure 1.15(c),

```
n = 15;  yy(1) = y(1);
for k = 1 : n;  yy(k + 1) = yy(k) + y(k + 1);  end;
stairs(x, yy)
```

(2) The *uniform distribution* $f(x) = \begin{cases} 1/(b-a), & x \in [a,b], \\ 0, & x \notin [a,b] \end{cases}$ on $[a,b]$ has the continuous distribution $F(x)$. We plot the graphs.

```
syms a b x;
f = (heaviside(x - a) - heaviside(x - b))/(b - a)
f0 = subs(subs(f, a, 1), b, 3)
ezplot(f0, [-1, 4])                    % Figure 1.33(a)
F0 = int(f0, x),  ezplot(F0, [-1, 4])   % Figure 1.33(b)
```

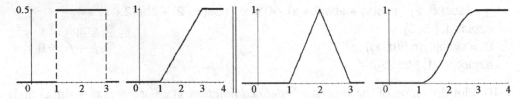

Fig. 1.33 Uniform and Simpson distributions.

(3) The *Simpson distribution* (triangle) on the segment $[a,b]$.

$$f(x) = \begin{cases} \frac{2}{b-a} - \frac{2}{(b-a)^2}|a+b-2x|, & x \in [a,b], \\ 0, & x \notin [a,b]. \end{cases}$$

```
syms a b x;
f = 2/(b - a) - 2/(b - a)^2*abs(a + b - 2*x);
fab = f*(heaviside(x - a) - heaviside(x - b));
f0 = subs(subs(fab, a, 1), b, 3);
ezplot(f0, [0, 4]), axis equal          % Figure 1.33(c)
F0 = simple(int(f0, x))
```

ezplot(F0, [0, 4])   % MATLAB gives wrong answer:
                                    % F_0 is discontinuous and $F_0(3) = 5 > 1$.

We represent and plot the right F_0 using an additional construction:

    x2 = 1 : 0.1 : 2; y2 = 1/2*x2.^2 - x2 + 1/2;
    x3 = 2 : 0.1 : 3; y3 = -1/2*x3.^2 + 3*x3 - 7/2;
    plot([x2 x3], [y2 y3]), grid on;   % Figure 1.33(d)

Section 1.4   1. Check that roses belong to the family of cycloidal curves: epitrochoids for $k > 1$ and hypotrochoids for $k < 1$.

Hint. Substitute $h = R + mR$ in the parametric equations of trochoids (see Section 5.1) and obtain the *trochoidal roses* $\rho = 2R(1+m)\sin(\frac{\varphi + \pi/2}{2m+1})$.

2. Plot (approximately) the crosses by their polar equations, Figure 1.25.

(2) *St. Andrew's cross.*

    f = ' min(1/(2*abs(cos(2*t))), 2)';
    hold on; axis equal; ezpolar('2.05'); ezpolar(f);   % Figure 1.25(2)

(3) *Catacomb cross.*

    f = ' min(2/(3*abs(sin(2*t))), 2)';
    hold on; axis equal; ezpolar('2.07'); ezpolar(f, [-.01, 2*pi]); % Figure 1.25(3)

(4) *St. George's cross*, two examples.

    f1 = ' min((9/(10*abs(sin(2*t))))^5, 2)';
    hold on; axis equal; ezpolar('2.05'); ezpolar(f1, [-.01, 2*pi]);% Figure 1.25(4)
    f2 = 'min(4/((2*sin(2*t)))^4, 2)';
    hold on; axis equal; ezpolar('2.05'); ezpolar(f2, [-.01, 2*pi]); % Figure 1.25(5)

(5) *On-Bread cross.*

    f = ' min(1/(10*abs(sin(2*t))), 1)';
    hold on; ezpolar('1.02'); ezpolar('1.2');
    ezpolar(f, [-.1, 2*pi]); axis equal;   % Figure 1.25(6)

(6) *St. George's cross (sharp).* Explain the program. First, calculate t_0:

    syms t; g = 4/(2*sin(2*t)).^4;
    sol = solve(g - 2, t);   % $-\frac{1}{2}\operatorname{asin}(\frac{1}{2}2^{\frac{1}{4}}); \frac{1}{2}\operatorname{asin}(\frac{1}{2}2^{\frac{1}{4}}); -\frac{1}{2}\operatorname{asin}(\frac{i}{2}2^{\frac{1}{4}}); \frac{1}{2}\operatorname{asin}(\frac{i}{2}2^{\frac{1}{4}})$
    t1 = sol(1); t0 = double(t1);
    hold on; ezplot('.5'); ezplot(1/g, [-t0, 2*pi - t0]);   % Figure 1.25(8)

$t_1 = (1/2)\arcsin(2^{1/4}/2)$ $t_0 = 0.3184$ % Answer

Next, create the function M-file:

```
function f = George_cross(t, t0)
  g = 4/(2*sin(2*t)).^4;   h = @(t) 2*(t^2 - t0^2) + 2;
  f1 = h(t)*(t < t0);
  f2 = g*((t < pi/2 - t0) - (t < t0));
  f3 = h(t - pi/2)*((t < pi/2 + t0) - (t < pi/2 - t0));
  f4 = g*((t < pi - t0) - (t < pi/2 + t0));
  f5 = h(t - pi)*((t ¡ pi + t0) - (t < pi - t0));
  f6 = g*((t < 3*pi/2 - t0) - (t < pi + t0));
  f7 = h(t - 3*pi/2)*((t < 3*pi/2 + t0) - (t < 3*pi/2 - t0));
  f8 = g*(1 - (t < 3*pi/2 + t0));
  f = f1 + f2 + f3 + f4 + f5 + f6 + f7 + f8;
end
```

Finally, plot the figure:

```
t0 = 0.3184;   ezpolar(@(t) George_cross(t, t0), [-t0, 2*pi - t0])
```
                                                           % Figure 1.25(7)

3. Plot the polygon through some points of Archimedes' spiral.

```
i = 0 : 20;   polar(i, i)
```
                                                           % Figure 1.16(a)

4. Plot the stopwatch with *moving* arrow using pause.

```
r1 = @(n) cosd(30*(3 - n));  r2 = @(n) sind(30*(3 - n));
hold on;  axis equal;  ezpolar('1');  ezpolar('.05');
text(-.1, -.3, 'Cosmos');
for i = 1 : 12;  text(.9*r1(i), .9*r2(i), num2str(i));  end;
for n = 0 : 12;
  compass(.8*r1(n), .8*r2(n));    pause(0.2);
  compass(.8*r1(n), .8*r2(n), 'w');
end;
```
                                                           % Figure 1.16(d)

5. Plot the *sunflower* $\rho = 3 + |\cos(n\varphi)|$, Figure 1.34(a).

```
n = 7;  t = linspace(-pi, pi, 101);
hold on;  axis equal;  ezpolar('3');  polar(t, 3 + abs(cos(n*t)), 'y')
```

6. Plot the *loop coupling* $\rho = 2\cos(2\varphi) + 1$ where $\varphi \in [0, 2\pi]$.

```
ezpolar('2*cos(2*t) + 1')
```
                                                           % Figure 1.34(b)

7. Consider a generalization $\rho = 2a\cos(n\varphi) - b$ $(a, b > 0)$ of the cardioid and limaçon and plot them, for example, for $n = 3$, $a = 0.5$, and $b = 2$.

8. Plot the *leaf of a Japanese maple* (*Acer palmatum*) by the equation $\rho = (1 + \sin(t))(1 + 0.3\cos(8t))(1 + 0.1\cos(24t))$.

```
ezpolar('(1 + sin(t)).*(1 + .3*cos(8*t)).*(1 + .1*cos(24*t))')
```
                                                           % Figure 1.35(a)

Plot another leaf by the following equation in polar coordinates:

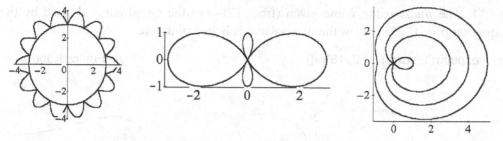

Fig. 1.34 Sunflower, loop coupling, Pascal's limaçon and cardioid.

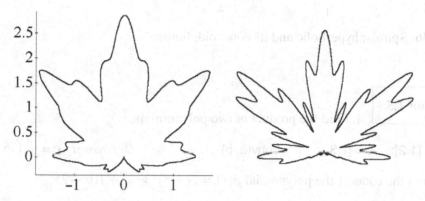

Fig. 1.35 A leaf of a Japanese maple and another leaf.

```
syms t;
g = 100/(100 + (t - pi/2).^8);                              % For scaling
ezpolar(g.*(2 - sin(7*t) - cos(30*t)/2),  [-pi/2, 3*pi/2])       % Figure 1.35(b)
```

9. The equations $\rho = 2a\cos\varphi$ and $\rho = a/\cos(\varphi - \varphi_0)$ define in polar coordinates a circle and a line, respectively.

What is the geometrical sense of the parameters a and φ_0 ?

Both equations are particular cases of the equation $\rho^m = a^m \cos(m\varphi)$, which defines in polar coordinates the family of curves called *sinusoidal spirals*. Check that in addition to the line and the circle, this family includes the rectangular hyperbola ($m = -2$), the parabola ($m = -1/2$), the cardioid ($m = 1/2$), the lemniscate of Bernoulli ($m = 2$), and the sinusoidal spiral ($m = \pm 4$).

10. The *hyperbolic spiral* $\rho = a/\varphi$ was studied by P. Varignon in 1704. The pole plays the role of an asymptotic (limit) point. The asymptote is the straight line parallel to the polar axis at the distance a from it. Plot this curve and its conchoid.

```
ezpolar('2/t',  [.5, 6*pi])                                % Figures 1.36(a,b)
```

11. The *lituus* is the name given (from 1714) to the spiral curve defined by the equation $\rho = a/\sqrt{\varphi}$. Show that the polar axis is its asymptote.

ezpolar('2/sqrt(t)', [.2, 16*pi])   % Figure 1.36(c)

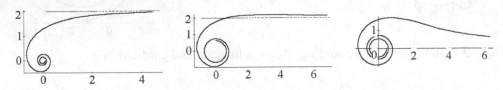

Fig. 1.36 Spirals: hyperbolic and its conchoid, lituus.

Section 1.5   1. Find the product of two polynomials,

a = [1 2];   b = [1 3];   c = conv(a, b)   % Answer: c =   1 5 6

2. Find the roots of the polynomial $p(x) = (x+5)^2 = x^2 + 10x + 25$.

p = [1 10 25];   rp = roots(p)   % Answer: $r_p = -5, -5$
P = poly(rp)   % verify

3. Find the derivative of a rational polynomial using polyder:

[nd, dd] = polyder(n, d)   % the numerator and denominator of $(n/d)'$
$nd = -1\ 10\ -25\ -250\ -625$,
$dd = 1\ 10\ 75\ 250\ 625\ 0\ 0$ % Answer

4. Show that
(a) the choice of similar (nonuniform) distribution of nodes

$$x_i = -\cos\left(\frac{2i+1}{2n}\pi\right), \quad \text{where} \quad i = 0,\ldots,n, \tag{1.29}$$

shares the same convergence properties of the Chebyshev nodes;
  (b) the nodes (1.29) are actually zeros of the Chebyshev polynomial

$$T_n(x) = \cos(n \arccos x) \quad \text{for} \quad |x| \le 1. \tag{1.30}$$

Find $T_n(x)$ for up to $n = 7$ and plot their graphs. *Hint:*

syms x;   n = 2 : 7;   T = simplify(expand(cos(n.*acos(x))))

(c) One may write (1.30) via the trigonometry in a recursive form as

$$T_0(x) = \cos 0 = 1, \quad T_1(x) = \cos(\arccos x) = x,$$
$$T_{n+1}(x) = 2x T_n(x) - T_{n-1}(x). \tag{1.31}$$

The leading coefficient (of x^n) in $T_n(x)$ is 2^{n-1} when $n > 0$.

(d) *Symmetry*: $T_{2n+1}(x)$ is an odd function, $T_{2n}(x)$ is an even function.

5. Verify that $L_{1,0}(x) = \frac{x-x_2}{x_1-x_2}$, $L_{1,1}(x) = \frac{x-x_1}{x_2-x_1}$. Show that $L_{n,i}(x) = \frac{\varphi(x)}{(x-x_i)\,\varphi'(x_i)}$.

6. Plot $L_6(x)$ over the range $[1,9]$ and look at its strange behavior outside of the segment containing the given nodes, Figure 1.26(a).

```
x = 2 : 8;  y = [32, 34, 36, 34, 39, 40, 37];   n = length(y);
xi = 1 : .01 : 9;
f = polyfit(x, y, n - 1),   p = polyval(f, xi);

f =  0.0972 -2.933 35.47 -219.1 725.4 1.214. 831.0        % Answer
```

Hence $f = \frac{7}{72}x^6 - \frac{44}{15}x^5 + \frac{1277}{36}x^4 - \frac{2629}{12}x^3 + \frac{52231}{72}x^2 - \frac{72839}{60}x + 831.$

```
plot(xi, p,  x, y, 'ro');                          % Figure 1.26(a)
```

7. Alternatively, you can use symbolic polynomials:

```
P = poly2sym(f, x)
ezplot(P, [1, 9]);
```

Conclude from this that interpolation polynomials cannot be used for the *extrapolation* of functions.

8. Plot the graph of $L_6(x)$ on the segment $[1.8, 8.2]$, Figure 1.26(b), and look at its oscillations. Use the range xi = 1.8 : .01 : 8.2.

9. Find interpolation polynomials for the data from Example, p. 49.

```
x = 2 : 8;  y = [32, 34, 36, 34, 39, 40, 37];
x2 = linspace(0, 10, 25);                          % extended segment
P1 = polyfit(x, y, 1),  f1 = polyval(P1, x2);      % linear
P2 = polyfit(x, y, 2),  f2 = polyval(P2, x2);      % quadratic
P3 = polyfit(x, y, 3),  f3 = polyval(P3, x2);      % cubic, etc.

P1 = 1.0714   30.6429       P2 = -0.1905   2.9762   26.6429
P3 = -0.1111   1.4762   - 4.5794   36.6429              % Answer
```

In classical notation the polynomials are

$$P_1 = 1.0714x + 30.6429, \quad P_2 = -0.1905x^2 + 2.9762x + 26.6429,$$
and $P_3 = -0.1111x^3 + 1.4762x^2 - 4.5794x + 36.6429.$

Plot polynomials on the extended segment to compare with the data:

```
plot(x, y, 'o',  x2, f1,  x2, f2,  x2, f3)          % Figure 9.2(a)
```

10. Extract 15 values of the function $y = \sin(x)$ ($|x| \leq \pi$), approximate these values, and visualize them on the segment $|x| \leq 4$. *Hint:*

```
x1= linspace(-pi, pi, 15);    y1 = sin(x1);                    % points
n = 3;    f1 = polyfit(x1, y1, n);                 % polynomial of degree n
x2 = -4 :.4 : 4;    y2 = sin(x2);                     % extended segment
fp = polyval(f1, x2);                                % values of polynomial
plot(x2, y2, 'ro', x2, fp);
```

11. Consider a polynomial $h(x) = H_3 x^3 + H_2 x^2 + H_1 x + H_0$ matching just two points (x_1, y_1), (x_2, y_2) and having the specified first derivatives y_1', y_2' at these points. To obtain H_i you must solve the linear system

$$h(x_1) = H_3 x_1^3 + H_2 x_1^2 + H_1 x_1 + H_0 = y_1, \quad h(x_2) = H_3 x_2^3 + H_2 x_2^2 + H_1 x_2 + H_0 = y_2,$$
$$h'(x_1) = 3H_3 x_1^2 + 2H_2 x_1 + H_1 = y_1', \qquad h'(x_2) = 3H_3 x_2^2 + 2H_2 x_2 + H_1 = y_2'.$$

Use the data $\{(0,0), (1,1), (2,4), (3,5)\}$ and $y_1' = 0$, $y_2' = 1$. Alternatively, approximate the specified derivatives at the data points by their differences: $y_1' = y_2 - y_1$, $y_2' = y_2 - y_4$.

12. Compute the *Legendre polynomials* through the following relations:

$$\hat{L}_0(x) = 1, \qquad \hat{L}_1(x) = x,$$
$$\hat{L}_k(x) = \frac{2k-1}{k} x \hat{L}_{k-1}(x) - \frac{k-1}{k} \hat{L}_{k-2}(x), \qquad k = 2, 3, \ldots$$

Show that (a) every polynomial of degree n can be obtained by a linear combination of $\hat{L}_0, \ldots, \hat{L}_n$; (b) $\hat{L}_{n+1}$ is orthogonal to all the polynomials of degree less than or equal to n, i.e., $\int_{-1}^{1} \hat{L}_{n+1}(x) \hat{L}_j(x) \, dx = 0$ for all $j \leq n$. Plot the graphs of the first few Legendre polynomials for $-1 \leq x \leq 1$.

13. Another series of orthogonal polynomials defined over $[-1, 1]$ are the Chebyshev polynomials, $T_i(x)$, see (1.30), (1.31). The reader is invited to change the previous codes, both for least-squares fitting and for the Legendre polynomials, into codes that use Chebyshev polynomials.

Section 1.6   1. Plot "triangular waves" using a linear spline and then smooth them by a cubic spline.

```
x = 0 : 11;   y = [-1 1 -1 1 -1 1 -1 1 -1 1 -1 1];
t = linspace(0, 11, 201);    v = spline(x, y, t);
plot(x, y, '-r', t, v);   axis equal,  grid;                   % Figure 1.28(c)
```

2. Check that for the function $f(x) = \frac{1}{1+25x^2}$ the cubic spline with $n = 6$ nodes contains an error of approximation of the same order as that of the interpolation polynomial $L_5(x)$, and on the net with $n = 21$ nodes this error just cannot be seen on the scale of usual figures in a book (in this case $L_{20}(x)$ gives a huge error).

3. Plot spline eight based on 11 points of the lemniscate $\rho = \sqrt{2\cos(2\varphi)}$, see Figure 1.21(c), given in polar coordinates. *Hint:*

```
x = [-45 -40, -33, -20, 0, 20, 33, 40, 45]*pi/180;
r1 = sqrt(cos(2.*x));    r2 = sqrt(cos(2.*(x + pi)));
x1 = linspace(-45, 45, 100);   xx = [x1  x1 + 180]*pi/180;
s = spline([x  x + pi], [r1  r2], xx)
polar([x  x + pi], [r1  r2], 'o'), hold;    polar(xx, s)
```

4. Apply the M-file **cubic_spline.m** for deriving a cubic spline $S(x)$ (say, specifying $S''(x)$ at each endpoint):

```
m_in = 0;    m_fin = 0;                        % specify end conditions
x = 2 : 8;   y = [32, 34, 36, 34, 39, 40, 37];
syms t;   S = cubic_spline(x, y, m_in, m_fin)
```

Then plot a composed curve:

```
hold on;   plot(x, y, 'r o - -');
for i = 1 : length(x) - 1;
  ti = linspace(x(i), x(i + 1), 20);     plot(ti, subs(S(i), t, ti));
end
```

Section 1.7	1. Plot two spline surfaces in $\mathbb{R}^3$, using interp2.
	(a) Prepare the data for a *torus*, Figure 1.30(a):

```
R = 1;   n = 5;    m = 5;
[uu, vv] = meshgrid(0 : 2*pi/(n - 1) : 2*pi,  0 : 2*pi/(m - 1) : 2*pi);
X = (3+R*cos(uu)).*cos(vv);   Y = (3+R*cos(uu)).*sin(vv);   Z = R*sin(uu);
[UI, VI] = meshgrid(0 : .2 : 2*pi,  0 : .2 : 2*pi);
```

(b) Prepare the data for a *helicoid*, Figure 1.30(c):

```
b = .7;   n = 2;   m = 5;
[uu, vv] = meshgrid(0 : 1/(n - 1) : 3,  0 : 2*pi/(m - 1) : 4*pi);
X = uu.*cos(vv);   Y = uu.*sin(vv);   Z = b*vv;
[UI, VI] = meshgrid(0 : .3 : 3, 0 : .2 : 4*pi);
```

Then calculate and plot the surface similarly as in Example, p. 63.

2. Represent the least-squares approximation by a *bilinear function* $y = c_0 + c_1 x_1 + c_2 x_2 + c_3 x_1 x_2$. Let the data be given on a rectangular net by $y = \sin(x_1 x_2)$. Estimate the coefficients using the program below:

```
i = 0 : 10;   j = -1 : 9;
[X1 X2] = meshgrid(i, j);
Y = sin(X1.*X2);                              % enter your function
[m n] = size(Y);
x1 = reshape(X1, m*n, 1);   x2 = reshape(X2, m*n, 1);
```

```
y = reshape(Y, m*n, 1);                                        % reshape of data
A = [ones(size(x1)) x1 x2 x1.*x2];
c = A \ y;                                                     % solution
x1 = reshape(x1, m, n);   x2 = reshape(x2, m, n);
y = reshape(y, m, n);                                          % output
hold on;  view(3);   plot3(x1, x2, y, 'o');        % Figure 1.37(a), data and
syms u v;
ezsurf(u, v, c(1) + c(2)*u + c(3)*v + c(4)*u*v, [0, 10, -1, 9]);   % surface
```

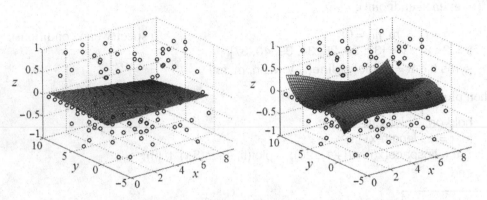

Fig. 1.37 Bilinear and bicubic surface approximations.

3. Modify two lines of the program above

```
A = [ones(size(x1)) x1 x2 x1.*x2 x1.^2 x2.^2 (x1.^2).*x2 (x2.^2).*x1
x1.^3 x2.^3];
...
ezsurf(u, v,  c(1) + c(2)*u + c(3)*v + c(4)*u*v + c(5)*u^2 + c(6)*v^2
+ c(7)*u^2*v + c(8)*u*v^2 + c(9)*u^3 + c(10)*v^3,  [0, 10, -1, 9]);
```

for the least-squares approximation by a *bicubic function*, Figure 1.37(b).

$$y = c_0 + c_1 x_1 + c_2 x_2 + c_3 x_1 x_2 + c_4 x_1^2 + c_5 x_2^2 + c_6 x_1^2 x_2 + c_7 x_1 x_2^2 + c_9 x_1^3 + c_9 x_2^3.$$

Estimate the 10 coefficients of y by a similar procedure.

2

Rigid Motions (Isometries)

Section 2.1 discusses vectors and their use in analytic geometry. In Section 2.2 we study rigid motions of $\mathbb{R}^n$ ($n \geq 2$), including the complex and quaternionic approaches. Section 2.3 is devoted to geometry on a sphere (induced by Euclidean geometry of space), and also introduces stereographic projection. The study of polyhedra is organized in Section 2.4 into a sequence of themes: the notion of a polyhedron, Platonic solids, symmetries of geometrical figures, star-shaped polyhedra, and Archimedean solids. Section 2.5 (Appendix) surveys matrices and groups.

2.1 Vectors

A vector in $\mathbb{R}^n$ (as a point of $\mathbb{R}^n$) is an ordered set $\mathbf{x} = (x_1, x_2, \ldots, x_n)$ of n real numbers. The *origin O* in $\mathbb{R}^n$ is the zero vector.

2.1.1 *Vectors in* $\mathbb{R}^3$

If $\mathbf{a}_1, \ldots, \mathbf{a}_n$ are vectors, and if $\lambda_1, \ldots, \lambda_n$ are real numbers, then the *linear combination* $\lambda_1 \mathbf{a}_1 + \cdots + \lambda_n \mathbf{a}_n$ *of vectors* is again a vector. The *directed line segment from the point* $\mathbf{a}$ *to the point* $\mathbf{b}$ is the set of points

$$[\mathbf{a}, \mathbf{b}] = \{\mathbf{a} + t(\mathbf{b} - \mathbf{a}) : 0 \leq t \leq 1\}$$

with initial point $\mathbf{a}$ and final point $\mathbf{b}$.

A nonempty set $X \subset \mathbb{R}^3$ is called *convex* if X contains, with any two points $\mathbf{x}, \mathbf{y}$, also the line segment $[\mathbf{x}, \mathbf{y}]$.

V. Rovenski, *Modeling of Curves and Surfaces with MATLAB®*,
Springer Undergraduate Texts in Mathematics and Technology 7, DOI 10.1007/978-0-387-71278-9_2,
© Springer Science+Business Media, LLC 2010

Triangles, rectangles, and disks in a plane are examples of convex sets.

Recall that for all $\mathbf{x} \in \mathbb{R}^3$, $\mathbf{x} = x_1\mathbf{i} + x_2\mathbf{j} + x_3\mathbf{k}$, where $\mathbf{i} = (1,0,0)$, $\mathbf{j} = (0,1,0)$, $\mathbf{k} = (0,0,1)$ are the unit vectors; these points lie at a unit distance along the three coordinate axes.

The *scalar product* of the vectors $\mathbf{x}$ and $\mathbf{y}$ is $\mathbf{x} \cdot \mathbf{y} := x_1y_1 + x_2y_2 + x_3y_3$. We call $\|\mathbf{x}\| = \sqrt{x_1^2 + x_2^2 + x_3^2}$ the *norm* (*length*) of $\mathbf{x}$.

The following properties of the scalar product are immediate:

(1) $(\lambda_1\mathbf{x} + \lambda_2\mathbf{y}) \cdot \mathbf{z} = \lambda_1(\mathbf{x} \cdot \mathbf{z}) + \lambda_2(\mathbf{y} \cdot \mathbf{z})$;
(2) $\mathbf{x} \cdot \mathbf{y} = \mathbf{y} \cdot \mathbf{x}$;
(3) $\|\mathbf{x} - \mathbf{y}\|^2 = \|\mathbf{x}\|^2 + \|\mathbf{y}\|^2 - 2(\mathbf{x} \cdot \mathbf{y})$;
(4) $\mathbf{i} \cdot \mathbf{j} = \mathbf{j} \cdot \mathbf{k} = \mathbf{k} \cdot \mathbf{i} = 0$, and $\mathbf{i} \cdot \mathbf{i} = \mathbf{j} \cdot \mathbf{j} = \mathbf{k} \cdot \mathbf{k} = 1$.

We calculate the *angle* θ between the vectors $\mathbf{x}$, $\mathbf{y}$ by $\cos\theta = \mathbf{x} \cdot \mathbf{y}/(\|\mathbf{x}\| \cdot \|\mathbf{y}\|)$.

The triangle inequality. $\|\mathbf{x} - \mathbf{z}\| \le \|\mathbf{x} - \mathbf{y}\| + \|\mathbf{y} - \mathbf{z}\|$, $\mathbf{x}$, $\mathbf{y}$, $\mathbf{z} \in \mathbb{R}^3$.

The *vector product* of the two vectors $\mathbf{x}$ and $\mathbf{y}$ is

$$\mathbf{x} \times \mathbf{y} := (x_2y_3 - x_3y_2)\mathbf{i} + (x_3y_1 - x_1y_3)\mathbf{j} + (x_1y_2 - x_2y_1)\mathbf{k}.$$

The following properties of the vector product are immediate:

(1) $\mathbf{x} \times \mathbf{y}$ is orthogonal to $\mathbf{x}$ and to $\mathbf{y}$;
(2) $(\lambda_1\mathbf{x} + \lambda_2\mathbf{y}) \times \mathbf{z} = \lambda_1(\mathbf{x} \times \mathbf{z}) + \lambda_2(\mathbf{y} \times \mathbf{z})$;
(3) $\mathbf{x} \times \mathbf{y} = -\mathbf{y} \times \mathbf{x}$;
(4) $\mathbf{x} \times \mathbf{y} = 0$ if and only if $\mathbf{x}$, $\mathbf{y}$ are collinear. In particular, $\mathbf{x} \times \mathbf{x} = 0$.

The *scalar triple product* of the vectors $\mathbf{x}$, $\mathbf{y}$, and $\mathbf{z}$ is $(\mathbf{x}, \mathbf{y}, \mathbf{z}) := \mathbf{x} \cdot (\mathbf{y} \times \mathbf{z})$. Hence, the scalar triple product is the determinant $(\mathbf{x}, \mathbf{y}, \mathbf{z}) = \begin{vmatrix} x_1 & x_2 & x_3 \\ y_1 & y_2 & y_3 \\ z_1 & z_2 & z_3 \end{vmatrix}$. The *vector triple product* of the vectors $\mathbf{x}$, $\mathbf{y}$, and $\mathbf{z}$ is

$$[\mathbf{x}, \mathbf{y}, \mathbf{z}] := \mathbf{x} \times (\mathbf{y} \times \mathbf{z}).$$

Consider two nonzero vectors $\mathbf{a}$ and $\mathbf{b}$, in the plane $x_3 = 0$. Let $\theta \in (0, \pi)$ be the angle between $\mathbf{a}$ and $\mathbf{b}$, measured in the counterclockwise direction. We can easily see that $\Delta(\mathbf{a}, \mathbf{b}) := u_1v_2 - u_2v_1 = \|\mathbf{a}\| \cdot \|\mathbf{b}\| \sin\theta$.

Following the case of two variables, we now use the determinant $\Delta(\mathbf{a}, \mathbf{b}, \mathbf{c}) = (\mathbf{a}, \mathbf{b}, \mathbf{c})$ for vectors $\mathbf{a}$, $\mathbf{b}$, and $\mathbf{c}$ in $\mathbb{R}^3$.

Two vectors $\mathbf{a}$ and $\mathbf{b}$ lying in the plane $x_3 = 0$ are *positively (negatively) oriented* if $\Delta(\mathbf{a}, \mathbf{b}) > 0$ ($\Delta(\mathbf{a}, \mathbf{b}) < 0$).

Three vectors $\mathbf{a}$, $\mathbf{b}$, $\mathbf{c}$ in $\mathbb{R}^3$ are said to be *positively oriented* if $\Delta(\mathbf{a}, \mathbf{b}, \mathbf{c}) > 0$, and *negatively oriented* if $\Delta(\mathbf{a}, \mathbf{b}, \mathbf{c}) < 0$.

Example. Show that $[\mathbf{x},\mathbf{y},\mathbf{z}] = (\mathbf{x}\cdot\mathbf{z})\,\mathbf{y} - (\mathbf{x}\cdot\mathbf{y})\,\mathbf{z}$ for all $\mathbf{x}$, $\mathbf{y}$, $\mathbf{z} \in \mathbb{R}^3$.

Solution. Execute the following code:

```
syms  x1 x2 x3 y1 y2 y3 z1 z2 z3  real;
x = [x1 x2 x3];    y = [y1 y2 y3];    z = [z1 z2 z3];
xyz = cross(x, cross(y, z))
F = (x*z')*y - (x*y')*z              % compare LHS and RHS of the identity
simplify(xyz - F)                    % Answer: [0,0,0].
```

2.1.2 Lines and conics

The key principle of analytic geometry is to link algebra to the study of geometry. There are two main problems studied in two-dimensional analytic geometry:

(1) given the equation of a curve, determine its shape, location, etc.;
(2) given a description of the plot of a curve, determine its equation.

A *line* in $\mathbb{R}^2$ with the normal $\mathbf{n} = (A,B) \neq 0$ is given by the equation

$$Ax + By + C = 0.$$

A *plane* in $\mathbb{R}^3$ with the normal $\mathbf{n} = (A,B,C) \neq 0$ is given by the equation

$$Ax + By + Cz + D = 0.$$

For example, the equations $2x + 3y - 4 = 0$ and $x + y + z - 1 = 0$ are entered using the commands:

```
eq1 = 2*x + 3*y - 4;    eq2 = x + y + z - 1 = 0;
```

The line $l(P,\mathbf{a}) \subset \mathbb{R}^3$ through a point $P = (p_1, p_2, p_3)$ in the direction of a vector $\mathbf{a} = (a_1, a_2, a_3)$ is given by the *parametric* equations

$$\mathbf{r}_P + t\,\mathbf{a} = (p_1 + a_1 t,\ p_2 + a_2 t,\ p_3 + a_3 t), \qquad t \in \mathbb{R}.$$

The quadratic equation

$$Ax^2 + Bxy + Cy^2 + Dx + Ey + F = 0$$

represents *conics* (circles, ellipses, hyperbolas, parabolas) and pairs of lines in $\mathbb{R}^2$; see Section 1.4 and the table in what follows.

The three-dimensional analogs of lines and conics are planes and quadrics.

N	Curve	Equation	MATLAB® Commands
1a	circle	$(x-x_c)^2 + (y-y_c)^2 = R^2$ $\mathbf{r} = (x_c, y_c) + R(\cos t, \sin t)$	syms t; x = 1 + 3*cos(t) y = 2 + 3*sin(t); ezplot(x, y)
1b	arc	$\ldots, \varphi \in [2,5]$	... ezplot(x, y, [2, 5])
2	parabola	$y^2 = p(x - x_0)$	ezplot('t', 't^2/5', [-2 2])
3	ellipse	$\frac{(x-x_c)^2}{a^2} + \frac{(y-y_c)^2}{b^2} = 1$ $\mathbf{r} = (x_c, y_c) + (a\cos t, b\sin t)$	x = 1 + 4*cos(t); y = 2 + 3*sin(t); ezplot(x, y)
4	hyperbola	$\frac{(x-x_c)^2}{a^2} - \frac{(y-y_c)^2}{b^2} = 1$ $\mathbf{r} = (x_c, y_c) \pm (a\cosh t, b\sinh t)$	x = 2 + 4*cosh(t); y = 3*sinh(t); ezplot(x, y)
5	helix	$x = R\cos t,\ y = R\sin t,$ $z = vt$	t = linspace(0, 4*pi, 201) plot3(cos(t), sin(t), t/(2*pi))

Examples.

1. The generation of the trigonometric circle is immediate:

   t = linspace(0, 2*pi, 101);   plot(cos(t), sin(t)), axis equal;

In the *Argand presentation*, the equation of a circle is $z = (a + ib) + \rho e^{i\varphi}$.

   T = 0, .1 : 2*pi;   plot(3 + 5i + 4*exp(i*T))

2. Let $P = (1,2,3)$, $Q = (0,1,5)$. Enter the line $l(P,Q)$ using the commands:

   syms t;
   P = [1, 2, 3];   Q = [0, 1, 5];   % take $\mathbf{a} = Q - P$.
   r = P + t*(Q - P)   % obtain $\mathbf{r} = [1 - t,\ 2 - t,\ 3 + 2t]$

The plane through the 3 points $\mathbf{r}_i = P_i = (x_i, y_i, z_i)$ $(i \le 3)$ is given by

$$(\mathbf{r} - \mathbf{r}_1, \mathbf{r}_2 - \mathbf{r}_1, \mathbf{r}_3 - \mathbf{r}_1) = \det \begin{vmatrix} x - x_1 & y - y_1 & z - z_1 \\ x_2 - x_1 & y_2 - y_1 & z_2 - z_1 \\ x_3 - x_1 & y_3 - y_1 & z_3 - z_1 \end{vmatrix} = 0.$$

   syms x y z;   r = [x y z];
   r1 = [1 4 2];   r2 = [2 1 0];   r3 = [-1 2 3];
   det([r - r1; r2 - r1; r3 - r1])   % obtain $-7x + 3y - 8z + 11$

3. The *distance d from P to the line through Q and parallel to* $\mathbf{a}$ is $d = \|(\mathbf{r}_P - \mathbf{r}_Q) \times \mathbf{a}\| / \|\mathbf{a}\|$.

   p = [0 1 0];   q = [1 1 1];   a = [0 1 1];
   b = cross(p - q, a);   d = norm(b) / norm(a);   % obtain $d = 1.2247$

4. The *distance d between the parallel lines* given by $\mathbf{r} = \mathbf{r}_1 + t\mathbf{a}$ and $\mathbf{r} = \mathbf{r}_2 + t\mathbf{a}$ (by taking $\mathbf{r}_P = \mathbf{r}_2$, $\mathbf{r}_Q = \mathbf{r}_1$) is $d = \|(\mathbf{r}_2 - \mathbf{r}_1) \times \mathbf{a}\| / \|\mathbf{a}\|$.

```
r1 = [0 1 2];  r2 = [1 0 1];  a = [0 1 3];
b = cross(r2 - r1,  a);
d = norm(b) / norm(a);                          % obtain d = 1.1832
```

5. The *distance d between two skew lines* given by $\mathbf{r} = \mathbf{r}_1 + t\mathbf{a}_1$ and $\mathbf{R} = \mathbf{r}_2 + t\mathbf{a}_2$ is $d = |(\mathbf{r}_2 - \mathbf{r}_1, \mathbf{a}_1, \mathbf{a}_2)| / \|\mathbf{a}_1 \times \mathbf{a}_2\|$.

Hence, non-parallel lines given by $\mathbf{r} = \mathbf{r}_1 + t\mathbf{a}_1$ and $\mathbf{R} = \mathbf{r}_2 + t\mathbf{a}_2$ meet if and only if $\mathbf{a}_1$, $\mathbf{a}_2$, and $\mathbf{r}_2 - \mathbf{r}_1$ are coplanar: $(\mathbf{a}_1, \mathbf{a}_2, \mathbf{r}_2 - \mathbf{r}_1) = 0$.

```
r1 = [0 1 0];  a1 = [1 1 1];
r2 = [1 0 0];  a2 = [0 1 1];
f = cross(a1, a2);
g = det([ r2 - r1; a1; a2 ]);
d = abs(g) / sqrt(f*f')                         % obtain  d = 0.7071.
```

2.2   Rigid Motions in $\mathbb{R}^n$

The transformations we are concerned with are changes in the coordinates and shapes of objects upon the action of geometrical operations. We formulate both the problems and their solutions in the language of vectors and matrices.

Our next objective is to understand isometries of $\mathbb{R}^n$. A *rigid motion (isometry)* of $\mathbb{R}^n$ is a function $f\colon \mathbb{R}^n \to \mathbb{R}^n$ that preserves the distance between points; that is, it satisfies

$$\|f(\mathbf{x}) - f(\mathbf{y})\| = \|\mathbf{x} - \mathbf{y}\| \text{ for all } \mathbf{x} \text{ and } \mathbf{y}.$$

The most general rigid motion $f\colon \mathbb{R}^n \to \mathbb{R}^n$ is of the form

$$f(\mathbf{x}) = \mathbf{x}A + \mathbf{a}, \text{ where } \mathbf{a} = f(\mathbf{0}) \text{ and } A \text{ is an orthogonal matrix of order } n.$$

The *translation* $\mathbf{T_a}$ defined by a vector $\mathbf{a} = (a_1, \ldots, a_n)$, is a transformation which maps a point $P = (x_1, \ldots, x_n)$ to a point $P' = (x_1', \ldots, x_n')$ by adding a constant amount to each coordinate so that $x_1' = x_1 + a_1, \ldots, x_n' = x_n + a_n$.

2.2.1   *Rigid motions in two dimensions*

A *rigid motion (isometry)* of the plane, $f\colon \mathbb{R}^2 \to \mathbb{R}^2$, is given by

$$f(x,y) = (x\cos\theta \mp y\sin\theta + b_1,\ x\sin\theta \pm y\cos\theta + b_2) \tag{2.1}$$

for some constant real numbers θ (angle) and b_1, b_2. Using the matrix

$$A = \begin{pmatrix} \cos\theta & \mp\sin\theta \\ \sin\theta & \pm\cos\theta \end{pmatrix}$$

and vector $\mathbf{b} = (b_1, b_2)$, we equivalently write $f(\mathbf{x}) = \mathbf{x}A + \mathbf{b}$ $(\mathbf{x} \in \mathbb{R}^2)$. For example,

```
syms t;                                    % rigid motions: A₁ direct, A₂ indirect.
A1 = [cos(t) -sin(t); sin(t) cos(t)],   A2 = [cos(t) sin(t); sin(t) -cos(t)]
```

Consider particular cases of rigid motions.

A *translation* $\mathbf{T_b}$ is defined by a vector $\mathbf{b} = (b_1, b_2)$. It is a transformation which maps a point $P = (x,y)$ to a point $P' = (x', y')$ by adding a constant amount to each coordinate so that $x' = x + b_1$, $y' = y + b_2$.

A *rotation about the origin through an angle* θ has the effect that a point $P = (x,y)$ is mapped to a point $P' = (x', y')$ so that the initial point P and its image point P' are the same distance from the origin, and the angle between lines OP and OP' is θ:

$$x' = x\cos\theta - y\sin\theta, \qquad y' = x\sin\theta + y\cos\theta. \qquad (2.2)$$

(The new coordinates of a point in the XY-plane rotated by an angle θ around the z-axis can be directly derived through some elementary trigonometry). The matrix $\mathbf{R}_\theta = \begin{pmatrix} \cos\theta & \sin\theta \\ -\sin\theta & \cos\theta \end{pmatrix}$ is called the *rotation matrix*. The particular case of $\theta = \pi$ (*inversion about the origin*) has the matrix $\mathbf{R}_\pi = \begin{pmatrix} -1 & 0 \\ 0 & -1 \end{pmatrix}$ and changes the coordinates as follows: $x' = -x$, $y' = -y$. Using complex numbers, we have $z' = ze^{i\theta}$, hence

$$x' + iy' = (x + iy)(\cos\theta + i\sin\theta) = (x\cos\theta - y\sin\theta) + i(x\sin\theta + y\cos\theta).$$

Equating separately the real parts and the imaginary parts, we also deduce the effect of a rotation on the coordinates of a point.

Another example of a rigid motion is a *reflection in a line*. If the line is $l: ax + by = 0$ (and at least one of the coefficients is nonzero), the matrix of reflection in l is $\mathbf{P}_l = \dfrac{1}{a^2 + b^2} \begin{pmatrix} b^2 - a^2 & -2ab \\ -2ab & a^2 - b^2 \end{pmatrix}$. From (2.1) we see that $a = -\sin(\theta/2)$, $b = \cos(\theta/2)$. The reflections in the x- and y-axes have the matrices $\mathbf{P}_x = \begin{pmatrix} 1 & 0 \\ 0 & -1 \end{pmatrix}$ (i.e., $a = 0$) and $\mathbf{P}_y = \begin{pmatrix} -1 & 0 \\ 0 & 1 \end{pmatrix}$ (i.e., $b = 0$).

```
t = pi/6;                        % Example: let ∠(l,OX) = π/6
P_l = [cos(2*t) sin(2*t); sin(2*t) -cos(2*t)]    % Pₗ = [0.50,0.87; 0.87,-0.50]
```

A *glide reflection* $\Gamma_{l,\mathbf{a}}$ is a type of indirect isometry of $\mathbb{R}^2$: the combination of a reflection in a line l and a translation along that line.

Theorem 2.1. *Each nonidentity motion (isometry) of* $\mathbb{R}^2$ *is exactly one of the following: translation, rotation, reflection, or glide reflection* (see the table in what follows).

	translations	rotations	reflections	glides
fixed points	no	unique	a line	no
invariant lines	a direction of lines	no	a line and a direction of lines	a unique line
decomposition in reflections	two parallel lines	two secant lines	one line	three lines
writing in complex numbers	$z \mapsto z+b$	$z \mapsto az+b$ $\|a\|=1$ $a \neq 1$	$z \mapsto a\bar{z}+b$ $\|a\|=1$ $a\bar{b}+b=0$	$z \mapsto a\bar{z}+b$ $\|a\|=1$ $a\bar{b}+b\neq 0$

Now we look at rigid motions (isometries) using the complex numbers. The *direct* and *indirect plane isometries* of $\mathbb{C}$ are given by

$$fz = az+b \text{ (direct)}, \qquad fz = a\bar{z}+b \text{ (indirect)}, \qquad |a|=1, \, b \in \mathbb{C}. \qquad (2.3)$$

The maps (2.3) are isometries (for example, $|(az+b)-(aw+b)| = |a| \cdot |z-w| = |z-w|$), and each isometry of $\mathbb{C}$ is of one of these forms:

(a) Let $f: z \mapsto az+b$, where $|a|=1$. If $a=1$ then f is a translation by the vector b; if $a \neq 1$ then $f(z_0)=z_0$ when $z_0 = b/(a-1)$, and $f(z)-z_0 = a(z-z_0)$, so clearly f is a rotation about z_0 of angle $\arg a$.

(b) Let $f: z \mapsto a\bar{z}+b$, where $|a|=1$. By Theorem 2.1, $f = T \circ S$, where S is a reflection in a line l, and T is a translation along l. Then $f^2 = T^2 \circ S^2 = T^2$. Since $f^2 z = z + a\bar{b}+b$, we define maps T and S as

$$Tz = z+(a\bar{b}+b)/2, \qquad Sz = T^{-1}fz = a\bar{z}+(b-a\bar{b})/2.$$

If $a\bar{b}+b=0$ then $T=0$ and f is a reflection in the line $l = \{\frac{1}{2}b + s\exp(it/2) : s \in \mathbb{R}\}$ where $t = \arg a$ (for instance, $f: z \mapsto \bar{z}$ is a reflection in the x-axis); if $a\bar{b}+b\neq 0$ then f is a glide reflection (see [Beardon]).

Hence, *every isometry of $\mathbb{C}$ can be obtained as a rotation (through an angle $\arg a$) followed by a translation (by the vector b), possibly all preceded by a reflection in the real axis.*

Examples. 1. If $b \neq 0$ is real then
  a) $z \mapsto \bar{z}+b$ is a glide reflection along the x-axis, and
  b) $z \mapsto -\bar{z}+ib$ is a glide reflection along the y-axis.

2. If $f \colon z \mapsto a\bar{z} + b$ (where $a = e^{it}$) is a reflection in a line l, then
   a) $fz = a(\bar{z} - \bar{b})$,   b) l is given by $c\bar{z} - \bar{c}z = c\bar{b}$ where $c = e^{it/2}$.

3. Given $f = a\bar{z} + b$, plot the axis l of the glide and the image of $\{0, 1, i\}$.

```
a = 1 + 2i,  b = 3 + i                                    % enter your data
aa = a/abs(a)                                             % unit complex number
e = (aa*conj(b) + b)/2,   bb = (b - aa*conj(b))/2    % direction of l and b̃ of S
t = atan2(imag(a), real(a))                               % arg a
syms s real;   l = bb/2 + s*exp(i*t/2);             % parameterization of l
hold on;  ezplot(real(l), imag(l), [-1 5]);               % l
plot([0, 1, i], 'o-r');  plot([b, aa+b, -aa*i+b], 'o-g')   % 3 points f(0), f(1), f(i)
```

2.2.2 *Rotations and reflections in planes of* $\mathbb{R}^3$

Rotations in $\mathbb{R}^3$ form a very useful collection of mappings. Any rotation about a line (passing through the origin) can be produced by applying one or more rotations about the coordinate axes, $\mathbf{R}_x$, $\mathbf{R}_y$, and $\mathbf{R}_z$. Hence, the formulae for $\mathbf{R}_i$ ($i = x, y, z$) are usually sufficient in practice. These are easily derived from the formula of a rotation in the plane. Namely,

$$\mathbf{R}_x(\theta) = \begin{pmatrix} 1 & 0 & 0 \\ 0 & \cos\theta & \sin\theta \\ 0 & -\sin\theta & \cos\theta \end{pmatrix}, \; \mathbf{R}_y(\theta) = \begin{pmatrix} \cos\theta & 0 & -\sin\theta \\ 0 & 1 & 0 \\ \sin\theta & 0 & \cos\theta \end{pmatrix}, \; \mathbf{R}_z(\theta) = \begin{pmatrix} \cos\theta & \sin\theta & 0 \\ -\sin\theta & \cos\theta & 0 \\ 0 & 0 & 1 \end{pmatrix}$$

where θ is the angle of rotation. The *half-turn* $\mathbf{H}_l$ about line l (i.e., by the angle π) is the involute rotation about l.

A *screw motion* in $\mathbb{R}^3$ is a rotation about some line in $\mathbb{R}^3$, followed by a translation in the direction of that line.

Example. Let A be the matrix

$$\begin{pmatrix} \cos\theta + u^2(1-\cos\theta) & uv(1-\cos\theta) + w\sin\theta & wu(1-\cos\theta) - v\sin\theta \\ uv(1-\cos\theta) - w\sin\theta & \cos\theta + v^2(1-\cos\theta) & vw(1-\cos\theta) + u\sin\theta \\ wu(1-\cos\theta) + v\sin\theta & vw(1-\cos\theta) - u\sin\theta & \cos\theta + w^2(1-\cos\theta) \end{pmatrix}$$

where $\mathbf{a} = (u, v, w)$ is a unit vector, $u^2 + v^2 + w^2 = 1$. Show that A is orthogonal and represents a *rotation* $\mathbf{R}(\mathbf{a}, \theta)$ about $\mathbf{a}$ through an angle θ.

Solution. It is easy to check that A is orthogonal and $\mathbf{a}$ is fixed by A. If $\mathbf{b}$ is orthogonal to $\mathbf{a}$, then a simple calculation shows that

$$\mathbf{b}A = \cos\theta\,\mathbf{b} + \sin\theta\,(\mathbf{a} \times \mathbf{b}),$$

which is the required rotation. We prepare a function M-file:

```
function A = s_rot(u, v, w, t);
    A = [ cos(t) + u^2*(1-cos(t)) u*v*(1-cos(t)) + w*sin(t) w*u*(1-cos(t)) - v*sin(t);
    u*v*(1-cos(t)) - w*sin(t) cos(t) + v^2*(1-cos(t)) v*w*(1-cos(t)) + u*sin(t)
    w*u*(1-cos(t)) + v*sin(t) v*w*(1-cos(t)) - u*sin(t) cos(t) + w^2*(1-cos(t)) ]
end
```

For example,

```
syms u v w t real
A = s_rot(u, v, w, t)
simplify(subs(factor(det(A)),  w^2,  1 - u^2 - v^2))                    % det(A) = 1
```

We verify orthogonality, $AA' = \mathrm{Id}_3$, for particular values of the parameters.

```
u = 1/sqrt(6);  v = -1/sqrt(6);  w = 2/sqrt(6);  t = pi/6;
eval(A),    eval(A*A')
```

$$\begin{bmatrix} 0.8884 & 0.3859 & 0.2488 \\ -0.4306 & 0.8884 & 0.1595 \\ -0.1595 & -0.2488 & 0.9553 \end{bmatrix} \quad \begin{bmatrix} 1.0 & 0 & 0 \\ 0 & 1.0 & 0 \\ 0 & 0 & 1.0 \end{bmatrix} \qquad \% \text{ Answer.}$$

Example. (Euler's angles.) The *orientation of a rigid body* is determined by the angles subtended by a frame on the body relative to a fixed reference frame. A body can be positioned with any desired orientation by applying a rotation about each of the axes. Here, we shall adopt the usage of three *Euler's angles* ϕ, θ, ψ that represent rotations about the x, y, z directions, respectively. A generic orthogonal transformation is composed of a rotation by the angle ψ about the z-axis, followed by a rotation by the angle θ about the y-axis (in its new position x', y', z'), followed by a rotation by the angle ϕ about the x-axis (in its new position x'', y'', z''); see Figure 2.1. Thus, the final resulting system is $\bar{x}$, $\bar{y}$, $\bar{z}$. Rotations about the x, y, and z axes are referred to as *pitch*, *yaw*, and *roll*, respectively.

The matrix $\mathbf{R_a} = \mathbf{R}_x(\phi)\mathbf{R}_y(\theta)\mathbf{R}_z(\psi)$ is

$$\begin{pmatrix} \cos\theta\cos\psi & \cos\theta\sin\psi & -\sin\theta \\ \sin\phi\sin\theta\cos\psi - \cos\phi\sin\psi & \sin\phi\sin\theta\sin\psi + \cos\phi\cos\psi & \sin\phi\cos\theta \\ \cos\phi\sin\theta\cos\psi + \sin\phi\sin\psi & \cos\phi\sin\theta\sin\psi - \sin\phi\cos\psi & \cos\phi\cos\theta \end{pmatrix}.$$

For a given transformation matrix, the transformation angles may be determined as $\tan\psi = \mathbf{R}_{a;12}/\mathbf{R}_{a;11}$, $\sin\theta = -\mathbf{R}_{a;13}$, $\tan\phi = \mathbf{R}_{a;23}/\mathbf{R}_{a;33}$. Special attention should be devoted to the sign of the angles in the case where $\mathbf{R}_{a;11} = 0$ and/or $\mathbf{R}_{a;33} = 0$. We prepare the M-file **e_rot.m**:

```
function A = e_rot(phi, t, psi)                              % matrix R_a
    A = [ cos(t)*cos(psi) cos(t)*sin(psi) -sin(t);
    sin(phi)*sin(t)*cos(psi)-cos(phi)*sin(psi)
```

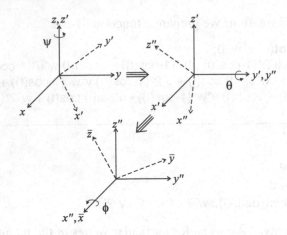

Fig. 2.1 Definition of Euler's angles.

sin(phi)*sin(t)*sin(psi)+cos(phi)*cos(psi)   sin(phi)*cos(t);
cos(phi)*sin(t)*cos(psi)+sin(phi)*sin(psi)
cos(phi)*sin(t)*sin(psi)-sin(phi)*cos(psi)   cos(phi)*cos(t)];
end

For example, we find the rotations about the coordinate axes, $\mathbf{R}_x$, $\mathbf{R}_y$, and $\mathbf{R}_z$:

syms t psi phi;
Rx = **e_rot**(phi, 0, 0),   Ry = **e_rot**(0, t, 0),   Rz = **e_rot**(0, 0, psi)

Next we verify the identity $\mathbf{R_a} = \mathbf{R}_x(\phi)\mathbf{R}_y(\theta)\mathbf{R}_z(\psi)$:

Ra = Rx*Ry*Rz
Ra - **e_rot**(phi, t, psi)   % obtain zero 3 × 3 matrix

Consider a plane ω given by $\mathbf{x} \cdot \mathbf{n} = d$, where $\mathbf{n} = (a,b,c) \neq 0$. The reflection in ω is given by the formula $\mathbf{P}_\omega(\mathbf{x}) = \mathbf{x} + 2\frac{d - \mathbf{x}\cdot\mathbf{n}}{\|\mathbf{n}\|^2}\mathbf{n}$ (see also Section 3.3). The matrix of *reflection* $\mathbf{P}_\omega$ for $d = 0$ is

$$\mathbf{P}_\omega = \frac{1}{a^2+b^2+c^2}\begin{pmatrix} b^2+c^2-a^2 & 2ab & 2ac \\ 2ab & a^2+c^2-b^2 & 2bc \\ 2ac & 2bc & a^2+b^2-c^2 \end{pmatrix}. \qquad (2.4)$$

It is known that *every isometry of $\mathbb{R}^3$ that fixes 0 is the composition of at most three reflections in planes that contain 0.* Hence, every isometry of $\mathbb{R}^3$ is the composition of at most four reflections. (The proof is based on the results of Exercise 10, p. 128).

Note that $\mathbf{P}_\omega = -\mathbf{R}(\mathbf{a}, \pi)$, where the plane ω (passing through the origin) is orthogonal to $\mathbf{a}$. Recall that $\mathbf{R}(\mathbf{a}, \theta)$ is the rotation about $\mathbf{a}$ through θ.

In order to carry out the computation of transformations of $\mathbb{R}^3$ by matrix multiplication, we increase their order by 1. Let $A \in O(3)$ and $\mathbf{a} = (a_1, a_2, a_3)$, $\mathbf{x} = (x_1, x_2, x_3) \in \mathbb{R}^3$. The rigid motion $f(\mathbf{x}) = R_A(\mathbf{x}) + \mathbf{a}$ is represented by the matrix

$$F := \begin{pmatrix} A & 0 \\ \mathbf{a} & 1 \end{pmatrix} = \begin{pmatrix} A_{11} & A_{12} & A_{13} & 0 \\ A_{21} & A_{22} & A_{23} & 0 \\ A_{31} & A_{32} & A_{33} & 0 \\ a_1 & a_2 & a_3 & 1 \end{pmatrix} \in GL_4(\mathbb{R}). \tag{2.5}$$

Let $(\mathbf{x}, 1) := (x_1, x_2, x_3, 1) \in \mathbb{R}^4$. Show that

$$(\mathbf{x}, 1)\, F = (R_A(\mathbf{x}) + \mathbf{a}, 1) \in \mathbb{R}^4.$$

In this way, F represents f. One may show that the composition of two isometries, like the ones represented by $F_1 = \begin{pmatrix} A_1 & 0 \\ \mathbf{a}_1 & 1 \end{pmatrix}$ and $F_2 = \begin{pmatrix} A_2 & 0 \\ \mathbf{a}_2 & 1 \end{pmatrix}$, is the isometry represented by the product:

$$\begin{pmatrix} A_1 & 0 \\ \mathbf{a}_1 & 1 \end{pmatrix} \cdot \begin{pmatrix} A_2 & 0 \\ \mathbf{a}_2 & 1 \end{pmatrix} = \begin{pmatrix} A_1 \cdot A_2 & 0 \\ R_{A_2}(\mathbf{a}_1) + \mathbf{a}_2 & 1 \end{pmatrix}. \tag{2.6}$$

This allows us to see that the isometry $\mathbf{x} \to R_{A_1}(\mathbf{x}) + \mathbf{a}_1$ followed by the isometry $\mathbf{x} \to R_{A_2}(\mathbf{x}) + \mathbf{a}_2$ is the isometry $\mathbf{x} \to R_{A_1 \cdot A_2}(\mathbf{x}) + R_{A_2}(\mathbf{a}_1) + \mathbf{a}_2$.

Example. We prepare the symbolic matrix F of (2.5):

```
syms t psi phi a1 a2 a3;
A = e_rot(phi, t, psi);
F1 = cat(2, A, [0; 0; 0]),    a = [a1 a2 a3];
F = cat(1, F1, [a 1])                              % extended matrix F
```

We verify (2.6) for numerical matrices:

```
F1 = double(subs(F, {t, phi, psi, a1, a2, a3}, {0, 1, 2, 5, 0, 0}))    % matrix F1
F2 = double(subs(F, {t, phi, psi, a1, a2, a3}, {1, 2, 0, 0, 2, 0}))    % matrix F2
F3 = F1*F2;
A1 = double(subs(A, {t, phi, psi}, {0, 1, 2}))
A2 = double(subs(A, {t, phi, psi}, {1, 2, 0}))
A3 = A1*A2;    a3 = [5 0 0]*A2 + [0, 2, 0];
G0 = cat(2, A3, [0; 0; 0]);    G3 = cat(1, G0, [a3 1]);
F3 - G3                                            % obtain zero 3 × 3 matrix !
```

Examples of transformations in $\mathbb{R}^4$ are:

(1) projection to the line along $\mathbf{u} \neq 0$, defined by $P(\mathbf{x}) = \frac{\mathbf{x} \cdot \mathbf{u}}{\mathbf{u} \cdot \mathbf{u}}\, \mathbf{u}$;
(2) reflection through the line $\mathbf{u} \neq 0$, defined by $S(\mathbf{x}) = 2P(\mathbf{x}) - \mathbf{x}$;
(3) projection to the hyperplane $\perp \mathbf{u} \neq 0$, defined by $Q(\mathbf{x}) = \mathbf{x} - P(\mathbf{x})$;

(4) projection to the coordinate plane P_{ij}, defined by $P_{ij}(\mathbf{x}) = (x_i, x_j)$;

(5) rotation in the x_1-x_2-plane by the angle θ, defined by

$$R_\theta^{12}(\mathbf{x}) = (\cos\theta\, x_1 - \sin\theta\, x_2, \sin\theta\, x_1 + \cos\theta\, x_2, x_3, x_4); \quad \text{etc.}$$

Example. Consider the 4-cube K^4 with 16 vertices with all coordinates either 1 or -1. The projection P_{12} of K^4 is the square with the vertices $(\pm 1, \pm 1)$. (Which four vertices are all sent to $(1,1)$ under P_{12}?).

In order to get more useful pictures, we first rotate K^4 before projecting to the 1-2-coordinate plane. If we rotate K^4 by $\theta = 30°$ in the 1-3-plane, we obtain $P_{12}R_\theta^{13}(\mathbf{x}) = (\cos\theta\, x_1 - \sin\theta\, x_3, x_2)$ and hence $P_{12}R_{30}^{13}(1,1,1,1) = (\sqrt{3}/2 - 1/2, 1)$. The picture is somewhat clearer, but we still see only 8 distinct vertex images.

If we first rotate in the 1-3-plane, then in the 2-4-plane, and then in the 1-4-plane, we obtain a cube in general position. Thus

$$P_{12}R_\psi^{14}R_\phi^{24}R_\theta^{13}(\mathbf{x})$$
$$= (\cos\psi(\cos\theta\, x_1 - \sin\theta\, x_3) - \sin\psi(\sin\phi\, x_2 + \cos\phi\, x_4), \cos\phi\, x_2 - \sin\phi\, x_4).$$

Then $P_{12}R_{30}^{14}R_{30}^{24}R_{30}^{13}(\mathbf{x}) = \left(\frac{3}{4}x_1 - \frac{1}{4}x_2 - \frac{\sqrt{3}}{4}x_3 - \frac{\sqrt{3}}{4}x_4, \frac{\sqrt{3}}{2}x_2 - \frac{1}{2}x_4\right)$. We obtain 16 different images for the 16 vertices of the 4-cube. Finally, if we rotate by 30 degrees in the 2-3-plane, we have

$$P_{12}R_{30}^{23}R_{30}^{14}R_{30}^{24}R_{30}^{13}(\mathbf{x}) = \left(\frac{3}{4}x_1 - \frac{1}{4}x_2 - \frac{\sqrt{3}}{4}x_3 - \frac{\sqrt{3}}{4}x_4, -\frac{1}{4}x_1 + \frac{3}{4}x_2 - \frac{\sqrt{3}}{4}x_3 - \frac{\sqrt{3}}{4}x_4\right).$$

Four pairs of vertices are brought into superposition by the final rotation (which ones?), and we now obtain only 12 different images for the 16 vertices of the 4-cube.

2.2.3 *Rotations and reflections using quaternions*

There are two linear transformations of $\mathbb{R}^4$ given by right and left multiplication by q, namely, $R_q(x) = xq$ and $L_q(x) = qx$; see Exercise 14, p. 129. The rows of the matrix R_q are determined by its action on 1, i, j, and k, viz.,

$$R_q = \begin{pmatrix} q_0 & q_1 & q_2 & q_3 \\ -q_1 & q_0 & -q_3 & q_2 \\ -q_2 & q_3 & q_0 & -q_1 \\ -q_3 & -q_2 & q_1 & q_0 \end{pmatrix}. \tag{2.7}$$

We shall see how quaternions provide an alternative algebraic way to express reflections and rotations.

Consider the map $C_q \colon \mathbb{H} \to \mathbb{H}$ given by $C_q(p) = -qpq^{-1}$, where $q = q_1\mathbf{i} + q_2\mathbf{j} + q_3\mathbf{k} \neq 0$ is a pure quaternion. (If q is a pure *unit* quaternion, then $C_q(p) = qpq$.) Clearly, C_q is a linear map, that is, $C_q(\lambda_1 p_1 + \lambda_2 p_2) = \lambda_1 C_q(p_1) + \lambda_2 C_q(p_2)$, and $C_q(0) = 0$, $C_q(q) = -q$. These facts suggest that C_q might be related to the reflection across the plane with normal $\mathbf{q}$. The 4×4 matrix C_q of an arbitrary quaternion $q = q_0 + q_1\mathbf{i} + q_2\mathbf{j} + q_3\mathbf{k}$ is

$$\begin{pmatrix} |q|^2 & 0 & 0 & 0 \\ 0 & q_0^2 + q_1^2 - q_2^2 - q_3^2 & 2q_1q_2 + 2q_0q_3 & 2q_1q_3 - 2q_0q_2 \\ 0 & 2q_1q_2 - 2q_0q_3 & q_0^2 - q_1^2 + q_2^2 - q_3^2 & 2q_2q_3 + 2q_0q_1 \\ 0 & 2q_1q_3 + 2q_0q_2 & 2q_2q_3 - 2q_0q_1 & q_0^2 - q_1^2 - q_2^2 + q_3^2 \end{pmatrix}.$$

Theorem 2.2. *For a pure quaternion q, C_q maps the set H_0 of pure quaternions into itself. When H_0 is identified with $\mathbb{R}^3$ and q with $\mathbf{q} = q_1\mathbf{i} + q_2\mathbf{j} + q_3\mathbf{k} \in \mathbb{R}^3$, C_q is the reflection* across the plane $q_1x + q_2y + q_3z = 0$.

We now use quaternions to describe rotations that fix the origin. Consider the rotation R obtained by the reflection across the plane with unit normal $\mathbf{p}$ followed by the reflection across the plane with unit normal $\mathbf{q}$. According to Theorem 2.2, $R(\mathbf{x}) = q(p\mathbf{x}p)q = (qp)\mathbf{x}(pq)$, where $p = (0, \mathbf{p})$ and $q = (0, \mathbf{q})$. This yields the following theorem.

Theorem 2.3. *Let $q = (\cos\frac{\theta}{2}, \sin\frac{\theta}{2}\mathbf{n})$ be a unit quaternion. Then $-C_q \colon \mathbf{x} \to q\mathbf{x}q^{-1} = q\mathbf{x}\bar{q}$ yields a rotation of $\mathbf{x}$ by an angle θ about the axis $\mathbf{n}$.*

Examples.

1. We create M-files **r_quat.m** and **r_quat.m** with $L_q(\mathbf{x})$ and $R_q(\mathbf{x})$.

```
function Rq = r_quat(q)
   q0 = q(1);  q1 = q(2);  q2 = q(3);  q3 = q(4);
   Rq = [q0 q1 q2 q3;  -q1 q0 -q3 q2;  -q2 q3 q0 -q1;  -q3 -q2 q1 q0];
end
```

```
function Lq = l_quat(q)
   q0 = q(1);  q1 = q(2);  q2 = q(3);  q3 = q(4);
   Lq = [q0 q1 q2 q3;  -q1 q0 q3 -q2;  -q2 -q3 q0 q1;  -q3 q2 -q1 q0];
end
```

Then we verify the equalities $pq = R_q(p)\ (= pR_q)$ and $qp = L_q(p)\ (= pL_q)$:

```
syms p0 p1 p2 p3 real;   p = [p0 p1 p2 p3]
quatmultiply(p, q) - p*r_quat(q)                            % obtain 0
quatmultiply(q, p) - p*l_quat(q)                            % obtain 0
```

2. We verify the equality $C_q = L_q R_{q^{-1}} = R_{q^{-1}} L_q$.

```
Cq1 = simple(l_quat(q)*r_quat(quatinv(q)))            % L_qR_{q^{-1}}
Cq2 = simple(r_quat(quatinv(q))*l_quat(q))            % R_{q^{-1}}L_q
Cq1 - Cq2                                  % obtain zero 4 × 4 matrix.
```

For future use prepare the function M-file **c_quat.m**:

```
function Cq = c_quat(q)
    Cq = -l_quat(q)*r_quat(quatinv(q));
end                        % Alternatively the reader can use the quatrotate command
```

3. The reflection across the plane $(\mathbf{x}, \mathbf{k}) = 0$ (i.e., coordinate plane $x_3 = 0$) is given by $(x_1, x_2, x_3) \to (x_1, x_2, -x_3)$ (see Theorem 2.2), or, in terms of quaternions, $\mathbf{x} \to -\mathbf{kxk}^{-1} = C_q(\mathbf{x})$, where $q = \mathbf{k} = [0, 0, 0, 1]$.

```
q = [0 0 0 1];  R3 = c_quat(q);  R3(2:4, 2:4)     % reflection across x_3 = 0
[1 0 0; 0 1 0; 0 0 -1]                            % Answer
```

We also note (in preparation for what follows) that the map $\mathbf{x} \to \mathbf{kxk}^{-1}$ is given by $(x_1, x_2, x_3) \to (-x_1, -x_2, x_3)$, and this is a rotation of angle π about the $\mathbf{k}$-axis.

4. One may prove Theorem 2.3 using a program.

```
syms t r1 r2 r3 real;
r = cos(t/2)*[1 0 0 0] + sin(t/2)*[0 r1 r2 r3]
Rr = -c_quat(r)                    % x → qxq^{-1} = -C_q(x)
R_r = s_rot(r1, r2, r3, t)         % rotation in R^3 by t about [r_1, r_2, r_3]
Q = Rr(2:4, 2:4) - R_r             % next we use r_1^2 + r_2^2 + r_3^2 = 1
D = simplify(subs(Q, r3^2, 1 - r1^2 - r2^2))   % obtain zero 3 × 3 matrix
```

5. Rotate $P = (3, 6, -5)$ by $\theta = 2\pi/3$ about the axis $\mathbf{r} = (-4, 2, 4)$.

```
t = 2*pi/3;  r = [-4 2 4];  P = [3 6 -5];           % enter data
r0 = r/norm(r);                   % normalize: r_0 = [-√3/3, √3/6, √3/3]
r1 = cos(t/2)*[1 0 0 0] + sin(t/2)*cat(2, 0, r0)    % r_1 ≈ [0.50, -0.58, 0.29, 0.58]
R0 = -c_quat(r1)                                    % matrix of rotation
[1 0 0 0; 0 0.17 0.24 -0.96; 0 -0.91 -0.33 -0.24; 0 -0.38 0.91 0.17]
P0 = cat(2, 0, P);  P0*R0          % Answer: ≈ [0, -3.07, -5.82, -5.16]
```

An animation of an object can be performed by determining a curve in the unit sphere of pure quaternions joining the two points (of start and end "orientations" q_{start}, q_{end}). Assume that the interpolating curve is an arc of a *great circle*; see Section 2.3. Let ϕ be the angle between q_{start} and q_{end} in $\mathbb{R}^4$. The expected formula is

$$q(t) = \frac{\sin((1-t)\phi)}{\sin\phi} q_{\text{start}} + \frac{\sin(t\phi)}{\sin\phi} q_{\text{end}}, \quad 0 \le t \le 1. \tag{2.8}$$

The locus of a point p is a curve parameterized by $C_q(t) = q(t) p q(t)^{-1}$.

Example. (Animation of a triangle using quaternions.)
Let $q_{\text{start}} = (\frac{1}{5}, \frac{2}{5}, \frac{4}{5}, \frac{2}{5})$ and $q_{\text{end}} = (\frac{2}{3}, 0, \frac{1}{3}, \frac{2}{3})$ be unit quaternions (verify!). Then $\cos\phi = q_{\text{start}} \cdot q_{\text{end}} = \frac{2}{3}$ (the scalar product).

```
qs = [1/5 2/5 4/5 2/5];   qe = [2/3 0 1/3 2/3];        % define two quaternions
phi = acos(qs*qe');                                    % obtain angle φ ≈ 0.8411
qt = @(t, ps, pe)(sin((1 - t)*phi))/sin(phi)*ps + sin(t*phi)/sin(phi)*pe;
syms t;  vpa(qt(t, qs, qe), 3)                          % obtain (2.8)
qt(2/5, qs, qe)                      % obtain [0.4250, 0.2595, 0.6666, 0.5547]
```

Apply the animation to the triangle $A = (5,5,0), B = (10,5,0), C = (7,10,0)$.

```
A = [5, 5, 0];   B = [10, 5, 0];   C = [7, 10, 0];          % define triangle
Aq = cat(2, 0, A);  Bq = cat(2, 0, B);  Cq = cat(2, 0, C);   % points in ℝ⁴
syms t real;
At = quatmultiply(quatmultiply(qt(t, qs, qe), Aq), quatinv(qt(t, qs, qe)));
                                           % Aₑ(t) = q(t)A_q q(t)⁻¹.
ezplot(At(4), [0,1])                 % graph of the z-coordinate of A(t)
ezplot3(At(2), At(3), At(4), [0, 1])       % trajectory of A in ℝ³, Section 5.3.
```

To see the whole picture (without animation), we prepare

```
N = 20;  for i = 1 : N;
AT(i, :) = quatmultiply(quatmultiply(qt(i/N,qs,qe),Aq), quatinv(qt(i/N,qs,qe)));
BT(i, :) = quatmultiply(quatmultiply(qt(i/N,qs,qe),Bq), quatinv(qt(i/N,qs,qe)));
CT(i, :) = quatmultiply(quatmultiply(qt(i/N,qs,qe),Cq), quatinv(qt(i/N,qs,qe)));
end
```

and then plot N triangles with (three) trajectories of their vertices:

```
hold on;   for k = 1 : N ;
 patch([AT(k, 2), BT(k, 2), CT(k, 2)],  [AT(k, 3), BT(k, 3), CT(k, 3)],
 [AT(k,4), BT(k,4), CT(k,4)], 'r');
end;                                                        % N triangles
plot3(AT(:, 2), AT(:, 3), AT(:, 4));                     % trajectory of A
plot3(BT(:, 2), BT(:, 3), BT(:, 4));                     % trajectory of B
plot3(CT(:, 2), CT(:, 3), CT(:, 4))                      % trajectory of C
```

2.3 Sphere

Spherical geometry is the study of geometry on the surface of a sphere of radius R:
$S^2(R) = \{(x,y,z) \in \mathbb{R}^3 : x^2 + y^2 + z^2 = R^2\}$. It is simplest to choose this sphere to have unit radius; we denote $S^2(1)$ by S^2.

2.3.1 *Spherical transformations*

A *great circle* (or *geodesic*) is the circle cut out on the sphere S^2 by a plane through the center of the sphere. The curve of intersection of S^2 with a plane not through the center is called a *little circle*. Any great circle through the North Pole N is called a *circle of longitude*. The part of the circle of longitude through $A = (1,0,0)$ that has positive x-coordinate is called the *Greenwich meridian*. The curve of intersection of S^2 with a plane parallel to the equator is called a *circle of latitude*.

Let γ be a great circle on S^2 through two points P and Q. Then either arc with endpoints P, Q is said to be a *line PQ*. For definiteness, we call the shorter of these two the *strict line segment PQ*. A *triangle* $\triangle PQR$ is defined by three points on S^2 that do not lie on a single great circle, and three lines PQ, QR, and RP. For definiteness, we call the triangle whose sides are strict line segments the *strict triangle*.

The *length* of a line on S^2 is the Euclidean length of the corresponding circular arc; the *(spherical) distance* between P and Q on S^2 is the length of the shorter of the two lines PQ. The distance d between any two points $P = (p_1, p_2, p_3)$ and $Q = (q_1, q_2, q_3)$ on S^2 is $\cos d = (P,Q) = p_1 q_1 + p_2 q_2 + p_3 q_3$.

Spherical triangles exist with any given angles; see exercises. The *angular excess* of a spherical triangle is the difference between the sum of the angles of the triangle and π. The area of a spherical triangle is equal to its angular excess: $S(\triangle ABC) = \alpha + \beta + \gamma - \pi$.

Vector products are useful in studying triangles on a sphere.

Let $\triangle ABC$ be a strict triangle in S^2 with the sides a, b, c and the angles α, β, γ. Then

(a) $\cos\alpha = \frac{(A\times B, A\times C)}{\|A\times B\|\,\|A\times C\|}$, $\cos\beta = \frac{(B\times A, B\times C)}{\|B\times A\|\,\|B\times C\|}$, $\cos\gamma = \frac{(C\times A, C\times B)}{\|C\times A\|\,\|C\times B\|}$,

(b) $\cos a = (B,C)$, $\cos b = (A,C)$, $\cos c = (A,B)$.

(c) (*Sine rule*) $\sin\alpha/\sin a = \sin\beta/\sin b = \sin\gamma/\sin c$.

(d) (*Cosine rule*) $\cos c = \cos a \cos b + \sin a \sin b \cos\gamma$ (for sides),
  $\cos\gamma = \sin\alpha \sin\beta \cos c - \cos\alpha\cos\beta$ (for angles).

An *isometry* (rigid motion) of S^2 is a mapping of the unit sphere to itself that preserves distances between points.

The isometries of S^2 form a group. The direct isometries of S^2 are simply rotations of S^2 that also form a group. Among them there are three *elementary rotations* of S^2 (i.e., rotations about coordinate axes) $\mathbf{R}_x(\theta)$, $\mathbf{R}_y(\theta)$, and $\mathbf{R}_z(\theta)$ that in general are not commutative.

A *reflection* of S^2 in a plane through the center of the unit sphere is not a direct isometry, but every composition of an even number of such reflections is a direct isometry of S^2. It follows from (2.4) that a reflection of S^2 in the plane $\alpha\colon ax + by + cz = 0$ $(a^2 + b^2 + c^2 = 1)$ is given by the mapping $\mathbf{x} \to \mathbf{x}A$, where

$$A = \begin{pmatrix} 1-2a^2 & -2ab & -2ac \\ -2ab & 1-2b^2 & -2bc \\ -2ac & -2bc & 1-2c^2 \end{pmatrix}.$$

One may show that any isometry (rigid motion) of S^2 is

(a) a product of at most three reflections in great circles;
(b) a reflection, a rotation, or a composite of a reflection and a rotation.

Any direct isometry of S^2 is a rotation, and vice versa.

Examples.

1. Find the sides of a spherical triangle $\triangle ABC$ with $\angle A = \angle B = \alpha$ and $\angle C = \pi/2$.

Solution. First note that $a = b$.

```
syms a c;   alpha = pi/3;
Eq1 = sin(alpha)/sin(a) - sin(pi/2)/sin(c);         % Sine rule
Eq2 = cos(c) - cos(a)^2;        % Cosine rule (Pythagoras' theorem, ∠C = π/2)
[a c] = solve(Eq1, Eq2, a, c);   mod(double([a, c]), pi)
[2.1863  1.9106]                 % Answer: a = 2.1863, c = 1.9106.
              % The second solution is π − a ≈ 0.9553, π − c ≈ 1.2310.
```

2. Show that a rotation of a sphere that maps $A = (1,0,0)$ to the point $P = (\cos\phi\sin\theta, \sin\phi\sin\theta, \cos\theta)$ is given by the product of $\mathbf{R}_z(\phi)$ and $\mathbf{R}_y(\theta - \pi/2)$.

```
syms t s real;  P = [cos(t)*sin(s), sin(t)*sin(s), cos(s)]
s1 = pi/2 - s;
Rz = e_rot(0, 0, t);   Ry = e_rot(0, -s1, 0);   R = Ry*Rz;
[1, 0, 0]*R - P                  % obtain (0,0,0). Hence P = (1,0,0) R_y R_z.
```

Alternatively, one may use the function M-file **s_rot.m**; see p. 91.

2.3.2 *Stereographic projection*

Identify the plane with $\mathbb{R}^2 \times \{0\}$ in $\mathbb{R}^3$. The *stereographic projection* π of $\mathbb{R}^2$ onto $S^2 \setminus \{e_3\}$ is defined by projecting $\mathbf{x}$ in $\mathbb{R}^2$ towards (or away from) e_3 until it meets the sphere S^2 in the unique point $\pi(\mathbf{x})$ other than e_3. As $\pi(\mathbf{x})$ is on the line passing through $\mathbf{x}$ in the direction of $e_3 - \mathbf{x}$, there is a scalar s such that $\pi(\mathbf{x}) = \mathbf{x} + s(e_3 - \mathbf{x})$. The condition $\|\pi(\mathbf{x})\|^2 = 1$ leads to the value $s = \frac{\|\mathbf{x}\|^2 - 1}{\|\mathbf{x}\|^2 + 1}$ and the explicit formula

$$\pi(\mathbf{x}) = \left(\frac{2x_1}{1 + \|\mathbf{x}\|^2}, \frac{2x_2}{1 + \|\mathbf{x}\|^2}, \frac{\|\mathbf{x}\|^2 - 1}{\|\mathbf{x}\|^2 + 1} \right). \tag{2.9}$$

The mapping π is a bijection of $\mathbb{C} \approx \mathbb{R}^2$ onto $S^2 \setminus \{\mathbf{e}_3\}$. The inverse mapping is given by

$$\pi^{-1}(x_1, x_2, x_3) = \frac{x_1}{1 - x_3} + \frac{x_2}{1 - x_3} i \quad \text{for all} \quad (x_1, x_2, x_3) \in S^2.$$

We prepare a function M-file **ster.m** to compute the projection π:

```
function f = ster(x)                              % for x = [x₁,x₂]
    f = [ 2*x(1)/(x*x' + 1) 2*x(2)/(x*x' + 1) (x*x' - 1)/(x*x' + 1) ];
end
```

Then we compute images of points, lines, circles (see Theorem 2.4), etc.

```
ster([1  2])                      % image of a point: [0.3333, 0.6667, 0.6667]
syms t real;
simplify(ster([2*cos(t) 2*sin(t)]))     % image of a circle: [⅘ cost, ⅘ sint, ⅗].
g = simplify(ster([3  t]))                        % g = [6/(10+t²), 2t/(10+t²), (8+t²)/(10+t²)]
simplify(g(1)/3 + g(3) - 1)                               % Answer: 0,
          % image of a line {x₁ = 3} belongs to the plane ⅓x₁ + x₃ = 1.
```

We define the *chordal distance* d on $\mathbb{C}$ (as it is the length of the chord of S^2 that joins $\pi(z)$ and $\pi(w)$) by the formula $d(z,w) = \|\pi(z) - \pi(w)\|$; see Exercise 12, p. 130. By definition, the map π is an isometry from $\mathbb{R}^2$, with the chordal metric, to S^2 with the Euclidean metric. It is geometrically evident that $d(z,w) \leq 2$. A simple calculation shows that

$$d(z,w) = \frac{2|z - w|}{\sqrt{(1 + |z|^2)(1 + |w|^2)}} \quad \text{and} \quad d(z,\infty) = \frac{2}{\sqrt{1 + |z|^2}} \quad \text{for } z, w \in \mathbb{C}.$$

The extended complex plane $\hat{\mathbb{C}} = \mathbb{C} \cup \{\infty\}$ (obtained from $\mathbb{C}$ by adjoining one point at infinity, see Section 4.2) is a compact metric space. The *one-point compactification* $\hat{\mathbb{C}}$ of the complex plane $\mathbb{C}$ is also called the *Riemann sphere*. We refer to the point $N = (0,0,1)$ of a 2-sphere $S^2 := \{x^2 + y^2 + z^2 = 1\}$ as the *North Pole* and $S = (0,0,-1)$ as the *South Pole*.

There are other kinds of compactification; for example, the projective plane $\mathbb{R}P^2$, which we will meet in Chapter 3, is obtained from the standard plane $\mathbb{R}^2$ by adding a line at infinity (actually a circle).

Theorem 2.4. *Under stereographic projection $\pi \colon \hat{\mathbb{C}} \to S^2$, generalized circles (i.e., circles and lines) map onto circles on the sphere. In particular, lines map onto circles on the sphere that pass through the North Pole N.*

Proof. A circle on S^2 is the intersection of the sphere with some plane $\omega \colon aX + bY + cZ = d$, where a, b, c are not all zero. It follows from substituting the expressions

$$X = \frac{2x_1}{x_1^2 + x_2^2 + 1}, \qquad Y = \frac{2x_2}{x_1^2 + x_2^2 + 1}, \qquad Z = \frac{x_1^2 + x_2^2 - 1}{x_1^2 + x_2^2 + 1}$$

into the equation of the plane that

$$\frac{2ax_1 + 2bx_2 + c\left(x_1^2 + x_2^2 - 1\right)}{x_1^2 + x_2^2 + 1} = d \iff (c-d)x_1^2 + (c-d)x_2^2 + 2ax_1 + 2bx_2 = c+d.$$
(2.10)

Hence, the inverse image (under π^{-1}) of the circle on the sphere is a generalized circle in $\hat{\mathbb{C}}$. The circle on the sphere passes through $N = (0,0,1)$ if the plane ω passes through N, whereupon $c = d$. In this case the circle's image in $\hat{\mathbb{C}}$ has an equation of the form $2ax_1 + 2bx_2 = c+d$, and so is an extended line. □

We deduce (2.10) using a simple program:

```
syms x1 x2 a b c d real;
f = ster([x1 x2]);
F = a*f(1) + b*f(2) + c*f(3) - d;
G = collect(simple(F*(x1^2 + x2^2 + 1)),  [x1, x2])
```

$$G = (c \quad d)x_1^2 + 2ax_1 + (c-d)x_2^2 + 2bx_2 - c - d \qquad \text{\% Answer.}$$

Remark 2.1. The notions of stereographic projection and chordal metric can be extended to arbitrary dimension. Let $S^n(0,1) = \{\mathbf{x} \in \mathbb{R}^{n+1} : \|\mathbf{x}\| = 1\}$. The *stereographic projection* $\pi\colon \mathbb{R}^n \to S^n(0,1) \setminus \{\mathbf{e}_{n+1}\} \subset \mathbb{R}^{n+1}$ (where $n \geq 1$) is derived by the formula

$$\pi(\mathbf{x}) = \left(\frac{2x_1}{1 + \|\mathbf{x}\|^2}, \dots, \frac{2x_n}{1 + \|\mathbf{x}\|^2}, \frac{\|\mathbf{x}\|^2 - 1}{\|\mathbf{x}\|^2 + 1} \right).$$
(2.11)

The *chordal distance* $d(\mathbf{x}, \mathbf{y}) = \|\pi(\mathbf{x}) - \pi(\mathbf{y})\|$ on $\mathbb{R}^n$ (the length of the chord of S^n that joins $\pi(\mathbf{x})$ and $\pi(\mathbf{y})$) is derived by the formula

$$d(\mathbf{x}, \mathbf{y}) = \frac{2\|\mathbf{x} - \mathbf{y}\|}{\sqrt{(1 + \|\mathbf{x}\|^2)(1 + \|\mathbf{y}\|^2)}}, \qquad d(\mathbf{x}, \infty) = \frac{2}{\sqrt{1 + \|\mathbf{x}\|^2}}.$$
(2.12)

The *one-point compactification* $\hat{\mathbb{R}}^n = \mathbb{R}^n \cup \{\infty\}$ (obtained from $\mathbb{R}^n$ by adjoining one point at infinity) is a compact metric space.

2.4 Polyhedra

We start from basic facts about polyhedra and simple programs for plotting the polygonal curves, prisms and pyramids. A typical program for plotting polyhedra has the following structure:

(a) the list of vertices with their coordinates,

(b) the lists of vertices in faces,

(c) the computations if required, and

(d) the visualizing command patch.

Next we study and visualize the Platonic solids and star-shaped polyhedra (Poinsot's polyhedra and Kepler's octahedron) in rotation about the coordinate axes. The reader can play with complicated polyhedra, putting objects through various transformations: rotation, coloring, various operations (stellate, etc.), change of parameters in coordinates of vertices and faces, intersections and other combinations of solids, projecting, discovering new polyhedra, and so on.

2.4.1 *What is a polyhedron?*

An acquaintance with polyhedra leaves a strong impression on the human mind, sometimes leading to creative work and discoveries. The importance of this theme in school geometry and for developing spatial imagination is well known. The theory of polyhedra goes back to antiquity: volume XII of Euclid's *Elements* is devoted to the regular polyhedra. Archimedes, in his work *"On polyhedra"* described the so-called semi-regular polyhedra.

A closed polygonal line without self-intersections that bounds a plane domain is called a *simple polygon*. A simple polygon is called *convex* if for each side of the polygon, all the vertices that do not lie on the side in question belong to one of the half-planes defined by this side.

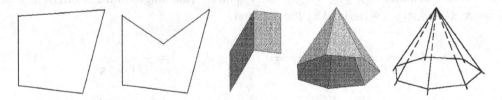

Fig. 2.2 Convex and non-convex polygons.   Adjacent faces. Solid angle.

A *solid angle* is a cone-like surface whose directrix is a simple polygon. A *polyhedron* is a figure in space that consists of a finite number of planar polygons (*faces* of the polyhedron) meeting the following conditions.

(1) Each side of any of these faces meets exactly one other face (said to be *adjacent* to the first one) along a side.

(2) For faces α, β one can find a sequence of faces α_1, α_2, ..., α_n such that face α is adjacent to α_1, face α_1 is adjacent to α_2, ..., and face α_n is adjacent to β.

(3) Let V be any vertex and let F_1, F_2, ..., F_n be the n faces that meet at V. It is possible to traverse the polygons F_i without passing through V.

A polyhedron is *convex* if all vertices that do not belong to an arbitrary face of this polyhedron lie in one of the half-spaces defined by this face.

Simple examples of polyhedra are *prisms* (in particular, a cube) and *pyramids* (in particular, a tetrahedron).

If all faces are simple polygons, then by definition, a polyhedron is a *surface* (see Section 7.1) in space. Moreover, condition (1) means that this surface is without a *boundary* and has no singular edges, as in Figures 7.1(c), 7.3(c), 8.26(b), and 8.5(b,c). Condition (2) means that it is *connected* (i.e., has exactly one component). Condition (3) excludes singular vertices; see Figures 7.16(a), 7.11, and 8.25(b).

To exclude self-intersecting polyhedra we give the following definition: A polyhedron is *simple* if

(i) all its faces are simple polygons;

(ii) any two non-adjacent faces have no (interior or boundary) points in common, except perhaps a single vertex;

(iii) any two adjacent faces have only one edge in common and do not have any other points in common.

The definition does not exclude polyhedra (solids) with *tunnels*. A simple polyhedron divides space into two regions, a finite one, called *inner*, and another, which contains lines, called *external*.

A closed curve in a surface is *non-separating* if the surface remains in one piece when it is cut along the curve. Some polyhedra and surfaces possess non-separating curves, Figures 7.1(b), 7.3, 8.15(a), and 8.16(d); in others (*simply connected*), every curve separates the surface or polyhedron into two pieces, Figures 8.1(b,c) and 8.2(c).

Non-separating curves (consisting of edges) in a polyhedron correspond to tunnels through a solid. One can define the number of tunnels as the maximum number of cuts (curves) which leave the polyhedron connected.

Examples. The commands patch(x, y, [r g b]) and patch(x, y, z, [r g b]) create plane and solid patches colored with the vector $[r\ g\ b]$.

Let us plot several figures.

(a) A plane triangle (one may similarly plot a polygon with arbitrary vertices) and a regular simple n-gon (inscribed in a circle).

```
patch([0 2 5], [0 4 3], [0 1 1])                % triangle in ℝ², Figure 2.3(a)
n = 6;  t = 0 : 2*pi/n : 2*pi;
X = sin(t);  Y = cos(t);
patch(X, Y, [0 1 1]);  axis equal               % hexagon
```

(b) A regular pyramid of height 2 with vertex $S = (0,0,2)$.

```
n = 6;  t = linspace(0, 2*pi, n);                    % pyramid, Figure 2.3(d)
X = sin(t);  Y = cos(t);  Z = 0*t;
V = [X' Y' Z'];   V0 = [0 0 1];   V(n + 1, :) = V0;
F = [n  1  n+1];
for i = 1 : n - 1;  F = cat(1, F, [i  i+1  n+1]);  end;
hold on;  axis equal;  view(3);  patch(X, Y, [0 1 1]);
patch('Vertices', V, 'Faces', F, 'FaceColor', [1 .5 1]);
```

(c) A regular prism of height 2 with vertical edge $[0,0,h]$.

```
R = 1;  h = 2;  n = 6;                               %  prism, Figure 2.3(c)
t = 0 : 2*pi/n : 2*pi;
X = R*sin(t);  Y = R*cos(t);  Z = 0*t;
X1 = cat(2, X(1 : n), X(1 : n));
Y1 = cat(2, Y(1 : n), Y(1 : n));
Z1 = cat(2, Z(1 : n), Z(1 : n) + h);
V = [X1' Y1' Z1'];  F = [n 1 n+1 2*n];
for i = 1 : n - 1;  F = cat(1, F, [i  i+1  n+i+1  n+i]);  end;
hold on;  axis equal;  view(3);
patch(X, Y, [0 1 1]);  patch(X, Y, Z + h, [0 1 1]);
patch('Vertices', V, 'Faces', F, 'FaceColor', [1 .5 1])
```

In what follows we plot regular convex polyhedra by different methods.

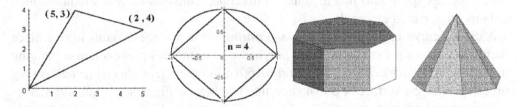

Fig. 2.3 Planar polygons.   Prism and pyramid.

The *development of a polyhedron* is the union of plane polygons according to a rule stating how one should glue them together along their sides and vertices to obtain the desired polyhedron.

Moreover, the following conditions must be satisfied:

(i) Each side of the polygon is attached to at most one side of another polygon.
(ii) Any two glued sides must have the same length.
(iii) One can reach any polygon by starting from any other polygon and moving along a sequence of glued polygons.

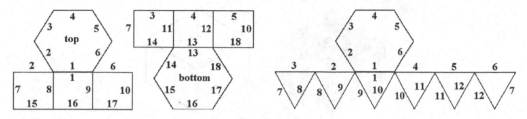

Fig. 2.4 Developments of 6-faced prism and pyramid.

For example, the development of the side of a cone is a circular sector.

One can apply the notion of development to curved surfaces (for instance, in descriptive geometry and drawing).

The following facts (A. Cauchy, G. Minkowski) are important in the theory of polyhedra.

(a) If two convex polyhedra are isometric to each other, then one can be obtained from the other by a motion in $\mathbb{R}^3$.

(b) There exist convex polyhedra with arbitrary area of the faces and arbitrary directions of the extrinsic normals to them, but the sum of the vectors in the directions of these normals with lengths equal to the areas of the corresponding faces must be zero, and not all the vectors are coplanar.

(c) A convex polyhedron is completely defined by the areas of its faces and the directions of the extrinsic normals to the faces.

The numbers of faces F, edges E, and vertices V of a convex polyhedron are related by *Euler's theorem* $F - E + V = 2$. However, this is not the only necessary condition. For a simple polyhedron M, the expression $F - E + V$ is called the *Euler characteristic* and is denoted by $\chi(M)$.

Suppose that we cut each edge of the polyhedron at its midpoint and then count the segments that are formed by this cutting. As each edge produces two segments, this count is clearly $2E$. On the other hand, every vertex has *at least* three segments ending there, and each segment ends at only one vertex; thus the count is at least $3V$. We conclude that $2E \geq 3V$. Similarly, imagine that we have separated the polyhedron into its faces, and that we now count the edges of all of these faces. Clearly this count is $2E$. On the other hand, each face has *at least* three sides so the count must be at least $3F$. We deduce that $2E \geq 3F$.

Theorem 2.5 (Existence). *There exists a convex polyhedron with F faces, E edges, and V vertices if and only if*

$$F - E + V = 2, \quad 2E \geq 3V, \quad 2E \geq 3F. \tag{2.13}$$

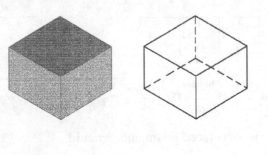

Fig. 2.5 Cube.

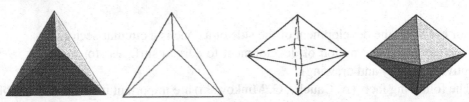

Fig. 2.6 Tetrahedron and octahedron.

2.4.2 *Platonic solids*

A *regular polyhedron* is a polyhedron whose faces are congruent regular polygons, with all solid angles equal.

As we can see from calculating the sum of the plane angles at a vertex, there are at most five regular convex polyhedra. These polyhedra, pictured in Figures 2.5–2.8, the so-called *Platonic solids* — the *tetrahedron,* the *cube,* the *octahedron,* the *dodecahedron,* and the *icosahedron* — were known in antiquity (Plato, around 400 BC), and their existence was proved by Euclid. The cube can be treated as a prism with $n = 4$ and $h = a$. The other four Platonic solids can be constructed using the cube.

- The cube and octahedron are combinatorially dual, i.e., one is obtained from the other if the centers of the faces of the first are considered the vertices of the second. Analogously, the dodecahedron and icosahedron are combinatorially dual.
- Any four vertices of a cube taken in pairs not adjacent along an edge are the vertices of a tetrahedron.
- The icosahedron can be inscribed by a special method in a cube if its opposite edges are fixed in pairs on parallel faces of the cube.
- The dodecahedron is obtained from the cube by building roofs over its faces (the method of Euclid).

The above polyhedra admit a number of *symmetries* (self-transformation by a rigid motion of space): cyclic transposition of vertices on any face and cyclic transposition of faces that meet at any vertex.

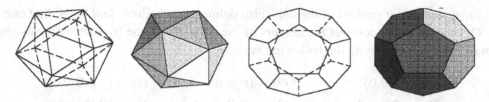

Fig. 2.7 Icosahedron and dodecahedron.

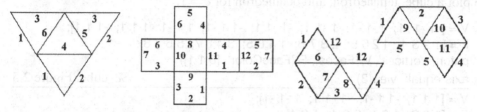

Fig. 2.8 Developments of tetrahedron, cube, and octahedron.

A polyhedron is denoted by the *symbol* $\{p,q\}$, or p^q, if every vertex is surrounded by q p-gons.

The radii R, r of the circumscribed and inscribed spheres and the volume of each regular convex polyhedron with edge of length a are given in the following table.

Name	Symbol	R	r	V
tetrahedron	$\{3,3\}$	$a\sqrt{6}/4$	$a\sqrt{6}/12$	$a^3\sqrt{2}/12$
cube	$\{4,3\}$	$a\sqrt{3}/2$	$a/2$	a^3
octahedron	$\{3,4\}$	$a\sqrt{2}/2$	$a\sqrt{6}/6$	$a^3\sqrt{2}/3$
dodecahedron	$\{5,3\}$	$\frac{a}{4}\sqrt{18+6\sqrt{5}}$	$\frac{a}{2}\sqrt{\frac{1}{10}(25+11\sqrt{5})}$	$\frac{a^3}{4}(15+7\sqrt{5})$
icosahedron	$\{3,5\}$	$\frac{a}{4}\sqrt{10+2\sqrt{5}}$	$\frac{a\sqrt{3}}{12}(3+\sqrt{5})$	$\frac{5a^3}{12}(3+\sqrt{5})$

Example. Consider the **cube** with edge $2c$ and the vertices

$$A_1 = (c,c,c), \qquad A_2 = (-c,c,c), \qquad A_3 = (-c,-c,c), \qquad A_4 = (c,-c,c),$$
$$A_5 = (c,c,-c), \qquad A_6 = (-c,c,-c), \qquad A_7 = (-c,-c,-c), \qquad A_8 = (c,-c,-c).$$

Let us find vertices of other regular convex polyhedra using their relation to the cube.

Tetrahedron. Take diagonals of the six faces of a cube in the role of edges of a tetrahedron. Then the vertices of the tetrahedron are the following:

$$A_1 = (c,c,c), \qquad A_3 = (-c,-c,c), \qquad A_6 = (-c,c,-c), \qquad A_8 = (c,-c,-c).$$

The remaining four vertices of the cube also define a tetrahedron *dual* to the first one.

Octahedron. We connect the centers of the faces of a cube by line segments and obtain an octahedron with the following vertices:

$$O_1 = (c,0,0), \qquad O_2 = (0,c,0), \qquad O_3 = (-c,0,0),$$
$$O_4 = (0,-c,0), \qquad O_5 = (0,0,c), \qquad O_6 = (0,0,-c).$$

We plot a cube, tetrahedron, and octahedron for $c = 1$.

```
V = [-1 -1 -1; 1 -1 -1; 1 -1 1; -1 -1 1; -1 1 -1; 1 1 -1; 1 1 1; -1 1 1];
F = [1 2 3 4; 1 2 6 5; 2 3 7 6; 1 4 8 5; 3 4 8 7; 5 6 7 8];
patch('Vertices', V, 'Faces', F, 'FaceColor', [0 1 1])
axis equal; view(3)                                    % cube, Figure 2.5
```
```
V = [1 1 1; -1 1 -1; -1 -1 1; 1 -1 -1];
F = [1 2 3; 1 2 4; 2 3 4; 1 3 4];
patch('Vertices', V, 'Faces', F, 'FaceColor', [0 1 1])
axis equal; view(3)                             % tetrahedron, Figure 2.6(a,b)
```
```
V = [-1 0 0; 0 1 0; 1 0 0; 0 -1 0; 0 0 -1; 0 0 1];
F = [1 2 5; 1 2 6; 2 3 5; 2 3 6; 3 4 5; 3 4 6; 1 4 5; 1 4 6];
patch('Vertices', V, 'Faces', F, 'FaceColor', [0 1 1])
axis equal; view(3)                             % octahedron, Figure 2.6(c,d)
```

Recall that the *golden section* (or *golden ratio*), $\tau = \frac{\sqrt{5}+1}{2} \approx 1.618$, is the division of a line segment AB by a point C, such that the ratio of the smaller part to the larger one is equal to the ratio of the larger part to the whole segment: $CB : AC = AC : AB$. Euclid made use of the golden section for constructing regular *pentagons* and *hexagons* and also for constructing the *icosahedron* and *dodecahedron*.

Icosahedron. We arrange six equal line segments with the ratio of their lengths to the edge of the cube equal to $1/\tau$ (see exercises) on the faces of a cube in such a way that each segment is symmetric to itself with respect to the center of a certain face and parallel to its opposite two sides, and such that the line segments in neighboring faces are mutually orthogonal. Then we connect the endpoints of these segments in a certain order and obtain an icosahedron inscribed in the cube. Hence, our icosahedron has the following vertices (where $p = c/\tau$):

$$B_1 = (p,0,c), \qquad B_2 = (-p,0,c), \qquad B_3 = (p,0,-c), \qquad B_4 = (-p,0,-c),$$
$$B_5 = (c,-p,0), \qquad B_6 = (c,p,0), \qquad B_7 = (-c,-p,0), \qquad B_8 = (-c,p,0),$$
$$B_9 = (0,c,p), \qquad B_{10} = (0,c,-p), \qquad B_{11} = (0,-c,p), \qquad B_{12} = (0,-c,-p).$$

Dodecahedron. As in the case of the icosahedron, we choose six line segments of length $2p = 2c/\tau$ on the faces of a cube. Then we make a parallel translation by the distance p (see exercise) of each of these segments in the direction of the extrinsic normal to the face. After connecting the twelve endpoints of the line segments so defined in a certain order with the eight vertices of the cube, we obtain a regular dodecahedron. Its vertices, in addition to the eight vertices of the cube, are

$$D_1 = (p,0,c+p), \qquad D_2 = (-p,0,c+p), \qquad D_3 = (p,0,-c-p),$$
$$D_4 = (-p,0,-c-p), \qquad D_5 = (c+p,-p,0), \qquad D_6 = (c+p,p,0),$$
$$D_7 = (-c-p,-p,0), \qquad D_8 = (-c-p,p,0), \qquad D_9 = (0,c+p,p),$$
$$D_{10} = (0,c+p,-p), \qquad D_{11} = (0,-c-p,p), \qquad D_{12} = (0,-c-p,-p).$$

Plot the icosahedron and dodecahedron using the coordinates of their vertices (do some calculations, see Exercises 5, 6, p. 130). For the icosahedron we directly describe its faces, but for the dodecahedron, we first describe its vertices, and then write sequences of vertices for each face.

```
c = 1;    tau = (sqrt(5) + 1)/2;   p = c/tau;              % use for both polyhedra
V = [p 0 c; -p 0 c; 0 c p; 0 -c p; -c p 0; -c -p 0; c p 0; c -p 0; p 0 -c; 0 c -p;
-p 0 -c;  0 -c -p];
F = [1 2 3; 1 2 4; 2 3 5; 2 5 6; 1 7 3; 1 8 7; 7 8 9; 7 10 3; 7 10 9; 3 10 5; 9 11
10; 1 8 4; 4 2 6; 4 12 6; 9 11 12; 12 11 6; 11 5 6; 8 12 9; 4 12 8; 5, 10, 11];
patch('Vertices', V, 'Faces', F, 'FaceColor', [0 1 1]);
axis equal;  view(3)                            % Icosahedron, Figure 2.7(a,b).
V = [p 0 c+p; -p 0 c+p; p 0 -c-p; -p 0 -c-p; c+p -p 0; c+p p 0; -c-p -p 0; -c-p
p 0; 0 c+p p; 0 c+p -p; 0 -c-p p; 0 -c-p -p; c c c; c c -c; c -c c; c -c -c; -c c c;
-c c -c; -c -c c; -c -c -c];
F = [1 13 9 17 2; 17 9 10 18 8; 8 18 4 20 7; 7 20 12 11 19; 19 11 15 1 2; 6 5
15 1 13; 6 13 9 10 14; 14 10 18 4 3; 3 4 20 12 16; 16 12 11 15 5; 5 6 14 3 16];
patch('Vertices', V, 'Faces', F, 'FaceColor', [0 1 1]);
axis equal;  view(3)                            % Dodecahedron, Figure 2.7(c,d).
```

There are other polyhedra besides the Platonic solids that are bounded by equal regular faces. In fact they are bounded by triangles, and together with the tetrahedron, the octahedron, and the icosahedron they form the family of eight (convex) *deltahedra*.

2.4.3 *Symmetries of regular polygons and polyhedra*

Symmetries preserve distances, angles, sizes, and shapes.

The *symmetry group of a subset* $X \subset \mathbb{R}^n$ is the group of all isometries of $\mathbb{R}^n$ which carry X onto itself: $\text{Symm}(X) := \{f \in \text{Isom}(X) : f(X) = X\}$. Further, $\text{Symm}(X) = \text{Symm}^+(X) \cup \text{Symm}^-(X)$, where the sets

$$\text{Symm}^\pm(X) := \{f \in \text{Symm}(X) : \det A = \pm 1\}$$

are respectively called *direct* and *indirect symmetries* of X.

For any $X \subset \mathbb{R}^n$, $\mathrm{Symm}^+(X) \subset \mathrm{Symm}(X)$ is a subgroup.

We now encounter some important finite symmetry groups.

The symmetry group of a regular m-gon (triangle, square, pentagon, hexagon, etc.) centered at the origin in $\mathbb{R}^2$ is called the *dihedral group* of order $2m$, denoted D_m (see also the definition in Section 2.5). The elements of D_m with determinant $+1$ are called *rotations*; they form a subgroup of index 2 which is isomorphic to the cyclic group C_m of order m. The elements of D_m with determinant -1 are called *flips*.

Symmetry groups of subsets of $\mathbb{R}^2$ are useful for studying objects which are essentially two-dimensional, like snowflakes and certain crystal structures.

Many subsets of $\mathbb{R}^2$, like the wallpaper tilings of $\mathbb{R}^2$ illustrated in some M.C. Escher prints, have infinite symmetry groups. The classification of such infinite "wallpaper groups" can be found in the literature. Surprisingly, *the only finite symmetry groups in dimension 2 are D_m and C_m*.

The proof involves two steps. First, when $\mathrm{Symm}(X)$ is finite, its elements must share a common fixed point, so it is isomorphic to a subgroup of $O(2)$. Second, D_m and C_m are the only finite subgroups of $O(2)$.

Examples.

1. Let P be the regular n-gon with vertices at the nth roots of unity. The symmetry group G of P acts on the complex plane $\mathbb{C}$. Each vertex v of P can mapped by a rotation in G to any other vertex. Thus $G * v = V$, where V is the set of vertices of P, and G_v contains two elements, namely the identity I and the reflection in the line through v and 0.

Next, consider the origin 0; here, $G * 0 = \{0\}$ and $G_0 = G$. Finally, if z does not lie on any line of symmetry of P, then $G_z = \{I\}$, and $G * z$ contains $2n$ points. In every case, $|G * z| \cdot |G_z| = |G|$.

2. We locate the *square* with the vertices $1, i, -1$ and $-i$, and relabel these vertices by the integers 1, 2, 3, and 4, respectively. Now any symmetry of the square can be written as a permutation of $\{1,2,3,4\}$. We have seen that D_4 is generated by r and s, where $r(z) = iz$ and $s(z) = \bar{z}$. In terms of permutations, $r = (1\,2\,3\,4)$, and $s = (2\,4)$. There are exactly four reflective symmetries of the square, namely, s, sr, sr^2, and sr^3; in terms of *permutations*, these are $(2\,4)$, $(1\,4)(2\,3)$, $(1\,3)$, and $(1\,2)(3\,4)$, respectively. Of course, $r^2 = (1\,3)(2\,4)$ and $r^3 = (1\,4\,3\,2)$. Although $s^2 = e$ and $r^4 = e$, D_4 is not isomorphic to $C_2 \times C_4$, as $C_2 \times C_4$ is abelian, whereas D_4 is not.

3. The command perms(v) returns all the permutations of the vector v.

   P = perms([1 3 5])   % all the permutations of the numbers 1, 3, and 5

The command p = randperm(n) returns a random permutation of the integers $1 : n$. The command below rearranges the elements of a vector.

    p = randperm(5);   % Answer: $p = [2,3,1,5,4]$ (permutation vector)
    a = intrlv(10 : 10 : 50, p)   % Answer: $a = [20,30,10,50,40]$.

The command deintrlv(data, elements) acts as an inverse of intrlv.

    b = deintrlv(a, p)   % Answer: $b = [10,20,30,40,50]$

Symmetry groups of subsets of $\mathbb{R}^n$ ($n > 2$) are even more interesting than those of $\mathbb{R}^2$. There are still very few finite symmetry groups in $\mathbb{R}^3$.

Denote by S_m the *group of permutations of a set with m elements*, and by $A_m \subset S_m$ the subgroup of even permutations (the *alternating group*).

Theorem 2.6. *For $X \subset \mathbb{R}^3$, if $\mathrm{Symm}^+(X)$ is finite, then it is isomorphic to D_m, C_m, A_4, S_4, or A_5.*

Like the $n = 2$ case, the proof involves verifying that all symmetries have a common fixed point and that the only finite subgroups of $SO(3)$ are D_m, C_m, A_4, S_4, and A_5.

The regular solids (or *Platonic solids*) provide examples of sets whose direct symmetry groups equal A_4, S_4, and A_5. These polyhedra were first listed by Theatus in about 400 BC, but they are often referred to as the Platonic solids. It turns out that

  A_4 *is the direct symmetry group of a tetrahedron*,
  S_4 *is the direct symmetry group of a cube or an octahedron*, and
  A_5 *is the direct symmetry group of a dodecahedron or an icosahedron*.

Each regular polyhedron is invariant by a reflection $\mathbf{P}_\omega$ across some plane ω passing through the origin. The group G of isometries that leave the polyhedron invariant has the coset decomposition $\mathbf{P}_\omega G + \cup G^+$ (direct and indirect isometries), so that $|G| = 2|G^+|$. Note that whereas each rotation (in G^+) is physically realizable as a rotation of $\mathbb{R}^3$, the reflections that leave the polyhedron invariant cannot be realized physically.

Computing the symmetry groups of the regular polyhedra may be approached as follows:

(a) **Cube and octahedron.** The symmetry group $G^+ = S_4$ has 24 elements. Clearly, the two symmetry groups of the octahedron coincide with those of the cube. There are three types of symmetry planes, and there are $3 + 6 = 9$ plane reflection symmetries. There are three types of symmetry axes, and there are $1 + 6 + 8 = 15$ rotary reflection symmetries. The 48 symmetries of the cube form a Coxeter group with presentation

$$\{x,y,z : x^2 = y^2 = z^3 = (xy)^4 = (yz)^2 = (xz)^3 = 1\}.$$

(b) **Tetrahedron.** The symmetry group $G^+ = A_4$ has 12 elements. The symmetry group $G = S_4$ has 24 elements.

(c) **Dodecahedron and icosahedron.** The dodecahedron and the icosahedron are dual solids; that is, the centroids of the faces of one of these solids are at the vertices of a solid of the other type. Accordingly, their symmetry groups are the same. The symmetry group $G^+ = A_5$ has 60 elements. The symmetry group $G = A_5 \times C_2$ has 120 elements.

2.4.4 Star-shaped polyhedra

One may represent star-shaped polyhedra (SPs) over *Platonic solids* using a *stellar construction* ("stellar" means "starlike") with several values of the height parameter h less or greater than 1. For particular values of h (calculate them) the polyhedra are regular SPs; see Figure 2.9.

Example. The reader can plot SPs over *Platonic solids*, see Example, p. 107, using *stellate construction*. First define the vertices V and faces F of a polyhedron, and h (positive or negative small); then call stellate(F, V, h).

```
function S = stellate(F, V, h)                          % M-file stellate.m
    NF = length(F);  NV = size(F, 2);
    hold on;  axis equal;  view(3)
    for i = 1 : NF;
     C(i, :) = sum(V(F(i, :), :)) / NV;                 % center of a face
     H(i, :) = cross(C(i, :) - V(F(i, 1), :), C(i, :) - V(F(i, 2), :));    % normal
     if H(i, :)*C(i, :)' < 0;
      H(i, :) = - H(i, :);
     end;
     NH = H(i, :)*H(i, :)';
     VH(i, :) = C(i, :) + h*H(i, :) / NH;               % vertex
     for j = 1 : NV;
     if j == NV;  j1 = 1;  else j1 = j + 1;  end;
     W(i, j, :, :) = [VH(i, :);  V(F(i, j), :);  V(F(i, j1), :)];   % triangle
      patch(reshape(W(i, j, :, 1), 1, 3),  reshape(W(i, j, :, 2), 1, 3),
      reshape(W(i, j, :, 3), 1, 3),  [i/NF/2 j/NV/2 1 - i/NF/2]);
     end;
    end;
end
```

For example, stellate constructions over a tetrahedron:

```
V = [1 1 1; -1 1 -1;  -1 -1 1;  1 -1 -1];
F = [1 2 3;  1 2 4;  2 3 4;  1 3 4];                    % define V, H
stellate(F, V, 1)                                       % h = 1
stellate(F, V, -.5)                                     % h = -0.5, etc.
```

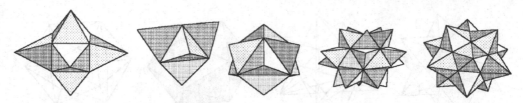

Fig. 2.9 Stars over Platonic solids.

Aside from the *regular convex solid angles* there also exist *regular star-shaped solid angles*. In both cases their planar angles are equal each to other; the solid angles are equal as well.

The regular star 5-faced angle is shown in Figure 2.11(c,d). Let O be the center of the star pentagon $A_1A_2A_3A_4A_5$. Plot the line OB perpendicular to the plane of pentagon. Then the five plane angles A_1BA_2, A_2BA_3, A_3BA_4, A_4BA_5, and A_5BA_1 form a stellated pentagonal solid.

In the 17th century J. Kepler found two regular SPs.

In 1810 L. Poinsot proved the existence of four such SPs, now called *Poinsot's polyhedra*, seen in Figures 2.11 and 2.12:

- The *small stellated dodecahedron* has 12 faces (regular star pentagons), 30 edges, 12 vertices.
- The *great stellated dodecahedron* has 12 faces (regular star pentagons), 30 edges, 20 vertices.
- The *great icosahedron* has 20 triangular faces, 30 edges, 12 vertices.
- The *great dodecahedron* has 12 faces (regular simple pentagons), 30 edges, 12 vertices.

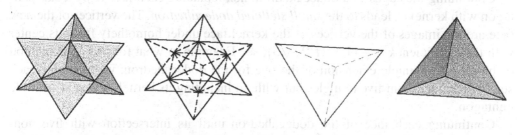

Fig. 2.10 The stellated octahedron consists of two tetrahedra.

In 1812 A. Cauchy proved that the set of all regular polyhedra contains only the five *Platonic solids*, the four *Poinsot bodies*, and the (*stellated*) *Kepler octahedron*, Figure 2.10. The last polyhedron has 8 faces and decomposes into two tetrahedra.

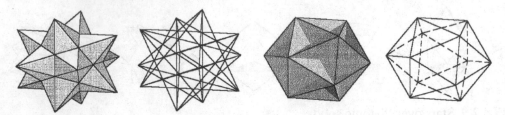

Fig. 2.11 Small stellated and great dodecahedra.

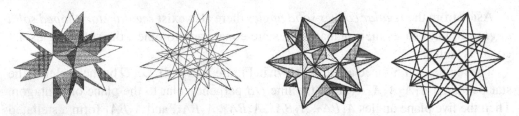

Fig. 2.12 Great stellated dodecahedron and great icosahedron.

The faces of an SP either are star polygons or mutually intersect in space. The supporting planes of the faces of a regular SP M divide space into some number of convex domains, one of which, the *kernel* of the polyhedron M, is a simple regular polyhedron M_0. The polyhedron M_0 has the same number of faces as M.

The *stellated octahedron* is obtained from the octahedron by continuation of each of its faces until it intersects three non-neighboring non-parallel faces. This polyhedron *decomposes into two tetrahedra*, "dual" to each other, inscribed in a cube. The method of plotting a stellated octahedron is an extension of the method for the tetrahedron; see Example, p. 107.

Continuing the edges of a dodecahedron, i.e., replacing each face α by a star pentagon with kernel α, leads to the *small stellated dodecahedron*. The vertices of the new face are the images of the vertices of the kernel face under homothety from its center with the coefficient $k = -(1 + 2\tau^{-1}/\sqrt{3})$; see Example in what follows. The method of plotting is a simple extension of the one for the dodecahedron: we assemble each star-shaped face from five triangles but without including its kernel, which is a simple pentagon.

Continuing each face of the dodecahedron until its intersection with five non-neighboring and non-parallel faces leads to two possible cases.

- If the new faces are *simple* pentagons, then we obtain the *great dodecahedron*. The MATLAB program given in Example below is a short continuation of the program for the dodecahedron on p. 109.
- If the new faces are *star* pentagons, then we obtain the *great stellated do-decahedron*. The MATLAB program is based on results of the program for the

small stellated dodecahedron, using homothety with negative coefficient $k = -(1 + 2\tau^{-1}/\sqrt{3})$; see exercises. Here we again assemble each star-shaped face from five triangles.

Examples.

1. Plot the small stellated dodecahedron starting from the coordinates of its vertices and lists of sequences of vertices in all faces. We complete the program for the regular dodecahedron; see Example, p. 107. Here $C(i, :)$ is the center of the ith face of the dodecahedron.

```
k = 1 + 2*cos(pi/5);
hold on;  axis equal;  view(3)                          % small stellated dodecahedron
for i = 1 : 12;
 C(i, :) = sum(V(F(i, :), :))/5;
 for j = 1 : 5;
 VV(j, :, i) = C(i, :) - k*(V(F(i, j), :) - C(i, :));
 j2 = mod(j + 2, 5);  j3 = mod(j + 3, 5);
 if j2 == 0;  j2 = 5;
  elseif j3 == 0;  j3 = 5;
  end;
  W(i, j, :, :) = [VV(j, :, i); V(F(i, j2), :); V(F(i, j3), :)];
  patch(reshape(W(i, j, :, 1), 1, 3),  reshape(W(i, j, :, 2), 1, 3),
  reshape(W(i, j, :, 3), 1, 3),  [i/15 j/15 1 - i/15]);
  end;
 end
```

2. Plot the great (stellated) icosahedron by using the coordinates of its vertices and their order in all the faces. The program is based on the one for the simple icosahedron. Here $C(i,:)$ is the center of the ith face of the icosahedron. We do not see on the display the complete picture of the intersections of all 20 of its triangular faces.

The program for plotting the *great icosahedron* given below is based on data of the program for the simple icosahedron. We extend the program for the icosahedron to plot a great stellated icosahedron.

```
k = (3 + sqrt(5))/(3 - sqrt(5));                        % great (stellated) icosahedron
hold on;  axis equal;  view(3)
for i = 1 : 20;
 C(i, :) = sum(V(F(i, :), :))/3;
 for j = 1 : 3;
  VV(j, :, i) = C(i, :) - k*(V(F(i, j), :) - C(i, :));
  j2 = mod(j + 2, 3);      j1 = mod(j + 1, 3);
  if j2 == 0;  j2 = 3;
  elseif j1 == 0;  j1 = 3;
  end;
  W(i, j, :, :) = [VV(j, :, i); V(F(i, j1), :); V(F(i, j2), :)];
```

```
    patch(reshape(W(i, j, :, 1), 1, 3),  reshape(W(i, j, :, 2), 1, 3),
    reshape(W(i, j, :, 3), 1, 3), [i/21 j/21 1 - i/21]);
    end;
end
```

A *compound polyhedron* is a set of distinct polyhedra, called the components of the compound, which are placed together so that their centers coincide. (Figure 2.10 shows the compound of two tetrahedra.)

Compounds of Platonic solids in which all components are the same have a high degree of symmetry which makes them very attractive. The idea of placing one polyhedron inside another in different ways is used in plotting compounds. Another method to plot compounds uses the matrix representation of finite symmetry groups in space.

2.4.5 *Archimedean solids*

We survey the construction of the Archimedean solids by transformations of the Platonic solids, and present programs to plot them by coordinates of vertices (in a similar way to the programs for the Platonic solids but with preparatory work for the lists of vertices in the faces).

An *isohedron* (*isogon*) is a convex polyhedron whose rotation group (of the first and second orders) moves any of its faces (vertices) to any other of its faces (resp., vertices).

Each isohedron corresponds to a dual isogon, and conversely. There are 13 different combinatorially special types and two infinite series of isogons. Each of them can be realized in space in such a way that all its faces are regular polygons, and one obtains semi-regular polyhedra from the following definition.

A polyhedron whose faces are regular polygons (perhaps of different sizes and types) and whose solid angles are equal is called a *semi-regular polyhedron*.

Simple examples of such polyhedra are the *regular prisms* $4^2 \cdot n$ (i.e., each vertex belongs to two squares and one n-gon; in the following discussion we use this classic designation for semi-regular polyhedra), whose bottom and top faces are regular simple n-gons and lateral faces are squares, and *antiprisms* $3^3 \cdot n$, whose bottom and top faces are regular simple n-gons with $n \geq 3$ and lateral faces are pairs of regular triangles. In particular, the 3-antiprism is the octahedron, and the 4-prism is the cube. Note that the n-antiprism can be obtained from the n-prism by rotating one of its bases about the center through an angle π/n in the same plane, while at the same time decreasing the height so that the distance between corresponding vertices is equal to their edges.

Long ago Archimedes proved that except for the two series of prisms and antiprisms, there exist 13 types of semi-regular convex polyhedra, the so-called *Archimedean solids*. The geometer Pappus of Alexandria tells about the work of Archimedes and gives a short description of the *Archimedean solids*. The complete theory of semi-regular polyhedra was discovered by J. Kepler in his book *"Harmonices mundilibri quinque..."* (1619).

Some restrictions on polyhedra follow directly from the definition:

- *There are no semi-regular polyhedra bounded by* more than three *different types of faces.* (In view of the equality of their solid angles, each of them contains at least one planar angle of each type of face, but the minimum angles come from polygons with 3, 4, 5 and 6 sides, i.e., we have the inequality $60° + 90° + 108° + 120° > 360°$.)

- *There are no semi-regular polyhedra whose vertices contain* more than five *faces.* (The assumption that a solid angle consists of at least six planar angles of regular polygons of two types leads to the inequality $5 \cdot 60° + 90° > 360°$.)

The list of possible types of semi-regular convex polyhedra can be obtained by topological (combinatorial) reasoning using Euler's formula for a sphere.

The names of the semi-regular convex polyhedra are the following:

(1) *snub cube* $3^4 \cdot 4$: 32 triangles, 6 squares

(2) *cuboctahedron* $(3 \cdot 4)^2$: 8 triangles, 6 squares

(3) *rhombicuboctahedron* $3 \cdot 4^3$: 8 triangles, 18 squares

(4) *snub dodecahedron* $3^4 \cdot 5$: 80 triangles, 12 pentagons

(5) *icosidodecahedron* $(3 \cdot 5)^2$: 20 triangles, 12 pentagons

(6) *truncated tetrahedron* $3 \cdot 6^2$: 4 triangles, 4 hexagons

(7) *truncated cube* $3 \cdot 8^2$: 8 triangles, 6 octagons

(8) *truncated dodecahedron* $3 \cdot 10^2$: 20 triangles, 12 decagons

(9) *truncated octahedron* $4 \cdot 6^2$: 6 squares, 8 hexagons

(10) *truncated icosahedron* $5 \cdot 6^2$: 12 pentagons, 20 hexagons

(11) *rhombicosidodecahedron* $3 \cdot 4 \cdot 5 \cdot 4$: 20 triangles, 30 squares, 12 pentagons

(12) *truncated cuboctahedron* $4 \cdot 6 \cdot 8$: 12 squares, 8 hexagons, 6 octagons

(13) *truncated icosidodecahedron* $4 \cdot 6 \cdot 10$: 30 squares, 20 hexagons, 12 decagons

(14) *Ashkinuze solid* $3 \cdot 4^3$: 8 triangles and 18 squares

In the 20th century, V. Ashkinuze (and J. Miller) found the fourteenth semi-regular polyhedron (of type $3 \cdot 4^3$), which differs from the *rhombicuboctahedron*, case (3), only by rotating the whole upper part of the polyhedron, consisting of five squares and four triangles, through the angle 45°. For each of the 13 *Archimedean solids*, prisms, and antiprisms, any two vertices can be translated one to the other by a symmetry of the polyhedron, but for the *Ashkinuze solid* this does not hold.

We now describe a method to plot the *Archimedean solids* using transformations of the five *Platonic solids* by cutting off the neighborhoods of vertices and edges by planes.

Let us start with the tetrahedron. We plot the planes at a distance $1/3$ from each edge meeting at a common vertex, cut off the resulting pyramid, and obtain the solid (6), bounded by regular triangles and hexagons.

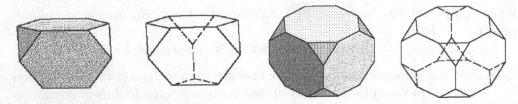

Fig. 2.13 (6) Truncated tetrahedron. (7) Truncated cube.

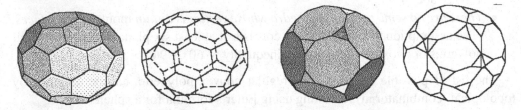

Fig. 2.14 (10) Truncated icosahedron. (8) Truncated dodecahedron.

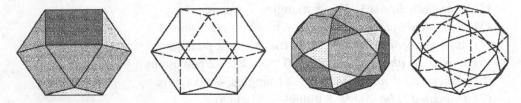

Fig. 2.15 (2) Cuboctahedron. (5) Icosidodecahedron.

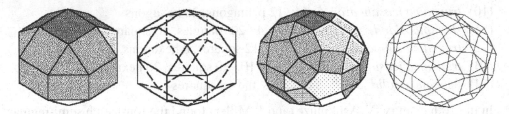

Fig. 2.16 (3) Rhombicuboctahedron. (11) Rhombicosidodecahedron.

An analogous transformation for the octahedron gives us the solid (9), and from the icosahedron we may obtain the solid (10).

Plotting the planes through the centers of certain edges of the cube or octahedron, we obtain the solid (2), i.e., the *cuboctahedron*.

By the same method we obtain (from the icosahedron or dodecahedron) solid (5), i.e., the *icosidodecahedron*.

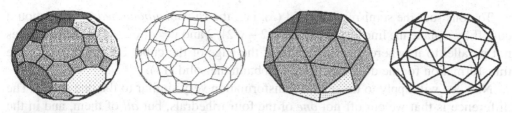

Fig. 2.17 (13) Truncated icosidodecahedron. (1) Snub cube.

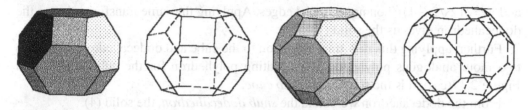

Fig. 2.18 (9) Truncated octahedron. (12) Truncated cuboctahedron.

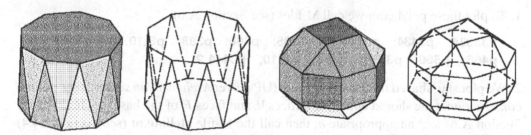

Fig. 2.19 n-antiprism. (14) Ashkinuze solid.

If we employ such cutting planes in the cube, so that its square faces are transformed into regular octagons, then we obtain the solid (7), i.e., the *truncated cube*. In an analogous way we obtain the solid (8) from the dodecahedron.

The transformations that proceed by *cutting off neighborhoods of both vertices and edges* are more complicated. If we cut off, one after another, the edges having a common vertex, by planes that define strips of equal height on the faces, we obtain four new trihedrals. Drawing this, we see that three of these vertices are located symmetrically around the fourth one. We cut off this last vertex by a plane that contains the first three vertices and thus obtain a triangle in the intersection.

Repeating the same procedure with the other edges and vertices, each time we obtain one triangle in place of a vertex of the cube, and between them, instead of the edges of the cube, we obtain squares; the faces of the cube also would also be transformed into new squares of smaller size.

To construct the semi-regular solid (3), i.e., the *rhombicuboctahedron*, one should cut off by a plane the line segments $x = a(2 - \sqrt{2})/2$ and $y = a(\sqrt{2} - 1)/2$ on the edges not parallel to it, where a is the length of the edge of the cube. Employing the same transformation for the dodecahedron, we obtain the solid (11).

Now we will apply to the cube a transformation very similar to the one above. The difference is that we cut off not *one* of the four trihedrals, but *all* of them, and in the intersection we obtain not a triangle, but a (regular) hexagon. The result will be the solid (12). Moreover, one can cut off by a plane the line segments $x = a(4 - \sqrt{2})/14$ and $y = a(2\sqrt{2} - 1)/7$ on non-parallel edges. Applying the same transformation to the dodecahedron gives us the solid (13).

Further, applying this new transformation to the cube and dodecahedron, we obtain two more analogous polyhedra. The resulting polyhedron for the cube is shown in Figure 2.17(c,d); it is the solid (1), a *snub cube*.

From the dodecahedron we obtain the *snub dodecahedron*, the solid (4).

Examples.

1. To plot these polyhedra we call M-files (see Section A.6):

   p33334; p3434; p3444-1; p3535; p366; p388; p3_10_10;
   p466; p566; p3454; p468; p4610; p3444_2;   % etc.

2. We plot star-shaped uniform polyhedra (UPs) over *Archimedean solids* using *stellate construction*. One should enter the vertices V and faces F of the base polyhedron (see Section A.6) and an appropriate h, then call the M-file **stellate.m** (see Section 2.4.4) for all types of faces. For example, plot a stellated **3444**:

   h = 1;   % enter h
   ·········   % enter F_3, F_4, V of **3444**
   hold on; **stellate**(F3, V, h);   **stellate**(F4, V, h)

   There also exist 75 *stellated semi-regular* (*uniform*) *polyhedra* (UPs). These are to the Archimedean solids what the regular SPs are to the Platonic solids. However, the faces of UPs can intersect each other.

   Half-regular polyhedra are defined as semi-regular ones, where we use half-regular (i.e., equilateral) polygons. Kepler restricted attention to half-regular 4-gons (rhombi) and found two such polyhedra. The first (*rhombic dodecahedron*) is bounded by twelve rhombi whose diagonals are in the ratio of $1 : \sqrt{2}$. The second rhombic polyhedron (*rhombic triacontahedron*) is bounded by 30 rhombi whose diagonals are in the golden ratio.

   A generalization of semi-regular polyhedra is the so-called *regular-faced polyhedra*, whose faces by definition are regular polygons. The constraints that restrict the number of such polyhedra are less topological and more metrical. Their classification requires length calculations.

Aside from the two series of prisms and antiprisms, there exist 92 regular-faced polyhedra, but only 28 of them are *indecomposable* (i.e., cannot be broken by a plane into two regular-faced polyhedra).

Examples of decomposable regular-faced polyhedra are the octahedron, icosahedron, Ashkinuze solid, and also among the *Archimedean solids*, the cuboctahedron, icosidodecahedron, rhombicuboctahedron, and rhombicosidodecahedron. Hence, only $8 \left(= \frac{28 - (5-2) - (13-4)}{2} \right)$ regular-faced polyhedra are not regular or semi-regular polyhedra or parts of them.

2.5 Appendix: Matrices and Groups

2.5.1 *Permutations and group actions*

We recall the necessary notations from algebra.

The pair $(G, \cdot)$ is called a *group* provided that the operation $\cdot$ satisfies

(i) *associativity*: $g_1 \cdot (g_2 \cdot g_3) = (g_1 \cdot g_2) \cdot g_3$ for all $g_1, g_2, g_3 \in G$;
(ii) *unit*: there is $e \in G$ (the *unit*) such that $e \cdot g = g \cdot e = g$ for every $g \in G$;
(iii) *inverse*: for every $g \in G$, there is $g^{-1} \in G$ (the *inverse* of g) such that
$g \cdot g^{-1} = g^{-1} \cdot g = e$.

Let $H \subset G$. We say that $(H, \cdot)$ is a *subgroup* of $(G, \cdot)$ if $g_1 \cdot g_2^{-1} \in H$ for all $g_1, g_2 \in H$.

A group G is called *cyclic* if there is some $g \in G$ such that $G = \{g^m : m \in \mathbb{Z}\}$, i.e., G is *generated* by g. The *cyclic group of order* $n > 1$ (that is generated by an element of order n) is denoted by C_n.

The *symmetric group* S_n is the group of permutations of $\{1, 2, \ldots, n\}$. The *alternating group* A_n is the group of all even permutations in S_n. The *dihedral group* D_n ($n > 1$) is the group of order $2n$ that is generated by an element a of order n and an element b of order two, where $aba = b$.

The *q-cycle* $(n_1 \ldots n_q)$ is the permutation that maps n_j to n_{j+1} when $1 \leq j < q$, maps n_q to n_1, and fixes all other integers in $\{1, 2, \ldots, n\}$. A 2-cycle is called a *transposition*.

Example. For each integer k, $k\mathbb{Z} = \{kn : n \in \mathbb{Z}\}$ is a subgroup of the group $\mathbb{Z}$ (with respect to addition). But the set $\{2n : n \in \mathbb{Z}\} \cup \{5\}$ is not a subgroup of $(\mathbb{Z}, +)$.

As every permutation is a product of cycles, every permutation is a product of transpositions (2-cycles). If α and β are permutations, then $\varepsilon(\alpha \cdot \beta) = \varepsilon(\alpha) \cdot \varepsilon(\alpha)$. The *sign* $\varepsilon(\sigma)$ of a permutation σ is -1^q, where σ can be expressed as a product of q transpositions.

The rest of this section is devoted to the idea of a group acting on a set. We have applied this idea in Section 2.4.3 above to the study of the symmetry groups of regular solids, and shall use it later to study hyperbolic geometry (Chapter 4).

An *action of a group* G (with *unit e*) on a set X is a mapping $\varphi \colon G \times X \to X$, i.e., $(g, x) \mapsto g * x$, such that $e * x = x$ and $(g_1 \cdot g_2) * x = g_1 * (g_2 * x)$ for all $x \in X$ and all $g_1, g_2 \in G$. The mapping $\varphi_g \colon x \to g * x$ is called the *transformation of X effected by g.*

We consider G as a group of transformations (or symmetries) of a set X, and refer to this by saying that G *acts* on X. We express the equation $y = g * x$ by saying that g *moves x to y.*

Suppose that G acts on X.

(a) Then x is a *fixed point* of g in G if $g * x = x$, and the set of fixed points of g is denoted by $\mathrm{Fix}(g)$. If $g * x = x$ for all $g \in G$ then x is called a *fixed point of the group action.*

(b) Given x in X, the subgroup $G_x := \{g \in G : g * x = x\}$ of elements of G that fix x is called the *stabilizer* of x.

(c) Given x in X, the subset $G * x := \{g * x : g \in G\}$ of X is called the *orbit* (or *trajectory*) of x under G.

(d) The group G is said to act *transitively* on X if there is only one orbit (i.e., every point of X can be moved to any other point).

Examples. Any group G acts on itself by:

1. *Left multiplication* (*left regular action*): There is only one orbit.

2. *Conjugation*: For $g \in G$, define $c_g \in \mathrm{Aut}(G)$ by $c_g(x) = gxg^{-1}$. The orbits of this action are called the *conjugacy classes* of G; two elements x and y are conjugate if there is a $g \in G$ such that $gxg^{-1} = y$. For example, two permutations (elements of the symmetric group S_n) are conjugate if and only if they have the same cycle structure. A subgroup H of G is called *normal* (or self-conjugate) if $gHg^{-1} = H$ for every $g \in G$. The dihedral group D_n has one conjugacy class of reflections if n is odd, and two conjugacy classes if n is even.

Proposition 2.1. *Suppose that G acts on X.*

(a) *The relation "$x \sim y \iff x$ can be moved to y" is an equivalence relation. The equivalence class of x is just the orbit $G * x$. Consequently, the orbits of the action form a partition of X.*

(b) *If $x \in X$ then G_x is a subgroup of G, and $G_{g*x} = gG_x g^{-1}$.*

We come now to a geometric form of Lagrange's theorem. The number of orbits is called the *index* of G_x in G and is denoted by $[G : G_x]$. (It can be a finite number or an infinite cardinal number.)

The orbit–stabilizer theorem. *Suppose that G acts on X. Then, for any $x \in X$, $|G * x| = [G : G_x]$. In particular, if G is finite then $|G * x|$ is a divisor of $|G|$:*

$$|G * x| \cdot |G_x| = |G|. \tag{2.14}$$

Therefore, a large stabilizer yields a small orbit, and conversely. For a finite G, (2.14) shows that $G * x$ and G_x are reciprocals of each other.

The last result in this section provides a formula for the number of orbits in a group action and it is proved in this way.

Burnside's lemma. *Let G be a finite group acting on a finite set X. Then there are N orbits, where*

$$N = \frac{1}{|G|} \sum_{g \in G} |\mathrm{Fix}(g)| = \frac{1}{|G|} \sum_{x \in X} |G_x|. \tag{2.15}$$

Hence, N is the average number of fixed points that an element of G has.

2.5.2 Matrices and linear transformations

Let $\mathbb{K}$ be a field, say $\mathbb{R}$ or $\mathbb{C}$, or a skew-field of quaternions $\mathbb{H}$. Denote by $M_{m,n}(\mathbb{K})$ (or $\mathbb{K}^{m,n}$) the set of all $m \times n$ matrices with entries in $\mathbb{K}$. For example, $M_{1,n}(\mathbb{K}) = \mathbb{K}^n$ is the set of *n-vectors*. If $A \in M_{m,n}(\mathbb{K})$, then A_{ij} is the element in row i and column j of A. Denote by $M_n(\mathbb{K}) = M_{n,n}(\mathbb{K})$ the space of *square matrices. Addition* of same-dimension matrices is defined component-wise, so that $(A + B)_{ij} = A_{ij} + B_{ij}$. The *product of* $A \in M_{m,n}(\mathbb{K})$ and $B \in M_{n,l}(\mathbb{K})$ is the element $AB \in M_{m,l}(\mathbb{K})$ defined by

$$(AB)_{ij} = (\text{row } i \text{ of } A) \cdot (\text{column } j \text{ of } B) = \sum_{s=1}^{n} A_{is} B_{sj}.$$

Use the term *diagonal matrix* $A = \mathrm{diag}(\lambda_1, \ldots, \lambda_n)$ for one with $A_{ij} = 0$ when $i \neq j$ and $A_{ii} = \lambda_i$. The *identity matrix* of order n is $I_n = \mathrm{diag}(1, \ldots, 1)$.

The *transpose of* $A \in M_{m,n}(\mathbb{K})$ is the matrix $A^t \in M_{n,m}(\mathbb{K})$ obtained by interchanging the rows and columns of A, so that $(A^t)_{ij} = A_{ji}$.

When $a \in \mathbb{K}$ and $A \in M_{n,m}(\mathbb{K})$, we define the *left scalar multiplication* $a \cdot A \in M_{n,m}(\mathbb{K})$ to be the result of *left*-multiplying the elements of A by a:

$$(a \cdot A)_{ij} := a \cdot A_{ij}.$$

The *trace* of a square matrix $A \in M_n(\mathbb{K})$ is defined as the sum of its diagonal elements:

$$\mathrm{trace}(A) = \sum_{i=1}^{n} A_{ii}.$$

Example. If $\mathbb{K} \in \{\mathbb{R}, \mathbb{C}\}$ then the *symmetric functions* $\sigma_j(A)$ of a quadratic matrix A of order n are defined by the polynomial equality

$$\det(I_n + tA) = \sum_j \sigma_j(A) t^j, \tag{2.16}$$

where I_n is a unit matrix and $t \in \mathbb{K}$. Prove that

$$\sigma_1(A) = \text{trace } A, \qquad \sigma_n(A) = \det A, \qquad 2\sigma_2(A) = \sigma_1^2(A) - \sigma_1(A^2),$$

$$3\sigma_3(A) = \frac{1}{2} \text{trace}^3(A) - \frac{3}{2} \text{trace}(A) \text{trace}(A^2) + \text{trace}(A^3). \tag{2.17}$$

Hint. We prepare an M-file to compute $\sigma_k(A)$:

```
function f = sigma1(A, k)
  syms t;
  n = length(A);  E = eye(n);
   if k > n;   f = 0;
   else f = subs( diff( det(E + t*A), t, k), t, 0)/factorial(k);
   end;
end
```

For example,

```
A = [1 2 2 3;  3 4 1 0;  2 5 0 1;  2 3 6 1]          % enter your matrix, n = 4
```

The following commands return 0:

```
sigma1(A, 1) - trace(A)                          % example for k = 1.
sigma1(A, 4) - det(A)                            % example for k = 4.
sigma1(A, 1)^2 - sigma1(A*A, 1) - 2*sigma1(A, 2)   % identity for σ2(A).
```

The *general linear group* over $\mathbb{K}$ is

$$GL_n(\mathbb{K}) := \{A \in M_n(\mathbb{K}) : \exists B \in M_n(K) \text{ with } AB = BA = I_n\}.$$

A function $f : \mathbb{K}^n \to \mathbb{K}^m$ is called a *linear mapping* if $f(\mathbf{u} + \mathbf{v}) = f(\mathbf{u}) + f(\mathbf{v})$ and $f(\lambda \mathbf{u}) = \lambda f(\mathbf{u})$ for all $\mathbf{u}, \mathbf{v} \in \mathbb{K}^n$ and $\lambda \in \mathbb{K}$. If $n = m$ then f is called a *linear isomorphism* of $\mathbb{K}^n$.

One may show that

$$GL_n(\mathbb{K}) = \{A \in M_n(\mathbb{K}) : R_A \text{ is a linear isomorphism of } \mathbb{K}^n\}.$$

If $\mathbb{K} \in \{\mathbb{R}, \mathbb{C}\}$ then $GL_n(\mathbb{K}) = \{A \in M_n(\mathbb{K}) : \det A \neq 0\}$.

Proposition 2.2. *For any $A \in M_n(\mathbb{K})$ and any $g \in GL_n(\mathbb{K})$, the matrix $g A g^{-1}$ represents R_A in the basis $V = \{e_1 g, \ldots, e_n g\}$.*

This key result requires only slight modification when representing linear transformations using *left* matrix multiplication when $\mathbb{K}$ is $\mathbb{R}$ or $\mathbb{C}$: for any $A \subset M_n(\mathbb{K})$ and any $g \in GL_n(\mathbb{K})$, the matrix $g^{-1}Ag$ represents L_A in the basis $\{ge_1, \ldots, ge_n\}$ (via *left* multiplication).

The *standard (Euclidean) inner product on* $\mathbb{K}^n$ is the function from $\mathbb{K}^n \times \mathbb{K}^n$ to $\mathbb{K}$ defined by $\langle (x_1, \ldots, x_n), (y_1, \ldots, y_n) \rangle_{\mathbb{K}} := x_1 \bar{y}_1 + \cdots + x_n \bar{y}_n$.

The *p-norm* of $\mathbf{x} \in \mathbb{K}^n$ is defined by $\|\mathbf{x}\|_p = \left(\sum_i |x_i|^p \right)^{1/p}$. The *Euclidean norm* ($p = 2$) is defined by $\|\mathbf{x}\| := \sqrt{\langle \mathbf{x}, \mathbf{x} \rangle_{\mathbb{K}}} = \left(\sum_i |x_i|^2 \right)^{1/2}$. For example:

```
v = [1 -2 3]; w = [i -2+i 3*i];                      % v ∈ ℝ³, w ∈ ℂ³
[norm(v, 1) norm(v) norm(v, inf)]           % Answer: 6.00, 3.74, 3.00
[norm(w, 1) norm(w) norm(w, inf)]           % Answer: 6.24, 3.87, 3.00.
```

If $\mathbb{K} = \mathbb{C}, \mathbb{H}$, the standard inner product is called *hermitian* or *symplectic*, respectively. It follows that $\langle \mathbf{x}, \mathbf{x} \rangle_{\mathbb{K}}$ is a real number ≥ 0, which equals 0 only when $\mathbf{x} = (0, \ldots, 0)$.

The most important subgroups of the general linear group GL_n are $O(n)$, $SO(n)$, $U(n)$, $SU(n)$, and $Sp(n)$.

Vectors $\mathbf{x}, \mathbf{y} \in \mathbb{K}^n$ are called *orthogonal* if $\langle \mathbf{x}, \mathbf{y} \rangle_{\mathbb{K}} = 0$. A basis $\{\mathbf{x}_1, \ldots, \mathbf{x}_n\}$ of $\mathbb{K}^n$ is called *orthonormal* if $\langle \mathbf{x}_i, \mathbf{x}_j \rangle$ equals 1 when $i = j$ and equals zero when $i \neq j$ (that is, the vectors have norm 1 and are mutually orthogonal). The *standard orthonormal basis* of $\mathbb{K}^n$ is $\mathbf{e}_1 = (1, 0, \ldots, 0), \ldots, \mathbf{e}_n = (0, \ldots, 0, 1)$. The *orthogonal group over* $\mathbb{K}$,

$$O_n(\mathbb{K}) := \{A \in GL_n(\mathbb{K}) : \langle \mathbf{x}A, \mathbf{y}A \rangle = \langle \mathbf{x}, \mathbf{y} \rangle \text{ for all } \mathbf{x}, \mathbf{y} \in \mathbb{K}^n\},$$

is denoted by $O(n)$ and called the *orthogonal group* for $\mathbb{K} = \mathbb{R}$;
is denoted by $U(n)$ and called the *unitary group* for $\mathbb{K} = \mathbb{C}$;
is denoted by $Sp(n)$ and called the *symplectic group* for $\mathbb{K} = \mathbb{H}$.

Its elements are called *orthogonal*, *unitary*, or *symplectic* matrices. The subgroups (in Chapter 4 we denote them by $SL_n(\mathbb{K})$)

$$SO_n := \{A \in O_n : \det(A) = 1\}, \qquad SU_n := \{A \in U_n : \det(A) = 1\}$$

are called the *special orthogonal group* and the *special unitary group*, respectively.

For $A \in GL_n(\mathbb{K})$ the following properties are equivalent:

(1) $A \in O_n(\mathbb{K})$.
(2) R_A preserves orthonormal bases; i.e., if $\{\mathbf{x}_1, \ldots, \mathbf{x}_n\}$ is an orthonormal basis of $\mathbb{K}^n$, then so is $\{R_A(\mathbf{x}_1), \ldots, R_A(\mathbf{x}_n)\}$.
(3) The rows of A form an orthonormal basis of $\mathbb{K}^n$.
(4) $A \cdot A^t = I_n$.

$O(n)$ is equal to the group of rigid motions of $\mathbb{R}^n$, $U(n)$ to the group of rigid motions of $\mathbb{C}^n$ that preserve the standard complex structure.

If $A \in O(n)$ then $\det(A) = \pm 1$; if $A \in U(n)$ then $\det(A) = e^{i\theta}$ for some $\theta \in [0, 2\pi)$; if $A \in Sp(n)$ then $\det(A) = 1$.

Example. Find an orthogonal basis (and its matrix B) for a set A of vectors. The number of columns of B is the rank of A.

```
a = [1 2 3]; b = [2 -1 0]; c = [-1 3 2]; d = [1 0 2];      % four vectors in R³
A = [a' b' c' d']; rank(A)           % Answer: 0, set of vectors, rank(A) = 3
B = orth(A)                          % an orthonormal basis for the range of A
[ -0.01 0.83 -0.55;  -0.66 -0.42 -0.62;  -0.75 0.36 0.56 ]
B'*B                                 % B is orthogonal matrix: B'B = eye(rank(A))
[ 1 0 0;  0 1 0;  0 0 1 ]                                              % Answer.
```

The group $\text{Isom}(\mathbb{R}^n) := \{ f \colon \mathbb{R}^n \to \mathbb{R}^n : f \text{ is an isometry} \}$ (under composition of functions) is equal to the subgroup of $GL_{n+1}(\mathbb{R})$

$$\text{Isom}(\mathbb{R}^n) \simeq \left\{ \begin{pmatrix} A & 0 \\ \mathbf{a} & 1 \end{pmatrix} : A \in O(n) \text{ and } \mathbf{a} \in \mathbb{R}^n \right\}.$$

The subgroup of $\text{Isom}(\mathbb{R}^n)$ given by

$$\text{Trans}(\mathbb{R}^n) := \left\{ \begin{pmatrix} I_n & 0 \\ \mathbf{a} & 1 \end{pmatrix} : \mathbf{a} \in \mathbb{R}^n \right\}$$

is equal to $(\mathbb{R}^n, +)$ ($\mathbb{R}^n$ under the group operation of vector addition). The isometries of $\text{Trans}(\mathbb{R}^n)$ only *translate* and do not rotate.

The affine group (under composition of functions)

$$\text{Aff}_n := \left\{ f \colon \mathbb{R}^n \to \mathbb{R}^n : f \text{ is an affine map} \right\}$$

is equal to the following subgroup of $GL_{n+1}(\mathbb{R})$:

$$\text{Aff}_n := \left\{ \begin{pmatrix} A & 0 \\ \mathbf{a} & 1 \end{pmatrix} : \det A \neq 0 \text{ and } \mathbf{a} \in \mathbb{R}^n \right\}.$$

2.6   Exercises

Section 2.1   1. Let $[0, \mathbf{a}]$, $[0, \mathbf{b}]$, $[0, \mathbf{c}]$ be segments that are not coplanar. Show that any $\mathbf{x}$ in $\mathbb{R}^3$ can be written as the linear combination of the three vectors $\mathbf{a}$, $\mathbf{b}$, and $\mathbf{c}$ given by

$$\mathbf{x} = \frac{(\mathbf{x}, \mathbf{b}, \mathbf{c})}{(\mathbf{a}, \mathbf{b}, \mathbf{c})} \mathbf{a} + \frac{(\mathbf{a}, \mathbf{x}, \mathbf{c})}{(\mathbf{a}, \mathbf{b}, \mathbf{c})} \mathbf{b} + \frac{(\mathbf{a}, \mathbf{b}, \mathbf{x})}{(\mathbf{a}, \mathbf{b}, \mathbf{c})} \mathbf{c}.$$

2. Prove the vector identities

$$[\mathbf{a},\mathbf{b},\mathbf{c}] + [\mathbf{b},\mathbf{c},\mathbf{a}] + [\mathbf{c},\mathbf{a},\mathbf{b}] = 0;$$
$$(\mathbf{a}\times\mathbf{b})\times(\mathbf{c}\times\mathbf{d}) = (\mathbf{a},\mathbf{b},\mathbf{d})\mathbf{c} - (\mathbf{a},\mathbf{b},\mathbf{c})\mathbf{d};$$
$$(\mathbf{a}\times\mathbf{b})\cdot((\mathbf{c}\times\mathbf{d})\times(\mathbf{e}\times\mathbf{f})) = (\mathbf{a},\mathbf{b},\mathbf{d})(\mathbf{c},\mathbf{e},\mathbf{f}) - (\mathbf{a},\mathbf{b},\mathbf{c})(\mathbf{d},\mathbf{e},\mathbf{f});$$
$$(\mathbf{b}\times\mathbf{c})\cdot(\mathbf{a}\times\mathbf{d}) + (\mathbf{c}\times\mathbf{a})\cdot(\mathbf{b}\times\mathbf{d}) + (\mathbf{a}\times\mathbf{b})\cdot(\mathbf{c}\times\mathbf{d}) = 0.$$

3. Show that for any vectors $\mathbf{a}_1, \mathbf{b}_1, \mathbf{c}_1$ and $\mathbf{a}_2, \mathbf{b}_2, \mathbf{c}_2$ in $\mathbb{R}^3$

$$(\mathbf{a}_1,\mathbf{b}_1,\mathbf{c}_1)(\mathbf{a}_2,\mathbf{b}_2,\mathbf{c}_2) = \begin{vmatrix} \mathbf{a}_1\cdot\mathbf{a}_2 & \mathbf{a}_1\cdot\mathbf{b}_2 & \mathbf{a}_1\cdot\mathbf{c}_2 \\ \mathbf{b}_1\cdot\mathbf{a}_2 & \mathbf{b}_1\cdot\mathbf{b}_2 & \mathbf{b}_1\cdot\mathbf{c}_2 \\ \mathbf{c}_1\cdot\mathbf{a}_2 & \mathbf{c}_1\cdot\mathbf{b}_2 & \mathbf{c}_1\cdot\mathbf{c}_2 \end{vmatrix}.$$

4. Show (and execute examples) that

(a) The line $(\mathbf{r} - \mathbf{R})\times\mathbf{a} = 0$ and the plane $(\mathbf{r} - \mathbf{r}_0)\cdot\mathbf{n} = 0$ are *parallel* if $\mathbf{a}\cdot\mathbf{n} = 0$. The *distance* from the point $\mathbf{R}$ to the plane $(\mathbf{r} - \mathbf{r}_0)\cdot\mathbf{n} = 0$ is

$$\|(\mathbf{R} - \mathbf{r}_0)\cdot\mathbf{n}\|/\|\mathbf{n}\|.$$

(b) The *projection* of the point $\mathbf{R}$ to the plane $(\mathbf{r} - \mathbf{r}_0)\cdot\mathbf{n} = 0$ is

$$\mathbf{R}_1 = \frac{(\mathbf{R} - \mathbf{r}_0)\cdot\mathbf{n}}{\|\mathbf{n}\|^2}\mathbf{n}.$$

(c) The *perpendicular* from the point $\mathbf{R}$ to the line $(\mathbf{r} - \mathbf{r}_0)\times\mathbf{a} = 0$ is

$$\mathbf{r} = \mathbf{R} + t\,[\mathbf{a}, \mathbf{R} - \mathbf{r}_0, \mathbf{a}].$$

5. Find the distance between two lines determined by $\mathbf{r}_1 = [0,1,0]$, $\mathbf{a}_1 = [1,1,1]$ and $\mathbf{r}_2 = [1,0,0]$, $\mathbf{a}_2 = [0,1,1]$.

Section 2.2

1. Show that

(a) $\mathbf{T}_\mathbf{b} \circ \mathbf{T}_\mathbf{b} = \mathbf{T}_{\mathbf{a}+\mathbf{b}}$ and $\mathbf{R}_\phi \circ \mathbf{R}_\theta = \mathbf{R}_{\theta+\phi}$.

(b) A reflection in a line other than the coordinate axes is a composition of $\mathbf{P}_x$ (or $\mathbf{P}_y$), translation and the rotation.

(c) The quantity $x^2 + y^2$ is invariant under the action of $\mathbf{P}_x$, $\mathbf{P}_y$, or $\mathbf{R}_\pi$.

(d) If $\boldsymbol{\Gamma}_{l,\mathbf{a}}$ is a glide reflection with axis l and f is an isometry, then $f \circ \boldsymbol{\Gamma}_{l,\mathbf{a}} \circ f^{-1}$ is a glide reflection with axis $f(l)$.

2. Prove that *every rigid motion of* $\mathbb{R}^2$ *is a composition of at most three reflections*.

3. Prove that an affine transformation of $\mathbb{C}$, see (2.3), takes lines to lines and circles to circles.

4. Show that

(a) The translation $\mathbf{T_a}$ is a composition of two reflections $\mathbf{T}_{l_1}$ and $\mathbf{T}_{l_2}$ where the lines l_1, l_2 are parallel, and $\|\mathbf{a}\| = d(l_1, l_2)$;

(b) The rotation $\mathbf{R}_\theta$ is a composition of two reflections $\mathbf{T}_{l_1}$ and $\mathbf{T}_{l_2}$ where the lines l_1, l_2 intersect at the origin, and $\theta = 2\angle(l_1, l_2)$;

(c) The glide reflection $\mathbf{\Gamma}_{l,\mathbf{a}}$ is a composition of three reflections $\mathbf{T}_{l_1}$, $\mathbf{T}_{l_2}$, $\mathbf{T}_{l_3}$ where the lines l_1, l_2 are perpendicular to l with $\|\mathbf{a}\| = d(l_1, l_2)$, and $l_3 = l$.

5. Show that the *reflections in the XY-, YZ- and XZ-planes* are

$$\mathbf{P}_{xy} = \begin{pmatrix} 1 & 0 & 0 \\ 0 & 1 & 0 \\ 0 & 0 & -1 \end{pmatrix}, \quad \mathbf{P}_{yz} = \begin{pmatrix} -1 & 0 & 0 \\ 0 & 1 & 0 \\ 0 & 0 & 1 \end{pmatrix}, \quad \mathbf{P}_{xz} = \begin{pmatrix} 1 & 0 & 0 \\ 0 & -1 & 0 \\ 0 & 0 & 1 \end{pmatrix}. \qquad (2.18)$$

Each reflection across a plane is an isometry, and every isometry is a composition of reflections.

6. Prove that every *direct* isometry of $\mathbb{R}^3$ is a screw motion.

7. Consider the composition of reflections $\mathbf{P}_j$ in two distinct planes ω_1 and ω_2 that intersect in a line l. Prove that $\mathbf{P}_2\mathbf{P}_1$ is a rotation of $\mathbb{R}^3$ about the axis l of twice the angle between the planes ω_j.

8. Consider two parallel planes, $\mathbf{x} \cdot \mathbf{n} = d_1$ and $\mathbf{x} \cdot \mathbf{n} = d_2$, where $\|\mathbf{n}\| = 1$, and let $\mathbf{P}_1$ and $\mathbf{P}_2$ denote the reflections in these planes. Then

$$\mathbf{P}_2\mathbf{P}_1(\mathbf{x}) = \mathbf{x} + 2(d_2 - d_1)\,\mathbf{n},$$

so that the composition of two reflections in parallel planes is a translation.

9. Show that the reflections of $\mathbb{R}^3$ in the planes $\omega(\mathbf{a}, t): \mathbf{a} \cdot \mathbf{x} = t$ and $\omega(\mathbf{b}, s): \mathbf{b} \cdot \mathbf{x} = s$ commute if and only if either $\omega(\mathbf{a}, t) = \omega(\mathbf{b}, s)$ or $\mathbf{a} \perp \mathbf{b}$.

10. Prove the following three simple results:

(a) Let f be an isometry with $f(0) = 0$. Then for all $\mathbf{x}$ and $\mathbf{y}$, $\|f(\mathbf{x})\| = \|\mathbf{x}\|$ and $f(\mathbf{x}) \cdot f(\mathbf{y}) = \mathbf{x} \cdot \mathbf{y}$ (that is, f preserves norms and scalar products).

(b) If an isometry f fixes 0, $\mathbf{i}$, $\mathbf{j}$, and $\mathbf{k}$ then $f = I_3$.

(c) Suppose that $\|\mathbf{a}\| = \|\mathbf{b}\| \neq 0$. Then there is a reflection R across a plane ω through 0 such that $R(\mathbf{a}) = \mathbf{b}$ and $R(\mathbf{b}) = \mathbf{a}$.

11. Let $\mathbf{u} = (1,1,1,1)$ in $\mathbb{R}^4$; then

(a) $P(\mathbf{x}) = \frac{1}{4}(x_1 + x_2 + x_3 + x_4)\,\mathbf{u}$,

(b) $S(\mathbf{x}) = 2P(\mathbf{x}) - \mathbf{x} = \frac{1}{4}(x_1 + x_2 + x_3 + x_4)\,\mathbf{u} - \mathbf{x}$.

12. The points $\mathbf{u} = (1,1,0,1)$, $\mathbf{v} = (1,0,1,1)$, $\mathbf{w} = (0,1,1,1)$ in $\mathbb{R}^4$ determine an equilateral triangle. The image points under P_{12} are $P_{12}(\mathbf{u}) = (1,1)$, $P_{12}(\mathbf{v}) = (1,0)$, $P_{12}(\mathbf{w}) = (0,1)$. (The image itself is not equilateral).

13. The same image as in Exercise 12 is seen for the tetrahedron determined by the vertices $\mathbf{u}_1 = (1,1,0,1)$, $\mathbf{u}_2 = (1,0,1,1)$, $\mathbf{u}_3 = (0,1,1,1)$, $\mathbf{u}_4 = (1,1,1,0)$ since $P_{12}(\mathbf{u}_1) = P_{12}(\mathbf{u}_4) = (1,1)$. Plot the 2-D image of the tetrahedron determined by the

vertices $\mathbf{v}_1 = (2,2,1,-1)$, $\mathbf{v}_2 = (1,-1,0,3)$, $\mathbf{v}_3 = (-1,1,6,4)$, $\mathbf{v}_4 = (-1/2,-1/2,7,7)$ under R_{12}.

14. Show that $L_q = \begin{pmatrix} q_0 & q_1 & q_2 & q_3 \\ -q_1 & q_0 & q_3 & -q_2 \\ -q_2 & -q_3 & q_0 & q_1 \\ -q_3 & q_2 & -q_1 & q_0 \end{pmatrix}$.

15. Show that

(a) $R_q: \mathbb{H} \to \mathbb{H}$ is $\mathbb{H}$-linear for each $q \in \mathbb{H}$.

(b) For $q \in \mathbb{H}$, $L_q: \mathbb{H} \to \mathbb{H}$ is not necessarily $\mathbb{H}$-linear (find an example).

(c) If q has unit length then both matrices R_q, L_q are orthogonal; each of them has two mutually orthogonal invariant planes on which they act by rotations.

16. Use Theorem 2.2 to show that if p and q are pure quaternions ($p = \mathbf{p}$ and $q = \mathbf{q}$), and if $\mathbf{p} \perp \mathbf{q}$, then $-qpq^{-1} = p$.

17. Use quaternions to find the image of $\mathbf{x}$ under a rotation about the $\mathbf{k}$-axis of angle $\pi/6$. Now verify your result by elementary geometry.

18. Let α be the plane $\mathbf{x} \cdot \mathbf{n} = 0$, where $\mathbf{n} = (1,1,0)$, and let R be the reflection across α. Write $\mathbf{x} = (a,b,c)$ and $\mathbf{y} = R(\mathbf{x})$. Verify, both using elementary geometry and quaternion algebra, that $\mathbf{y} = (b,a,c)$.

19. Find the quaternion q such that M_q is the rotation of $\mathbb{R}^3$ given in coordinate form by $\mathbf{x} \to \mathbf{x}M$, where $M = \begin{pmatrix} 0 & 1 & 0 \\ 0 & 0 & 1 \\ 1 & 0 & 0 \end{pmatrix}$.

20. Can every orthogonal linear map of $\mathbb{R}^4$ with determinant 1 be realized by $\mathbf{x} \to p\mathbf{x}q$ for some quaternions q, p?

21. Let R_r and R_s be the rotations associated with the quaternions

$$r = \cos(\theta/2) + \sin(\theta/2)\,\mathbf{n}, \quad s = \cos(\varphi/2) + \sin(\varphi/2)\,\mathbf{m}.$$

Show (using Theorem 2.3) that the composition $R_s R_r$ is the map $\mathbf{x} \to s(r\mathbf{x}\bar{r})\bar{s} = (sr)\mathbf{x}\overline{sr}$, and as sr is also a unit quaternion (which is necessarily of the form $\cos\frac{1}{2}\psi + \sin\frac{1}{2}\psi\,\mathbf{h}$ for some unit vector $\mathbf{h}$), hence $R_s R_r = R_{sr}$. By computing sr as a quaternion product, find the axis and angle of rotation of the composition $R_s R_r$.

Section 2.3 1. Prove (visually) that *given any three angles α, β, γ with $\alpha + \beta + \gamma > \pi$, there exists a spherical triangle with those angles.*

2. Find the sides of a spherical triangle $\triangle ABC$ with $\angle A = \angle B = \angle C = \alpha$.

3. Prove that the product of any two reflections of $\mathbb{R}^3$ in planes through O that meet in a common line is a rotation about that common line.

4. Given any two great circles on S^2, there are two great circles bisecting the angles between them, each at right angles to the other.

5. Find a spherical triangle of area $3\pi/4$.

6. Prove the Pythagorean theorem for a sphere:

Let $\triangle ABC$ be a triangle in S^2 in which the angle at C is a right angle. If a, b, and c are the lengths of BC, CA, and AB, then

$$\cos c = \cos a \cdot \cos b. \tag{2.19}$$

7. For small a, b, c the classical Pythagorean theorem follows from (2.19). *Hint:*

$$\cos x \approx 1 - \tfrac{1}{2}x^2 \ (x = a, b, c) \Longrightarrow \cos c - \cos a \cos b \approx -\tfrac{1}{2}(c^2 - a^2 - b^2).$$

The useful identities for a right angle triangle also are

$$\sin \alpha = \sin a / \sin c, \quad \tan \alpha = \tan a / \tan b.$$

8. Show that $\mathbf{R}_x(\tfrac{\pi}{2})\mathbf{R}_y(\tfrac{\pi}{2}) = \left(\begin{smallmatrix} 0 & 0 & -1 \\ 1 & 0 & 0 \\ 0 & -1 & 0 \end{smallmatrix}\right)$ and $\mathbf{R}_y(\tfrac{\pi}{2})\mathbf{R}_x(\tfrac{\pi}{2}) = \left(\begin{smallmatrix} 0 & 1 & 0 \\ 0 & 0 & 1 \\ 1 & 0 & 0 \end{smallmatrix}\right)$.

9. Determine the matrix A of a rotation $\mathbf{x} \to \mathbf{x}A$ of S^2 that maps $Q(\tfrac{1}{2}, -\tfrac{1}{2}, -\tfrac{1}{\sqrt{2}})$ to $P(\tfrac{1}{2\sqrt{2}}, \tfrac{\sqrt{3}}{2\sqrt{2}}, \tfrac{1}{2\sqrt{2}})$.

10. Let a triangle $\triangle PQR$ in S^2 have sides $QR = RP$. Prove (using a reflection) that the angles at the vertices P and Q coincide.

11. Let a point P in $\mathbb{R}^2$ with polar coordinates (r, θ) map under stereographic projection to a point P' on S^2. Then the spherical distance of P' from the South Pole S is $2\tan^{-1}(r)$.

12. The coordinates of $P \in S^2 \setminus \{\mathbf{e}_3\}$ under stereographic projection can be expressed in terms of complex numbers

$$\pi(z) = \left(\frac{z + \bar{z}}{1 + |z|^2}, \frac{i(z - \bar{z})}{1 + |z|^2}, \frac{z\bar{z} - 1}{z\bar{z} + 1} \right). \tag{2.20}$$

Section 2.4 | 1. Plot the pyramid and the prism each *with rotation* about its geometrical axis. Use the programs of Example, p. 103, and the following program:

```
N = 36;  axis off;
for n = 0 : N;  view(5*n, 30);                                % animation
    patch(X, Y,  [0 1 1]);
    patch('Vertices', V, 'Faces', F, 'FaceColor',  [1 .5 1]);
    pause(0.2)
end;
```

2. Plot an oblique pyramid and a prism using the vector $[x_0, y_0, z_0]$, modifying the program of Example, p. 103. Plot the tetrahedron and the cube as particular cases of a regular pyramid and a prism whose heights have special values.

3. Prove that the faces of a convex polyhedron are convex polygons.

4. Show that there is no convex polyhedron with $F = 4$, $E = 7$, and $V = 5$. (Use Theorem 2.5.)

5. Find the length of the line segment on a face of the cube that we use for plotting the icosahedron.

6. Find the height and an edge of the roof over the face of the cube that we use for plotting the dodecahedron. The resulting pentagons must be planar and regular.

7. Prove formulae for the radii of inscribed and circumscribed spheres and for the volume of *Platonic solids* as functions of their edges.

8. Plot five non-regular deltahedra. (Two of them are triangular and pentagonal bipyramids).

9. Show that (a) G acts transitively on X if and only if for each $x, y \in X$ there is some $g \in G$ such that $g * x = y$; (b) X is the union of its orbits, and any two orbits are either equal or disjoint.

10. Let Q be a plane quadrilateral and let $G(Q)$ be its symmetry group (that is, the group of Euclidean isometries mapping Q onto itself). Show that $G(Q)$ has at most 8 elements (so that Q has the largest symmetry group when it is a square). For each n in $\{1, 2, \ldots, 8\}$ determine whether or not there is a quadrilateral Q with $G(Q)$ of order n. Is it true that if $G(Q)$ has order eight then Q is a square?

11. Let $G = \langle \sigma \rangle$ be the subgroup of S_{10} generated by $\sigma = (0)(12)(345)(6789) \in S_{10}$. Then there are 4 orbits under the canonical action of G on $\{0, 1, \ldots, 9\}$, namely, the sets $\{0\}$, $\{1, 2\}$, $\{3, 4, 5\}$ and $\{6, 7, 8, 9\}$.

12. For any matrix $A \in M_2(\mathbb{R})$ we have an action of $(\mathbb{R}, +)$ on $\mathbb{R}^2$ via $t * (x, y) - (x, y) e^{tA}$. Show that if A equals

$$\begin{pmatrix} 1 & 0 \\ 0 & 1 \end{pmatrix}, \quad \begin{pmatrix} -1 & 0 \\ 0 & -1 \end{pmatrix}, \quad \begin{pmatrix} 1 & 0 \\ 0 & -1 \end{pmatrix}, \quad \begin{pmatrix} 0 & -1 \\ 1 & 0 \end{pmatrix}, \quad \begin{pmatrix} 0 & 1 \\ 0 & 0 \end{pmatrix},$$

then e^{tA} equals

$$\begin{pmatrix} e^t & 0 \\ 0 & e^t \end{pmatrix}, \quad \begin{pmatrix} e^{-t} & 0 \\ 0 & e^{-t} \end{pmatrix}, \quad \begin{pmatrix} e^t & 0 \\ 0 & e^{-t} \end{pmatrix}, \quad \begin{pmatrix} \cos t & \sin t \\ -\sin t & \cos t \end{pmatrix}, \quad \begin{pmatrix} 0 & 0 \\ t & 1 \end{pmatrix}, \quad \text{respectively.}$$

13. Show that
(a) every subgroup G of the symmetric group S_n acts on $\{1, \ldots, n\}$ via $\sigma * n := \sigma(n)$,
(b) every matrix group $G \subset GL(V)$ acts on the vector space V via $T * v := vT$. This is called the *natural action* of G on V.

14. Let $A \in GL(V)$ be an invertible matrix. Then $(\mathbb{Z}, +)$ acts on V via $m * v := vA^m$.

15. The transitive subgroups of S_3 are exactly S_3 and A_3. The transitive subgroups of S_4 are
(a) S_4,
(b) A_4, which is normal,
(c) D_4 (three conjugate copies),
(d) $G_1 = \{e, (12)(34), (13)(24), (14)(23)\}$, which is normal,
(e) $\mathbb{Z}_4$ (three conjugate copies).

16. Prove that the symmetry group of the sphere $S^n \subset \mathbb{R}^{n+1}$ equals the group of isometries of $\mathbb{R}^{n+1}$ with no translational component, which is isomorphic to the orthogonal group: $\mathrm{Symm}(S^n) = O(n+1)$.

17. Summarize the definition of the dihedral group as follows: D_m *is a group of order* $2m$ *that is generated by two elements* r *and* s *which are subject to the relations:* $r^n = e, s^2 = e, sr = r^{-1}s$. Show that

(a) The element of order n is the rotation by $2\pi/n$ degrees, and the element of order two is the reflection about any symmetry axis of the n-gon.

(b) The direct symmetries of a polygon form a cyclic group C_n.

18. Prove directly that the group of rotations of a regular polyhedron with E edges has order $2E$.

Hint. The vertices of each of these polyhedra lie on a sphere, which one may assume is centered at the origin in $\mathbb{R}^3$. Recall that, say, q regular p-gons meet at each vertex v, and we accept (again without proof) that there is a rotation of order q (about an axis through v and the center of the polyhedron) which fixes v and leaves the polyhedron invariant. The stabilizer of each vertex (which cyclically permutes the q edges emanating from v) is a cyclic group of order q.

Repeating rotations of the polyhedron, we can move any vertex to any other; thus the orbit of any given vertex v is the set of all V vertices (and G^+ acts transitively on V). Applying the orbit–stabilizer theorem, p. 123, one may conclude that G^+ has order qV; thus, $|G^+| = \frac{4pq}{2q+2p-pq} = 2E$. This makes $|G^+|$ equal to 12 for the tetrahedron, 24 for the cube and the octahedron, and 60 for the dodecahedron and the icosahedron.

19. Prove that the vertices of any face of the small stellated dodecahedron are the images under a homothety from the center of the "kernel" face with coefficient $k = -(1 + 2\tau^{-1}/\sqrt{3})$, but with different order.

Analogously, the vertices of the great stellated dodecahedron are the images under a homothety from the center of mass of the "kernel" face with the same coefficient k, this time in the same order.

20. Plot SPs over Platonic solids.

21. Some of the most fascinating models to play with are compounds. Plot compounds corresponding to:

(a) An octahedron and a cube coupled together so that their edges bisect each other at right angles. (The dodecahedron and the icosahedron can also be coupled together to form a compound polyhedron.)

(b) Five different ways to inscribe a cube in a dodecahedron.

(c) Three different ways to inscribe a cube in an octahedron.

22. Calculate the vertices of the *great icosahedron*.

Hint. Start from an icosahedron inscribed in a cube. The continuation of the faces of the icosahedron leads to only one case where the new polyhedron does not decompose. Each face must be continued until its intersection with the three faces neighboring its opposite face. By symmetry, each face of the great icosahedron is a triangle homothetic to the triangular face of the central icosahedron with some negative coefficient k. For calculating k, consider the cross section of the polyhedron cut by the plane XZ, which contains the center O_1 of the face $B_1 B_2 B_9$ and medians of this face, and the face $B_3 B_4 B_{10}$, which is neighbor to the parallel face $B_3 B_4 B_8$. Such medians $B_9 B_{1,2}$, $B_{10} B_{3,4}$, where $B_{1,2} = (B_1 + B_2)/2 = (0,0,c)$, $B_{3,4} = (B_3 + B_4)/2 = (0,0,-c)$, intersect at the point $F(0,y,0)$ on the axis OY, which is the center of an edge of the great icosahedron. Hence, F is the image of the point $B_{1,2}$ under homothety with the center O_1, i.e., $k = -\frac{O_1 E}{O_1 B_{1,2}}$. From similarity of triangles we find the value of y:

$$y/c = (y-c)/p \Longrightarrow y = c^2/(c-p) = 2c/(3-\sqrt{5}).$$

Then express the coefficient of homothety k through the ratio of projections of line segments: $k = -\frac{y-c/3}{c/3} = -\frac{3y-c}{c} = -\frac{3+\sqrt{5}}{3-\sqrt{5}} \approx -6.854$.

23. Plot a regular icosahedron combining a pentagonal antiprism with two regular pentagonal pyramids with the same edge length.

24. What regular stellated polyhedra are obtained from the *Platonic solids* by the stellate construction? Apply the stellate construction to the *Archimedean solids*.

25. Write a program for plotting two half-regular Kepler's polyhedra.

26. Write a program for plotting UP: prisms and antiprisms.

27*. Complete the programs of Section A.6 with a program for plotting the *snub dodecahedron*.

28. Plot prisms and antiprisms with star polygons as bases. Plot a decagon (sometimes called a *pentacle*) when the base for the prism is a pentagon.

Section 2.5 1. Show that the symmetric group S_n has $n!$ elements, and the alternating group A_n has $n!/2$ elements. *Hint*. Clearly, the dihedral group D_n is generated by two elements ab and b, both of order two.

2. Let $G = \{1, i, -1, -i\}$ be a subgroup of $\mathbb{C}$. Show that for each $g \in G$ the map $x \mapsto gx$ is a permutation of G.

3. Let $K = \{0,1,2\}$ be the field with 3 elements (defined in Section 1.1) and $A = A(K^2)$ be the affine plane over K. Thus in A there are exactly 9 points and 12 lines, each line has 3 points, and each point lies on 4 lines.

4. Show that matrix multiplication is not generally commutative.

5. A 3×3 matrix $\{a_{ij}\}$ is called a *magic square* if the 3 row sums, the 3 column sums and the 2 diagonal sums are all the same number s.

(a) Show that magic squares form a three-dimensional subspace of $M_3(\mathbb{R})$, and $A_1 = \left(\begin{smallmatrix} 1 & 1 & 1 \\ 1 & 1 & 1 \\ 1 & 1 & 1 \end{smallmatrix}\right)$, $A_2 = \left(\begin{smallmatrix} 1 & -1 & 0 \\ -1 & 0 & 1 \\ 0 & 1 & -1 \end{smallmatrix}\right)$, $A_3 = \left(\begin{smallmatrix} 0 & 1 & -1 \\ -1 & 0 & 1 \\ 1 & -1 & 0 \end{smallmatrix}\right)$ form a basis of this subspace.

(b) Show that $a_{22} = s/3$.

6. Check that $(A \cdot B)^t = B^t \cdot A^t$.

7. When $A, B \in M_n(\mathbb{K})$ and $\mathbb{K} \in \{\mathbb{R}, \mathbb{C}\}$, verify the following property: trace$(AB) =$ trace(BA). Since multiplication in $\mathbb{H}$ is not commutative, the above equality is false even in $M_1(\mathbb{H})$.

8. Show for the matrix $B = \begin{pmatrix} \cos\theta & \sin\theta \\ -\sin\theta & \cos\theta \end{pmatrix}$ that $R_B \colon \mathbb{R}^2 \to \mathbb{R}^2$ is a counterclockwise rotation through the angle θ. Compare this to the matrix $A = (e^{i\theta}) \in M_1(\mathbb{C})$. For this matrix $R_A \colon \mathbb{C} \to \mathbb{C}$ is also a counterclockwise rotation through the angle θ, since $R_A(re^{i\phi}) = re^{i(\theta+\phi)}$. Thus, $A \in M_1(\mathbb{C})$ and $B \in M_2(\mathbb{R})$ "represent the same motion."

9. Consider the complex matrices (called *pure quaternions*)

$$\mathbf{1} = \begin{pmatrix} 1 & 0 \\ 0 & 1 \end{pmatrix}, \quad I = \begin{pmatrix} i & 0 \\ 0 & -i \end{pmatrix}, \quad J = \begin{pmatrix} 0 & -1 \\ 1 & 0 \end{pmatrix}, \quad K = \begin{pmatrix} 0 & -i \\ -i & 0 \end{pmatrix}. \quad (2.21)$$

Verify the equations $I^2 = J^2 = K^2 = -\mathbf{1}$, $IJ = K$, $JK = I$, $KI = J$. Hence, the subgroup G of $GL(2,\mathbb{C})$ that is generated by the four elements (2.21) consists of 8 elements, $G = \{\pm\mathbf{1}, \pm I, \pm J, \pm K\}$.

10. Prove that quaternion multiplication corresponds to matrix multiplication:

$$M(q)M(q') = M(q \cdot q').$$

11. Prove the *Schwarz inequality* $|\langle \mathbf{x}, \mathbf{y} \rangle| \leq \|\mathbf{x}\| \cdot \|\mathbf{y}\|$ for all $\mathbf{x}, \mathbf{y} \in \mathbb{K}^n$.

12. Verify using sigma1.m the last identity of (2.17).

3

Affine and Projective Transformations

In addition to isometries, there are two kinds of mappings that preserve lines: *affine* (Section 3.1) and *projective* (Section 3.2) transformations.

Affine transformations f of $\mathbb{R}^n$ have the following property:

> *If l is a line then $f(l)$ is also a line, and if $l \parallel k$ then $f(l) \parallel f(k)$.*

A line in $\mathbb{R}^n$ means a set of the form $\{\mathbf{r}_0 + \mathbf{r} : \mathbf{r} \in W\}$, where $\mathbf{r}_0 \in \mathbb{R}^n$ and $W \subset \mathbb{R}^n$ is a one-dimensional subspace.

Projective transformations f of $\mathbb{R}^n$

> *map lines to lines, preserving the cross-ratio of four points.*

We also use homogeneous coordinates $\mathbf{x} = (x_1 : \ldots : x_{n+1})$ in $\mathbb{R}^{n+1}$. Section 3.3 describes transformation matrices in homogeneous coordinates.

3.1 Affine Transformations

An *affine transformation* $f \colon \mathbb{R}^n \to \mathbb{R}^m$ is a linear mapping followed by a translation, i.e., $f(\mathbf{v}) = A(\mathbf{v}) + \mathbf{b}$, where $\mathbf{b} \in \mathbb{R}^m$ and $A \colon \mathbb{R}^n \to \mathbb{R}^m$ is an invertible linear mapping. The *set* $\mathrm{Aff}_n(\mathbb{R})$ *of affine transformations* of $\mathbb{R}^n$ is a group under the operation of composition.

3.1.1 *Affine transformations in two and three dimensions*

An *affine transformation* of the plane, $f \colon \mathbb{R}^2 \to \mathbb{R}^2$, has the form

$$f(x,y) = (a_{11}x + a_{12}y + b_1, \; a_{21}x + a_{22}y + b_2) \tag{3.1}$$

V. Rovenski, *Modeling of Curves and Surfaces with MATLAB®*,
Springer Undergraduate Texts in Mathematics and Technology 7, DOI 10.1007/978-0-387-71278-9_3,
© Springer Science+Business Media, LLC 2010

for some real numbers a_{ij}, b_j satisfying $\Delta = a_{11}a_{22} - a_{12}a_{21} \neq 0$.

In matrix form we write $f(\mathbf{x}) = \mathbf{x}A + \mathbf{b}$ $(\mathbf{x} \in \mathbb{R}^2)$, where $A = \begin{pmatrix} a_{11} & a_{21} \\ a_{12} & a_{22} \end{pmatrix}$ is an invertible matrix, and $\mathbf{b} = (b_1, b_2) \in \mathbb{R}^2$.

One may use (3.1) to verify the properties of affine transformations:

(1) *mapping lines to lines*;
(2) *mapping parallel lines to parallel lines*;
(3) *preserving ratios of lengths along a given line*.

If $A, B, C, D \in \mathbb{R}^2$ are four collinear points (with $C \neq D$), the *ratio* $\overrightarrow{AB}/\overrightarrow{CD}$ is well defined; this is a scalar $\lambda \in \mathbb{R}$ such that $\overrightarrow{AB} = \lambda \overrightarrow{CD}$. The property (3) means that $\overrightarrow{AB}/\overrightarrow{CD} = \overrightarrow{f(A)f(B)} / \overrightarrow{f(C)f(D)}$.

Fundamental theorem of affine geometry. (For simplicity, consider the two-dimensional case.) *Let A, B, C and A', B', C' be two sets of three non-collinear points in $\mathbb{R}^2$. Then there is a unique affine transformation f that maps A, B, C to A', B', C', respectively.*

Next, we consider particular cases of affine transformations.

A *scaling about the origin* is a transformation that maps a point $P = (x, y)$ to a point $P' = (x', y')$ by multiplying the x and y coordinates by positive constant scaling factors λ_x and λ_y, respectively, to give $x' = \lambda_x x$ and $y' = \lambda_y y$. The matrix $\mathbf{S}(\lambda_x, \lambda_y) = \begin{pmatrix} \lambda_x & 0 \\ 0 & \lambda_y \end{pmatrix}$ is called the *scaling transformation matrix*. If $\lambda_x \neq 1$ and $\lambda_y = 1$, we have a *horizontal scaling* $\mathbf{S}_x(\lambda_x)$, while if $\lambda_y \neq 1$ and $\lambda_x = 1$, we have a *vertical scaling* $\mathbf{S}_y(\lambda_y)$:

$$\begin{cases} x' = \lambda_x x, \\ y' = y \end{cases} \quad \text{and} \quad \begin{cases} x' = x, \\ y' = \lambda_y y. \end{cases}$$

A scaling transformation is *uniform* (another name is a *dilation* about O) whenever $\lambda_x = \lambda_y = \lambda$. A scaling factor λ is said to be an *expansion* if $\lambda > 1$, and a *compression* if $\lambda < 1$. If the constant λ_x is allowed to be negative, a horizontal scaling by the factor $-\lambda_x$ is followed by a mirror reflection in the vertical axis, and similarly for negative values of the constant λ_y.

A slightly less obvious transformation of the plane is called *shear*. The *horizontal shear* and *vertical shear* transformations, $\hat{\mathbf{S}}_x(r)$ and $\hat{\mathbf{S}}_y(r)$, are given by

$$\begin{cases} x' = x + ry, \\ y' = y \end{cases} \quad \text{and} \quad \begin{cases} x' = x, \\ y' = rx + y. \end{cases} \tag{3.2}$$

In both cases r is a real constant other than zero.

Both scaling and shear can be applied in an arbitrary direction.

The *directional scaling* by a factor $\lambda > 0$ in the direction $\mathbf{v} = (a,b)$ is given by the matrix $\mathbf{S}_\mathbf{v}(\lambda) = \dfrac{1}{a^2+b^2}\begin{pmatrix} \lambda a^2+b^2 & (\lambda-1)ab \\ (\lambda-1)ab & a^2+\lambda b^2 \end{pmatrix}$, i.e.,

$$x' = \frac{(\lambda a^2+b^2)x+(\lambda-1)aby}{a^2+b^2}, \qquad y' = \frac{(a^2+\lambda b^2)y+(\lambda-1)abx}{a^2+b^2}. \tag{3.3}$$

The *directional shear* by a factor $r > 0$ in the direction $\mathbf{v} = (a,b)$ is given by the matrix $\hat{\mathbf{S}}_\mathbf{v}(r) = \dfrac{1}{a^2+b^2}\begin{pmatrix} a^2+b^2-rab & -rb^2 \\ ra^2 & a^2+b^2+rab \end{pmatrix}$, i.e.,

$$x' = \frac{(a^2+b^2-rab)x+ra^2y}{a^2+b^2}, \qquad y' = \frac{(a^2+b^2+rab)y-rb^2x}{a^2+b^2}. \tag{3.4}$$

Examples.

1. Einstein's theory of special relativity studies the relationship of the dynamics of a system, if described in two coordinate systems moving with constant speed v one from the other. The *Lorentz transformation* gives the rules for going from one coordinate system to the other. Einstein, on the basis of two postulates, derived the (affine) transformation $\mathbf{L}_x$ relating the coordinates of the two systems: $y' = y$, $z' = z$, and

$$\begin{cases} x' = \beta(x-vt), \\ t' = \beta\left(t-\dfrac{v}{c^2}x\right) \end{cases} \implies \mathbf{L}_x = \begin{pmatrix} \beta & -\beta v \\ -\beta\frac{v}{c^2} & \beta \end{pmatrix}.$$

Here $c = \mathrm{const} > 0$ is the velocity of light and $\beta = 1/\sqrt{1-v^2/c^2}$.

2. Let P, Q, R be points other than the vertices on the (possibly extended) sides of a triangle $\triangle ABC$. One may show that

(a) the points P, Q, R are collinear $\iff \dfrac{AR}{RB}\cdot\dfrac{BP}{PC}\cdot\dfrac{CQ}{QA} = -1$ (Menelaus);

(b) the lines AP, BQ, CR are concurrent $\iff \dfrac{AR}{RB}\cdot\dfrac{BP}{PC}\cdot\dfrac{CQ}{QA} = 1$ (Ceva).

We may use (b) in solving this problem: Let $A = (1,3)$, $B = (-1,0)$, $C = (4,0)$ and $P = (0,0)$, $Q = \left(\frac{8}{3},\frac{4}{3}\right)$, $R = \left(-\frac{2}{3},\frac{1}{2}\right)$ be the points. Determine whether the lines AP, BQ, and CR are concurrent.

```
A = [2 4];  B = [-2 0];  C = [1 0];  P = [5/2 0];  Q = [3/2 2];  R = [1 3];
det([A - B; R - B]),  det([A - C; Q - C]),  det([C - B; P - B])        % Answer: 0,0,0
AR = norm(A - R),   BP = norm(B - P),   CQ = norm(C - Q)               % verify
RB = norm(R - B),   PC = norm(P - C),   QA = norm(Q - A)
Ceva = (AR*BP*CQ)/(RB*PC*QA)                                  % obtain 1 (Yes!)
```

3. How is an affine transformation written in complex numbers? Affine mappings are of the form $fz = az + b\bar{z} + c$; the affine transformations are those for which $|a|^2 \neq |b|^2$. We confirm this using the determinant:

```
syms a1 a2 b1 b2 x y real
f = (a1 + i*a2)*(x + i*y) + (b1 + i*b2)*conj(x + i*y);      % one may assume c = 0
f1 = collect(real(f), [x, y]),  f2 = collect(imag(f), [x, y])
```
$f_1 = (a_1 + b_1)x + (b_2 - a_2)y \quad f_2 = (a_2 + b_2)x + (a_1 - b_1)y$ % Answer
```
A = [a1+b1  b2-a2;  a2+b2  a1-b1]      % the matrix of the linear system
det(A)                                 % Answer: a₁²+a₂²−b₁²−b₂²
```
det(A) % Answer: $a_1^2 + a_2^2 - b_1^2 - b_2^2$

An affine transformation $f\colon \mathbb{R}^3 \to \mathbb{R}^3$ of space has the form $f(\mathbf{x}) = \mathbf{x}A + \mathbf{b}$ ($\mathbf{x} \in \mathbb{R}^3$), where $A = \begin{pmatrix} a_{11} & a_{21} & a_{31} \\ a_{12} & a_{22} & a_{32} \\ a_{13} & a_{23} & a_{33} \end{pmatrix}$ is an invertible matrix, and $\mathbf{b} = (b_1, b_2, b_3) \in \mathbb{R}^3$. In coordinate form we have

$$f(\mathbf{x}) = (a_{11}x + a_{12}y + a_{13}z + b_1,\, a_{21}x + a_{22}y + a_{23}z + b_2,\, a_{31}x + a_{32}y + a_{33}z + b_3)$$

for some real numbers a_{ij}, b_j satisfying $\det A \neq 0$.

A *scaling about the origin* is a transformation that maps a point $P = (x, y, z)$ to a point $P' = (x', y', z')$, multiplying the x, y, and z coordinates by positive constant scaling factors λ_x, λ_y, and λ_z, respectively, to give $x' = \lambda_x x$, $y' = \lambda_y y$, and $z' = \lambda_z z$. The matrix $\mathbf{S}(\lambda_x, \lambda_y, \lambda_z) = \begin{pmatrix} \lambda_x & 0 & 0 \\ 0 & \lambda_y & 0 \\ 0 & 0 & \lambda_z \end{pmatrix}$ is called the *scaling transformation matrix*. A scaling transformation is *uniform* whenever $\lambda_x = \lambda_y = \lambda_z = \lambda$. A scaling factor λ is said to be an *expansion* if $\lambda > 1$, and a *compression* if $\lambda < 1$.

Examples. The command maketform (from the Image Processing Toolbox) creates spatial transformation structures of the following types:

'affine' – affine transformation in 2-D or n-D,
'projective' – projective transformation in 2-D or n-D, etc.

f = maketform('affine', A) builds a nonsingular $(n+1)$-by-$(n+1)$ matrix A for an affine transformation, the last column of A being [zeros(n, 1); 1].

f = maketform('projective', A) builds a nonsingular real $(n+1)$-by-$(n+1)$ matrix A for an n-dimensional projective transformation; the entry $A(n+1, n+1)$ cannot be 0. The matrix A defines a forward transformation such that tformfwd(U, f), where U is a 1-by-n vector, returns a 1-by-n vector $\mathbf{x}$ such that x = W(1 : n)/W(n+1), where W = [U 1]*A.

1. Make and apply an affine transformation of $\mathbb{R}^2$:

f = maketform('affine', [.5 0 0; .5 2 0; 1 2 1]);   % $f(\mathbf{x}) = \mathbf{x} \begin{pmatrix} .5 & 0 \\ .5 & 2 \end{pmatrix} + (1,2)$

tformfwd([10 20], f)   % Answer: $(16,42)$ (see also Section 3.3).

2. Apply an affine transformation of $\mathbb{R}^3$, $f(\mathbf{x}) = \mathbf{x}A$, $A = \begin{pmatrix} .5 & 0 & 0 \\ .5 & 2 & 0 \\ 0 & 0 & 1 \end{pmatrix}$:

F = maketform('affine', [.5 0 0 0; .5 2 0 0; 0 0 1 0; 0 0 0 1]);   % matrix A
tformfwd([10 20 5], F)   % Answer: $(15,40,5)$.

3.1.2 *Parallel projections*

We will see that a parallel projection (of one plane onto another one) is a special type of affine transformation.

There are several variants of parallel projections (see also Section 5.3.2), but they are all based on the following principle: Select a direction $\mathbf{d} \in \mathbb{R}^3$ and construct a ray that starts at a general point P in the object, and goes in the direction $\mathbf{d}$. The point P' where this ray intercepts the projection plane becomes the projection of P. The rays may be perpendicular to the projection plane (*orthogonal projection*), or they may strike at a different angle. This is why the latter method is called *oblique projection*.

Because the rays are parallel, we can imagine that they originate at a center of projection located at infinity. This interpretation unifies *parallel* and *perspective* projections (see also Section 3.2.3) and is in accordance with the general rule of projections which distinguishes between parallel and perspective projections by the location of the *center of projection*.

We create the parallel projection ourselves using calculations with affine maps. Assume the vector $\mathbf{d} = (a,b,c)$ is not parallel to the plane ω: $Ax + By + Cz = 0$ (with normal vector $\mathbf{n} = (A,B,C)$). Then the parallel projection of the point $P = (x,y,z)$ in the plane ω in the direction $\mathbf{d}$ is the point $Q = (x',y',z')$, where $\mathbf{r}_Q = \mathbf{r}_P - \frac{\mathbf{r}_P \cdot \mathbf{n}}{\mathbf{d} \cdot \mathbf{n}} \mathbf{d}$ (verify!) or

$$\begin{aligned} x' &= \frac{(bB+cC)x - aBy - aCz}{aA+bB+cC}, \\ y' &= \frac{-bAx + (aA+cC)y - bCz}{aA+bB+cC}, \\ z' &= \frac{-cAx - cBy + (aA+bB)z}{aA+bB+cC}. \end{aligned} \tag{3.5}$$

If the vector $\mathbf{d}$ is perpendicular to the plane ω (i.e., $\mathbf{d}$ and $\mathbf{n}$ are parallel), the term *orthogonal projection* or *orthographic projection* is used.

We will now study the relations of parallel projections and affine transformations of $\mathbb{R}^n$. For concreteness we will restrict ourselves to $n = 2, 3$.

Let α_1 and α_2 be two planes of $\mathbb{R}^3$ and let $\mathbf{d}$ be a vector not parallel to either plane. Imagine rays of light shining parallel to $\mathbf{d}$ through these planes. Each point $P_1 \in \alpha_1$ has a unique ray passing through it that also passes through a point P_2, say, in the plane α_2. We call the function π that maps each point P_1 in α_1 to the corresponding point P_2 in α_2 a *parallel projection from* α_1 *onto* α_2. If the roles of the planes are reversed, then we obtain the inverse function $\pi^{-1} \colon \alpha_2 \to \alpha_1$, a *parallel projection from* α_2 *onto* α_1.

We can imagine α_1 and α_2 as copies of $\mathbb{R}^2$, each equipped with Cartesian coordinates (due to isometries $f_i \colon \mathbb{R}^2 \rightleftarrows \alpha_i$). Hence, a parallel projection $\pi \colon \alpha_1 \to \alpha_2$ becomes a transformation of $\mathbb{R}^2$, i.e., $f_2^{-1} \cdot \pi \cdot f_1 \colon \mathbb{R}^2 \mapsto \mathbb{R}^2$.

Note the following properties:

(a) *Each parallel projection* $\pi \colon \alpha_1 \to \alpha_2$ *is an affine transformation.*
(b) *An affine transformation is not necessarily a parallel projection.*
(c) *An affine transformation of* $\mathbb{R}^2$ *can be expressed as the composite of two parallel projections.*

Hence, certain properties of figures, such as length and angle, are not preserved under a parallel projection. The difference arises because the affine group Aff_2 is larger than the group Isom_2 (of isometries).

Examples.

1. A parallel projection is either between two parallel planes, in which case all lengths are unchanged, or between two intersecting planes, in which case distances along the line of intersection are unchanged. Hence, the "doubling map" $\mathbf{x} \to 2\mathbf{x}$ is not a parallel projection, but it is possible to realize it by following one parallel projection with another: the first doubles all horizontal lengths, and the second doubles all vertical lengths.

2. Consider a parallel projection $\pi \colon \omega_2 \to \omega_1$ of coordinate plane $\omega_2 \colon x = 0$ onto $\omega_1 \colon z = 0$ (i.e., $A = B = 0$, $C = 1$) in the direction $\mathbf{d} = (\cos\phi, \sin\phi, -1)$. In this view, (3.5) becomes $x' = x + \cos\phi z$, $y' = x + \sin\phi z$, $z' = 0$. Let us find the image of the triangle $\triangle PQR \subset \omega_2$:

```
t = pi/3;  d = [cos(t) sin(t) -1];
P = [0; 0; 1];  Q = [0; 1; 2];  R = [0; 2; 1];
T = [1 0 cos(t);  0 1 sin(t);  0 0 0]          % projection in the direction d
P1 = T*P;  Q1 = T*Q;  R1 = T*R;                % images of P, Q, R
hold on;  view(3);  arrow3(R, d, 'r');
patch([P(1) Q(1) R(1)], [P(2) Q(2) R(2)], [P(3) Q(3) R(3)], [0 0 1]);
patch([P1(1) Q1(1) R1(1)], [P1(2) Q1(2) R1(2)], [P1(3) Q1(3) R1(3)], [0 1 0]);
```

3.1.3 *Convex hull and Delaunay triangulation*

Let $A_1, \ldots, A_k$ be distinct points in $\mathbb{R}^n$ and $\alpha_1, \ldots, \alpha_k$ be scalars. The set $(A_1, \alpha_1), \ldots, (A_k, \alpha_k)$ is called a system of *weighted points*.

If (A_i, α_i) $(1 \le i \le k)$ are weighted points such that $\sum \alpha_i \ne 0$, there exists a unique point P with the property $\sum \alpha_i \overrightarrow{PA_i} = 0$. (Verify!) Moreover, for any point Q, we have $(\sum \alpha_i) \overrightarrow{QP} = \sum \alpha_i \overrightarrow{QA_i}$.

The unique point P defined by this claim is called the *barycenter* of the system. Notice that the barycenter of the system $(A_1, \alpha_1), \ldots, (A_k, \alpha_k)$ is, by definition and for any nonzero scalar λ, the same as that of the system $(A_1, \lambda \alpha_1), \ldots, (A_k, \lambda \alpha_k)$.

A subset S of $\mathbb{R}^n$ is *convex* if for all points A and B of X the segment $[A, B]$ is contained in S. The intersection of all convex subsets containing $X \subset \mathbb{R}^n$ is a convex subset $C(X) \subset \mathbb{R}^n$, called the *convex hull* of X.

A convex polygon bounds a convex plane domain. New convex subsets can be constructed using the following properties, whose proofs are easy:

(a) The intersection of convex subsets is convex.
(b) The convex hull of X is the set of barycenters of the points of X endowed with positive masses.

Several properties of conics are affine invariant. Let $f \colon \mathbb{R}^2 \to \mathbb{R}^2$ be an affine transformation, and C a conic.

(a) *If R is the center of C, then $f(R)$ is the center of $f(C)$.*
(b) *If l is the asymptote of C, then $f(l)$ is the asymptote of $f(C)$.*
(c) *If l is a tangent to C, then $f(l)$ is a tangent to $f(C)$.*

The convex hull of a finite planar or spatial set $\mathbf{P} = \{P_1, P_2, \ldots, P_n\}$ of points is useful in spline constructions; see Chapter 9. The derivation and plotting of the convex hull, shown in Figure 3.1(a), are the key to a number of problems of *computational geometry*, in which algorithms of geometrical problems and their computational complexity are studied. There are many well-known algorithms for the convex hull. We will consider one of them, based on the *gift wrapping* method. The idea is the following:

1. *Find the P_{i_1} with the minimal y-coordinate (P_{i_1} lies on the convex hull).*
2. *Rotate a ray around P_{i_1} until the last point, say P_{i_2}, of the given set lies on it. (Obviously, P_{i_2} belongs to the convex hull.)*
3. *Repeat this rotation for P_{i_2} and find P_{i_3}, etc., until we return to the initial point P_{i_1}.*

Example. We find the convex hull using the command convhulln.

```
x = randn(1, 35);   y = randn(1, 35);
k = convhulln( [x; y]', {'Qt', 'Pp'} )
plot(x, y, 'o',  x(k), y(k), '+r')          % The vertices of CH(X) are in red
```

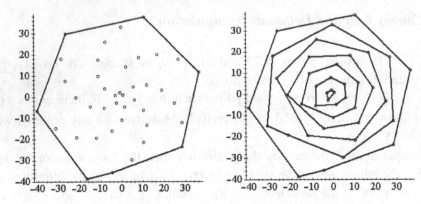

Fig. 3.1 The convex hull. The depth (layers) of the planar point set.

Alternatively, we create an M-file **CH_my**:

```
function [L, m] = CH_my(x, y)                    % first m points of L form CH(P)
   n = length(x);
   P = cat(1, x, y);   L = P;               % P array of given points; P_i = (x_i, y_i)
   Y = L(2, 1);   a = 1;
   for i = 2 : n;   if L(2, i) <= Y;
   a = i;   Y = L(2, i);
   end;   end
   L(:, n + 1) = L(:, a);   B = [1; 0];
   for j = 1 : (n + 1);
   C = -1;   T = L(:, a);   L(:, a) = L(:, j);   L(:, j) = T;
   if (a == n + 1) | (j == n + 1)                      % condition "or"
   m = j;
   j = n + 1;   return
   end;
   for i = (j + 1) : (n + 1);
   r = L(:, i) - L(:, j);
   if r'*r > 0
   Y = r'*B/(sqrt(r'*r)*sqrt(B'*B));
   if Y >= C;
   C = Y;   a = i;
   end;   end;   end;
   B = L(:, a) - L(:, j);
   end
end
```

Now we apply **CH_my**:

```
N = 35;  x = randn(1, N);  y = randn(1, N);      % random definition of points
[L, m] = CH_my(x, y);                                   % compute CH(P)
```

```
plot(L(1, 1 : m), L(2, 1 : m), '-r');              % plot CH(P), Figure 3.1(a)
hold on;  plot(x, y, 'bo')                         % plot all points P (inside CH(P))
```

There are several related problems, for example, finding the intersection of two given convex polygons. Below we consider a problem having applications in statistics.

The *depth of a point A* in a finite planar set **P** is the number of convex hulls (we will say *convex layers*) that bound this point and that must be deleted before the point A can be deleted (Figure 3.1). The *depth of a finite planar set* is the number of convex hulls (i.e., convex layers) obtained by the above procedure. For example, the depth of a triangle and its center (i.e., four points) is equal to two.

The *Delaunay triangulation* returns a set of triangles connecting the data points in such a way that no data point is contained within any triangle.

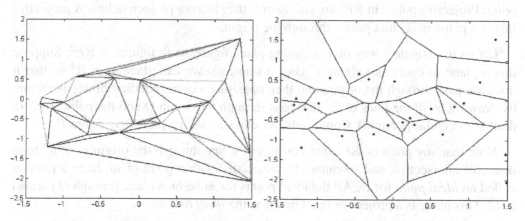

Fig. 3.2 Delaunay triangulation. Lines that separate regions.

Example. Given 2-D data, apply Delaunay triangulation. Obtain the index number of the triangle containing a given point by using the function tsearch.

```
x = randn(1, 12);  y = randn(1, 12);
T = delaunay(x, y)
tsearch(x, y, T,  [.2 .7])
```

Plot the triangles by using trimesh.

```
trimesh(T, x, y)                                   % Delaunay triangulation
```

Compute and plot the lines that separate regions in the plane closest to a particular data point by using the function voronoi.

```
voronoi(x, y, T)                                   % lines that separate regions
```

3.2 Projective Transformations

3.2.1 *Homogeneous coordinates*

The space $\mathbb{R}^3$ is used to develop the theory of two-dimensional projective geometry.

A *projective point* is a line in $\mathbb{R}^3$ through the origin. The *real projective plane* $\mathbb{R}P^2$ is the set of all such points.

The expression $(a:b:c)$, in which the numbers a, b, c are not all zero, represents the point Q in $\mathbb{R}P^2$ corresponding to the line in $\mathbb{R}^3$ through $(0,0,0)$ and (a,b,c). We refer to $(a:b:c)$ as *homogeneous coordinates* of Q. The points $(1:0:0)$, $(0:1:0)$, $(0:0:1)$ are known as the *triangle of reference*. The point $(1:1:1)$ is called the *unit point*. Projective points in $\mathbb{R}P^2$ are *collinear* if they lie on a projective line. A *projective line* is a plane in $\mathbb{R}^3$ that passes through the origin.

Let us investigate a way of associating plane figures with figures in $\mathbb{R}P^2$. Suppose that a plane ω contains a figure F (say, a triangle). We can place ω in $\mathbb{R}^3$ so that it does not pass through the origin, and then construct a corresponding projective figure by drawing in all the lines of $\mathbb{R}^3$ that pass through the origin O and the points of F (a double triangular pyramid). Conversely, we can start with a projective figure F.

Note that any point of $\mathbb{R}P^2$ that consists of a line through the origin *parallel* to ω does not intersect ω, and so cannot be associated with a point of ω. Such a point is called an *ideal point* for ω. All the ideal points for ω lie on a plane through O parallel to ω. This plane is a projective line known as the *ideal line* for ω.

An *embedding plane* is a plane ω that does not pass through the origin, together with the set of all ideal points for ω. The simplest approach is to designate ω to be the *standard embedding plane* $z = 1$. (Two other possible standard choices are $x = 1$ and $y = 1$.)

A polynomial $p(x_1,\ldots,x_n)$ in n variables is *homogeneous of degree k* if

$$p(\lambda x_1,\ldots,\lambda x_n) = \lambda^k p(x_1,\ldots,x_n) \qquad \forall \lambda \in \mathbb{R}.$$

This means that p is a sum of *kth-degree monomials* $ax_1^{t_1}\cdots x_n^{t_n}$, where $t_1 + \cdots + t_n = k$.

A *projective figure* is a subset of $\mathbb{R}P^2$. Figures in $\mathbb{R}P^2$ can be determined by homogeneous polynomials in the variables x, y, z.

A homogeneous polynomial of degree k in the variables x, y, z defines a figure $p_k(x,y,z) = 0$ in $\mathbb{R}P^2$, which is called an *algebraic set of kth order*. The *algebraic set of second order* $p_2(x,y,z) = 0$ is called a *projective conic* in $\mathbb{R}P^2$. A projective conic is *nondegenerate* if it can be represented by a nondegenerate conic in the standard embedding plane.

Example. Show that $p = x^3y + xy^3$ is a homogeneous polynomial.

```
syms x y k;
p = @(x, y) x^3*y + x*y^3;                    % enter polynomial
p2 = p(k*x, k*y);   simplify(p2/p(x, y))      % obtain k⁴
```

3.2.2 *Duality and cross-ratio*

We define cross-ratio in terms of projective geometry.

Let A, B, C, D be four collinear points in $\mathbb{R}P^2$ represented by position vectors $\mathbf{a}$, $\mathbf{b}$, $\mathbf{c}$, $\mathbf{d}$, and let $\mathbf{c} = \alpha\mathbf{a} + \beta\mathbf{b}$ and $\mathbf{d} = \gamma\mathbf{a} + \delta\mathbf{b}$. Then the *cross-ratio* of A, B, C, D is

$$(A,B,C,D) = (\beta/\alpha):(\delta/\gamma).$$

Four collinear points A, B, C, D are called *harmonic* if $(A,B,C,D) = -1$. In this case, A, B are *harmonically conjugate with respect to* C, D.

The following properties are easy to prove:

(a) The cross-ratio (A,B,C,D) is independent of the homogeneous coordinates that are used to represent the collinear points A, B, C, D.

(b) Projective transformations preserve the cross-ratio.

In affine plane geometry, if two points A, B are given, then the ratio AC/CB uniquely determines a third point C on the line AB.

We now explore the analogous result for projective geometry.

Unique fourth point. *Let* A, B, C, X, Y *be collinear points in* $\mathbb{R}P^2$ *such that*

$$(A,B,C,X) = (A,B,C,Y).$$

Then $X = Y$.

Hence, the cross-ratio is like a coordinate determining the fourth member from a given triple. Any number, as well as ∞, can occur.

Example. Let A, B, C, D and A, E, F, G be two sets of collinear points (on different lines) in $\mathbb{R}P^2$ such that $(A,B,C,D) = (A,E,F,G)$. To show that the lines BE, CF, and DG are concurrent, we will use symbolic arguments. We may assume $A = (0,0,1)$, $B = (1,0,1)$, $C = (c,0,1)$, and $D = (d,0,1)$.

```
syms c d f g;
A = [0 0 1];  B = [1 0 1];
C = [c 0 1];  D = [d 0 1];
E = [0 1 1];  F = [0 f 1];  G = [0 g 1];
```

```
CR1 = (A(1)-C(1))*(B(1)-D(1))/(A(1)-B(1))/(C(1)-D(1));        % 1st cross-ratio
CR2 = (A(2)-F(2))*(E(2)-G(2))/(A(2)-E(2))/(F(2)-G(2));        % 2nd cross-ratio
f2 = factor(solve(CR1 - CR2, f))          % condition (A,B,C,D) = (A,E,F,G)
BE = cross(B, E);  CF = cross(C, F);  DG = cross(D, G);
BE_CF = cross(BE, CF);
BE_DG = cross(BE, DG);
R1 = factor(subs(BE_CF, f, f2));
R2 = factor(subs(BE_DG, f, f2));
h = R1(2)/R2(2);  simplify(R1/h - R2)                        % obtain zero vector
```

Duality is the bijective correspondence between points and lines of $\mathbb{R}P^2$ that respects *incidence*: a point and a line are incident if the point belongs to the line, or the line passes through the point. The *duality principle* of projective geometry states:

> *If a valid statement is posed only in terms of points, lines, and the incidence relations, then the dual proposition, obtained from the original statement by interchanging the terms point and line, is also true.*

A set of points on a line is called a *range* and the line itself is called the *axis* of the range. The dual of a range is called a *pencil* of lines. The common point of the lines of the pencil is called the *apex* of the pencil. Any line not passing through the apex of a pencil is called a *transversal*.

If a range has axis l, then the lines of the dual pencil pass through the point dual to the line l. Since the dual coordinates of four lines of a pencil are dependent, a cross-ratio can be defined for a pencil of four lines in the same way one was defined for four collinear points.

Examples.

1. In homogeneous coordinates, the *dual line* of the point $P = (a : b : c)$ is the line $l: ax + by + cz = 0$. The dual of the line containing two points is the point where the dual lines meet. If $P = (a : b : c)$ and $X = (x : y : z)$ then $ax + by + cz = 0$ can be written as $X \cdot P = 0$.

```
syms x y z;
P = [1 2 5];
l = [x y z]'*P                              % obtain l = x + 2y + 5z
```

2. Prove that two distinct points in $\mathbb{R}P^2$ lie on a unique line.

Solution. An equation for the line in $\mathbb{R}P^2$ through the points $Q = (a : b : c)$ and $R = (d : e : f)$ is the determinant $\begin{vmatrix} x & y & z \\ a & b & c \\ d & e & f \end{vmatrix} = \begin{vmatrix} b & c \\ e & f \end{vmatrix} x - \begin{vmatrix} a & c \\ d & f \end{vmatrix} y + \begin{vmatrix} a & b \\ d & e \end{vmatrix} z = 0$. We compute this equation for the points $Q = (1 : 3 : 5)$ and $R = (-1 : 2 : 0)$:

```
syms x y z;
S = [x y z];    Q = [1 3 5];    R = [-1 2 0];
det([Q; R; S])                                          % obtain equation.
```

3. Show that the four lines

$$x + y + 2z = 0, \qquad 3x - y + 4z = 0, \qquad 5x + y + 8z = 0, \qquad 2x + 3z = 0,$$

are concurrent and find their cross-ratio. *Hint.* Use duality and the definition of cross-ratio.

```
L1 = [1 1 2];  L2 = [3 -1 4];  L3 = [5 1 8];  L4 = [2 0 3];
M = [L1; L2; L3; L4];   rank(M)                  % verify that rank(M) = 2
syms a1 a2 b1 b2;
Eq1 = a1*L1 + a2*L2 - L3;
Eq2 = b1*L1 + b2*L2 - L4;
[a1 a2] = solve(Eq1(1), Eq1(2), Eq1(3), a1, a2)
[b1 b2] = solve(Eq2(1), Eq2(2), Eq2(3), b1, b2)
CR = (a2/a1)/(b2/b1)                             % Answer: the cross-ratio 1/2.
```

3.2.3 *Projective transformations of* $\mathbb{R}P^2$

A *projective transformation* of $\mathbb{R}^{n+1}$ is a function of the form $f: (x_1 : \ldots : x_{n+1}) \rightarrow (x_1 : \ldots : x_{n+1})A$, where A is an invertible $(n+1) \times (n+1)$ matrix. We say A is a matrix *associated* with f. The set $P(n)$ of projective transformations of $\mathbb{R}P^n$ is a group under the operation of composition.

We define the transformations of $\mathbb{R}P^2$ algebraically, and then we give their geometrical interpretation using the ideas of perspectivity.

A *projective transformation* of $\mathbb{R}P^2$ is a function of the form $f : (x : y : z) \rightarrow (x : y : z)A$, where A is an invertible 3×3 matrix. We say A is a matrix *associated* with f.

If R is a (real) eigenvector of the matrix A corresponding to an eigenvalue λ, then $RA = \lambda R$ and so R is a *fixed point* of the corresponding projective transformation.

Strategy:

1. To *compose* two projective transformations f_1 and f_2:

   (a) write down matrices A_1 and A_2 associated with f_1, f_2;
   (b) calculate $A_1 A_2$;
   (c) write down the composite $f_1 \circ f_2$ with which $A_1 A_2$ is associated.

2. To find the *inverse* of a projective transformation f:

   (a) write down a matrix A associated with f;

(b) calculate $\text{adj}(A)$;

(c) write down the inverse f^{-1} with which $\text{adj}(A)$ is associated.

3. To find the *image of a line* $ax + by + cz = 0$ under the projective transformation $f: (x:y:z) \rightarrow (x:y:z)A$:

   (a) write the equation of the line $(x:y:z)L = 0$ where $L = (a,b,c)$;
   (b) find a matrix B associated with f^{-1};
   (c) write down the equation of the image $(x:y:z)LB = 0$.

Example. Find the fixed points of the projective transformation corresponding to the

matrix $A = \begin{pmatrix} 2 & 1 & 0 \\ 0 & 1 & -1 \\ 0 & 2 & 4 \end{pmatrix}$.

Solution.

    A = [2 1 0; 0 1 -1; 0 2 4];
    [V, E] = eig(A)

A projective transformation $f: (x:y:z) \rightarrow (x:y:z)A$ of $\mathbb{R}P^2$ is called a *perspective* if there are planes ω_1 and ω_2 such that the linear transformation $f: \mathbf{x} \rightarrow \mathbf{x}A$ maps ω_2 rigidly onto ω_1 in $\mathbb{R}^3$.

Perspectivity theorem. *Every projective transformation of $\mathbb{R}P^2$ is the composite of at most three perspective transformations. Hence, every property preserved by perspective transformations is also preserved by projective transformations.*

For example, the projective transformation associated with the matrix

$$A = [2\,0\,0;\; 0\,3\,0;\; 0\,0\,4]$$

cannot be a perspective transformation. Because the linear transformation associated with (any multiple of) A stretches distances in each direction of $\mathbb{R}^3$ by different factors, it cannot map any plane rigidly onto another.

A *quadrilateral* is a set of four points A, B, C, D no three of which are collinear. The four points determine six lines, which fall into three pairs of opposite sides, the lines $\{AC, BD\}$, $\{AD, BC\}$, $\{AB, CD\}$ meeting respectively in three diagonal points X, Y, Z.

Fundamental theorem of projective geometry. (For simplicity, consider the two-dimensional case.) *Let A, B, C, D and A', B', C', D' be two quadrilaterals in $\mathbb{R}P^2$. Then there is a unique projective transformation f mapping A, B, C, D to A', B', C', D'.*

Strategy. To determine the projective transformation f that maps the vertices of the quadrilateral $ABCD$ to the corresponding vertices of the quadrilateral $A'B'C'D'$:

(a) find the projective transformation f_1 that maps the triangle of reference and unit point to the points A, B, C, D, respectively;

(b) find the projective transformation f_2 that maps the triangle of reference and unit point to the points A', B', C', D', respectively;

(c) calculate $f = f_2 \circ f_1^{-1}$.

Examples. (*Using the fundamental theorem.*)

1. **Desargues' theorem.** *Let* $\triangle ABC$ *and* $\triangle A'B'C'$ *be two triangles in* $\mathbb{R}^2$ *such that the lines* AA', BB', *and* CC' *meet at a point* U. *Let* BC *and* $B'C'$ *meet at* P, CA *and* $C'A'$ *meet at* Q, *and* AB *and* $A'B'$ *meet at* R. *Then* P, Q, *and* R *are collinear.*

Proof. Let A', B' be points that do not lie on the line $z = 0$. By the fundamental theorem, one may assume $A = (1:0:0)$, $B = (0:1:0)$, $C = (0:0:1)$, and $C' = (1:1:1)$.

```
syms p1 p2 q1 q2 r1 r2;
A = [1 0 0];  B = [0 1 0];  C = [0 0 1];
A1 = [p1 p2 1];  B1 = [q1 q2 1];  C1 = [1 1 1];
AA1 = cross(A, A1);
BB1 = cross(B, B1);
CC1 = cross(C, C1);
D1 = det([AA1; BB1; CC1])              % obtain D₁ = p₂ − q₁
       % D₁ = 0 means that the lines AA′, BB′, and CC′ meet at a point U.
BC = cross(B, C);
B1C1 = cross(B1, C1);
P = cross(BC, B1C1);
CA = cross(C, A);
C1A1 = cross(C1, A1);
Q = cross(CA, C1A1);
AB = cross(A, B);
A1B1 = cross(A1, B1);
R = cross(AB, A1B1);
D2 = factor(det([P; Q; R]))            % find D₁ as a factor
R2 = solve(D1, r2);
D2 = simplify(subs(det([P; Q; R]), r2, R2))           % obtain 0
```

$\square$

2. **Pappus' theorem.** *Let* A_1, A_2, A_3 *and* B_1, B_2, B_3 *be two sets of collinear points lying on distinct lines. Let* $C_1 = A_2 B_3 \cap A_3 B_2$, $C_2 = A_3 B_1 \cap A_1 B_3$, *and* $C_3 = A_1 B_2 \cap A_2 B_1$. *Then* C_1, C_2, *and* C_3 *are collinear.*

Proof. We apply a projective transformation to simplify the algebra. Take $A_1 A_2 A_3$ to be the line $y = 0$ and $B_1 B_2 B_3$ to be the line $z = 0$. Let $A_1 = (p:0:1)$, $A_2 = (q:0:1)$, $A_3 = (r:0:1)$, $B_1 = (l:1:0)$, $B_2 = (m:1:0)$, $B_3 = (n:1:0)$. Then the line $A_2 B_3$ has dual coordinates $A_2 \times B_3 = (-1:n:q)$, and the line $A_3 B_2$ has dual coordinates

$A_3 \times B_2 = (-1:m:r)$. These lines meet at $c_1 = (-1:n:q) \times (-1:m:r) = (nr-qm:r-q:n-m)$. Similarly, $c_2 = (lp-rn:p-r:l-n)$ and $c_3 = (mq-pl:q-p:m-l)$.

Since $[C_1,C_2,C_3] = \begin{vmatrix} nr-qm & r-q & n-m \\ lp-rn & p-r & l-n \\ mq-pl & q-p & m-l \end{vmatrix} = 0$ (the rows sum to zero), it follows that C_1, C_2, C_3 are collinear. We verify the proof using the program below.

```
syms p q r l m n;
A1 = [p 0 1]; A2 = [q 0 1]; A3 = [r 0 1];
B1 = [l 1 0]; B2 = [m 1 0]; B3 = [n 1 0];
A2B3 = cross(A2, B3);
A3B2 = cross(A3, B2);
c1 = cross(A2B3, A3B2)
A1B3 = cross(A1, B3);
A3B1 = cross(A3, B1);
c2 = cross(A3B1, A1B3)
A1B2 = cross(A1, B2);
A2B1 = cross(A2, B1);
c3 = cross(A1B2, A2B1)
det([c1; c2; c3])                                    % obtain 0
```

□

Note that the vanishing of the determinant depends on the fact that $nr = rn$, $qm = mq$, $pl = lp$. With more general non-commutative algebraic systems such as the quaternions, Pappus' theorem may fail to hold.

3. The *Fano plane*, $\mathbb{Z}_2 P^2$, is the projective plane defined by 3 homogeneous $\mathbb{Z}_2$ coordinates. An element of $\mathbb{Z}_2 P^2$ is a triple $(x:y:z)$, where x, y, z are 0 or 1. Since $(0:0:0)$ is excluded, $\mathbb{Z}_2 P^2$ has seven points. By duality there are seven lines. The *fundamental theorem of plane projective geometry* is true for $\mathbb{Z}_2 P^2$. The set of all projective transformations of $\mathbb{Z}_2 P^2$ consists of $7 \cdot 6 \cdot 4 = 168$ elements.

3.3 Transformations in Homogeneous Coordinates

When $X_{n+1} \neq 0$, the homogeneous coordinates $(X_1 : \ldots : X_{n+1})$ represent the Cartesian point $[x_1,\ldots,x_n] = (X_1/X_{n+1},\ldots,X_n/X_{n+1})$ in $\mathbb{R}^n$. A point of the form $(X_1 : \ldots : X_n : 0)$ does not correspond to a Cartesian point, but represents the point at infinity in the direction of the vector $(X_1,\ldots,X_n)$.

3.3.1 *Transformations in two dimensions*

When $W \neq 0$, the homogeneous coordinates (X, Y, W) represent the Cartesian point $(x, y) = (X/W, Y/W)$ in $\mathbb{R}^2$. A point of the form $(X, Y, 0)$ represents the point at infinity in the direction of the two-dimensional vector (X, Y).

In homogeneous coordinates, as shown in Section 3.2, the composition of transformations f_1 and f_2, denoted by $f_1 \circ f_2$, can be performed with matrix multiplications alone. The homogeneous transformation matrices for translations, scalings, and rotations are presented in what follows.

Examples.

1. A rotation $\mathbf{R}(\theta)$ about the origin followed by a translation $\mathbf{T_a}$ is denoted by $\mathbf{R}(\theta) \circ \mathbf{T_a}$, and has the transformation matrix

$$\begin{pmatrix} \cos\theta & \sin\theta & 0 \\ -\sin\theta & \cos\theta & 0 \\ 0 & 0 & 1 \end{pmatrix} \begin{pmatrix} 1 & 0 & 0 \\ 0 & 1 & 0 \\ a_x & a_y & 1 \end{pmatrix} = \begin{pmatrix} \cos\theta & \sin\theta & 0 \\ -\sin\theta & \cos\theta & 0 \\ a_x & a_y & 1 \end{pmatrix}.$$

```
syms t ax ay;   Rz = e_rot(0, 0, t);   Ta = [1 0 0; 0 1 0; ax ay 1];
Rz*Ta                                  % obtain above R(θ) ∘ Ta
```

A *rotation through an angle* θ *about an arbitrary point* (x_0, y_0) is obtained by performing a translation which maps (x_0, y_0) to the origin, followed by a rotation through the angle θ about the origin, followed by a translation which maps the origin to (x_0, y_0). The rotation matrix is

$$\mathbf{R}_{(x_0, y_0)}(\theta) = \mathbf{T}_{-(x_0, y_0)} \mathbf{R}(\theta) \mathbf{T}_{(x_0, y_0)}$$

$$= \begin{pmatrix} 1 & 0 & 0 \\ 0 & 1 & 0 \\ -x_0 & -y_0 & 1 \end{pmatrix} \begin{pmatrix} \cos\theta & \sin\theta & 0 \\ -\sin\theta & \cos\theta & 0 \\ 0 & 0 & 1 \end{pmatrix} \begin{pmatrix} 1 & 0 & 0 \\ 0 & 1 & 0 \\ x_0 & y_0 & 1 \end{pmatrix}$$

$$= \begin{pmatrix} \cos\theta & \sin\theta & 0 \\ -\sin\theta & \cos\theta & 0 \\ -x_0\cos\theta + y_0\sin\theta + x_0 & -x_0\sin\theta - y_0\cos\theta + y_0 & 1 \end{pmatrix}.$$

```
syms t ax ay;   Rz = e_rot(0, 0, t);   Ta = [ 1 0 0; 0 1 0; ax ay 1];
Rt = inv(Ta)*Rz*Ta                      % obtain above R(x0,y0)(θ)
```

2. A square has vertices $A = (1, 1), B = (2, 1), C = (2, 2), D = (1, 2)$. Find the coordinates of the vertices when the figure is rotated about B through the angle $\pi/4$.

```
R1 = e_rot(0, 0, pi/4)      % obtain ≈ [0.71, 0.71, 0; −0.71, 0.71, 0; 0, 0, 1.00]
Tb = [ 1 0 0; 0 1 0; 2 1 1];
Rt = inv(Tb)*R1*Tb          % ≈ [0.71, 0.71, 0; −0.71, 0.71, 0; 1.29, −1.12, 1.00]
```

```
S = [1 1 1; 2 1 1; 2 2 1; 1 2 1];
Q = S*Rt                           % rows of Q are images of A, B, C, D

1.2929 0.2929 1.0000
2.0000 1.0000 1.0000
1.2929 1.7071 1.0000
0.5858 1.0000 1.0000                          % Answer: matrix Q.
```

3. A reflection P_l in an arbitrary line l: $ax + by + c = 0$ (see Section 2.2) may be obtained by transforming the line to one of the axes, reflecting in that axis, and then applying the inverse of the first transformation. Reflections P_x and P_y correspond to the x- and y-axes. The general reflection matrix in homogeneous coordinates is

$$\mathbf{P}_l = \begin{pmatrix} b^2 - a^2 & -2ab & 0 \\ -2ab & a^2 - b^2 & 0 \\ -2ac & -2bc & b^2 + a^2 \end{pmatrix}. \tag{3.6}$$

To show this, let $b \neq 0$. The line l intersects the y-axis in $(0, -c/b)$.

Strategy:

(a) Apply a translation mapping $(0, -c/b)$ to the origin, thus mapping l to a line l' through the origin with an identical gradient to l.

(b) The gradient of l' is $\tan(\theta) = -a/b$, where θ is the angle that l makes with the x-axis. Rotate l' about the origin through the angle $-\theta$. The line is now mapped to the x-axis.

(c) Apply a reflection in the x-axis.

(d) Apply the inverse of the rotation of step (b), followed by the inverse of the translation of step (a).

The composition of the transformations above is

$$\begin{pmatrix} 1 & 0 & 0 \\ 0 & 1 & 0 \\ 0 & c/b & 1 \end{pmatrix} \begin{pmatrix} \cos\theta & -\sin\theta & 0 \\ \sin\theta & \cos\theta & 0 \\ 0 & 0 & 1 \end{pmatrix} \begin{pmatrix} 1 & 0 & 0 \\ 0 & -1 & 0 \\ 0 & 0 & 1 \end{pmatrix} \begin{pmatrix} \cos\theta & \sin\theta & 0 \\ -\sin\theta & \cos\theta & 0 \\ 0 & 0 & 1 \end{pmatrix} \begin{pmatrix} 1 & 0 & 0 \\ 0 & 1 & 0 \\ 0 & -c/b & 1 \end{pmatrix}$$

$$= \begin{pmatrix} \cos 2\theta & \sin 2\theta & 0 \\ \sin 2\theta & -\cos 2\theta & 0 \\ (c/b)\sin 2\theta & -(c/b)(1 + \cos 2\theta) & 1 \end{pmatrix}. \tag{3.7}$$

```
syms a b c t;
Rt = e_rot(0, 0, t);
T1 = [1 0 0; 0 1 0; 0 c/b 1];  T2 = [1 0 0; 0 -1 0; 0 0 1];
Pl = simple(T1*inv(Rt)*T2*Rt*inv(T1))        % obtain result of (3.7)
```

Since $\tan\theta = -a/b$, it follows that $\cos 2\theta = (b^2 - a^2)/(b^2 + a^2)$ and $\sin 2\theta = -2ab/(b^2 + a^2)$. Finally, substitution for the trigonometric functions in (3.7) gives

(3.6) with the factor $1/(b^2 + a^2)$. (In homogeneous coordinates multiplication by a factor does not affect the result.)

3.3.2 Transformations in three dimensions

Homogeneous coordinates in three-dimensional space are derived similarly to homogeneous coordinates of the plane. A point (x, y, z) in $\mathbb{R}^3$ is represented in the four-dimensional space $\mathbb{R}^4$ by the vector $(x, y, z, 1)$, or by any multiple (rx, ry, rz, r) with $r \neq 0$.

When $W \neq 0$, the homogeneous coordinates $(X : Y : Z : W)$ represent the Cartesian point $(x, y, z) = (X/W, Y/W, Z/W)$. A point of the form $(X, Y, Z, 0)$ does not correspond to a Cartesian point, but represents the point at infinity in the direction of the three-dimensional vector (X, Y, Z). The set of all homogeneous coordinates $(X : Y : Z : W)$ is called *three-dimensional projective space* and denoted by $\mathbb{R}P^3$. An *embedding space* in $\mathbb{R}P^3$ is defined similarly to an embedding plane in $\mathbb{R}P^2$.

A *projective transformation* of $\mathbb{R}P^3$ is a mapping $f \colon \mathbb{R}P^3 \to \mathbb{R}P^3$ of the form $f(x : y : z : w) = (x, y, z, w)A$, where $A \in GL(4, \mathbb{R})$ (the *homogeneous transformation matrix of f*) is a nondegenerate matrix of order 4.

The homogeneous *translation* matrix is $\mathbf{T_a} = \begin{pmatrix} 1 & 0 & 0 & 0 \\ 0 & 1 & 0 & 0 \\ 0 & 0 & 1 & 0 \\ a_x & a_y & a_z & 1 \end{pmatrix}$, where $\mathbf{a} = (a_x, a_y, a_z)$.

Then

$$(x, y, z, 1)\, \mathbf{T_a} = (x + a_x, y + a_y, z + a_z, 1).$$

The homogeneous *scaling* matrix is $\mathbf{S}(s_x, s_y, s_z) = \begin{pmatrix} s_x & 0 & 0 & 0 \\ 0 & s_y & 0 & 0 \\ 0 & 0 & s_z & 0 \\ 0 & 0 & 0 & 1 \end{pmatrix}$. Then

$$(x, y, z, 1)\, \mathbf{S}(s_x, s_y, s_z) = (s_x x, s_y y, s_z z, 1).$$

The composition of transformations can be performed with matrix multiplications alone. The rotations about the coordinate axes (see Section 2.2) are called the *primary rotations*. Reflections in the XY-, YZ-, and XZ-planes are derived in (2.18). A *reflection in an arbitrary plane* $\omega \colon ax + by + cz = 0$ is represented by (2.4).

Example. Let m be a line with direction $\mathbf{a} = (u, v, w)$ through a point $Q = (x_0, y_0, z_0)$. A rotation about m through an angle θ can be represented by the product $\mathbf{R}_m(\theta) = \mathbf{T}_{-Q}\mathbf{R_a}(\theta)\mathbf{T}_Q$, where $\mathbf{R_a}(\theta)$ is the *rotation about the axis* $\mathbf{a}$ through the angle θ, given (see also Example, p. 90) by the matrix

$$\begin{pmatrix} \cos\theta + u^2(1-\cos\theta) & uv(1-\cos\theta) + w\sin\theta & wu(1-\cos\theta) - v\sin\theta & 0 \\ uv(1-\cos\theta) - w\sin\theta & \cos\theta + v^2(1-\cos\theta) & vw(1-\cos\theta) + u\sin\theta & 0 \\ wu(1-\cos\theta) + v\sin\theta & vw(1-\cos\theta) - u\sin\theta & \cos\theta + w^2(1-\cos\theta) & 0 \\ 0 & 0 & 0 & 1 \end{pmatrix}.$$

To prove the claim, compute $\mathbf{R}_m(\theta)$ symbolically (using Q and $\mathbf{a}$).

```
syms u v w t x0 y0 z0;
Q = [x0 y0 z0];    a = [u v w];
Ra = [cos(t)+u^2*(1-cos(t)) u*v*(1-cos(t))+w*sin(t) w*u*(1-cos(t))-v*sin(t) 0;
u*v*(1-cos(t))-w*sin(t) cos(t)+v^2*(1-cos(t)) v*w*(1-cos(t))+u*sin(t) 0;
w*u*(1-cos(t))+v*sin(t) v*w*(1-cos(t))-u*sin(t) cos(t)+w^2*(1-cos(t)) 0;
0 0 0 1]
TQ = [1 0 0 0; 0 1 0 0; 0 0 1 0; Q 1]; TQ2 = [1 0 0 0; 0 1 0 0; 0 0 1 0; -Q 1];
Rm = TQ2*Ra*TQ
```

Compute $\mathbf{R}_m(\theta)$ for $\mathbf{a} = (0,1,1)$ through a point $Q = (1,2,0)$ with $\theta = \pi/4$.

```
a1 = [0 1 1]; Q1 = [1 2 0]; t0 = pi/4;
subs(Rm, {u v w x0 y0 z0 t}, {a1(1) a1(2) a1(3) Q1(1) Q1(2) Q1(3) t0})
```

[0.71 0.71 -0.71 0; -0.71 1 0.29 0; 0.71 0.29 1 0; 1.71 -0.71 0.12 1] % Answer

A homogeneous polynomial of degree k in x, y, z, t defines an *algebraic set of kth order*, $p_k(x,y,z,t) = 0$, in $\mathbb{R}P^3$. An *algebraic set of order* 2, $\{p_2(x,y,z,t) = 0\}$, is called a *projective quadric* in $\mathbb{R}P^3$. A projective quadric is *nondegenerate* if it can be represented by a nondegenerate quadric in the standard embedding space.

Lagrange theorem. A projective quadric has the form $\sum_i A_{ii}x_i^2 + \sum_{i<j} A_{ij}x_i x_j = 0$ ($1 \leq i, j \leq 4$), where not all coefficients are zero. By replacing the coordinate system, the following canonical expression of a quadric in KP^3 over a field with char $K \neq 2$ may be obtained: $\lambda_1 x_1^2 + \cdots + \lambda_r x_r^2 = 0$, where $r \leq 4$ is the *rank* of a quadratic form.

3.4 Exercises

Section 3.1

1. Let $a_{11}B - a_{12}A$ and $a_{21}B - a_{22}A$ be not both zero. Prove that the transformation f given by (3.1) maps the line $Ax + By + C = 0$ (A and B not both zero) to the line

$$(a_{22}A - a_{21}B)x + (a_{11}B - a_{12}A)y + (a_{12}f - d_1 a_{22})A$$
$$- (a_{11}d_2 - d_1 a_{21})B + (a_{11}a_{22} - a_{12}a_{21})C = 0.$$

2. Prove that the inverse of the affine transformation $f(\mathbf{x}) = \mathbf{x}A + \mathbf{b}$ is given by $f^{-1}(\mathbf{x}) = \mathbf{x}A^{-1} - \mathbf{b}A^{-1}$.

3. Given n points $A_1, \ldots, A_n$ in the plane, find n points $B_1, \ldots, B_n$ such that $A_1, \ldots, A_n$ are the midpoints, respectively, of $B_1 B_2, \ldots, B_n B_1$. Consider, in particular, the cases $n = 3, 4$.

4. On the three sides of $\triangle ABC$, put three points A', B', C' in such a way that $AC' = \frac{2}{3} AB$, $BA' = \frac{2}{3} BC$, $CB' = \frac{2}{3} CA$. The lines AA', BB', and CC' draw a small $\triangle A''B''C''$. Compute the ratio of the area of $\triangle A''B''C''$ to that of $\triangle ABC$.

5. Prove that all triangles are affine-congruent.

6. Find the inverse matrices for $\mathbf{S}_x(\lambda_x)$ and $\mathbf{S}_y(\lambda_y)$.

7. Under what conditions do the two systems of equations

$$a_i x + b_i y + c_i z = d_i, \quad i = 1, 2 \quad \text{and} \quad a_i' x + b_i' y + c_i' z = d_i', \quad i = 1, 2$$

describe (a) (parallel) lines in $\mathbb{R}^3$? (b) the same line in $\mathbb{R}^3$ (for $d_i = d_i' = 0$)?

8. Illustrate the fundamental theorem of 3-D affine geometry.

9. Prove that affine mappings preserve barycenters.

10. Let P be the barycenter of $(A, \alpha), (B, \beta), (C, \gamma)$, where $\alpha + \beta + \gamma \neq 0$ and $\beta + \gamma \neq 0$. Prove that the intersection point A' of AP and BC is the barycenter of $(B, \beta), (C, \gamma)$. As a consequence of this, in a triangle, the three medians (lines from the vertices to the midpoints of the opposite sides) are concurrent at the *centroid* of the triangle.

11. Let A, B, $C \in \mathbb{R}^2$ be non-collinear points. Prove that any point M of $\mathbb{R}^2$ is a barycenter of A, B, C for some masses α, β, γ. The system (α, β, γ) is called a system of *barycentric coordinates* of the point M in the frame A, B, C. What points correspond to nonnegative α, β, γ?

12. In Example 2(a), p. 137, let $A = (2, 4)$, $B = (-2, 0)$, $C = (1, 0)$ and the points $P = (\frac{5}{2}, 0)$, $Q = (\frac{3}{2}, 2)$, $R = (1, 3)$. Determine whether the points P, Q, R are collinear.

13. Prove that the image of a convex set by an affine mapping is convex.

14. Prove that
(a) any ellipse is affine-congruent to the unit circle $x^2 + y^2 = 1$;
(b) any parabola is affine-congruent to the parabola $y^2 = x$;
(c) any hyperbola is affine-congruent to the rectangular hyperbola $xy = 1$.

15. (Tangents to conics in standard form.) The equation of the tangent to a standard conic at the point $P(x_0, y_0)$ is

$$x x_0 + y y_0 = 1 \quad \text{(unit circle)}, \qquad x y_0 + y x_0 = 2 \quad \text{(rectangular hyperbola)},$$
$$2 y y_0 = x + x_0 \quad \text{(unit parabola)}.$$

16. An ellipse touches the sides AB, BC, and CA of $\triangle ABC$ at the points R, P, and Q, respectively. Prove that $\frac{AR}{RB} \cdot \frac{BP}{PC} \cdot \frac{CQ}{QA} = 1$ (and hence the lines AP, BQ, and CR are concurrent).

17. Write a program that uses the procedure convhulln and returns the set of layers of the finite planar set $\mathbf{P}$, shown in Figure 3.1(b).

18. Write a program to derive the *Lorentz transformation* (Example 1, p. 137).

19. Use **CH_my** to plot the set of layers of the planar point set.

Section 3.2

1. Prove that:

(a) A line in $\mathbb{R}P^2$ is an algebraic set of 1st order: $ax + by + cz = 0$, where a, b, $c \in \mathbb{R}$ and not all zero. If $\lambda \neq 0$ then $\lambda ax + \lambda by + \lambda cz = 0$ is also an equation of the same line.

(b) Any two distinct lines in $\mathbb{R}P^2$ intersect in a unique point.

(c) To determine whether or not three points $Q = (a : b : c)$, $R = (d : e : f)$, and $S = (g : h : i)$ are collinear (i.e., belong to a line), one should evaluate the determinant

$$\Delta = \begin{vmatrix} a & b & c \\ d & e & f \\ g & h & i \end{vmatrix};$$ the points are collinear if and only if $\Delta = 0$. Determine whether $Q = (1 : 2 : -3)$, $R = (1 : 2 : 0)$, and $S = (1 : 1 : 1)$ are collinear.

Hint:

    Q = [1 2 -3]; R = [1 2 0]; S = [1 1 1];
    det([Q; R; S])

2. A projective conic has an equation $Ax^2 + Bxy + Cy^2 + Dxz + Eyz + Fz^2 = 0$, where not all coefficients are zero; see p. 144. Show (by changing the coordinate system) that a projective conic over $\mathbb{R}$, $\mathbb{C}$ has an equation that may be brought into one of the canonical forms

$$
\begin{array}{llll}
x^2 + y^2 \pm z^2 = 0, & x^2 \pm y^2 = 0, & x^2 = 0 & (\mathbb{R}P^2), \\
x^2 + y^2 + z^2 = 0, & x^2 + y^2 = 0, & x^2 = 0 & (\mathbb{C}P^2).
\end{array}
$$

3. Show that the cross-ratio satisfies the identities of Exercise 8, p. 193.

4. Four distinct points are harmonic (see the definition of cross-ratio) if and only if $(A, B, C, D) = (B, A, C, D)$.

5. The line through the points A, $B \in \mathbb{R}P^2$ has dual coordinates $A \times B$. Dually, the lines $A \cdot X = 0$ and $B \cdot X = 0$ meet at $A \times B$.

6. Prove that

(a) Three points P, Q, and R are incident with the same line if and only if the determinant of their coordinates vanishes.

(b) Three lines l, m, and n are incident with the same point if and only if the determinant of their coordinates vanishes.

7. Any transversal of a pencil of four lines will meet the pencil in four collinear points, and their cross-ratio is the same as that of the pencil.

8. Give an alternative proof of Pappus' theorem (see Example 2, p. 149) from the properties of the cross-ratio.

9. The set of projective transformations of $\mathbb{R}P^2$ forms a group under the operation of composition of functions.

10. Collinearity and incidence are preserved by projective transformations.

11. Show that the projective transformation associated with the diagonal matrix $A = [2,3,4]$ cannot be a perspective transformation.

12. Prove that all quadrilaterals are projectively congruent.

13. What is the maximum number of points in $\mathbb{Z}_2 P^2$, no three of which are collinear? How many (nondegenerate) triangles, quadrilaterals are there in $\mathbb{Z}_2 P^2$?

Section 3.3 1. Show that the homogeneous *translation* matrix is $\mathbf{T_a} = \begin{pmatrix} 1 & 0 & 0 \\ 0 & 1 & 0 \\ a_x & a_y & 1 \end{pmatrix}$, where $\mathbf{a} = (a_x, a_y)$. Then $(x, y, 1)\mathbf{T_a} = (x + a_x, y + a_y, 1)$.

2. Show that in homogeneous coordinates the matrix for a *rotation* $\mathbf{R}(\theta)$ about the origin through an angle θ is $\mathbf{R}(\theta) = \begin{pmatrix} \cos\theta & \sin\theta & 0 \\ -\sin\theta & \cos\theta & 0 \\ 0 & 0 & 1 \end{pmatrix}$. Hence, $(x, y, 1)\mathbf{R}(\theta) = (x\cos\theta - y\sin\theta, x\sin\theta + y\cos\theta, 1)$.

3. Show that the homogeneous *scaling* matrix is $\mathbf{S}(s_x, s_y) = \begin{pmatrix} s_x & 0 & 0 \\ 0 & s_y & 0 \\ 0 & 0 & 1 \end{pmatrix}$. Then $(x, y, 1)\mathbf{S}(s_x, s_y) = (s_x x, s_y y, 1)$.

4. Show that the matrix for a rotation $\mathbf{R}_z(\theta)$ about the z-axis through an angle θ is $\mathbf{R}_z(\theta) = \begin{pmatrix} \cos\theta & \sin\theta & 0 & 0 \\ -\sin\theta & \cos\theta & 0 & 0 \\ 0 & 0 & 1 & 0 \\ 0 & 0 & 0 & 1 \end{pmatrix}$.

Hence, $(x, y, z, 1)\mathbf{R}_z(\theta) = (x\cos\theta - y\sin\theta, x\sin\theta + y\cos\theta, z, 1)$. The matrices for rotations $\mathbf{R}_x(\theta)$ and $\mathbf{R}_y(\theta)$ about the x- and y-axes, respectively, through the angle θ are $\mathbf{R}_x(\theta) = \begin{pmatrix} 1 & 0 & 0 & 0 \\ 0 & \cos\theta & \sin\theta & 0 \\ 0 & -\sin\theta & \cos\theta & 0 \\ 0 & 0 & 0 & 1 \end{pmatrix}$, $\mathbf{R}_y(\theta) = \begin{pmatrix} \cos\theta & 0 & -\sin\theta & 0 \\ 0 & 1 & 0 & 0 \\ \sin\theta & 0 & \cos\theta & 0 \\ 0 & 0 & 0 & 1 \end{pmatrix}$.

5. Show that the reflection in the plane $\omega_0: ax + by + cz = 0$ is given by

$$\mathbf{P}_{\omega_0} = I_4 - \begin{pmatrix} \frac{2}{a^2+b^2+c^2}\tilde{\mathbf{P}} & 0 \\ 0 & 0 \end{pmatrix}, \text{ where } \tilde{\mathbf{P}} = \begin{pmatrix} a^2 & ab & ac \\ ab & b^2 & bc \\ ac & bc & c^2 \end{pmatrix};$$

see (2.4).

Hint. Let $\omega: ax + by + cz = d$ be a plane with $c \neq 0$. Then a reflection in ω can be represented by the product $\mathbf{P}_\omega(\theta) = \mathbf{T}_{-Q}\mathbf{P}_{\omega_0}\mathbf{T}_Q$, where $Q = (0, 0, -d/c)$.

6. A rotation $\mathbf{R}_y(\theta)$ followed by a translation $\mathbf{T}_\mathbf{a}$ has the homogeneous transformation matrix

$$\begin{pmatrix} \cos\theta & 0 & \sin\theta & 0 \\ 0 & 1 & 0 & 0 \\ -\sin\theta & 0 & \cos\theta & 0 \\ 0 & 0 & 0 & 1 \end{pmatrix} \begin{pmatrix} 1 & 0 & 0 & 0 \\ 0 & 1 & 0 & 0 \\ 0 & 0 & 1 & 0 \\ a_x & a_y & a_z & 1 \end{pmatrix} = \begin{pmatrix} \cos\theta & 0 & \sin\theta & 0 \\ 0 & 1 & 0 & 0 \\ -\sin\theta & 0 & \cos\theta & 0 \\ a_x & a_y & a_z & 1 \end{pmatrix}.$$

7. Show that

(a) A plane in $\mathbb{R}P^3$ is an algebraic set of 1st order: $ax_1 + bx_2 + cx_3 + dx_4 = 0$, where $a, b, c, d \in \mathbb{R}$ and not all zero. If $\lambda \neq 0$ then $\lambda ax_1 + \lambda bx_2 + \lambda cx_3 + \lambda dx_4 = 0$ is also an equation of the plane.

(b) Any three distinct points in $\mathbb{R}P^3$ lie on a unique plane. An equation for the plane in $\mathbb{R}P^3$ through the points $Q_i = (a_i : b_i : c_i : d_i)$ $(i = 1, 2, 3)$ is the determinant

$$\begin{vmatrix} x_1 & x_2 & x_3 & x_4 \\ a_1 & b_1 & c_1 & d_1 \\ a_2 & b_2 & c_2 & d_2 \\ a_3 & b_3 & c_3 & d_3 \end{vmatrix} = 0.$$

(c) Any three distinct planes in $\mathbb{R}P^3$ intersect. When is the intersection a point, a line?

(d) To determine whether or not four points $Q_i = (a_i : b_i : c_i : d_i)$ $(i = 1, 2, 3, 4)$ are coplanar (i.e., belong to a plane) one should evaluate the determinant $\Delta = \begin{vmatrix} a_1 & b_1 & c_1 & d_1 \\ a_2 & b_2 & c_2 & d_2 \\ a_3 & b_3 & c_3 & d_3 \\ a_4 & b_4 & c_4 & d_4 \end{vmatrix}$; the points are coplanar if and only if $\Delta = 0$.

8. Quadrics in a projective 3-space are surfaces $\sum_{i,j \leq 4} a_{ij} x_i x_j = 0$, where not all coefficients are zero; see p. 154. Show (by changing the coordinate system) that a projective quadric over $\mathbb{K} = \mathbb{R}, \mathbb{C}$ has one of the following canonical forms:

$$\begin{aligned} x_1^2 + x_2^2 \pm x_3^2 \pm x_4^2 = 0, \quad & x_1^2 + x_2^2 \pm x_3^2 = 0, \quad x_1^2 \pm x_2^2 = 0, \quad x_1^2 = 0 \quad (\mathbb{R}P^3), \\ x_1^2 + x_2^2 + x_3^2 + x_4^2 = 0, \quad & x_1^2 + x_2^2 + x_3^2 = 0, \quad x_1^2 + x_2^2 = 0, \quad x_1^2 = 0 \quad (\mathbb{C}P^3). \end{aligned}$$

4

Möbius Transformations

In Section 4.1 we review reflections in a sphere and the inversive geometry of a plane. Sections 4.2 and 4.3 are devoted to Möbius transformations and their applications in spherical and hyperbolic geometries. In Section 4.4 we develop several MATLAB® procedures to solve problems and visualize solutions in a half-plane and half-space (the Poincaré model).

4.1 Inversive Geometry

In this section we introduce an *inversion*, which is a generalization of the reflection in a plane (or a line for the two-dimensional case); see Section 2.2.

Just as a reflection in a plane (line) maps points on one side of it to points on the other, so inversion in a sphere (circle) maps points inside of it to points outside, and vice versa.

4.1.1 *Reflection in a sphere*

The sphere of radius r centered at Q in $\mathbb{R}^3$ is the set (see Section 2.3)

$$S^2(Q,r) = \{\mathbf{x} \in \mathbb{R}^3 : \|\mathbf{x} - Q\| = r\}.$$

The *reflection* (or *inversion*) of $\mathbb{R}^3$ in the sphere $S^2(Q,r)$ is the transformation $\mathbf{P}_{Q,r}$ of the space that maps any point M onto the point M' on the same ray QM such that the product of the distances QM and QM' is equal to r^2, which is called the *degree of the inversion*. In other words,

V. Rovenski, *Modeling of Curves and Surfaces with MATLAB®*,
Springer Undergraduate Texts in Mathematics and Technology 7, DOI 10.1007/978-0-387-71278-9_4,
© Springer Science+Business Media, LLC 2010

$$\mathbf{P}_{Q,r}(\mathbf{x}) = Q + \frac{r^2}{\|\mathbf{x} - Q\|^2}(\mathbf{x} - Q). \tag{4.1}$$

Let $f(\mathbf{x}) = Q + r\mathbf{x}$ $(r > 0)$. Observe that $\mathbf{P}_{Q,r} = f \circ \mathbf{P}_{O,1} \circ f^{-1}$:

$$\mathbf{P}_{Q,r}(\mathbf{x}) = f \circ \mathbf{P}_{O,1}\big((\mathbf{x} - Q)/r\big) = f \circ \mathbf{P}_{O,1} \circ f^{-1}(\mathbf{x}).$$

We prepare an M-file to derive the reflection in the unit sphere.

```
function P = refl3(A)                          % works for any A ∈ ℝⁿ \ {0}
    P = A / norm(A, 2)^2;
end
```

For example, let $A = (1,3,4) \in \mathbb{R}^3$:

```
A = [1 3 4];   refl3(A)                        % Answer: 0.0385  0.1154  0.1538
```

The *angle between two curves at a point of intersection* is defined as the angle between the tangent lines at that point (see Section 6.1). We will show that *a reflection in any sphere preserves the magnitude of angles between curves but reverses their orientation: the angle between two curves is equal to the angle between the images of these two curves under inversion.*

Let $U \subset \mathbb{R}^3$ be an open subset, $f: U \to \mathbb{R}^3$ be a differentiable map, and $df(\mathbf{x}) = \{\frac{\partial f_i}{\partial x_j}(\mathbf{x})\}$ be the matrix of partial derivatives of f. The function f is said to be *conformal* if and only if there is a function $\mu: U \to \mathbb{R}_+$, called the *scale factor* of f, such that $\mu^{-1}(\mathbf{x})\,df(\mathbf{x})$ is an orthogonal matrix for each $\mathbf{x} \in U$.

Note that the scale factor μ of a conformal function f is uniquely determined by f, since $\mu^3(\mathbf{x}) = \det|df(\mathbf{x})|$.

Theorem 4.1. *Let U be an open subset of $\mathbb{R}^3$ and let $f: U \to \mathbb{R}^3$ be a differentiable map. Then f is conformal if and only if f preserves angles between differentiable curves in U.*

Proposition 4.1. *Every inversion of $\mathbb{R}^3$ in a sphere is conformal and reverses orientation.*

Proof. Let $\mathbf{P}_{O,1}(\mathbf{x}) = \mathbf{x}/\|\mathbf{x}\|^2$ be the reflection in the unit sphere $S^2(O,1)$. We have $\frac{\partial}{\partial x_j}(x_i/\|\mathbf{x}\|^2) = \delta_{ij}/\|\mathbf{x}\|^2 - 2x_i x_j/\|\mathbf{x}\|^4$. Hence,

$$d\mathbf{P}_{O,1}(\mathbf{x}) = \{\delta_{ij}/\|\mathbf{x}\|^2 - 2x_i x_j/\|\mathbf{x}\|^4\} = (\mathrm{Id} - 2B)/\|\mathbf{x}\|^2,$$

where B is the matrix $\{x_i x_j/\|\mathbf{x}\|^2\}$. One may show that $\mathrm{Id} - 2B$ is orthogonal, and so $\mathbf{P}_{O,1}$ is conformal. Let $\mathbf{a}_i$ be the rows of the matrix $\mathrm{Id} - 2B$. Then

$$\mathbf{a}_i \cdot \mathbf{a}_i = (1 - 2x_i^2)^2 + \sum_{k \neq i}(-2x_ix_k)^2 = (1 - 2x_i^2)^2 + 4x_i^2(1 - x_i^2) = 1,$$

$$\mathbf{a}_i \cdot \mathbf{a}_j = (1 - 2x_i^2)(-2x_ix_j) + (1 - 2x_i^2)(-2x_ix_j) + \sum_{k \neq i,j}(-2x_ix_k)(-2x_jx_k)$$

$$= 4x_ix_j(1 - x_i^2x_j^2) - 4x_ix_j(1 - x_i^2x_j^2) = 0.$$

Since $|d\mathbf{P}_{O,1}(\mathbf{x})| = -1/\|\mathbf{x}\|^6 < 0$, $\mathbf{P}_{O,1}$ reverses orientation.   □

The following program verifies the claim:

```
syms x y z  real
A = [x^2 x*y x*z; x*y y^2 y*z; x*z y*z z^2];
Q = eye(3) - 2*A/(x^2 + y^2 + z^2)
simplify(Q'*Q)                                    % identity matrix
det(Q)                         % Answer: −1 (reverses orientation).
```

4.1.2 *Inversive geometry of a plane*

Note that the restriction of $\mathbf{P}_{Q,r}$ to a plane α through Q is the *reflection in the circle* $S^1(Q,r) = S^2(Q,r) \cap \alpha$ in this plane.

We briefly consider the two-dimensional (2-D) case (the 3-D case is similar). *Inversion* (symmetry) with respect to the circle of radius R with center O is the transformation of the plane that maps any point M onto the point M' on the same ray OM, such that the product of distances OM and OM' is equal to R^2, the *degree of the inversion*.

Example. Let $\gamma \colon \rho = f(\varphi)$ be a curve in $\mathbb{R}^2$ given in polar coordinates. Then the inverse curve with respect to the circle $\omega = S^1(O,r)$ of radius r has the equation $\gamma' \colon \rho = r^2/f(\varphi)$. There are different mechanisms (*inversors*) that help us avoid long calculations when plotting the image under inversion of a given plane curve.

The reflections of roses $\rho = R\cos(k\varphi)$ with respect to the base circle of radius R are the *ear* curves $\rho = \frac{R}{\cos(k\varphi)}$ outside of this circle (verify!). Show visually that the inversion of the *three-leafed rose* $\rho = R\cos(3\varphi)$ (see Section 1.4) is the *trisectrix of Longchamps* $\rho = \frac{R}{\cos(3\varphi)}$.

```
syms t;
a = 1;  e = .07;  f = a/cos(3*t);
for i = 0 : 2;
   ezpolar(f, [(2*i - 1)*pi/6 + e, (2*i + 1)*pi/6 - e]);
   hold on;  axis equal;
end                                               % Figure 4.1(b)
```

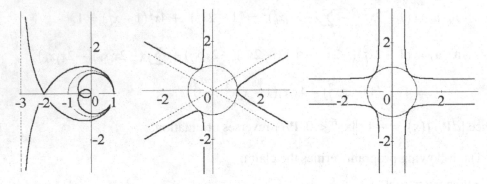

Fig. 4.1 Trisectrices of Maclaurin and Longchamps, cross-shaped curve.

We will imagine the plane as the union of $\mathbb{R}^2 \approx \mathbb{C}$ and one extra point, the *point at infinity*, denoted by ∞. The symbol ∞ does not represent a complex number, so it should not be used in association with arithmetic operations that act on complex numbers.

The *extended complex plane* is the union $\hat{\mathbb{C}} = \mathbb{C} \cup \{\infty\}$. A line $l \in \mathbb{C}$ may be extended to $\hat{l} = l \cup \{\infty\}$; such a set is called an *extended line*. A *generalized circle* in $\hat{\mathbb{C}}$ is either a circle or an extended line.

Example. The *reciprocal function* is defined by $\mathbf{T}(z) = 1/z$. This function may be expanded into the composite $\mathbf{T}_2 \circ \mathbf{T}_1$, where $\mathbf{T}_1$ is the inversion $z \to 1/\bar{z}$ and $\mathbf{T}_2$ is the conjugation $z \to \bar{z}$.

The next theorem (one may reformulate it for space of any dimension) says that an inversion maps a line or a circle again onto a line or a circle.

Theorem 4.2. *Under inversion in a circle* $\omega = S^1(O, r)$:

(a) *a* line *that does not pass through O maps onto a* circle *punctured at O;*
(b) *a* line *that is punctured at O maps onto the same* line;
(c) *a* circle *that does not pass through O maps onto a* circle;
(d) *a* circle *punctured at O maps onto a* line *that does not pass through O.*

Proof. To determine an equation for the *image of a curve under inversion* in the unit circle $\omega = S^1(O, 1)$:

(i) write down an equation of the curve lying in the xy-plane;
(ii) replace x by $\frac{x}{x^2+y^2}$, y by $\frac{y}{x^2+y^2}$, and simplify the resulting equation.

(a) Let a line be given by $l\colon ax + by + c = 0$, where $a^2 + b^2 \neq 0$ and $c \neq 0$; see (i). Applying (ii), we see that $\mathbf{T}(l)$ is a circle given by the equation

$$a\frac{x}{x^2+y^2} + b\frac{y}{x^2+y^2} + c = 0 \iff ax + by + c(x^2+y^2) = 0.$$

The proof of (b)–(d) is similar.   □

```
syms a b c  x y  X Y;              % use this code for (a) to show the cases (b)–(d)
  rr = x^2 + y^2;
  eq = simplify(rr*subs(a*X + b*Y + c, [X Y], [x/rr y/rr]))
```

One can obviously extend the notions above to the extended complex plane. Hence, Theorem 4.2 says that *inversions of the extended plane map generalized circles onto generalized circles.*

Each geometry is used to study those properties of figures in its space that are preserved by its transformations. For example, Euclidean geometry is used to study those properties of figures in $\mathbb{R}^2$ ($\mathbb{R}^n$), such as angle and distance, that are preserved by the isometries of $\mathbb{R}^2$. Since each isometry can be decomposed into a composite of reflections, we think of the group associated with Euclidean geometry as the group of all possible composites of reflections.

A transformation $\mathbf{T}\colon \hat{\mathbb{C}} \to \hat{\mathbb{C}}$ is an *inversive transformation* if it can be expressed as a composite of (a finite number of) inversions. *Inversive geometry* studies those properties of figures (circles, lines, angles, etc.) in $\hat{\mathbb{C}}$ that are preserved by inversive transformations.

One may notice that the affine group of transformations contains the Euclidean group but overlaps the inversive group. Because the Euclidean group is smaller than the inversive group, it follows that Euclidean geometry has more properties than does inversive geometry.

Example. There is a nice interpretation of the stereographic projection in terms of inversive geometry. Let $\mathbf{T}$ be the inversion of $\mathbb{R}^3$ in the sphere $S^2(\mathbf{e}_3, \sqrt{2})$. Then $\mathbf{T}(\mathbf{x}) = \mathbf{e}_3 + \frac{2(\mathbf{x}-\mathbf{e}_3)}{\|\mathbf{x}-\mathbf{e}_3\|^2}$. If $\mathbf{x} \in \mathbb{R}^2$, then

$$\mathbf{T}(\mathbf{x}) = \mathbf{e}_3 + \frac{2}{1+\|\mathbf{x}\|^2}(x_1, x_2, -1) = \left(\frac{2x_1}{1+\|\mathbf{x}\|^2}, \frac{2x_2}{1+\|\mathbf{x}\|^2}, \frac{\|\mathbf{x}\|^2-1}{\|\mathbf{x}\|^2+1}\right).$$

Thus, the restriction of $\mathbf{T}$ to $\mathbb{R}^2$ is the stereographic projection. From this and Theorem 4.1 it follows that *stereographic projection preserves the magnitude of angles (between differentiable curves)*.

4.2   Möbius Transformations

4.2.1   *Möbius transformations in two dimensions*

A *Möbius transformation* is a linear-fractional function $f(z) = \frac{az+b}{cz+d}$ ($z \in \hat{\mathbb{C}}$) depending on four parameters $a, b, c, d \in \mathbb{C}$, satisfying the condition $ad - bc \neq 0$. It is easy to see that $f(z) - f(w) = \frac{(ad-bc)(z-w)}{(cz+d)(cw+d)}$; hence, f is injective. The domain of this function is $\hat{\mathbb{C}} = \mathbb{C} \cup \{\infty\}$. If $c = 0$, f is simply a linear function. One may extend the definition of f when $c \neq 0$:

$$f(-d/c) = \lim_{z \to -d/c} f(z) = \infty, \qquad f(\infty) = \lim_{z \to \infty} f(z) = a/c. \qquad (4.2)$$

By Theorem 4.1, *Möbius transformations preserve the magnitude and orientation of angles, and map generalized circles onto generalized circles.*

Every inversion of $\mathbb{C}$ has the form $F(z) = f(\bar{z})$, where f is a Möbius transformation (verify!).

Theorem 4.3. *Every inversive transformation of $\mathbb{C}$ can be represented in $\hat{\mathbb{C}}$ by one of the formulae*

$$f(z) = \frac{az+b}{cz+d} \quad (direct) \quad or \quad f(z) = \frac{a\bar{z}+b}{c\bar{z}+d} \quad (indirect), \qquad (4.3)$$

where $a, b, c, d \in \mathbb{C}$ and $ad - bc \neq 0$.

We usually represent a Möbius transformation by a 2×2 matrix.

Let $f(z) = \frac{az+b}{cz+d}$, where $a, b, c, d \in \mathbb{C}$. Then $A = \begin{pmatrix} a & b \\ c & d \end{pmatrix}$ is the matrix *associated* with f. We will write $f = f_A$. We say that two matrices are *equivalent*, $A \equiv A'$, if $f_A = f_{A'}$ (as functions).

It is easy to show that *the composition of Möbius transformations is related to matrix multiplication by the following result:*

$$f_{A_1 A_2}(z) = f_{A_1}(f_{A_2}(z)).$$

The Möbius transformation related to the identity matrix $I = \begin{pmatrix} 1 & 0 \\ 0 & 1 \end{pmatrix}$ is just the identity function $f_I(z) = \frac{z+0}{0z+1} = z$. Hence, $f_A(f_{A^{-1}}(z)) = f_I(z) = z$. Thus $f_A^{-1} = f_{A^{-1}}$. So, f_A^{-1} is represented by

$$\begin{pmatrix} d & -b \\ -c & a \end{pmatrix} \equiv \frac{1}{ad-bc} \begin{pmatrix} d & -b \\ -c & a \end{pmatrix} = A^{-1}.$$

Strategy. To compose two Möbius transformations f_1 and f_2:

  (i)   write down the matrices A_1 and A_2 associated with f_1 and f_2;
  (ii) calculate $A_1 A_2$;
  (iii) write down the Möbius transformation $f_1 \circ f_2$ with which $A_1 A_2$ is associated.

An important tool when working with Möbius transformations is the cross-ratio. We will see that it is invariant under linear-fractional functions.

Let z_1, z_2, z_3, z_4 be elements (numbers) of $\mathbb{C} \cup \{\infty\}$, at least three of which are distinct. We define the *cross-ratio* of z_1, z_2, z_3, z_4 to be $[z_1, z_2, z_3, z_4] = \frac{(z_1-z_3)(z_2-z_4)}{(z_1-z_2)(z_3-z_4)}$. The value of the cross-ratio when, say, $z_j = \infty$ is the limiting value of $[z_1, z_2, z_3, z_4]$ as z_j tends to ∞.

We are free to permute the indices, but all possible choices of cross-ratio are closely related. We prepare an M-file to derive the cross-ratio.

```
function w = prod4(z1, z2, z3, z4)
        if z4 == inf;  w = simplify((z1-z3)/(z1-z2));
    else if z3 == inf;  w = simplify(-(z2-z4)/(z1-z2));
    else if z2 == inf;  w = simplify(-(z1-z3)/(z3-z4));
    else if z1 == inf;  w = simplify((z2-z4)/(z3-z4));
    else   w = (z1-z3)*(z2-z4)/(z1-z2)/(z3-z4);
end
```

The unique Möbius transformation with the property that $f(z_1) = 1$, $f(z_2) = 0$, and $f(z_3) = \infty$ is $f(z) = [z, z_1, z_2, z_3]$. This observation can explain the definition of the cross-ratio.

Examples.

1. Find the Möbius transformation which sends $1, -i, -1$ to $1, 0, \infty$.

Solution. We obtain $w = [1, -i, -1, z] = \frac{2z+2i}{(1+i)(z+1)} \Leftrightarrow w = \begin{pmatrix} 2 & 2i \\ 1+i & 1+i \end{pmatrix} z$.

Alternatively, we use the M-file **prod4.m**:

```
syms z w;
Z = prod4(1, -i, -1, z);    W = prod4(1, 0, inf, w);          % enter data
solve(Z - W, w)                                                % compare answers.
```

2. Show that any Möbius transformation of the form $f(z) = k\dfrac{z - z_1}{z - z_3}$ maps z_1 to 0 and z_3 to ∞. We can choose k so that $f(z_2) = 1$. Hence, $k = \dfrac{z_2 - z_3}{z_2 - z_1}$.

3. Similarly to affine geometry, we obtain the **fundamental theorem of inversive geometry**, mapping via auxiliary points 0, 1, and ∞:

Let z_1, z_2, z_3 and w_1, w_2, w_3 be triples of distinct points in $\hat{\mathbb{C}}$. Then there is a unique Möbius transformation f with $f(z_j) = w_j$ for $j = 1, 2, 3$.

We find this f using the M-file **prod4.m**:

```
syms z1 z2 z3 w1 w2 w3 z f
Eq = prod4(z1, z2, z3, z) - prod4(w1, w2, w3, f)
solve(Eq, f)                                          % obtain expression for f.
```

Hence, if a Möbius transformation has *three fixed points*, then it is the identity map.

A *fixed point* of a Möbius transformation $f(z) = \frac{az+b}{cz+d}$, $ad \neq bc$, is a point z for which $f(z) = z$, i.e., $cz^2 + z(d - a) - b = 0$. Since a quadratic equation can have at most two roots, it follows that a non-trivial Möbius transformation can have at most two fixed points.

If a Möbius transformation $f \colon \hat{\mathbb{C}} \to \hat{\mathbb{C}}$ has exactly

(a) *two fixed points*, then it is conjugate to a map $z \to az$, where $a \neq 0$;
(b) *one fixed point*, then it is conjugate to $z \to z + 1$.

The points $z_1, \ldots, z_n$ are called *concyclic* if they lie on some circle or line in $\mathbb{C}$.

Strategy. To determine the image of a generalized circle ω under an inversive transformation f:

(i)   take three points z_1, z_2, z_3 on $\mathbb{C}$;
(ii) determine the images $f(z_1), f(z_2), f(z_3)$;
(iii) the image $f(\omega)$ is the generalized circle through $f(z_1), f(z_2), f(z_3)$.

Example. (Apollonian circle.) Let $A \neq B$ be two points in the plane, and let $k \neq 1$ be a positive real number. Then the locus of points P that satisfy $PA : PB = k : 1$ is a circle (called the *Apollonian circle*) whose center lies on the line through A and B. *Hint.* Two proofs are possible. The first uses coordinates:

```
syms a K x y real;                                          % K = k²
P = [x y];  A = [-a 0];                                     % B = −A
S = simplify( (P - A)*(P - A)' - K*(P + A)*(P + A)' )
Eq =collect(simplify( S/(1 - K)), [x y])                    % equation of a circle
```

$$x^2 + y^2 + 2ax(1 + K)/(1 - K) + a^2 \qquad \text{\% Answer.}$$

The second uses inversive geometry (setting up a one-to-one correspondence between the family of Apollonian circles and a family of concentric circles). Plot Apollonian circles for several values of k.

```
syms x y  real;
Eq = @(k) x^2 + y^2 + (4 - 4*k)/(1 - k) + x*(4*k + 4)/(1 - k)
axis equal;  hold on;    for j = 1 : 5,   j = j + .001;
  ezplot(Eq(j), [-12, 12, -7, 7]);  ezplot(Eq(1/j), [-12, 12, -7, 7]);
end;                                                        % see Figure 4.2(1).
```

The correspondence in the example above enables us to characterize *Apollonian families of circles* in terms of inversive transformations.

Theorem 4.4. *Let A, B be two distinct points in the plane, and let f be inversion in the circle of unit radius centered at A. Then the Apollonian family of circles defined by A, B is mapped by f to the family of all concentric circles centered at f(B).*

A *coaxal family of circles* in the plane is a family of (generalized) circles of one of the following types (see Figure 4.2):

(1) an Apollonian family (in which no two circles intersect);
(2) a family that intersect at one particular point;
(3) a family that intersect at two particular points.

The extended line in each family is called the *radical axis* of the family.

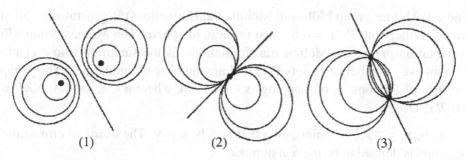

         (1)   (2)   (3)

Fig. 4.2 Coaxal families of circles.

Proposition 4.2.

(a) *Let $A \neq B$ be two points in the plane. Let F be the Apollonian family defined by A, B, and let $F^{\perp}$ be the family of all generalized circles through A, B. Then every member of F is orthogonal to every member of $F^{\perp}$.*

(b) *Points A, $B \in \hat{\mathbb{C}}$ are inverse points with respect to a generalized circle ω if and only if every generalized circle through A, B meets ω at right angle.*

(c) *Let F be the family of all generalized circles that have A, B as inverse points. Then F is either a concentric family of circles centered at A or B, or an Apollonian family of circles with point circles A and B.*

Obviously, two circles in a coaxal family of type (2) or (3) determine the whole family. The next proposition shows that the same is true for an Apollonian family:

(a) Let ω_1 and ω_2 be two disjoint circles in the plane. Then there is a Möbius transformation that maps ω_1, ω_2 onto a pair of concentric circles.

(b) Let ω_1 and ω_2 be two non-concentric circles in the plane. Then there is a unique Apollonian family of circles that contains ω_1 and ω_2.

4.2.2 *Möbius transformations in three dimensions*

Denote by $\hat{\mathbb{R}}^3 = \mathbb{R}^3 \cup \{\infty\}$ the one-point compactification of the space; see Section 2.3. A plane α may be extended by $\hat{\alpha} = \alpha \cup \{\infty\}$. A *generalized sphere* in $\hat{\mathbb{R}}^3$ is either a sphere or an extended plane. A Möbius transformation of $\hat{\mathbb{R}}^3$ is a finite composition of reflections of $\hat{\mathbb{R}}^3$ in generalized spheres.

In order to simplify notation, we shall not use a hat to denote the extension of a map to $\hat{\mathbb{R}}^3$.

The full Möbius group Möb$_3$ (all Möbius transformations) is generated by all similarities together with $\mathbf{P}_{O,1}$ = reflection in the unit sphere. The Möbius group Möb$_3^+$ (all orientation-preserving Möbius transformations) is the subgroup whose elements contain an even number of factors $\mathbf{P}_{O,1}$ and any number of similarities. The *group of similarities* of $\mathbb{R}^3$ consists of mappings $\mathbf{x} \to \mathbf{x}A + \mathbf{a}$, where $\mathbf{a} \in \mathbb{R}^3$ and $A = \lambda B$ with $\lambda > 0, B \in O(n)$.

Let $\mathbf{a}, \mathbf{b}, \mathbf{x}, \mathbf{y} \in \mathbb{R}^3$ be points such that $\mathbf{a} \neq \mathbf{b}, \mathbf{x} \neq \mathbf{y}$. The *absolute cross-ratio* of these points is defined to be the real number

$$[\mathbf{a}, \mathbf{b}, \mathbf{x}, \mathbf{y}] := \frac{d(\mathbf{a}, \mathbf{x})\, d(\mathbf{b}, \mathbf{y})}{d(\mathbf{a}, \mathbf{b})\, d(\mathbf{x}, \mathbf{y})}.$$

The absolute cross-ratio is a continuous function of four variables, since the distance $d \colon \hat{\mathbb{R}}^3 \times \hat{\mathbb{R}}^3 \to \mathbb{R}$ is a continuous function (verify!).

Theorem 4.5. *A function* $f \colon \mathbb{R}^3 \to \mathbb{R}^3$ *belongs to* Möb$_3$ *if and only if it preserves absolute cross-ratios.*

Proof. Indeed, $d(\mathbf{x}, \mathbf{y}) = \|\mathbf{x} - \mathbf{y}\|$. We may assume that f is a reflection in $S^2(Q, r)$. We have $\|f(\mathbf{x}) - f(\mathbf{y})\| = \|\mathbf{x} - \mathbf{y}\| / (\|\mathbf{x}\| \cdot \|\mathbf{y}\|)$ (see exercises). From this the claim follows.   □

Example. To illustrate Theorem 4.5, prepare an M-file. (It can be easily extended for the ∞ cases similarly to the way this was done in **prod4.m**.)

```
function f = prod43(A1, A2, A3, A4)              % the absolute cross-ratio
    f = norm(A1 - A3, 2)*norm(A2 - A4, 2)/norm(A1 - A2, 2)/norm(A3 - A4, 2);
end
```

For instance,

```
A1 = [1 3 4];  A2 = [0 1 0];  A3 = [1 0 0];  A4 = [0 0 1];
prod43(A1, A2, A3, A4)              % absolute cross-ratio of A1,A2,A3,A4
1.0911                                           % Answer.
```

Then we verify the property using the M-file **refl3** (Section 4.1):

prod43(refl3(A1), refl3(A2), refl3(A3), refl3(A4)) - prod43(A1, A2, A3, A4)
2.2204e-016   % Answer: 0.

4.2.3 *The sphere and Möbius transformations*

Stereographic projection enables us to interpret the theory of coaxal circles on the sphere S^2. We now study the correspondence between rotations of S^2 and Möbius transformations of $\mathbb{C}$.

Proposition 4.3. *Two points z, w map to diametrically opposite points on the sphere S^2 under the stereographic projection π if and only if $w = -1/\bar{z}$.*

Proof. The points $\pi(z)$ and $\pi(w)$ are diametrically opposite on S^2 if and only if $\pi(z) = -\pi(w)$. One may check from (2.20) that this is so if and only if $w = -1/\bar{z} = -z/|z|^2$. One may show this using the M-file **ster.m**.

```
syms x1 x2 y1 y2 real;
Eq = ster([x1 x2]) + ster([y1 y2]);
S = solve(Eq(1), Eq(2), Eq(3), y1, y2)
S.y1, S.y2              % Answer: −x1/(x2² + x1²),   −x2/(x2² + x1²)
z = x1 + i*x2;  w = S.y1 + i*S.y2;
simplify(w + 1/conj(z))            % Answer: 0. Hence w + 1/z̄ = 0.
```

$\hspace{11cm}\square$

Every map $f \colon \hat{\mathbb{C}} \to \hat{\mathbb{C}}$ corresponds to a map $\tilde{f} \colon S^2 \to S^2$ defined by $\tilde{f} = \pi \circ f \circ \pi^{-1}$, and vice versa.

Let $\mathbf{R}$ be a rotation of S^2. The transformation $\pi^{-1}\mathbf{R}\pi$ of $\hat{\mathbb{C}}$ is called the *conjugate transformation* of the rotation $\mathbf{R}$. One may show that

Every Möbius transformation $\mathbf{T}$ of the form

$$\mathbf{T}(z) = \frac{az + b}{-\bar{b}z + \bar{a}}, \quad \text{where } a, b \in \mathbb{C} \text{ and } |a|^2 + |b|^2 = 1, \qquad (4.4)$$

is conjugate to a rotation of S^2, and every rotation of S^2 arises in this way.

The Möbius transformations of the form (4.4) form a group, which we denote by Möb_2^0. The same argument leads to the following result:

The set $U(2) = \left\{ \begin{pmatrix} a & b \\ -\bar{b} & \bar{a} \end{pmatrix} : |a|^2 + |b|^2 = 1 \right\}$ of 2×2 complex matrices is a group, called the *unitary group*.

Example. To show the above claim, we verify that the product $L = M \cdot N$ of matrices keeps the property.

```
syms a b c d;
M = [a b;  -conj(b) conj(a)];  N = [c d;  -conj(d) conj(c)];
L = simple(expand(M*N))                          % product of matrices
```

$[ac - b\,\text{conj}(d),\ ad + b\,\text{conj}(c);$
$-\text{conj}(b)c - \text{conj}(ad),\ \text{conj}(ac) - \text{conj}(b)d]$ % Answer

```
r = simple(expand(L(2, 1) + conj(L(1, 2))))      % Answer: 0, i.e., L21 = -L12
s = simple(expand(L(2, 2) - conj(L(1, 1))))      % Answer: 0, i.e., L22 = L11
t = expand(L(1,1)*conj(L(1,1)) + L(1,2)*conj(L(1,2)))
t2 = expand(simple(subs(t, a*conj(a), 1 - b*conj(b))))
simple(subs(t2,  abs(b)^2, 1-abs(a)^2))          % hence |L11|^2 + |L12|^2 = 1
```

$\text{abs}(c)^2 + \text{abs}(d)^2$ % we assume $|a|^2 + |b|^2 = |c|^2 + |d|^2 = 1$.

4.3 Hyperbolic Geometry

Non-Euclidean geometry has had great methodological importance in the development of mathematics. The Cayley–Klein (in a disk or a ball) and Poincaré (in a half-plane or a half-space) models of hyperbolic geometry are studied in the foundations of geometry and have various applications.

The geometry induced on the half-space $\mathbb{R}_+^n = \{\mathbf{x} \in \mathbb{R}^n : x_n > 0\}$ by the *Poincaré metric* $ds^2 = \frac{k^2}{x_n^2}(dx_1^2 + \cdots + dx_n^2)$, see (4.7), is the *hyperbolic geometry* of Lobachevsky.

In this section we study the distance, the volume (area), and the hyperbolic rigid motions (isometries) of the Poincaré model.

4.3.1 *The hyperbolic length*

First, recall necessary facts about metric spaces and path metric spaces.

A *metric* (distance) on a set X is a function $\rho: X \times X \to \mathbb{R}$ satisfying three conditions:

(1) $\rho(x,y) \geq 0$ for all $x, y \in X$, and $\rho(x,y) = 0$ if and only if $x = y$;
(2) $\rho(x,y) = \rho(y,x)$ for all $x, y \in X$;
(3) $\rho(x,z) \leq \rho(x,y) + \rho(y,z)$ for all $x, y, z \in X$ (the *triangle inequality*).

If ρ is a metric on X, we often refer to the *metric space* (X, ρ).

An *isometry* of a metric space (X, ρ) is a homeomorphism f of X that preserves distance: $\rho(x,y) = \rho(f(x), f(y))$ for all $x, y \in X$.

In a metric space (X, ρ), we define the *open disc* of radius $\varepsilon > 0$ centered at a point x to be $U_\varepsilon(x) = \{y \in X : \rho(x, y) < \varepsilon\}$. A subset A of X is *open* if for every $x \in A$ there exists an $\varepsilon > 0$ such that $U_\varepsilon(x) \subset A$. A subset B of X is *closed* if its complement $X \setminus B$ is open.

A sequence $\{x_n\}$ of points of X *converges to a point x of X* if for every $\varepsilon > 0$ there exists some $N > 0$ such that $x_n \in U_\varepsilon(x)$ for all $n > N$.

If (X, ρ_X) and (Y, ρ_Y) are two metric spaces and if $f: X \to Y$ is a function, then f is *continuous at a point* $x \in X$ if, given $\varepsilon > 0$, there exists $\delta > 0$ such that $f(U_\delta(x)) \subset U_\varepsilon(f(x))$. We say that f is *continuous* if it is continuous at every point of X. For example, the function $f_z: X \to \mathbb{R}$ given by $f_z(x) = \rho(z, x)$ is continuous.

Consider the relations between the length of a path and distance.

A path $\varphi: [a, b] \to \mathbb{R}^n$ is *piecewise C^1* if φ is continuous and there is a partition of $[a, b]$ into subintervals $[a = a_0, a_1], [a_1, a_2], \ldots, [a_m, a_{m+1} = b]$ such that φ is a C^1 path when restricted to each subinterval $[a_k, a_{k+1}]$. In coordinates, we can write $\varphi(t) = (x_1(t), \ldots, x_n(t))$, where $x_i(t)$ have piecewise continuous derivative on (a, b). The image of each subinterval (a_k, a_{k+1}) under φ is a C^1 curve in $\mathbb{R}^n$.

Unless otherwise stated, we assume that all paths are piecewise C^1.

The *Euclidean length* of φ is given by (see also Section 6.3)

$$l(\varphi) = \int_a^b dl = \int_a^b \sqrt{(x_1'(t))^2 + \cdots + (x_n'(t))^2} \, dt, \tag{4.5}$$

and $dl = \sqrt{(x_1'(t))^2 + \cdots + (x_n'(t))^2} \, dt$ is the *element of arc length* in $\mathbb{R}^n$.

Let $\mu > 0$ be a continuous function on a domain $U \subset \mathbb{R}^n$. The element of arc length $dl_\mu = \mu(\mathbf{x}) \, dl$ on U is a conformal distortion of the Euclidean element of arc length dl, for which the length of a piecewise C^1 path $\varphi: [a, b] \to U$ is given by the integral

$$l_\mu(\varphi) = \int_a^b dl_\mu = \int_a^b \mu(\varphi(t)) \sqrt{(x_1'(t))^2 + \cdots + (x_n'(t))^2} \, dt. \tag{4.6}$$

In particular, $\mu(\mathbf{x}) = 1/x_n$ is a continuous function on $\mathbb{R}_+^n = \{\mathbf{x} \in \mathbb{R}^n : x_n > 0\}$. The *hyperbolic length* of a piecewise C^1 path $\varphi: [a, b] \to \mathbb{R}_+^n$ is

$$l_H(\varphi) = \int_a^b (1/x_n(t)) \sqrt{(x_1'(t))^2 + \cdots + (x_n'(t))^2} \, dt. \tag{4.7}$$

Examples.

1. The Euclidean distance on $\mathbb{K}^n$ is given by

$$\rho(\mathbf{x}, \mathbf{y}) = \sqrt{|x_1 - y_1|^2 + \cdots + |x_n - y_n|^2}.$$

The distance on the Riemann sphere $\hat{\mathbb{C}}$ (see Section 2.3) is given by

$$s(z,w) = \frac{2|z-w|}{\sqrt{(1+|z|^2)(1+|w|^2)}}, \quad s(z,\infty) = \frac{2}{\sqrt{1+|z|^2}} \quad \text{for} \quad z,w \in \mathbb{C}.$$

2. A natural example of a piecewise C^1 path that is not a C^1 path comes from considering absolute value. Specifically, let $f \colon [-1,1] \to \mathbb{C}$ be defined by $\varphi(t) = t + i|t|$. Since $|t|$ is not differentiable at $t = 0$, this is not a C^1 path. However, on $[-1,0]$ ($[0,1]$), we have that $|t| = -t$ (resp., $|t| = t$) and hence that $\varphi(t) = t - it$ (resp., $\varphi(t) = t + it$), which is a C^1 path. So, φ is a piecewise C^1 path on $[-1,1]$.

3. Consider the C^1 path $\varphi \colon [0,2] \to \mathbb{R}^2$ given by $\varphi(t) = (2+t, 1+t^2/2)$. The length of φ is $l(\varphi) = \int_0^2 \sqrt{1+t^2}\, dt = \sqrt{5} - \frac{1}{2}\ln(\sqrt{5}-2)$.

```
syms t;
f = [2 + t  1 + t^2/2];
df = diff(f, t);   v = sqrt(df(1)^2 + df(2)^2)
s = int(v, t, 0, 2),  eval(s);        % Answer: s = √5 − ½ log(√5 − 2), 2.9579.
```

4. Let the C^1 path $\varphi_h \colon [0,2] \to \mathbb{R}^2$ be given by $\varphi_h(t) = (1+t, h+t^2/2)$ for $h > 0$. One may show that $l(\varphi_h) \to \infty$ for $h \to 0$. For $h = 0.01$ we obtain $l(\varphi_{0.01}) = \int_0^2 \sqrt{1+t^2}/(0.01+t^2/2)\, dt = 22.6611$.

```
h = 0.01                                          % define h > 0
syms t;
f = [1 + t  h + t^2/2];   df = diff(f, t);   v = sqrt(df(1)^2 + df(2)^2)
s = int(v/f(2), t, 0, 2),    eval(s)
s = 14 atan(14/√5) − 2 log(√5 − 2),  22.6611                % Answer.
```

The problem we are interested in is how the *hyperbolic length* determines the *hyperbolic distance*.

Let X be a set in which we know how to measure the lengths of paths. Specifically, for each pair $x, y \in X$, suppose there is a nonempty collection $\Gamma[x,y]$ of paths $\varphi \colon [a,b] \to X$ satisfying $\varphi(a) = x$ and $\varphi(b) = y$, and assume that to each path φ in $\Gamma[x,y]$ we can associate in a "reasonable way" a nonnegative real number $\text{length}(\varphi)$, which we refer to as the length of φ. Consider the function $\rho_l \colon X \times X \to \mathbb{R}$ defined by taking the infimum

$$\rho_l(x,y) = \inf\{\text{length}(\varphi) : \varphi \in \Gamma[x,y]\}.$$

For $X = \mathbb{R}^n$, because the shortest Euclidean distance between two points is along a Euclidean line, which can be parameterized by a C^1 path, we see that $\rho_l(\mathbf{x},\mathbf{y}) = \|\mathbf{x} - \mathbf{y}\| = \rho(\mathbf{x},\mathbf{y})$.

Here is an example that illustrates one of the difficulties that can arise.

Let $X = \mathbb{C} \setminus \{0\}$ be the *punctured plane*, and for each pair of points $x, y \in X$, let $\Gamma[x,y]$ be the set of all piecewise C^1 paths $\varphi \colon [a,b] \to X$ with $\varphi(a) = x$ and $\varphi(b) = y$. The construction of a function on $X \times X$ by taking the infimum of the lengths of paths gives rise to the metric $\rho(x,y) = |x-y|$ on X. However, it is no longer true that there necessarily exists a path in $\Gamma[x,y]$ realizing the Euclidean distance between x and y. Specifically, consider the two points 1 and -1: The Euclidean line segment in $\mathbb{C}$ joining 1 to -1 passes through 0, and so it is not a path in X. Every other path joining 1 to -1 has length strictly greater than $\rho(1,-1) = 2$.

Definition 4.1. A metric space (X,d) is a *path metric space* if

$$\rho(x,y) = \inf\{\text{length}(\varphi) : \varphi \in \Gamma[x,y]\}$$

for each pair of points $x, y \in X$, and if there exists a distance-realizing path φ in $\Gamma[x,y]$ satisfying $\rho(x,y) = \text{length}(\varphi)$.

A *minimizing geodesic* in a path metric space (X,d) is any curve $\varphi \colon I \to X$ such that $\rho(\varphi(t), \varphi(t')) = |t - t'|$ for each $t, t' \in I$. A *geodesic* in X is any curve $\varphi \colon I \to X$ whose restriction to any sufficiently small subinterval in I is a minimizing geodesic.

The "reasonable way" in the above discussion means the following.

A *length structure* on a set X consists of a family $C(I)$ of mappings $\varphi \colon I \to X$ for each interval I and a map l of $C = \bigcup C(I)$ into $\mathbb{R}$ having the following four properties:

Positivity. $l(\varphi) \geq 0$ for each $\varphi \in C$, and $l(\varphi) = 0$ if and only if φ is constant (we assume of course that the constant functions belong to C).

Restriction, composition. If $I \subset J$, then the restriction to I of any member of $C(J)$ is contained in $C(I)$. If $\varphi \in C([a,b])$ and $\psi \in C([b,c])$, then the function h obtained by composition (or juxtaposition) of φ and ψ lies in $C([a,c])$ and $l(h) = l(\varphi) + l(\psi)$.

Invariance under change of parameter. If φ is a homeomorphism from I onto J and if $\psi \in C(J)$, then $\psi \circ \varphi \in C(I)$ and $l(\psi \circ \varphi) = l(\psi)$.

Continuity. For each $I = [a,b]$ and $\varphi \in C(I)$, the map $t \to l(\varphi|_{[a,t]})$ is continuous.

4.3.2 *The hyperbolic distance*

For simplicity, we restrict ourselves to the two-dimensional case, the *hyperbolic plane* $\mathbb{R}_+^2 = \{(x,y) \in \mathbb{R}^2 : y > 0\}$, where $ds^2 = \frac{k^2}{y^2}(dx^2 + dy^2)$.

Since the Gaussian curvature of the metric $ds^2 = f(x,y)(dx^2 + dy^2)$ is given by the formula $K = -\frac{1}{2f} \Delta \ln f$, the curvature of the Poincaré metric is a negative constant $K = -k^2$.

Lemma 4.1. *Each of the following maps ψ is an isometry of $\mathbb{R}_+^2$ onto itself (a motion of the metric).*

(1) $\psi(x,y) = (x+a, y)$ *are parallel translations along the axis OX,*

(2) $\psi(x,y) = (\lambda x, \lambda y)$ *are dilations (with center at O),*

(3) $\psi(x,y) = (-x, y)$ *is a symmetry with respect to the axis OY,*

(4) $\psi(x,y) = (x/(x^2+y^2), y/(x^2+y^2))$ *is an inversion with respect to the unit circle with center at O.*

Proof. We check that $(\mathbf{a},\mathbf{b})_L := \frac{k^2}{y^2}(\mathbf{a},\mathbf{b})$ is a scalar product. In turn,

(1), (3): $(d\psi(\mathbf{a}), d\psi(\mathbf{b}))_L = \frac{k^2}{y^2}(d\psi(\mathbf{a}), d\psi(\mathbf{b})) = \frac{k^2}{y^2}(\mathbf{a},\mathbf{b}) = (\mathbf{a},\mathbf{b})_L;$

(2): $(d\psi(\mathbf{a}), d\psi(\mathbf{b}))_L = \frac{k^2}{(\lambda y)^2}(d\psi(\mathbf{a}), d\psi(\mathbf{b})) = \frac{k^2}{\lambda^2 y^2}(\lambda\mathbf{a}, \lambda\mathbf{b}) = (\mathbf{a},\mathbf{b})_L;$

(4): for the basic vectors $\mathbf{i}, \mathbf{j}$ at the point (x,y), $d\psi(\mathbf{i}) = \left(\frac{y^2-x^2}{(x^2+y^2)^2}, \frac{-2xy}{(x^2+y^2)^2}\right)$, $d\psi(\mathbf{j}) = \left(\frac{-2xy}{(x^2+y^2)^2}, \frac{x^2-y^2}{(x^2+y^2)^2}\right)$. Thus $(d\psi(\mathbf{i}), d\psi(\mathbf{i}))_L = \frac{k^2}{(x^2+y^2)^2} \Big/ \left(\frac{y}{x^2+y^2}\right)^2 = \frac{k^2}{y^2} = (\mathbf{i},\mathbf{i})_L$, and

analogously for $\mathbf{j}$. Moreover, $(d\psi(\mathbf{i}), d\psi(\mathbf{j}))_L = 0 = (\mathbf{i},\mathbf{j})_L$ holds.   □

The group generated by all motions (1)–(4) is $\text{Möb}(\mathbb{R}_+^2)$.

Theorem 4.6. *The lines (geodesics) of the Poincaré metric on the half-plane $\mathbb{R}_+^2$ are* Euclidean *vertical rays and* Euclidean *half-circles with centers on the x-axis, with corresponding parameterization.*

Proof. Let $A_1, A_2 \in \mathbb{R}_+^2$ be arbitrary points. The circle ω that passes through both points and whose center lies on the x-axis intersects the x-axis at B_1 and B_2. Let φ be the composition of horizontal translation by $-B_1$ followed by inversion in the unit circle. This map is an isometry since it is a composition of isometries. Note that B_1 is first sent to the origin O and then to ∞ by the inversion. Thus, the image of ω is a vertical Euclidean line. Since vertical rays are lines in the model, and isometries preserve arc length, it follows that half of ω is a line (geodesic) through A_1 and A_2.   □

Notice that any of the curves in Theorem 4.6 coincides with a set of fixed points of some transformation from $\text{Möb}(\mathbb{R}_+^2)$. But the curve of fixed points of an isometry (in the Riemannian metric) is always a geodesic.

One may define the *hyperbolic distance* by the rule

$$\rho_H(A_1, A_2) = |\ln([A_2, A_1, B_1, B_2])|.$$

For every element γ of $\text{Möb}(\mathbb{R}_+^2)$ and for any pair of points $A_1, A_2 \in \mathbb{R}_+^2$ we have that $\rho_H(\gamma(A_1), \gamma(A_2)) = \rho_H(A_1, A_2)$.

Alternatively, one may derive the distance between the points $A_1 = (x_1, y_1)$ and $A_2 = (x_2, y_2)$ by integrating the hyperbolic arc length dl_H, see (4.6), along the segment (of the hyperbolic line):

$$\rho_H(A_1, A_2) = \begin{cases} k \left| \ln \dfrac{y_2}{y_1} \right| & \text{if } x_2 = x_1, \\[2mm] k \left| \ln \dfrac{\tan(t_1/2)}{\tan(t_2/2)} \right| & \text{if } x_2 \neq x_1. \end{cases} \tag{4.8}$$

Here t_i are the angles between the vectors $O'A_i$ and the x-axis, and O' is the center of the Euclidean circle through the points A_1 and A_2 which represents the line A_1A_2. If the points are given in the form z_1, z_2 (of complex numbers $z = x + iy$), then the formula (4.8) takes the form

$$\rho_H(z_1, z_2) = k \ln \frac{1+Z}{1-Z}, \quad \text{where} \quad Z = \frac{z_1 - z_2}{z_1 - \overline{z_2}}.$$

4.3.3 *The hyperbolic area*

The area element for the Poincaré upper half-plane is derived by taking a small Euclidean rectangle with sides oriented horizontally and vertically. The sides approximate hyperbolic line segments, since the rectangle is small. The area is therefore the product of the height and width (measured with the Euclidean arc-length element). The vertical sides have Euclidean length Δy, and since y is essentially unchanged, the hyperbolic length is $\frac{1}{y}\Delta y$. The horizontal sides have Euclidean length Δx, and hence hyperbolic length $\frac{1}{y}\Delta x$. Thus, the *area element* is given by $dS = \frac{1}{y^2}\,dx\,dy$.

The *hyperbolic volume* (area, when $n = 2$) of a set X in $\mathbb{R}^n_+$ is given by the integral

$$S(X) = \int_X (1/x_n^2)\,dx_1 \cdots dx_n.$$

The hyperbolic volume is invariant under the action of $\text{Möb}(\mathbb{R}^n_+)$. That is, $S(f(X)) = S(X)$ for any element f of $\text{Möb}(\mathbb{R}^n_+)$.

Example. First, look at the doubly asymptotic triangle with one vertex at $P = e^{i(\pi - t)}$ in $\mathbb{R}^2_+$, and vertices at infinity of 1 and ∞, as in Figure 4.3(a).

The area of a doubly asymptotic triangle $\triangle PMN$ with M, N at infinity and with $\angle MPN = P$ (measured in radians) is $S(\triangle PMN) = \pi - P$.

To prove this, let the angle at P have measure $t \in (0, \pi)$. Then $\triangle PMN$ in Figure 4.3(b) is similar to the doubly asymptotic triangle in Figure 4.3(a) and hence is congruent to it. But the area of that triangle is given by

$$S(t) = \int_{-\cos t}^{1} \int_{\sqrt{1-x^2}}^{\infty} \frac{dx\,dy}{y^2} = \int_{-\cos t}^{1} \frac{dx}{\sqrt{1-x^2}} = \pi - t.$$

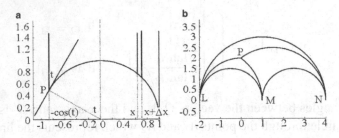

Fig. 4.3 The area of asymptotic triangles.

```
syms t x y;
int( int(1/y^2, y, sqrt(1 - x^2), inf), x, -cos(t), 1)        % Answer: π − acos(cos t).
```

Note that in Euclidean geometry a triangle with one infinite vertex can be represented as the half-strip between two parallel lines, a triangle with two infinite vertices — as the angle between two rays, and a triangle with three infinite vertices has no analogous Euclidean figure.

Theorem 4.7. *Let* $\triangle ABC$ *be a triangle in* $\mathbb{R}^2_+$, *with angles* $A, B,$ *and* C. *Then the area of* $\triangle ABC$ *is* $S(\triangle ABC) = \pi - A - B - C$.

Proof. Continue the edges of the triangle indefinitely as rays AB, BC, and CA. Let the points at infinity on these rays be, respectively, L, M, and N. Find the common parallels LM, MN, and NL, as in Figure 4.4. These lines form a triply asymptotic triangle, whose area is π. Thus

$$
\begin{aligned}
S(\triangle ABC) &= \pi - S(\triangle ALN) - S(\triangle BLM) - S(\triangle MCN) \\
&= \pi - (\pi - (\pi - A)) - (\pi - (\pi - B)) - (\pi - (\pi - C)) \\
&= \pi - A - B - C.
\end{aligned}
$$

□

In the exercises we will show that the area of an asymptotic triangle is finite. We will also show that all triply asymptotic triangles are congruent; hence, their area is constant.

One can generalize Theorem 4.7 to all reasonable hyperbolic polygons. By *hyperbolic polygon* in $\mathbb{R}^2_+$ we mean an ordered set of vertices $P_1, \ldots, P_m$ (some of them may be ideal) and a set of segments (of hyperbolic lines) connecting neighbor vertices in such a way that they bound a connected (one-component) domain. (For a more precise definition see [Anderson].)

Let Q be a hyperbolic polygon in $\mathbb{R}^2_+$, with interior angles α_k. Then the area of Q is $S(Q) = (n-2)\pi - \sum_{k=1}^{n} \alpha_k$.

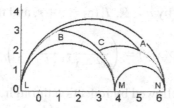

Fig. 4.4 The area of a triangle.

(To prove the claim, decompose Q into hyperbolic triangles, use Theorem 4.7 to calculate the hyperbolic area of each triangle in this decomposition, and then by summing get the hyperbolic area of Q.)

The following facts from hyperbolic trigonometry help us to solve problems with triangles.

Sine theorem: $\dfrac{\sinh(a)}{\sin(\alpha)} = \dfrac{\sinh(b)}{\sin(\beta)} = \dfrac{\sinh(c)}{\sin(\gamma)}.$

Cosine theorem: $\cosh(c) = \cosh(a)\cosh(b) - \sinh(a)\sinh(b)\cos(\gamma).$

4.3.4 *Hyperbolic motions in two dimensions*

The group of direct isometries of the hyperbolic half-plane (i.e., preserving orientation) is easy to describe — they are all *Möbius transformations* of a particular type. Recall notations for various sets of matrices:

$$M_2(\mathbb{K}) = M_{2\times 2}(\mathbb{K}) = \left\{ \begin{pmatrix} a & b \\ c & d \end{pmatrix} : a,b,c,d \in \mathbb{K} \right\},$$
$$GL_2(\mathbb{K}) = \{A \in M_2(\mathbb{K}) : \det A \neq 0\},$$
$$SL_2(\mathbb{K}) = \{A \in M_2(\mathbb{K}) : \det A = 1\}$$

(see Section 2.5), where $\mathbb{K}$ can be any of the sets $\mathbb{Z}$ (integers), $\mathbb{Q}$ (rationals), $\mathbb{R}$ (reals), $\mathbb{C}$ (complex numbers), $\mathbb{H}$ (quaternions), etc.

The set $GL_2(\mathbb{K})$ is called the *general linear group* over $\mathbb{K}$, and the set $SL_2(\mathbb{K})$ is called the *special linear group* over $\mathbb{K}$, both in dimension two.

Note that the group $GL_2(\mathbb{K})_{/\equiv}$ (the group $GL_2(\mathbb{K})$ modulo the equivalence relation $\lambda A \equiv A$), called the *projective special linear group*, is isomorphic to the group of Möbius transformations.

The horizontal translation by $a \in \mathbb{R}$, $T_r(z) = z + r$, and the map

$$\varphi(x,y) = \left(\frac{-x}{x^2 + y^2}, \frac{y}{x^2 + y^2} \right)$$

(inversion in the unit circle followed by reflection through $x = 0$), are Möbius transformations which are represented by elements $\tau_r = \begin{pmatrix} 1 & r \\ 0 & 1 \end{pmatrix}$ and $\sigma = \begin{pmatrix} 0 & -1 \\ 1 & 0 \end{pmatrix}$ of $SL_2(\mathbb{R})$.

To show the claim, note that $\mathbf{T}_r(x,y) = (x+r,y)$ is generated by τ_r. As a function of complex numbers, the map φ is just

$$\varphi(z) = \frac{-x + iy}{x^2 + y^2} = \frac{-(x - iy)}{(x + iy)(x - iy)} = \frac{-1}{z}.$$

This map is represented by σ.

Example. We create the M-file **taur.m** to compute the matrix τ_r:

```
function f = taur(r)
   f = [1 r;  0 1];
end
```

For instance, we compute

$$\sigma\tau_r = \begin{pmatrix} 0 & -1 \\ 1 & 0 \end{pmatrix} \begin{pmatrix} 1 & r \\ 0 & 1 \end{pmatrix} = \begin{pmatrix} 0 & -1 \\ 1 & r \end{pmatrix} \text{ and } \sigma\tau_s\sigma\tau_r = \begin{pmatrix} 0 & -1 \\ 1 & s \end{pmatrix} \begin{pmatrix} 0 & -1 \\ 1 & r \end{pmatrix} = \begin{pmatrix} -1 & -r \\ s & rs - 1 \end{pmatrix}.$$

```
syms r s;
sigma = [0 -1;  1 0]                           % we define the matrix σ
sigma*taur(r)
sigma*taur(s)*sigma*taur(r)        % Answer:  [0, -1;  1, r]    [-1, -r;  s, sr - 1]
```

Theorem 4.8. *The group $SL_2(\mathbb{R})$ is generated by σ and τ_r, for $r \in \mathbb{R}$.*

Hence, $SL_2(\mathbb{R})$, *when thought of as a subgroup of the Möbius transformations, is a subgroup of the isometries of the Poincaré model in a half-plane.* To prove this corollary, consider $A \in SL_2(\mathbb{R})$ with $\det A > 0$. Note that $\frac{1}{\sqrt{\det A}} A \in SL_2(\mathbb{R})$, and $(kA)z = Az$. Hence, A is an isometry of the Poincaré half-plane.

As with inversions, *the image of a circle or a line in $\mathbb{C}$ under the action of a Möbius transformation $A \in SL_2(\mathbb{R})$ is again a circle or line* (verify!).

The model of hyperbolic geometry must satisfy the following axiom:

For any two points P, Q, there is an isometry f such that $f(P) = Q$.

Let $P = a + ib$ and $Q = c + id$. One may verify that $Q = f(P)$, where

$$f(z) = \frac{d(z-a)}{b} + c = \begin{pmatrix} d & bc - ad \\ 0 & b \end{pmatrix} z. \tag{4.9}$$

Since this matrix has real coefficients and positive determinant ($b > 0$ and $d > 0$), it is an isometry of the Poincaré upper half-plane.

Recall that *translations* are direct isometries that have no fixed points (and *rotations* are direct isometries that have exactly one fixed point). We call z_0 a *fixed point* of a map A if $Az_0 = z_0$.

To see that the map (4.9) is a translation, we solve for z and get

$$\frac{dz_0 + bc - ad}{b} = z_0 \Longleftrightarrow z_0 = \frac{ad - bc}{d - b}.$$

Note that a, b, c, d are all real so if $b \neq d$, then z_0 is real and hence is not in the upper half-plane. If $b = d$, then $z_0 = \infty$, and again there are no solutions in the upper half-plane. So the map is a translation.

In hyperbolic geometry we classify translations according to how many fixed points there are on the line at infinity (that is, in $\mathbb{R} \cup \{\infty\}$).

Let $A = \begin{pmatrix} a & b \\ c & d \end{pmatrix}$. Then $A(z) = z$ if $cz^2 + (d-a)z - b = 0$. If $c \neq 0$ then this is a quadratic with discriminant $\Delta = (d-a)^2 + 4bc$. Hence, A has a fixed point in $\mathbb{R}_+^2$ if $\Delta < 0$, and no fixed points if $\Delta > 0$. If $\Delta > 0$, we call A a *hyperbolic translation*. If $\Delta = 0$, then A has exactly one fixed point on the line at infinity; we call A a *parabolic translation*.

One can find a Möbius transformation that fixes $P = a + ib$ and rotates counterclockwise through an angle t (see exercises). The set of isometries that do not preserve orientation in $\mathbb{R}_+^2$ also has a nice description.

Theorem 4.9. *Every isometry f of $\mathbb{R}_+^2$ that is not direct can be written in the form* $f(z) = A(-\bar{z})$ *for some* $A \in SL_2(\mathbb{R})$. *Furthermore, if* $A = \begin{pmatrix} a & b \\ c & d \end{pmatrix}$, *then* $f(z)$ *is a reflection if and only if* $a = d$.

4.3.5 *Hyperbolic motions in three dimensions*

Under the identification of $\mathbb{R}^2$ with the plane $\mathbb{R}^2 \times \{0\}$ in $\mathbb{R}^3$, a point $\mathbf{x} = (x, y)$ of $\mathbb{R}^2$ corresponds to the point $\tilde{\mathbf{x}} = (x, y, 0)$ of $\mathbb{R}^3$. Similarly to the two-dimensional case (see Theorem 4.6), the *lines* of the Poincaré half-space $\mathbb{R}_+^3 = \{(x, y, t) : t > 0\}$ are Euclidean *vertical rays* and Euclidean *vertical half-circles* with centers on the xy-plane, with corresponding parameterization. The *planes* of $\mathbb{R}^3_+$ are Euclidean *vertical half-planes* and Euclidean *upper half-spheres* with centers on the xy-plane. Hence, a hyperbolic plane $\hat{\omega}$ (upper half-sphere) or $\hat{l}$ (vertical half-plane) "intersects" the xy-plane in a circle ω or a line l.

We will show that the action of Möb$_2$ can be lifted to $\mathbb{R}^3_+$ (or if one prefers, to all $\mathbb{R}^3$). In fact, any $f \in$ Möb$_2$ is a product of reflections in circles (or lines), the circle (line) determines a hemisphere (or half-plane) orthogonal to $t = 0$, and the reflection in a circle extends to a reflection in the hemisphere. It remains to show that the result does not depend on the particular decomposition of f into a product.

Let f be a Möbius transformation of $\mathbb{R}^2$. We shall extend f to a Möbius transformation $\tilde{f}$ of $\mathbb{R}^3$ (called the *Poincaré extension* of f) as follows:

(i) If f is the reflection of $\mathbb{R}^2$ in ω (a circle) or l (a line), then $\tilde{f}$ is the reflection of $\mathbb{R}^3$ in $\hat{\omega}$ or $\hat{l}$, respectively. In both these cases $\tilde{f}(\mathbf{x},0) = (f(\mathbf{x}),0)$ for all $\mathbf{x} \in \mathbb{R}^2$. Thus $\tilde{f}$ extends f and leaves $\mathbb{R}^3_+$ invariant.

(ii) If $f = f_1 \circ \cdots \circ f_m$ is the composition of $m \geq 2$ reflections, set $\tilde{f} = \tilde{f}_1 \circ \cdots \circ \tilde{f}_m$. Then $\tilde{f}$ extends f and leaves $\mathbb{R}^3_+$ invariant.

Let $\tilde{f}_1$ and $\tilde{f}_2$ be two such extensions of f. Then $\tilde{f}_1 \tilde{f}_2^{-1}$ fixes each point of $\mathbb{R}^2$ and leaves $\mathbb{R}^3_+$ invariant. By Exercise 13, p. 194, we have that $\tilde{f}_1 \tilde{f}_2^{-1} = \mathrm{Id}$ and so $\tilde{f}_1 = \tilde{f}_2$. Thus $\tilde{f}$ depends only on f and not on the decomposition $f = f_1 \circ \cdots \circ f_m$.

One may show that *a Möbius transformation f of $\mathbb{R}^3$ leaves the upper half-space $\mathbb{R}^3_+$ invariant if and only if f is the Poincaré extension of a Möbius transformation of $\mathbb{R}^2$.*

We shall pay special attention to the *group* Möb$(\mathbb{R}^3_+)$ of Möbius transformations acting on the *upper half-space*.

In the special case of $\mathbb{R}^3_+$ one can make use of quaternions. It is convenient to identify the point $(x,y,t) \in \mathbb{R}^3_+$ with the quaternion $\mathbf{q} = x + y\mathbf{i} + t\mathbf{j}$, and also to identify the quaternion $\mathbf{i}$ with the complex number i. Thus we can write (x,y,t) as $\mathbf{q} = z + t\mathbf{j}$, where $z = x + iy$ is a complex number. In this notation we have the formula

$$z\mathbf{j} = (x + y\mathbf{i})\mathbf{j} = x\mathbf{j} + y\mathbf{i}\mathbf{j} = x\mathbf{j} - y\mathbf{j}\mathbf{i} = \mathbf{j}(x - y\mathbf{i}) = \mathbf{j}\bar{z}.$$

The conjugate of a quaternion $\mathbf{q} = z + t\mathbf{j}$ is $\bar{\mathbf{q}} = \bar{z} - t\mathbf{j}$, and the absolute value $|\mathbf{q}|$ is given by $|\mathbf{q}|^2 = \mathbf{q}\bar{\mathbf{q}} = |z|^2 + |t|^2$.

Suppose that $f(z) = (az + b)/(cz + d)$, where $ad - bc \neq 0$, is a Möbius transformation of $\hat{\mathbb{C}}$. We can let f act on $\mathbb{R}^3_+$ by the rule

$$f: z + t\mathbf{j} \to [a(z + t\mathbf{j}) + b][c(z + t\mathbf{j}) + d]^{-1}, \tag{4.10}$$

where this computation is to be carried out in the algebra of quaternions.

The following equality holds:

$$f(z+t\mathbf{j}) = \frac{(az+b)(\bar{c}\bar{z}+\bar{d}) + a\bar{c}t^2 + |ad-bc|t\mathbf{j}}{|cz+d|^2 + |c|^2 t^2}. \qquad (4.11)$$

If we put $t = 0$ in (4.11), we obtain the formula for the action of f on $\mathbb{C}$.

Consider the important particular cases of (4.11).

(a) $f(z) = z+b$ is a translation. We find that $f(z+t\mathbf{j}) = (z+b)+t\mathbf{j}$; thus f is just the horizontal translation of $\mathbb{R}^3_+$ by b.

(b) $f(z) = az$. We find that $f(z+t\mathbf{j}) = az + |a|t\mathbf{j}$.

   If $|a| = 1$, so that $f(z)$ is a rotation of a complex plane, then f acts on $\mathbb{R}^3_+$ as a rotation about a vertical axis through the origin.

   If $a > 0$ (a positive real), so that $f(z)$ is a *dilation* (a scale change by positive a) from the origin in $\mathbb{C}$, then f also acts as a dilation (from the origin and with the same factor) in $\mathbb{R}^3_+$.

(c) The more interesting case is when $f(z) = 1/z$; here $f(z+t\mathbf{j}) = \dfrac{\bar{z}+t\mathbf{j}}{|z|^2 + t^2}$.

The *cross-ratio* for quaternions is defined by

$$[\mathbf{q}_1, \mathbf{q}_2, \mathbf{q}_3, \mathbf{q}_4] = (\mathbf{q}_1 - \mathbf{q}_3)(\mathbf{q}_1 - \mathbf{q}_4)^{-1}(\mathbf{q}_2 - \mathbf{q}_4)(\mathbf{q}_2 - \mathbf{q}_3)^{-1}.$$

Proposition 4.4. *Four pairwise distinct points* $\mathbf{q}_1$, $\mathbf{q}_2$, $\mathbf{q}_3$, $\mathbf{q}_4 \in \mathbb{H}$ *lie on the same (one-dimensional) circle if and only if their cross-ratio is real. The two pairs of points* $\mathbf{q}_1$, $\mathbf{q}_2$, $\mathbf{q}_3$, $\mathbf{q}_4$ *lying on the same circle separate each other if and only if* $[\mathbf{q}_1, \mathbf{q}_2, \mathbf{q}_3, \mathbf{q}_4] < 0$.

One may show that *every Möbius map acts on* $\mathbb{R}^3_+$ *by* (4.10) *as a hyperbolic motion, and every hyperbolic motion of* $\mathbb{R}^3_+$ *that preserves orientation is a Möbius map.*

Example. We prepare the M-file **prod4q.m**. (It can be easily extended for the ∞ cases as in **prod4.m**.)

```
function w = prod4q(q1, q2, q3, q4)          % the cross-ratio for quaternions
    w = simplify( quatmultiply( quatmultiply(q1 - q3, quatinv(q1 - q4)),
        quatmultiply(q2 - q4, quatinv(q2 - q3)) ) );
end
```

Then we apply it.

```
q1 = [0 0 3 0]; q2 = [2 1 2 0]; q3 = [2 -1 3 0]; q4 = [-1 -2 4 0];
prod4q(q1, q2, q3, q4)                        % define four points in ℝ³₊

1.2333 1.3000 -0.5667 0.3667                  % Answer
p1 = [0 0 3 0]; p2 = [0 0 9 0]; p3 = [0 0 2 0]; p4 = [0 0 6 0];
prod4q(p1, p2, p3, p4)       % points on a hyperbolic line, see Proposition 4.4
-0.1429 0 0 0                                 % Answer
```

4.4 Examples of Visualization

Hyperbolic geometry in the half-plane $\mathbb{R}^2_+$ can be thought of as the negative-curvature analogue of the geometry in the 2-sphere S^2.

We plot basic objects of $\mathbb{R}^2_+$ using the procedures of Section A.7: **segment, distance, triangle, perimeter, angle, lambert, parallel, perpendicular, biorthogonal, transversal, equidistant, horocycle, fifth**, etc.

In some visualizations the reader should add the axis equal option.

Examples. (Lines, polygons, circles.)

1. We plot the line segment through two given points A, B.

```
A = [2, 3];  B = [3, 2];  segment(A, B);          % segment ([2, 3], [2, 2])
for ii = -4 : 4
    segment([0, 1], [ii/4, 0]);  hold on;         % sheaf of lines, Figure 4.5(a)
end
```

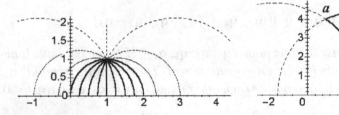

Fig. 4.5 Sheaf of lines. Triangle.

2. Find the distance between two points on the basis of the formula (4.8) and using the procedure **distance**.

```
distance([0, 20], [0, .05]);                      % Answer:  5.99146
```

3. Plot a triangle with given vertices using the statements

```
A = [1, 1];  B = [5,2];  C = [4, 4];              % Input three vertices
hold on;  segment(A,B);  segment(B,C);  segment(A,C);
```

or create a new M-file **triangle.m**.

```
triangle([1, 1], [5, 2], [4, 4])                  % Figure 4.5(b)
```

Calculate its perimeter using the M-file **perimeter.m**.

    perimeter([1, 1], [2, 1], [5, 4]);   % Answer: 1.21036

4. Calculate the angles of a triangle using the M-file **angle.m**.

    angle([1, 1], [1, 2], [2, 3]);   % Answer: 0.588003

Use **sum_angles** to compute the sum of its angles and the area.

    sum_angles([1, 1], [3, 1], [2, 1]);   % Answer: 1.24905

5. Next we visualize a *square* (right quadrilateral) with edge a. The *Lambert quadrilateral* (with angles $\frac{\pi}{2}, \frac{\pi}{2}, \frac{\pi}{2}, \varphi$ and edges a_1, a_2 between right angles) has the property $\sinh(a_1)\sinh(a_2) = \cos(\varphi)$.

If we reflect this quadrilateral with respect to an edge that connects the right angles, then we obtain the *Saccheri quadrilateral* with angles $\frac{\pi}{2}, \frac{\pi}{2}, \varphi, \varphi$. One may plot the quadrilaterals using the following procedure.

    lambert(0.7, 1);   % Lambert, Figure 4.6(a)
    lambert(0.7,1); hold on; **lambert**(-0.7,1));   % Saccheri, Figure 4.6(b)

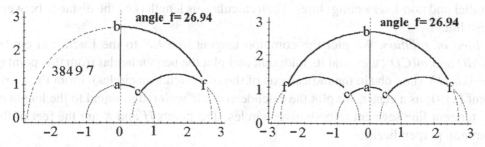

Fig. 4.6 Quadrilaterals of Lambert and Saccheri.

Example. Let us deduce the formula for the *angle of parallelism* (i.e., the angle between two rays with a common vertex that are *left* and *right parallel to a given line*), using Figure 4.7(a).

```
syms t k d;
F = k*log(tan(t/2));  alpha = solve(subs(F, t, pi/2) - F - d)
```
$$\alpha = 2\arctan\left(\exp(-d/k)\right)$$

We plot right and left parallels to a given line through a given point using the procedure **parallel**. We also calculate the angle between them and the angle of parallelism.

parallel([3, 5], [5, 2], [2, 4]);   % Figure 4.7(b)

Using the program, one can check that the limit angle as the point approaches the given line is equal to $\frac{\pi}{2}$, and the limit angle as the point moves away from the given line decreases to zero.

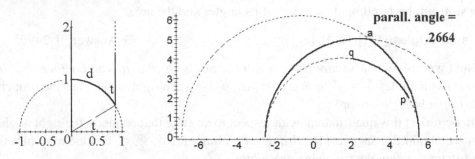

Fig. 4.7 The angle of parallelism. Left and right parallel lines.

Example. Let us plot the common perpendicular to given "*super-parallel*" (non-parallel and non-intersecting) lines. Then calculate its length (i.e., the distance between these lines).

Idea of solution. We plot the common tangent line EF to the Euclidean circles $\omega(AB)$ and $\omega(CD)$; then find its midpoint and plot the perpendicular from this point to the axis OX (i.e., obtain the radical axis of the two Euclidean circles). From the above point $(x_3, 0)$ as a center, we plot the Euclidean circle with radius equal to the length of the tangent line segment to both given circles. The points H and K are the feet of the common perpendicular.

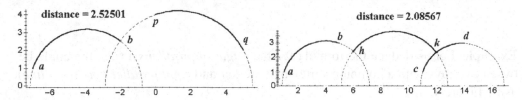

Fig. 4.8 Perpendicular to a line. The common perpendicular.

In the program **biorthogonal.m** we assume that the lines AB and CD are "super-parallel."

biorthogonal([1,1], [5,3], [11,1], [14,3]);   % Figure 4.8(b)

The following generalizes the notion of a common perpendicular.

The line L is called a θ-*transversal* for L_1 and L_2 if it makes the same angle θ with each of these lines. The common perpendicular L to L_1, L_2 is the unique $\frac{\pi}{2}$-transversal. For every other value of θ there are four θ-transversals (two *alternate* and two *adjacent*). The length t_θ of the segment of the transversal is equal to

$$\sinh\left(\frac{1}{2}d(L_1,L_2)\right) = \sinh\left(\frac{1}{2}t_\theta\right)\sin\theta, \qquad \text{the alternate transversal,}$$

$$\cosh\left(\frac{1}{2}d(L_1,L_2)\right) = \cosh\left(\frac{1}{2}t_\theta\right)\sin\theta, \qquad \text{the adjacent transversal.}$$

Examples.

1. Plot the perpendicular to a given line $l(P,Q)$ through a given point A and calculate its length (i.e., the distance from the point to the line), using the procedure **perpendicular.m**.

   perpendicular([-7, 1], [5, 2], [2, 4]);   % Figure 4.8(a)

   Idea of the program. Let $P_1 = Q_1 = x_1$. Then the perpendicular is represented by the Euclidean half-circle $(x - x_1) + y^2 = R^2$, where $R = \sqrt{(A_1 - x_1)^2 + A_2^2}$. Now let $P_1 \neq Q_1$. Then $l(P,Q)$ is represented by the Euclidean half-circle ω of radius R_1 centered at $C_1 = (x_1,0)$, where $R_1 = \sqrt{(P_1 - x_1)^2 + P_2^2}$ and x_1 is such that $(P - C_1)^2 = (Q - C_1)^2$.

   If $x_1 = A_1$, then the perpendicular is represented by the vertical ray through A. If $x_1 \neq A_1$, the perpendicular is represented by the Euclidean half-circle of radius R_2 centered at $C_2 = (x_2,0)$, where $R_2 = \sqrt{(A_1 - x_2)^2 + A_2^2}$ and x_2 is such that $(A - C_2)^2 = (C_1 - C_2)^2 - R_1^2$.

2. Plot the θ-transversals to lines L_1, L_2, using the statements

   transversal([-4,0], [-1,0], [1,0], [3,1], pi/3);
   transversal([2,2], [5,1], [7,5], [7,1], pi/3);   % Figure 4.9

3. **Pencil of lines and its orthogonal trajectories.** Plot a family of circles:

   for i = 1 : 5;
     circle([3, 1], i / 5); hold on;
   end   % Figure 4.10(a)

Next, plot two families of circles:

   n = 15; hold on;
   for i = 1 : n;
     circle([.3, 1], i / n); **circle**([-.3, 1], i / n);
   end   % Figure 4.10(b)

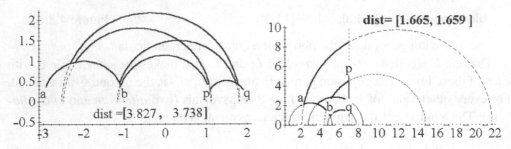

Fig. 4.9 Four θ-transversals to "super-parallel" lines.

The pencil (sheaf) of

(1) lines through a given point,
(2) lines orthogonal to a given line, or
(3) parallel lines (in some direction)

is called, respectively, an *elliptic,* a *hyperbolic,* or a *parabolic* pencil of circles.

The curves (in the hyperbolic plane) that are orthogonal to these sheaves are called a *circle,* an *equidistant,* and a *horocycle;* they are analytic curves.

(1) The *circle* is drawn as a Euclidean circle that does not intersect the axis OX. The pencil of lines is represented by a family of Euclidean circles through the given point and its symmetric point with respect to the axis OX. The orthogonal trajectories of this elliptic pencil are seen to be the family of Euclidean circles from the problem of Apollonius of Perga (with constant ratio of distances from given points).

If we throw a pebble in a pool of water, concentric circular waves (ripples) appear. In "hyperbolic water" the circles (ripples) are not concentric (in the Euclidean sense); they are orthogonal to a pencil of "hyperbolic lines" of type (1), through a given point. The figure is similar to the net of *bipolar coordinates.*

One can prove this using calculations. The area of the circle of radius R and the circumference of the circle are given by the formulae $S(R) = 4\pi \sinh^2(R/2)$ and $L(R) = S(R)' = 2\pi \sinh^2(R)$. One can check that the angle subtended by the diameter of the circle is less than $90°$.

Idea of the program. Let the circle of radius R with center at $A = (a_1, a_2)$ intersect the vertical ray $\{x = a_1\}$ at (diametric) points $y_1 < y_2$. Then by the formula of distance we have $R = \ln \frac{a_2}{y_1} = \ln \frac{y_2}{a_2}$. From this it follows that $y_2 = a_2 \exp(R)$ and $y_1 = a_2 \exp(-R)$. The Euclidean center of the circle lies at the altitude $y_E = (y_1 + y_2)/2 = a_2 \cosh R$, and its Euclidean radius is equal to $R_1 = (y_2 - y_1)/2 = a_2 \sinh R$.

(2) The *equidistant* is represented on the half-plane as an arc of a Euclidean circle that intersects the axis OX at two points in a manner improper for the given line

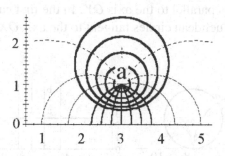 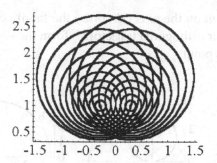

Fig. 4.10 Circles on non-Euclidean water.

through those points. It is a simple non-closed curve. The pencil of lines orthogonal to the given line when the given line is the axis OY is represented as the family of Euclidean concentric half-circles with center at O. The orthogonal trajectories of this pencil are the Euclidean rays through the point O. The figure is similar to the net of *polar coordinates*.

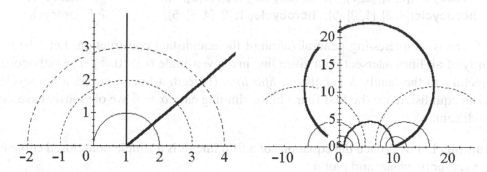

Fig. 4.11 Equidistant (2 cases).

Using inversion with respect to the Euclidean circle of radius R that is tangent to the axis OY, one maps the family of given rays (equidistants of OY) to the arcs of the hyperbolic pencil of Euclidean circles through the points $(0,0)$ and $(R,0)$. One can prove this using calculations.

> **equidistant**([1,1], [1,3], 1, 4); **equidistant**([1,1], [2,3], 1.6, 4);% Figure 4.11(a)
> **equidistant**([1,1], [2,3], 2, 4); **equidistant**([1,1], [2,3], -2, 4); % Figure 4.11(b)

(3) The *horocycle* is represented on the half-plane by a Euclidean circle that is tangent to the axis OX. It is a simple non-closed curve. A family of lines parallel to a given line is either represented by the parabolic pencil of Euclidean circles through the

point on the axis OX, or by the family of rays parallel to the axis OY. In the first case, their orthogonal trajectories form the set of Euclidean circles tangent to the axis OX at the point.

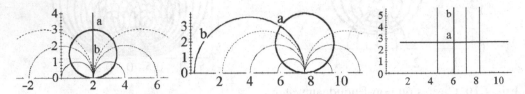

Fig. 4.12 Horocycle through A in the direction AB (3 cases).

In the second case, obviously, their orthogonal trajectories form the family of Euclidean lines parallel to the axis OX. Try to prove this by calculation. The *broken horocycle* (obtained as the union of two parts of a mutually tangent pair of horocycles) is only a C^1-regular curve. See Figure 4.12.

    horocycle([2, 3], [2, 1], 1); **horocycle**([6, 3], [1, 2], 1);   % 1 horocycle
    horocycle([4, 3], [1, 2], 5); **horocycle**([1, 2], [4, 3], 5);   % 5 horocycles

There is an interesting generalization of the equidistant construction. Let S be the family of all lines intersecting a given line in a given angle $\alpha \in (0, \frac{\pi}{2})$. The orthogonal trajectory to the family S is called a *fifth line*. (Indeed, a line, a circle, a horocycle, and an equidistant are the first four.) In the limiting case $\alpha = \frac{\pi}{2}$ we obviously have the equidistant.

Example. Let us deduce the equations of a fifth line (it is a simple non-closed curve in the hyperbolic plane) and plot it.

Solution (*implemented in the M-file* **fifth.m**). We deduce the ordinary differential equations (ODEs) of the fifth line in the case where the basic line coincides with the axis OY. The general case is obtained using inversion and translation of the hyperbolic plane.

A Euclidean circle intersecting the axis OY at the point $(0, t)$ with angle $a \in (0, \pi)$ has radius $R = t / \sin a$ and center at $C = (t \cot a, 0)$. The equation of the circle is

$$(x - t \cot a)^2 + y^2 = t^2 / \sin^2 a \iff x^2 - 2xt \cot a + y^2 = t^2. \qquad (4.12)$$

The parameter t of such a circle ω passing through the point (x, y) is the positive solution of the quadratic equation (4.12),

$$t_1 = \left(-x \cos a + \sqrt{x^2 + y^2 \sin^2 a} \right) \Big/ \sin a.$$

We calculate the normal vector $\mathbf{n}(x,y) = [x - t \cot a, y]$ to ω at (x,y),

$$\mathbf{n}(x,y) = \left[\frac{x - \cos a \sqrt{x^2 + y^2 \sin^2 a}}{\sin^2 a}, y \right].$$

Hence, the differential equations of a fifth line with the basic line OY are $[x'(s), y'(s)] = \mathbf{n}(x,y)$. A numerical solution of this system is possible; see the M-file **fifth.m**. One may also solve this system analytically. Assume x to be a function of y and obtain a *homogeneous* ODE

$$x' = \frac{dx}{dy} = \frac{x/y - \cos a \sqrt{(x/y)^2 + \sin^2 a}}{\sin^2 a}. \tag{4.13}$$

The geometric reason for the homogeneity is that the family of a-lines relative to the y-axis is invariant under the dilation $(x,y) \to (\lambda x, \lambda y)$ for all $\lambda > 0$. Hence, orthogonal trajectories are also invariant under these direct isometries. One can build one fifth line of the family, say, through the point $(0,1)$, and then produce the others by dilations. To solve (4.13), one uses the standard substitution $x = u(y)\, y$, and hence $x' = u'y + u$ and one obtains $u'y = A^2 u - A\sqrt{(1 + A^2)u^2 + 1}$, where $A = \cot a \in \mathbb{R}$. Using separation of variables, one gets

$$\int \frac{du}{A^2 u - A\sqrt{(1 + A^2)u^2 + 1}} = \ln y + c.$$

Denote the integral on the left-hand side by $F(u)$ for short. Then $y = Ce^{F(x/y)}$.

fifth([5, 0], [5, 5], pi/2 - 0.4, 8);   % 8 fifth lines, Figure 4.13(a)

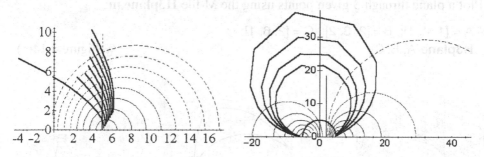

Fig. 4.13 Families of *fifth lines* (two cases).

A partition of a metric space by a family of congruent and mutually equidistant sets is called a *metric fibration*. Examples are parallel lines in the Euclidean plane, and generators or parallels on a circular cylinder. What are all the metric fibrations (by congruent and mutually equidistant curves) of the hyperbolic plane?

The metric fibrations of $\mathbb{R}^2_+$. *There are exactly three non-trivial metric fibrations of the hyperbolic plane:* (1) *the fibration into horocycles,* (2) *the fibration into broken horocycles, and* (3) *the fibration into fifth lines.*

Hyperbolic geometry in $\mathbb{R}^3_+$ can be thought of as the negative-curvature analogue of the positive-curvature geometry in the 3-sphere. We develop several procedures (see Section A.7) and plot the basic objects of $\mathbb{R}^3_+$.

Examples.

1. Plot the line segment through the points $A, B \in \mathbb{R}^3_+$:

```
A = [1, 4, 1];  B = [2, 3, 2];
% A = [1, 4, 1];  B = [1, 4, 2];                    % vertical segment
H3segment(A, B);                          % segment ([1,4,1], [2,3,2])
for j = -4 : 4;
  H3segment([0, 1, 1], [j/4, 0, 2]);  hold on;    % Sheaf of lines, Figure 4.14(a)
end
```

2. Compute the distance between the points $A_1 = (x_1, y_1, z_1)$ and $A_2 = (x_2, y_2, z_2)$:

```
A = [1, 4, 1];  B = [1, 4, 2];                           % z_1, z_2 > 0
H3distance(A, B)                                  % Answer: 0.6931.
```

3. Plot a triangle with given vertices using the M-file **H3triangle.m**:

```
A = [1, 1, 2];  B = [5, 2, 1];  C = [4, 4, 3];       % Input three vertices
H3triangle(A, B, C);                                 % Figure 4.14(b)
```

4. Plot a plane through 3 given points using the M-file **H3plane.m**:

```
A = [1, 4, 1];  B = [2, 3, 2];  C = [2, 6, 1];
H3plane(A, B, C);                                    % Figure 4.14(c)
```

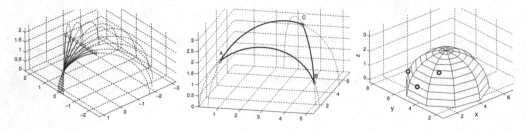

Fig. 4.14 Lines, triangle and plane in hyperbolic space.

5. Complute the distance between a point P and a plane $\alpha(A,B,C)$, and the projection, then plot its image, using the M-file **H3perp02.m**:

```
P = [4, 4, 2];  A = [1, 4, 1];  B = [2, 3, 2];  C = [2, 6, 1];
H3perp02(P, A, B, C);                              % obtain a Figure.
```

6. Plot two lines (through a point A) parallel to a given one $l(P,Q)$ in three dimensions using the M-file **H3parallel.m**:

```
P = [0, 1, 1];  Q = [-1, 0, 2];  A = [2, 4, 1];
H3parallel(A, P, Q);                               % obtain a Figure.
```

4.5 Exercises

Section 4.1

1. Let A be a real 3×3 matrix. Then there is a $\mu \in \mathbb{R}_+$ such that $\mu^{-1}A$ is an orthogonal matrix if and only if either of the following holds:

   (a) A preserves angles between nonzero vectors;

   (b) $\|\mathbf{x}A\| = \mu\|\mathbf{x}\|$ for all $\mathbf{x} \in \mathbb{R}^3$.

2. The *reflection* in the sphere $S^2(O,1) \subset \mathbb{R}^3$ is given by $\mathbf{P}_{O,1}(\mathbf{x}) = \mathbf{x}/\|\mathbf{x}\|^2$.

3. Show that $f(\mathbf{x}) = \mathbf{a} + r\mathbf{x}$ $(r > 0)$ is a conformal map.

4. Show that if $\mathbf{P}_{Q,r}$ is the inversion of $\mathbb{R}^3$ in the sphere $S^2(Q,r)$, then

   (i) $\mathbf{P}_{Q,r}(\mathbf{x}) = \mathbf{x}$ if and only if $\mathbf{x} \in S^2(Q,r)$;

   (ii) $\mathbf{P}^2_{Q,r}(\mathbf{x}) = \mathbf{x}$ for all $\mathbf{x} \neq Q$ (a self-inverse transformation);

   (iii) $\|\mathbf{P}_{Q,r}(\mathbf{x}) - \mathbf{P}_{Q,r}(\mathbf{y})\| = \frac{r^2}{\|\mathbf{x}-Q\|\cdot\|\mathbf{y}-Q\|}\|\mathbf{x} - \mathbf{y}\|$ for all $\mathbf{x}, \mathbf{y} \neq Q$.

5. The equation $\rho = \frac{R}{\cos(\varphi/3)}$ defines the *trisectrix of Maclaurin*, shown in Figure 4.1(a); it is useful for the problem of the *trisection of an angle*.

```
e = .5;  f = cos(t/3);
t = -3*pi/2 + e : .01 : 3*pi/2 - e;
hold on;  axis equal
polar(t, 1./f);  polar(t, f, ': g');
```

6. Inversion of the four-leafed rose $\rho = R\cos(2\varphi)$ leads to the *cross-shaped curve* $\rho = \frac{R}{\cos(2\varphi)}$ (see other *crosses* in Section 1.4).

```
syms t;
a = 1;  k = 2;  e = .15;  f = sin(k*t);
for i = 0 : 2*k;
  s = pi/2/k;
  hold on;  axis equal;
  ezpolar(1/f, [(2*i)*s + e, (2*i+2)*s - e]);
  ezpolar('1');  ezpolar(f);
end;                                               % Figure 4.1(c)
```

7. Check that these pairs of curves are mutually inverse: Archimedes' and hyperbolic spirals, the rectangular hyperbola and the lemniscate of Bernoulli, the parabola and the cardioid, the cochleoid and the quadratrix.

8. Show that

(a) the extended reciprocal function and the extended affine function are composites of inversions;

(b) all inversive transformations form a group under the operation of composition of functions;

(c) inversive transformations preserve the magnitude of angles and map generalized circles to generalized circles.

9. Show that

(a) the reflection of $\mathbb{R}^2 \approx \mathbb{C}$ in the unit circle $\omega = S^1(O,1)$ is the function
$$\mathbf{T}\colon z \to 1/\bar{z} \text{ or, equivalently, } \mathbf{T}\colon (x,y) \to \left(\frac{x}{x^2+y^2}, \frac{y}{x^2+y^2}\right);$$

(b) the reflection in a circle $\omega = S^1(c,r)$ may be represented in $\mathbb{C}$ by

$$\mathbf{T}(z) = r^2/(\bar{z}-\bar{c})+c, \quad z \in \mathbb{C}\setminus\{c\}. \tag{4.14}$$

10. **Steiner's porism.** Let the circle ω' be inside the circle ω. Insert a circle c_1 externally tangent to ω' and internally tangent to ω. Insert another circle c_2 tangent to c_1, ω, and ω', and so on, continuing around ω'. Suppose that for some n we have c_n tangent to c_1. Prove that if this happens for some n and some pair of circles ω' and ω, then no matter where we first draw c_1, we will always have c_n tangent to c_1.

Section 4.2	1. Prove that:

(a) The map $f^{-1}(z) = \frac{dz-b}{-cz+a}$ is the Möbius transformation inverse to $f = \frac{az+b}{cz+d}$. *Hint.* To verify directly, set $g = f_A(z)$ and isolate z:

$$(cz+d)g = az+b \Longleftrightarrow czg-az = b-dg \Longleftrightarrow z = (dg-b)/(-cg+a).$$

(b) Every Möbius transformation of $\mathbb{C}$ can be expressed as the composition of at most four maps, each of which is of one of the forms $z \to e^{it}z$, $z \to \lambda z$, $z \to z+b$, $z \to 1/z$, where $t, \lambda \in \mathbb{R}$ (i.e., rotations, dilations, translations, and the reciprocal function).

2. Show that:

(i) Every matrix associated with a Möbius transformation is invertible because $ad - bc \neq 0$. Moreover, $\begin{pmatrix} d & -b \\ -c & a \end{pmatrix}$ is associated with f^{-1}.

(ii) The matrix representation for f is not unique because f is also represented by $kA = \begin{pmatrix} ka & kb \\ kc & kd \end{pmatrix}$ for any $k \neq 0$.

3. Find the image of the circle $\omega = S^1(0,1)$ under the inversive transformation $f = \frac{\bar{z}+i}{\bar{z}-1}$. *Hint.* The image is an extended line through the origin.

4. Determine whether or not the four points i, $1 + 4i$, 3, $4 + 3i$ lie on a generalized circle. *Hint.* First we determine the Möbius transformation that maps i, $1 + 4i$, 3 to 0, 1, ∞, namely: $f(z) = k\frac{z-i}{z-3}$ where $k = 1 + i$. It follows that $f(4 + 3i) = 2$. Since this is real, it follows that i, $1 + 4i$, 3, $4 + 3i$ lie on a generalized circle.

5. Let ω_1 and ω_2 be generalized circles in $\hat{\mathbb{C}}$. Prove that there is a Möbius transformation that maps ω_1 to ω_2.

6. Show that

$$[\infty, z_2, z_3, z_4] = \frac{z_2 - z_4}{z_3 - z_4}, \qquad [z_1, \infty, z_3, z_4] = -\frac{z_1 - z_3}{z_3 - z_4},$$

$$[z_1, z_2, \infty, z_4] = -\frac{z_2 - z_4}{z_1 - z_2}, \qquad [z_1, z_2, z_3, \infty] = -\frac{z_1 - z_3}{z_1 - z_2}.$$

7. Prove that
(a) the linear maps $z \rightarrow az + b$ preserve the ratio $\lambda = \frac{z_2 - z_1}{z_4 - z_3}$;

(b) the ratio $\lambda = \frac{z_2 - z_1}{z_4 - z_2}$ is real if and only if the three points are collinear.

8. Prove that:
(a) Exchanging any two pairs of points preserves the cross-ratio:

$$[z_1, z_2, z_3, z_4] = [z_2, z_1, z_4, z_3] = [z_3, z_4, z_1, z_2] = [z_4, z_3, z_2, z_1].$$

(b) There are 6 possible values of the cross-ratio, depending on the order in which the points are given. These are:

$$[z_1, z_2, z_3, z_4] = \lambda, \qquad [z_1, z_2, z_4, z_3] = \frac{1}{\lambda}, \qquad [z_1, z_3, z_4, z_2] = \frac{1}{1 - \lambda},$$

$$[z_1, z_3, z_2, z_4] = 1 - \lambda, \qquad [z_1, z_4, z_3, z_2] = \frac{\lambda}{1 - \lambda}, \qquad [z_1, z_4, z_2, z_3] = \frac{\lambda - 1}{\lambda}.$$

9. Show that the non-constant linear-fractional function $f(z) = \frac{az+b}{cz+d}$ preserves the cross-ratio. Namely,

$$[z_1, z_2, z_3, z_4] = [f(z_1), f(z_2), f(z_3), f(z_4)] \quad \text{for all} \quad z_i \in \mathbb{C}.$$

10. Given two triples of distinct points z_1, z_2, z_3 and w_1, w_2, w_3 in $\mathbb{C}$, there is a linear-fractional function $f(z) = \frac{az+b}{cz+d}$ such that $w_i = f(z_i)$ ($i = 1, 2, 3$). *Hint.* Use the relation $[z_1, z_2, z_3, z] = [w_1, w_2, w_3, f(z)]$.

11. Show that
(a) the distinct points z_1, z_2, z_3, z_4 are concyclic if and only if their cross-ratio $[z_1, z_2, z_3, z_4]$ is real. *Hint.* Use Exercise 5 when z_1, z_2, z_3, z are real.

(b) $P(4 + 3i)$ belongs to the circle $\omega(i, 1 + 4i, 3)$. *Hint.* Use the cross-ratio.

prod4(i, $1 + 4*i$, 3, $4 + 3*i$)   % obtain $-1 \in \mathbb{R}$, hence $P \in \omega$.

12. A Möbius transformation $f : \mathbb{R}^3 \to \mathbb{R}^3$

(a) preserves extended spheres;

(b) fixes ∞ if and only if f is a similarity.

13. A Möbius transformation which fixes each point of a sphere S^2 is either the identity map of $\mathbb{R}^3$ or the inversion in S^2.

14. Show that

$$[\infty, \mathbf{b}, \mathbf{x}, \mathbf{y}] = d(\mathbf{b}, \mathbf{y})/d(\mathbf{x}, \mathbf{y}), \qquad [\mathbf{a}, \infty, \mathbf{x}, \mathbf{y}] = d(\mathbf{a}, \mathbf{x})/d(\mathbf{x}, \mathbf{y}),$$
$$[\mathbf{a}, \mathbf{b}, \infty, \mathbf{y}] = d(\mathbf{b}, \mathbf{y})/d(\mathbf{a}, \mathbf{b}), \qquad [\mathbf{a}, \mathbf{b}, \mathbf{x}, \infty] = d(\mathbf{a}, \mathbf{x})/d(\mathbf{a}, \mathbf{b}).$$

15. Use Theorem 4.5 to verify that *if a direct Möbius transformation* $f : \hat{\mathbb{R}}^3 \to \hat{\mathbb{R}}^3$ *has three fixed points, then it is the identity map.*

16. Prove that a Möbius transformation $\mathbf{T}$ that is conjugate

(a) to a rotation $\mathbf{R}_y(\theta)$ of S^2 is of the form $\mathbf{T}(z) = \frac{az-b}{bz+a}$,

(b) to a rotation $\mathbf{R}_x(\theta)$ of S^2 is of the form $\mathbf{T}(z) = \frac{az+ib}{ibz+a}$,

where $a, b \in \mathbb{R}$ are not both zero.

17. Show that:

(a) The map $f(z) = e^{2i\theta}z$, $\theta \in \mathbb{R}$, can be written in the form (4.4).

(b) The projected map $\tilde{f} = \pi(f(z)) = \left(\frac{2x\cos 2\theta - 2y\sin 2\theta}{|z|^2+1}, \frac{2x\sin 2\theta + 2y\cos 2\theta}{|z|^2+1}, \frac{|z|^2-1}{|z|^2+1} \right)$ for

$z = x + iy$, and $\tilde{f}$ is a rotation of the sphere by the angle 2θ about the vertical axis.

(c) If $f(z) = \frac{z\cos\theta + i\sin\theta}{iz\sin\theta + \cos\theta}$, then $\tilde{f}$ is a rotation of the sphere about the real axis. What is the angle of rotation of $\tilde{f}$? *Hint.* Consider $\pi(i)$ and $\pi(f(i))$.

18. Show that for any complex number $k \neq 0$:

(a) the Möbius transformation $f(z) = \frac{(k-2)z-2(k-1)}{(k-1)z-(2k-1)}$ fixes the points 1 and 2;

(b) the matrix $A = \begin{pmatrix} k-2 & k-1 \\ 2-2k & 1-2k \end{pmatrix}$ represents f so that A has eigenvectors $(1,1)$ and $(2,1)$, and the corresponding eigenvalues are $-k$ and -1.

19. Prove that a circle in $\mathbb{R}^2$ with equation $x^2 + y^2 + 2\alpha x + 2\beta y + \gamma = 0$ corresponds under stereographic projection to the circle cut out on S^2 by the plane with equation $2\alpha X + 2\beta Y + (1-\gamma)Z + (1+\gamma) = 0$.

Section 4.3   1. Show that the hyperbolic distance $\rho_H(A_1, A_2)$ (below Theorem 4.6) does not depend on which endpoint we call B_1 and which we call B_2: $|\ln([A_2, A_1, B_1, B_2])| = |\ln([A_2, A_1, B_2, B_1])|$. Show also that $\rho_H(A_1, A_2) = \rho_H(A_2, A_1)$.

2. Find the area in $\mathbb{R}_+^2$ of the doubly asymptotic triangle with vertices i, 1, and $1 + \sqrt{2}$.

3. Show that the area of a triply asymptotic triangle is π. *Hint.* See Figure 4.3(b):

$$S(\triangle LMN) = S(\triangle PLM) + S(\triangle PMN) + S(\triangle PNL)$$
$$= (\pi - \angle MPL) + (\pi - \angle MPN) + (\pi - \angle NPL) = 3\pi - 2\pi = \pi.$$

4. Show that for each $n \geq 3$ and $\alpha \in (0, \frac{n-2}{n}\pi)$, there is a regular hyperbolic n-gon whose interior angle is α. Plot examples.

5. Show that the dilation $T_\lambda(z) = \lambda z$ by $\lambda > 0$ can be thought of as a Möbius transformation which is represented by $\delta_\lambda = \begin{pmatrix} \lambda & 0 \\ 0 & 0 \end{pmatrix}$ of $SL_2(\mathbb{R})$.

6. Find a map $A \in SL_2(\mathbb{R})$ which sends $1 + i$ to i and ∞ to 1. Prove that A represents a rotation of $\mathbb{R}_+^2$.

Hint. An isometry A sends the line $l(1, \infty)$ to a Euclidean circle $\omega(i, 1)$. Hence, A sends 1 to -1. Thus

$$(z, i+1; 1, \infty) = (w, i; -1, 1) \iff \frac{z-1}{z-\infty} : \frac{i+1-1}{i+1-\infty} = \frac{w+1}{w-1} : \frac{i+1}{i-1}$$

$$\iff \begin{pmatrix} -1 & 1 \\ 0 & 1 \end{pmatrix} z = \begin{pmatrix} 1 & 1 \\ 1 & -1 \end{pmatrix} w \iff w = \begin{pmatrix} -1 & -1 \\ -1 & 1 \end{pmatrix} \cdot \begin{pmatrix} -1 & 1 \\ 0 & 1 \end{pmatrix} z = \begin{pmatrix} 1 & -2 \\ 1 & 0 \end{pmatrix} z.$$

To see whether $A = \begin{pmatrix} 1 & -2 \\ 1 & 0 \end{pmatrix}$ has a fixed point, solve

$$Az = z \iff (z-2)/z = z \iff z^2 - z + 2 = 0 \iff z = (1 \pm i\sqrt{7})/2.$$

```
syms z w;
S = solve(prod4(z, i + 1, 1, inf) - prod4(w, i, -1, 1), w)        % Answer: S = z-2/z
solve(S - z, z)                    % Answer: 1/2 - (i/2)√7,  1/2 + (i/2)√7
```

Since $z = \frac{1}{2}(1 + i\sqrt{7}) \in \mathbb{R}_+^2$, this map has a fixed point. Thus A is a rotation.

7. Prove Theorem 4.8. *Hint.* Using results of Example, p. 178, we obtain

$$\sigma\tau_t \, \sigma\tau_s \, \sigma\tau_r = \begin{pmatrix} 0 & -1 \\ 1 & t \end{pmatrix} \begin{pmatrix} -1 & -r \\ s & rs-1 \end{pmatrix} = \begin{pmatrix} -s & 1-rs \\ st-1 & rst-r-t \end{pmatrix}.$$

```
syms r s t a b c d;
sigma = [0 -1; 1 0]
B = sigma*taur(t)*sigma*taur(s)*sigma*taur(r)
```
$B = [-s, \ -sr+1; \ ts-1, \ (ts-1)r-t]$ % Answer
```
A = [a b; c d];    S = A - B
Sol = solve(S(1, 1), S(2, 1), S(1, 2), s, r, t);
Sol.r, Sol.s, Sol.t
```
$(b-1)/a, \ -a, \ -(c+1)/a$ % Answer
```
simplify(subs(S(2, 2), {r, s, t, d}, {Sol.r, Sol.s, Sol.t, (1 + b*c)/a} ))    % check d
0                                              % Answer
```

Thus, if $A = \begin{pmatrix} a & b \\ c & d \end{pmatrix} \in SL_2(\mathbb{R})$ and $a \neq 0$, then set $s = -a$ and solve $b = 1 - rs = 1 + ra$ and $c = st - 1 = -at - 1$. This gives $r = (b-1)/a$, $t = -(1+c)/a$. Since $\det A = 1$, this forces $d = rst - r - t$. Thus, if $a \neq 0$, then A can be written as a product involving only σ and translations. If $a = 0$, then $c \neq 0$, since $ad - bc \neq 0$, and hence $\sigma A = \begin{pmatrix} -c & -d \\ a & b \end{pmatrix}$, which can be written as a suitable product. So, $SL_2(\mathbb{R})$ is generated by translations and σ.

8. Find a Möbius transformation which fixes $P = a + ib$ and rotates counterclockwise through an angle t.

Hint. Find the Euclidean line through A_1 which makes an angle t with the vertical line. Find the perpendicular to this line and its intersection with the x-axis. The circle centered at this intersection and through A_1 is the image of the vertical line under the rotation. Let this circle intersect the x-axis at B_1 and B_2. The rotation is given by $[w, A_1, B_2, B_1] = [z, A_1, a, \infty]$. Let, for instance, $A_1 = i$. Let the center of a half-circle be $-x$, and the Euclidean radius of the circle be r. Then $r\cos t = x$, $r\sin t = 1$ and $B_1 = -r - x$, $B_2 = r - x$. After algebraic manipulation, find $w = \rho_t z$, where $\rho_t = \begin{pmatrix} \cos\frac{t}{2} & \sin\frac{t}{2} \\ -\sin\frac{t}{2} & \cos\frac{t}{2} \end{pmatrix}$. To find the rotation by an angle t about A_1, one should translate A_1 to i, then rotate, and translate back.

9. Prove (4.11). *Hint.* Use

$$(a\mathbf{q} + b)(c\mathbf{q} + d)^{-1} = (a\mathbf{q} + b)(\overline{c\mathbf{q} + d})(c\mathbf{q} + d)^{-1}(c\mathbf{q} + d)^{-1} = \frac{(a\mathbf{q} + b)(\overline{c\mathbf{q} + d})}{|c\mathbf{q} + d|^2}.$$

10. Show that the group $\text{Möb}(\mathbb{R}_+^3)$ is transitive, and that $\mathbf{T} = \begin{pmatrix} v^{1/2} & 0 \\ uv^{-1/2} & v^{-1/2} \end{pmatrix}$ transforms $\mathbf{j}$ to $u + v\mathbf{j}$.

Section 4.4   1. Find the Poincaré distance between $A_1 = 12 + 5i$ and $A_2 = 5 + 12i$. Answer in the form $\log(a/b)$, where a, b are positive integers.

2. Find the image of i under reflection through the line whose endpoints are 1 and 3.

3. What is the radius of the largest circle that can be inscribed in (a) a triangle, (b) a quadrilateral, and (c) a pentagon?

4. Plot the triangle
(a) with 1, 2 or 3 infinite vertices;
(b) symmetric to a given one with respect to (1) an axis, (2) a point.

5. With the help of the procedure **distance**, plot
(a) the central point of a segment,
(b) a doubling of a given segment,
(c) a given segment on a given ray.

6. Use the above programs to show that for "small" triangles the sum of their angles is close to π. Plot the bisector of an angle, and double an angle.

7. Plot the perpendicular to a segment through its midpoint using the procedure **perpendicular.m**.

8. Calculate the edge of, and then plot, the *regular 2n-polygon* the sum of whose angles is equal to 2π. (Gluing pairs of its edges will create a model for a surface of constant negative curvature.)

9. Plot the *circumcircle* for a given triangle. Note that a circumcircle does not always exist (for a given triangle), because through three points (in the hyperbolic half-plane) that do not belong to the same line, we can plot one of three curves: a circle, an equidistant, or a horocycle.

10. Plot the *inscribed circle* for a given triangle. This triangle always exists; its center lies on the intersection of the three angle bisectors.

11. Prove the **Pythagorean theorem**: *Let T be a triangle with angles α, β, $\pi/2$. Then* $\cosh(c) = \cosh(a)\cosh(b)$.

12. Let T be a triangle with angles $\alpha, 0, \frac{\pi}{2}$. Show that

    (1) $\sinh(b)\tan(\alpha) = 1$,
    (2) $\cosh(b)\sin(\alpha) = 1$,
    (3) $\tanh(b)\sec(\alpha) = 1$.

13. Let T be a triangle with angles $\alpha, \beta, 0$. Show that

    (4) $\cosh(c) = \dfrac{1 + \cos(\alpha)\cos(\beta)}{\sin(\alpha)\sin(\beta)}$,

    (5) $\sinh(c) = \dfrac{\cos(\alpha) + \cos(\beta)}{\sin(\alpha)\sin(\beta)}$.

14. Plot several lines of each of the metric fibrations, discussed on p. 190.

15. Calculate the angles in a triangle by developing an elementary geometry approach, by the method in **angle.m**, or by using the cosine theorem.

16. Verify the geometric constructions in the programs of Section A.7.

17*. Develop M-files to plot spheres, horospheres, equidistant surfaces, and tetrahedra in $\mathbb{R}^3_+$.

Part II
Curves and Surfaces

5

Examples of Curves

Section 5.1 starts from the basic notion of a regular curve. Then we investigate cycloidal curves and other remarkable parametric curves, as well as curves given implicitly as level curves of functions in two variables. The level sets are useful in solving problems with conditional extrema. We study the variational calculus for functions of one variable and employ it to derive Euler's spiral. In Section 5.2 we study some fractal curves. Section 5.3 discusses the basic MATLAB® capabilities for plotting space curves, surveys the parallel and perspective projections, and presents curves with shadows on planar, cylindrical, or spherical displays. Section 5.4 introduces helix-type curves on surfaces of revolution and studies curves obtained by the intersection of pairs of surfaces.

5.1 Plane Curves

Intuitively, according to Euclid, a *curve* is the trajectory of a moving point, a boundary of a surface, a one-dimensional figure.

The mathematically correct definition of a curve is based on notions from *topology*, but it starts from the key notion of an *elementary curve*, which can be imagined as an *interval* $I = (a, b)$ (or a *segment* $\bar{I} = [a, b]$) of the line after continuous deformation.

A set γ in $\mathbb{R}^n$ is called a *curve* if it can be covered by a finite or countable number of elementary curves.

One can distinguish *self-intersecting, simple* (i.e., without self-intersections; for example, graphs), *closed,* and *connected* curves.

There are many methods of classifying curves. One method is to divide them into *algebraic* and *transcendental* curves. An algebraic (plane) curve is given by a polynomial equation $P(x, y) = 0$. Its degree $n = \deg P$ is called the *order of the curve. Curves of order* $n = 2$ are studied in analytical geometry. The first classification of *curves of*

V. Rovenski, *Modeling of Curves and Surfaces with MATLAB®*,
Springer Undergraduate Texts in Mathematics and Technology 7, DOI 10.1007/978-0-387-71278-9_5,
© Springer Science+Business Media, LLC 2010

order $n = 3$ was obtained by Newton. The case $n > 3$ is more difficult. But among easily obtained curves there are many that are nonalgebraic, for example, the cycloid and the spiral of Archimedes; we study them by using parametric or implicit equations or by plotting them in polar coordinates (see Section 1.4).

The names of curves, like geographical names, contain much interesting information. We may group curves according to the meaning of their names:

By the name of a scientist — Dinostratus' quadratrix, conchoid of Nicomedes, Pascal's limaçon, lemniscate of Bernoulli, cissoid of Diocles, etc.

By the method of construction — caustics, equidistant (parallel) curves, pedal curves, evolutes, evolvents, etc.

By an important property — trisectrices, quadratrices, tractrices, etc.

By an essential property of their shape — astroid (star-shaped), deltoid (Greek letter Δ), cardioid (heart-shaped), nephroid (kidney-shaped), etc.

By historical factors — ellipse, parabola, hyperbola.

By the structure of the formula — (semi)cubical parabola, logarithmic spiral, exponential and logarithmic curves, sine (cosine) integral, etc.

5.1.1 *Parametric plane curves*

Let us fix rectangular coordinates (with the orthonormal basis $\{\mathbf{i}, \mathbf{j}\}$) in the plane with origin O. The coordinates x, y of a point P on the elementary curve γ are functions $x = x(t)$, $y = y(t)$ of $t \in I$. In other words, the position vector $\mathbf{r} = \overrightarrow{OP}$ of the point $P = (x, y)$ on the curve γ is the vector-valued function (the *parameterization*)

$$\mathbf{r}(t) = x(t)\mathbf{i} + y(t)\mathbf{j}, \qquad t \in I.$$

Example. A line in the plane is given by a linear vector-valued function in the variable t: $\mathbf{r}(t) = [a_1 t + b_1, a_2 t + b_2]$, or, in complex form, $z(t) = (a_1 + ia_2)t + (b_1 + ib_2)$.

We define functions $x(t)$, $y(t)$ and then introduce $\mathbf{r}(t)$ and $z(t)$.

```
x = @(t) 2*t + 3;   y = @(t) t - 1;        % define functions x(t), y(t)
r = [x(t), y(t)];                          % vector-function
z = x(t) + i*y(t)                          % complex parametric equation
```

One may derive in MATLAB some operations of calculus on vector-valued functions $\mathbf{r}(t) = [x(t), y(t)]$ of class C^k ($k \geq 0$).

The *limit* $\mathbf{r}(t_0) = \lim_{t \to t_0} \mathbf{r}(t)$ and the *derivative* $\mathbf{r}'(t) = \lim_{\Delta t \to 0} \frac{\mathbf{r}(t + \Delta t) - \mathbf{r}(t)}{\Delta t}$:

```
r0 = subs(r, t, 1),   z0 = subs(z, t, 1)                      % t_0 = 1
rt = diff(r, t);   zt = diff(z, t)
```

Similarly, we compute the derivatives $\mathbf{r}^{(k)}$ and *Taylor series* (Section 1.2).

A point P on a simple curve $\gamma \in \mathbb{R}^n$ is called *regular of class* C^k (C^∞ or C^ω) if some neighborhood of P admits a parameterization $\mathbf{r}(t) = [x_1(t), \ldots, x_n(t)]$, $t \in I$ ($\mathbf{r}(t) = [x(t), y(t)]$ for $n = 2$), where the functions $x_i(t)$ belong to class C^k (C^∞ or C^ω) and the vector $\mathbf{r}'(t)$ is nonzero at P. For the opposite case, a point P is called a *singular point* (see Figure 1.4(c) with a tractrix). A curve consisting of regular points is called a *regular curve of class* C^k (C^∞, C^ω) (*smooth* when $k = 1$).

Next we consider some remarkable parametric curves.

The trajectory of a point on the circle of radius R traveling (without sliding) along another circle of radius R' (or along a line) is a *cycloidal curve* (circle-shaped). Its parametric equations are

$$x = Rt - hR\sin(t), \qquad y = R - hR\cos(t). \tag{5.1}$$

If a circle drives along and inside of the motionless circle, then such a curve is a *hypocycloid*; if outside, then the curve is an *epicycloid*. If we follow a point not on the border of the wheel but on a spoke or on its continuation, then we obtain a *curtate* or *prolate trochoid* (wheel-shaped). Parametric equations of *trochoids* (*prolate* for $h > 1$ and *curtate* for $0 < h < 1$), with *modulus* $m = R/R'$, are the following:

$$\begin{cases} x = (R + mR)\cos(mt) - hmR\cos(t + mt), \\ y = (R + mR)\sin(mt) - hmR\sin(t + mt). \end{cases} \tag{5.2}$$

If m is represented in the form of an irreducible fraction a/b, then the cycloidal curve is closed with b branches; its period with respect to the parameter t is $2\pi b$. If the modulus m is irrational, then the curve consists of an infinite number of branches.

In the case $h = 1$ we have cycloidal curves: *epicycloids* for $m > 0$ and *hypocycloids* for $m < 0$. See Figures 5.1–5.3 for some tables of cycloidal curves.

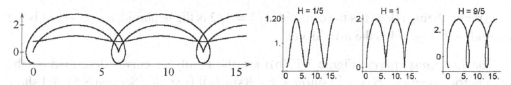

Fig. 5.1 Common cycloid, curtate and prolate cycloids.

Examples.

1. Let us plot the *cycloid* for $h = 1$ and its *trochoids: prolate* for $h > 1$ and *curtate* for $0 < h < 1$ (Figure 5.1).

```
hold on;  for i = 1 : 3;  h = 1 + (i - 2).*8/10;
 ezplot(t - h*sin(t), 1 - h*cos(t), [0 5*pi]);
end                                                        % Figure 5.1(a)
syms t;
hold on;  for i = 1 : 3;
 h = 1 + (i - 2).*8/10;
 subplot(1, 3, i);    ezplot(t - h*sin(t), 1 - h*cos(t), [0 5*pi]);
end                                                        % Figure 5.1(b)
```

2. One may generate tables of cycloidal curves (Figures 5.2, 5.3) with varying parameters a, b in the modulus. For an *epicycloid* the program is:

```
syms t;  h = 1;  hold on;  axis equal;
for a = 1 : 3;  for b = 3 : 5;
 x = (1 + a/b)*cos(a/b*t) - h*a/b*cos(t + a/b*t);
 y = (1 + a/b)*sin(a/b*t) - h*a/b*sin(t + a/b*t);
 subplot(3, 3, 3*(a - 1) + b - 2);  ezplot(x, y, [0  2*b*pi]);
end;  end
```

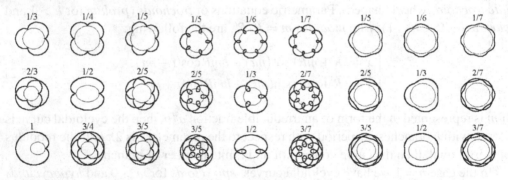

Fig. 5.2 Epicycloids.   Prolate epicycloids.   Curtate epicycloids.

Among them, $a/b = 1$ is the cardioid, $a/b = 1/2$ is the nephroid, $a/b = -1/3$ is the deltoid, $a/b = -1/4$ is the astroid, etc.

3. The *Lissajous curves* (Figure 8.17(b)) are the family of curves described by the parametric equations $\mathbf{r}(t) = [A\sin(at + \varphi), B\sin(bt)]$ (see also Section 8.5) and show a complex harmonic motion (Lissajous figure on an oscilloscope). Their appearance is highly sensitive to the ratio a/b (for a ratio of 1, the figure is an ellipse). Lissajous figures are sometimes used in graphic design as logos.

```
a = 2;  b = 1;  phi = 0;  A = 1;  B = 1;                   % enter your data
syms t;  ezplot(A*sin(a*t + phi), B*sin(t), [0 4*pi]);
```

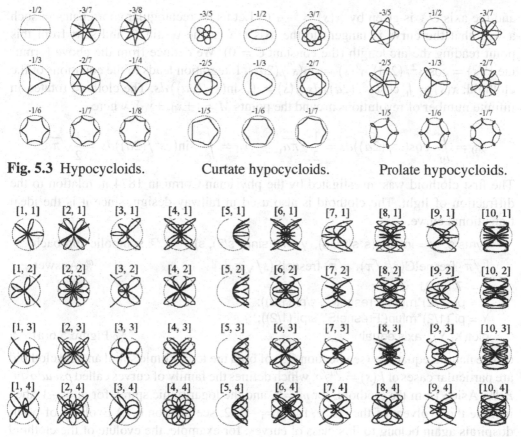

Fig. 5.3 Hypocycloids.   Curtate hypocycloids.   Prolate hypocycloids.

Fig. 5.4 Experiment with polar coordinates: $k = 5$.

4. Plot a number of interesting figures by inserting into the formulae relating *polar* and *rectangular* coordinates $[\rho \cos(t), \rho \sin(t)]$ (see Section 1.4) two additional parameters A, B.

Now substituting in the equations $[\rho \cos(At), \rho \sin(Bt)]$ the relationship $\rho = \sin(kt)$ (for *roses*, see Section 1.4) for $k = 5$, and then changing A and B, you will obtain a splendid view: see Figure 5.4.

```
N = 4; k = 5;                                    % define parameters
hold on; axis equal;
for a = 1 : 4; for b = 1 : N;
 x = sin(k*t)*cos(a*t); y = sin(k*t)*sin(b*t);
 subplot(4, N, 4*(a - 1) + b); ezplot(x, y, [0 2*pi]);
end; end
```

5. The *clothoid* (Cornu spiral) is the curve whose *curvature* (see Section 6.4) is proportional to its arc length: $k(s) = s/a$, and hence the angle between the tangent line

and the axis OX is given by $\alpha(s) = \frac{s^2}{2a} + C$. Let us fix rectangular coordinates in such a way that the curve is tangent to the axis OX at the origin O and start from this point reading the arc length (the constant $C = 0$). We deduce from the above formulae $x'(s) = \cos(s^2/(2a))$, $y'(s) = \sin(s^2/(2a))$. Integration leads to the equations of the clothoid: $x(s) = \int_0^s \cos(s^2/(2a))\,ds$, $y(s) = \int_0^s \sin(s^2/(2a))\,ds$. The clothoid rotates an infinite number of revolutions around the points $M = (\pm x_0, \pm y_0)$, where

$$x_0 = \int_0^\infty \cos(s^2/(2a))ds = \frac{1}{2}\sqrt{\pi a}, \qquad y_0 = \int_0^\infty \sin(cs^2/(2a))ds = \frac{1}{2}\sqrt{\pi a}.$$

The first clothoid was investigated by the physician Cornu in 1874 in relation to the diffraction of light. The clothoid is also used in railway design, since it is the ideal transitional curve.

```
syms s;  x = int(cos(s*s/2), s),  y = int(sin(s*s/2), s)        % symbolic approach
```
$$[\sqrt{\pi}\cdot\mathrm{fresnelC}(s/\sqrt{\pi}),\ \sqrt{\pi}\cdot\mathrm{fresnelS}(s/\sqrt{\pi})] \qquad \text{\% Answer}$$
```
s = -7 : 0.01 : 7;
X = pi (1/2)*mfun('FresnelC', s/pi^(1/2));
Y = pi^(1/2)*mfun('FresnelS', s/pi^(1/2));
plot(X, Y);  axis equal;                                       % Figure 5.6(a)
```

6. The natural equations (see Section 6.3) of both the logarithmic spiral and the clothoid are particular cases of $k(s) = s^m/a$, which defines the family of curves called *pseudospirals*. (Aside from the clothoid, for $m = 1$, and the logarithmic spiral, for $m = -1$, they include the evolvent of the circle, for $m = -1/2$; see Section 6.1.) Evolutes of pseudospirals again belong to this class of curves; for example, the evolute of the clothoid for $m = 3$.

One can study a more general relationship $k(s) = f(s)$. The curves having natural equations $k(s) = \frac{1}{a}\cos s$ are called *patterned curves*. These curves have a similar shape to *elastica* (Example, p. 215) studied in mechanics.

We plot the patterned curves when $a = \frac{\pi}{2}$, π, 2π (Figure 5.5).

```
syms s t n;  n = 25;
X = @(t)int(cos(2*pi*sin(s)), 0, t);  Y = @(t)int(sin(2*pi*sin(s)), 0, t);
tt = -2*pi : 2*pi/n : 2*pi;
hold on;   for m = 1 : n;
x(m) = double(X(tt(m)));    y(m) = double(Y(tt(m)));
plot(x(m), y(m), '.');
end
```

5.1.2   *Level lines and extremal problems*

An implicitly given curve $f(x,y) = c$ can be plotted using ezplot.

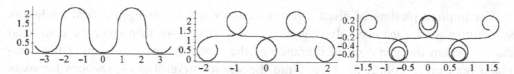

Fig. 5.5 Patterned curves (pseudospirals): $a = \frac{\pi}{2}, \pi, 2\pi$.

The command contour allows one to obtain several implicitly given curves $f(x,y) = c$ for the values $c = c_1, \ldots, c_n$.

Example. Plot some implicitly given curves by using ezplot

    ezplot('x.^3 + y.^3 - 3*x.*y'), axis equal   % 0-level line of $f = x^3 + y^3 - 3xy$

or by using meshgrid and then contour.

    x = -pi : .01 : pi;   [X, Y] = meshgrid(x);   % $f = \sin(xy)$
    contour(X, Y, sin(X.*Y), [-0.8, -1/2, 0, 1/2, 0.8])   % Figure 5.7(a).
    xy = -2 : .01 : 2;   [X, Y] = meshgrid(xy);   % folium of Descartes,
    contour(X, Y, X.^3 + Y.^3 - 3*X.*Y, [0 0])   % Figure 5.6(c).

Plot a perturbed folium of Descartes with equation $x^3 + y^3 - 3xy \pm 0.01$.
Let us find parametric equations for the folium of Descartes.
Set $t = y/x$ and substitute in the implicit equation:

$$x^3 + t^3 x^3 - 3x(tx) = 0 \iff x = 3t/(1+t^3), \qquad \text{where } t \neq -1.$$

Hence $y = 3t^2/(1+t^3)$. Then we plot (two parts of) the curve:

    syms t;   x = 3*t/(1 + t^3);   y = t*x;
    hold on;   ezplot(x, y, [-50, -1.3]);   ezplot(x, y, [-.8, 50]);

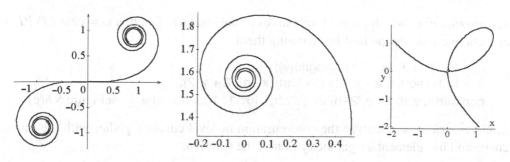

Fig. 5.6 Clothoid.   *SICI* spiral.   Folium of Descartes.

For implicit plotting of given *algebraic curves*, the following practical method is sometimes used. One can solve equations (of degree 2–4), then enter the command plot to obtain the graph of each branch of the function, and finally enter the command hold on to collect the pieces into the whole curve. The expressions for roots are often complicated, but their graphs are plotted exactly, and the obtained curve is glued together from different colored branches. We recommend that the reader use both methods to plot some curves of third and fourth degree: *folium of Descartes, cissoid, strophoid, trisectrix of Maclaurin, cardioid, Nicomedian conchoid, lemniscate of Bernoulli,* and *kappa.* Some of these curves are also successfully plotted in polar coordinates, as shown in Section 1.4.

Level sets of functions are useful in problems with extrema. The problem of *relative extrema* is formulated as follows: *What is the maximum (minimum) of the function f defined in the plane along the given curve γ?*

Theorem 5.1. *Let the differentiable function F be given in the domain $G \subset \mathbb{R}^2$ and assume the gradient ∇F is nonzero along γ_c: $F(x,y) = c$ (i.e., the curve γ_c of level c for F is a regular curve). Assume also that $g(x,y)$ is a smooth function on G and that $P \in \gamma_c$ is an extremal point of g on γ_c. Then the tangent line (see Section 6.1) to γ_c at the point P is orthogonal to the gradient $\nabla g(P)$; i.e., there exists a real number λ such that $\nabla g(P) = \lambda \nabla F(P)$ (λ is called a* Lagrange multiplier*).*

The equation $\nabla g = \lambda \nabla F$ is equivalent to the system

$$F(x,y) = c, \qquad F'_x(x,y) = \lambda g'_x(x,y), \qquad F'_y(x,y) = \lambda g'_y(x,y).$$

Example. An excursion bus travels along a highway (the line *l*). There is a castle (the line segment $P_1 P_2$ of length *d*) at some distance and elevated at some angle from the highway. What is the optimal stopping point of the bus on the highway (the point $P \in l$) from which the front of the castle can be viewed from *P* under the maximal angle?

Let us use this problem to study and compare different possibilities of MATLAB for solving problems of conditional extremum.

1. *Experimental:* we plot some level curves of the *objective function* $G = \cos(\angle P_1 P P_2)$ and choose visually the best from among them.

```
v = -2 : .01 : 3;  [x, y] = meshgrid(v);
d = 1;  f = sqrt(x.^2 + y.^2);  g = sqrt((x - d).^2 + y.^2);
contour(x, y, (f.^2 + g.^2 - d.^2)./(2*f.*g), [0 : .1 : 1]), axis equal      %Figure 5.8(c)
```

2. *Geometrical:* we analyze the configuration of level curves together with the given curve and use elementary geometry to find the solution.

3. *Analytical:*

   (a) we reduce the problem to the case of an unconditional extremum; or

(b) we apply the method of Lagrange multipliers.

Hint. **Method 2.** Let the function $f(P) = \angle P_1PP_2$ reach its maximum at $P \in l$. The value of the angle of view φ of the segment P_1P_2 is constant on arcs of circles through the endpoints of the segment, i.e., level curves of the function $f(P)$. A local maximum is reached at a point on the line l that is *tangent* to one of arcs of the family of level curves.

The problem has two solutions. Let us fix orthogonal coordinates in the plane such that $P_1 = (-d/2, 0)$ and $P_2 = (d/2, 0)$; then write down the equation of the line $l: ax + by = 1$. Find the center $O' = (0, y)$ of the arc tangent to l from the equation $|O'P_1| = d(O', l)$: $\sqrt{y^2 + d^2/4} = \frac{|by-1|}{\sqrt{a^2+b^2}}$ (the second root corresponds to the best view of the back side of the castle). We find the optimal angle φ from the equation $\tan \varphi = d/(2|y|)$.

```
a = 1/2; b = -1/2; d = 1;      % Answers in what follows are for symbolic a, b, d
syms y;
f1 = sqrt(y^2 + d^2/4); f2 = (b*y - 1)/sqrt(a^2 + b^2);
solve(f1^2 - f2^2, y)
```

$$\frac{-4b+2\sqrt{4b^2-a^4d^2-a^2d^2b^2+4a^2}}{4a^2}, \quad \frac{-4b-2\sqrt{4b^2-a^4d^2-a^2d^2b^2+4a^2}}{4a^2} \qquad \text{% Answer.}$$

Method 3a. Let us fix orthogonal coordinates such that $P_1 = (0,0)$, $P_2 = (d, 0)$, and write down the equation of the highway $l: ax + by = 1$. Represent the point on the line l in the form $P = (x = \frac{1-bt}{a}, y = t)$. Let us find $F = \cos(\angle P_1PP_2)$ by the cosine theorem and substitute, in place of the variables f, g, h, the sides of $\triangle(P_1PP_2)$ expressed in terms of coordinates. We differentiate the function $F(t)$ and find three roots.

```
f = sqrt(t^2 + ((1 - b*t)/a)^2): g = sqrt(t^2 + (d - (1 - b*t)/a)^2):
F = simple((f^2 + g^2 - d^2)/(2*f*g))
```

$$F = \frac{t^2a^2+b^2t^2-2bt+1+dabt-da}{a^2\sqrt{\frac{t^2a^2+b^2t^2-2bt+1}{a^2}}\sqrt{\frac{t^2a^2+d^2a^2+2dabt-2da+b^2t^2-2bt+1}{a^2}}} \qquad \text{% Answer.}$$

```
dF = simple(diff(F, t))
```

$$dF = \frac{d^2a^2t(b^2t^2+t^2a^2+da-1)}{(t^2a^2+b^2t^2-2bt+1)\sqrt{(t^2a^2+b^2t^2-2bt+1)/a^2}} \times$$

$$\times \frac{1}{(t^2a^2+d^2a^2+2dabt-2da+b^2t^2-2bt+1)\sqrt{\frac{t^2a^2+d^2a^2+2dabt-2da+b^2t^2-2bt+1}{a^2}}} \qquad \text{% Answer.}$$

```
T = [solve(dF, t)];  P = simple([(b*T(2) - 1)/a, T(2)])
```

$$T = [0, \frac{\sqrt{(b^2+a^2)(1-da)}}{b^2+a^2}, -\frac{\sqrt{(b^2+a^2)(1-da)}}{b^2+a^2}]$$

$$P = [-\frac{-b\sqrt{(a^2+b^2)(1-da)}+a^2+b^2}{(a^2+b^2)a}, \frac{\sqrt{(a^2+b^2)(1-da)}}{a^2+b^2}] \qquad \text{% Answer.}$$

Method 3b. For deriving the maximum of the function

$$G = \cos(\angle P_1PP_2) = (f^2 + g^2 - d^2)/(2fg),$$

where $f = \sqrt{x^2 + y^2}$, $g = \sqrt{(x-d)^2 + y^2}$ with the restriction $F = ax + by - 1 = 0$, it is sufficient to solve the system grad $G = \lambda$ grad F with respect to x, y, and λ under the given restriction. Since the line segment $[P_1, P_2]$ does not intersect l, the function F has the same sign at the points P_1 and P_2, that is, $ad < 1$.

```
a = 1/2;  b = -1/2;  d = 1;
syms x y lambda;
f = sqrt(x^2+y^2);  g = sqrt((x - d)^2 + y^2);
F = a*x + b*y - 1;  G = simplify((f^2 + g^2 - d^2)/(2*f*g));
GradG = [diff(G, x), diff(G, y)];  GradF = [diff(F, x), diff(F, y)];
Eq1 = simplify(GradG(1) - lambda*GradF(1));
Eq2 = simplify(GradG(2) - lambda*GradF(2));
[X Y Lambda] = solve(F, Eq1, Eq2, x, y, lambda);
x0 = X(1),  y0 = Y(1),  lambda0 = Lambda(1)
```

$$x_0 = \left(\frac{b\sqrt{-(b^2+a^2)(-1+da)}}{b^2+a^2} - 1 \right) a^{-1}, \quad y_0 = \frac{\sqrt{-(b^2+a^2)(-1+da)}}{b^2+a^2} \qquad \% \text{ Answer.}$$

```
G0 = subs(G, [x y lambda], [x0 y0 lambda0])
```

Conclusion: The point of the best view with $y > 0$ has coordinates

$$x_0 = \left(\frac{b\sqrt{(b^2+a^2)(1-da)}}{b^2+a^2} - 1 \right) a^{-1}, \quad y_0 = \frac{\sqrt{(b^2+a^2)(1-da)}}{b^2+a^2}.$$

5.1.3 *Trajectories of a vector field and ODEs*

The trajectories of a vector field $V = (x, F(x,y))$ and solutions of the ordinary differential equation (ODE) $y' = F(x,y)$ are the same. They can be approximated by the Euler or *tangent line method*. If the step size has a uniform value, this method is expressed by the equation

$$y_{i+1} = y_i + F(x_i, y_i)h, \quad i = 0, 1, \dots. \tag{5.3}$$

Euler's method consists of repeatedly evaluating (5.3), using the result of each step to execute the next step. We create a short M-file with an ODE solver for $\frac{dy}{dx} = F(x,y)$ with $y(x_0) = y_0$ on the interval $x_0 \le x \le x_1$.

```
function ode1(F, x0, x1, y0)        % illustrative program uses Euler's method
   global Y;       % to solve dy/dx = F(x,y) with y(x0) = y0 on x0 ≤ x ≤ x1.
   h = (x1 - x0)/50;   x = x0;   y = y0;   Y = y0;     % number of steps N = 50
   while x < x1
   y = y + h*F(x, y);   x = x + h;
   Y = cat(2, Y, y);
   end;
```

```
    plot(x0 : h : x1, Y)
end
```

To obtain the integral curves for the vector field $V = (f_1(x,y), f_2(x,y))$, we solve an ODE using the command dsolve and then plot the graphs of the solutions. Several procedures in MATLAB help us to study vector fields.

Examples.

1. Find the trajectory of a vector field $V = (x, x^2 + y^2)$ through the point $(0, 1)$ using **ode1**.

```
F = @ (x, y) x^2 + y^2;   ode1(F, 0, 1, 1);          % Obtain a trajectory of V
Y                                                     % the values {y_i}
```

2. The numerical gradient of a function is derived by using the command gradient. The graph of the gradient vector field is obtained in Figure 5.7(b) by using the command quiver.

```
v = -pi : .3 : pi;  [x, y] = meshgrid(v);  z = sin(x.*y);
[fx, fy] = gradient(z, .2, .2);  quiver(v, v, fx, fy)
```

3. Symbolic solutions to ODEs.

```
dsolve('Dx = -x/a', 'x(0) = 1')                          % Answer 1/ exp(t/a)
dsolve('D2x + 5*Dx + 6*x = 10*sin(t)', 'x(0) = 0', 'Dx(0) = 0')
```
$2/\exp(2t) - 1/\exp(3t) - \cos(t) + \sin(t)$ % Answer
```
[x, y] = dsolve('Dx=x^2', 'Dy=y^2', 'x(0)=1, y(0)=1')      % 2-D system
```
$x = -1/(t-1), \quad y = -1/(t-1)$ % Answer

Numerical solutions to ODEs.

```
ODE1 = inline('x(1)*(.1 - .01*x(1))', 't', 'x');                    % enter ODE
[t, x1] = ode45(ODE1, [0 100], 50);
plot(t, x1(:, 1))                          % graph of a solution, Figure 5.8(a)
ODE2 = inline('[.1*x(1) + x(2);  -x(1) + .1*x(2)]', 't', 'x');      % enter ODEs
[t, x2] = ode45(ODE2, [0 50], [.01, 0]);                            % 2-D system
plot(x2(:, 1), x2(:, 2))                                            % Figure 5.8(b)
```

4. Let us study the effect of the interaction coefficients, A and B, in the Lotka–Volterra predator–prey model. We create an M-file, **lotka.m**.

```
function yp = lotka(t, y)
    global A B;
    yp = [y(1) - A*y(1)*y(2);  -y(2) + B*y(1)*y(2)];
end
```

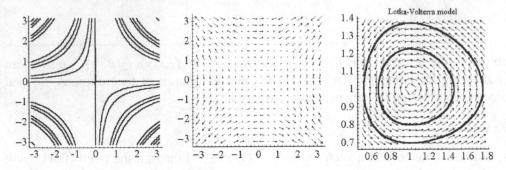

Fig. 5.7 Level curves, and   vector fields.

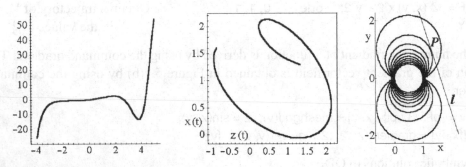

Fig. 5.8 Trajectories of ODEs in the plane. Level curves   and   extremum.

We plot two integral curves and also the direction field for the following system of ODEs (Figure 5.7(c)):

```
global A B;    A = 1;  B = 0.3;
[t, y1] = ode23(@lotka, [0,7], [1; 1]);                        % 1st curve
[t, y2] = ode23(@lotka, [0,7], [1; 0.7]);                      % 2nd curve
vx = 1 : .3 : 8.5;  vy = 0 : .2 : 2.5;  [xx, yy] = meshgrid(vx, vy);
fx = xx - A*xx.*yy;  fy = -yy + B*xx.*yy;
hold on;  plot(y1(:,1), y1(:,2));  plot(y2(:,1), y2(:,2), 'r');
quiver(vx, vy, fx, fy);
title('Lotka-Volterra Predator-Prey Population Model', 'FontWeight', 'bold')
```

The command quiver3(X, Y, Z, U, V, W) (the 3-D analog of quiver) draws arrows to indicate a gradient or arbitrary 3-D vector field. Here X, Y, Z are matrices that describe some surface in $\mathbb{R}^3$ (Chapters 7, 8). U, V, W are the matrices of the components of vectors that are to be attached at each of the points (x, y, z) that lie on the surface.

Examples.

1. We plot the vector field $\mathbf{V} = [zy, zx, xy]$ given for $-1 \leq x, y, z \leq 1$.

```
v = -1 : .3 : 1;   [X, Y, Z] = meshgrid(v);
quiver3(X, Y, Z, Z.*Y, Z.*X, X.*Y);                              % Figure 5.9(a)
```

The gradient of a scalar function $f(x,y,z)$ is $\nabla f = \left[\frac{\partial f}{\partial x}, \frac{\partial f}{\partial y}, \frac{\partial f}{\partial z}\right]$. The command gradient computes the gradient of a vector field. For example,

```
v = -2: .5 : 2;   vz = -8 : 2 : 0;   [X, Y, Z] = meshgrid(v, v, vz);
V = X.^2 + Y.^2 + Z;                                             % enter function
[fx, fy, fz] = gradient(V, .5, .5, 2);   quiver3(X, Y, Z, fx, fy, fz);
```

2. Plot the space curve $\mathbf{r}(t) = (x_1(t), x_2(t), x_3(t))$ that is the solution of a system of three ODEs (Figure 5.9(c)).

```
ODE3 = inline('[x(2) - x(3); x(3) - x(1); x(1) - 2*x(2)]', 't', 'x');
[t, x3] = ode45(ODE3, [-2 2], [1, 2, 2]);              % 3-D system, 1st curve
[t, x4] = ode45(ODE3, [-2 2], [1, 2, 0]);              % 3-D system, 2nd curve
hold on;  grid on;  view(20, 70);
plot3(x3(:, 1), x3(:, 2), x3(:, 3));   plot3(x4(:, 1), x4(:, 2), x4(:, 3))
```

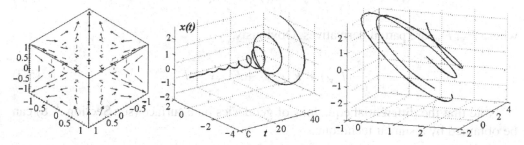

Fig. 5.9 Vector fields and trajectories in space.

5.1.4 *Euler's equations*

In this section we shall employ techniques of the *calculus of variations*, which deals with minimization of functionals. A *functional* is a quantity that depends on a set of functions. A fundamental result of the calculus of variations is that the extreme values of a functional must satisfy an associated differential equation (or a set of differential equations) over the discussed domain. These are generally termed the *Euler equations*. The notion "extreme values" stands for local minima, maxima, or inflection points. The calculus of variations is a fundamental analytical tool in many areas of general physics and engineering.

Functional with functions of one variable. We find extreme values of an integral functional J whose integrand F contains one or several functions associated with the admissible function $y(x) \in C^2$, $x \in [x_0, x_1]$.

Example. Consider the problem

$$J(y) = \int_{x_0}^{x_1} F(x, y, y') \, dx \to \min, \tag{5.4}$$

where F is a continuous function of three arguments (the problem of determining a maximum may be dealt with by replacing F with $-F$). The boundary values of $y(x)$ are generally given as

$$y(x_0) = y_0, \qquad y(x_1) = y_1. \tag{5.5}$$

The minimization of the functional $J(y)$ leads to the *Euler equation* for its integrand (a necessary condition that $J(y)$ possess a stationary value)

$$F_{,y} - \frac{d}{dx}(F_{,y'}) = 0, \tag{5.6}$$

where $F_{,y}$, $F_{,y'}$ are partial derivatives. Obviously,

$$\frac{d}{dx}(F_{,y'}) = F_{,y'y'}y'' + F_{,y'y}y' + F_{,y'x}. \tag{5.7}$$

Note that the differential equations of geodesics on a surface (see Section A.8) can be obtained by a similar technique.

Example. We compute the Euler equation (5.6) for a given function.

```
F = x^2 + y^2 + dy^2;          % define F (Here F = (y'(x))^2 + x^2 + (y(x))^2).
syms x y dy;
part1 = subs(diff(F, y), {y, dy}, {'y(x)', diff('y(x)', x)});
dFdy = diff(F, dy);
dFddy = subs(dFdy, {y, dy}, {'y(x)', 'dy(x)'});
p2 = diff(dFddy, x);
part2 = subs(p2, 'dy(x)', diff('y(x)',x));
simplify(part1 - part2)        % Euler's equation 2y(x) - 2\frac{d^2}{dx^2}y(x).
```

Variational problems with constraints. Consider a case where *Lagrange multipliers* should be employed within a variational problem (5.4) with the *isoperimeter conditions*

$$J_k(y) = \int_{x_0}^{x_1} F_k(x, y, y') \, dx = g_k, \qquad 1 \le k \le n, \tag{5.8}$$

where g_k are constants and F_k are continuous functions of three arguments. The corresponding Lagrange functional has the form

$$J_L(y) = \int_{x_0}^{x_1} \left[F(x,y,y') + \sum_{k=1}^{n} \lambda_k F_k(x,y,y') \right] dx. \tag{5.9}$$

In this case we solve the variational problem $J_L(y) \to \min$ by considering the Lagrange multipliers λ_k to be constants.

Example. (Elastica.) The analysis described in this example deals with large deformations of an elastic rod. Problems of this kind are traditionally termed "elastica."

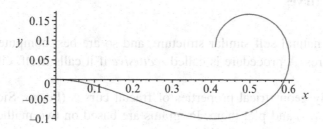

Fig. 5.10 Euler's spiral for $l = 1$, $\theta_e = \frac{3}{2}\pi$, $\tilde{g} = 1$.

The *bending energy* of an *elastic rod* of length l that is deformed along a plane curve $\mathbf{r}(s) = [x(s), y(s)]$ is assumed, for simplicity, to be proportional to the integral of the squared curvature over the length of the curve, i.e., $\int_0^l \kappa^2(s)\,ds$. Here $s \in [0,l]$ is the natural length parameter (see Section 6.3). The *curvature* $\kappa(s)$ of a plane curve is $\frac{d\theta}{ds}$ (see Section 6.4), where $\theta(s)$ is the angle between the local tangent line and the x-axis. One may therefore ask the following question: *what shape will the curve take if the total turning of its tangent is given and the turning is zero at one endpoint and θ_e at the other?* As a constrained variational problem, we write

$$J = \int_0^l (\theta')^2 ds \to \min, \quad J_1 = \int_0^l \theta\,ds = \tilde{g} \text{ with } \theta(0) = 0, \quad \theta(l) = \theta_e.$$

The Lagrange functional of (5.9) takes the form $J_L(y) = \int_0^l [(\theta')^2 + \lambda\theta]\,ds$, and F depends on two of the three variables s, θ, θ' and on the parameter λ. The Euler equation for the constrained problem is of the form of (5.6),

$$\lambda - 2\theta'' = 0. \tag{5.10}$$

Using the boundary conditions, one may find the solution $\theta = \frac{\lambda}{4}s^2 - \frac{\lambda l^2 - 4\theta_e}{4l}s$ depending on the parameter λ, which with the constraint $J_1 = \tilde{g}$ gives

$$\widetilde{g} = \left(\frac{\lambda}{12} s^3 - \frac{\lambda l^2 - 4\theta_e}{8l} s^2 \right) \Bigg|_0^l = \frac{l}{24} \left(12\theta_e - \lambda l^2 \right).$$

Hence,

$$\lambda = 12(\theta_e l - 2\widetilde{g})/l^3, \quad \theta = (3\theta_e - 6\widetilde{g}/l)(s/l)^2 + (-2\theta_e + 6\widetilde{g}/l)(s/l).$$

The resulting curve is known as *Euler's spiral* (or the *spiral of Cornu*), and an example of it is shown in Figure 5.10.

5.2 Fractal Curves

Fractals have a natural self-similar structure, and so are best computed by using recursive procedures. A procedure is called *recursive* if it calls itself, either directly or indirectly.

We will study geometrical properties of fractal curves (Peano, Sierpiński, Koch, and Menger curves) and plot them. Programs are based on two methods of deriving and plotting self-similar (fractal) objects:

(1) *symmetry and periodicity*, and (2) *recursion*.

Example. The *Cantor set* is well known from analysis. The following program plots "Cantor stairs" (i.e., the graph of a continuous function on $[0, 1]$) by recursion.

```
function [x, y] = cantor_stairs(n)
    [x0, y0] = cantor_stairs(n - 1);               % recursion
    x = (1/3)*[x0 x0+2];    y = .5*[y0 1+y0];
    axis square;  plot(x, y, '-');
end
```
```
    [x, y] = cantor_stairs(3);                      % example: n = 3, Figure 5.11
```

A geometrical figure (set) that can be broken into a finite number of equal figures similar to the given one is called a *self-similar figure*.

Simple examples are the segment, the triangle, the square, and the cube, as shown in Figure 5.13. Other self-similar figures in Figures 5.11–5.15 look more complicated, but they can be plotted easily.

5.2.1 *Peano curves*

If one starts from parametric equations of the curve in the form of a continuous vector-valued function $\mathbf{r}(t)$, where t ranges over the segment $[a, b]$, but considers only the

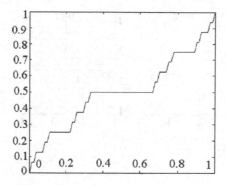

Fig. 5.11 Cantor stairs ($n = 3$).

image, i.e., the set of points without taking account of their order, then one arrives at the notion of a curve formed in the nineteenth century by C. Jordan. Moreover, such continuous images of the segment can fill the square, the cube, etc. On other hand, there exist objects similar to curves that are not continuous images of the segment.

Example.

Plot the graph of the function $\sin \frac{1}{x}$ ($0 < x \le 1$) and the segment $\{(0,y) : -1 \le y \le 1\}$.

```
T = -1 : 0.1 : 1;
hold on;  plot(T - T,  T,  '-r');   ezplot('sin(1/x)',  [-1 1])
```

The *Peano curve* (discovered by G. Peano in 1890) is a continuous image of the segment filling the interior of a square (or triangle). The Peano curve is related to the existence of simple curves in space whose projection onto the plane is in the form of filled areas, such as, for example, the curve $[f_1(t), f_2(t), t]$ where the first two functions define the Peano curve. Although this curve would serve nicely as a roof to keep out the rain, it cannot be identified with any continuous surface.

Example. We use recursion for plotting *Hilbert's curve*. This polygon H_1 is shaped like the letter Π, plotted in the form of three edges of a square. The polygon H_2 can be considered as a letter Π four parts of which are replaced by the same letters Π of one-third size, etc. The coefficient of similarity of the curve H_n is equal to $2^n - 1$.

```
function [x,y] = hilbert(n)
    if n <= 0
    x = 0;  y = 0;
    else
    [x0, y0] = hilbert(n - 1);                          % recursion
    x = .5*[-.5+y0 -.5+x0 .5+x0 .5-y0];  y = .5*[-.5+x0 .5+y0 .5+y0 -.5-x0];
end
```

```
    [x, y] = hilbert(3); line(x, y);            % example: n = 3; Figure 5.12(a–c)
```

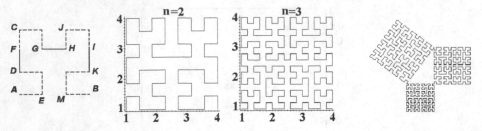

Fig. 5.12 Hilbert's curve.   Pythagorean triangle.

5.2.2 *Koch snowflake*

Starting from an equilateral triangle with edge a, one may repeat (recursively) the following process. Divide each segment connecting the vertices of the polygon into three parts and replace the middle part by two segments of length $a_1/3$, where a_1 is the length of the given segment. For the nth step, we obtain a polygon similar to a snowflake, which is called *Koch's snowflake* (in honor of its discoverer, the Swedish mathematician Helge von Koch.)

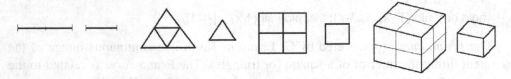

Fig. 5.13 The simplest self-similar objects.

Example. One may use a program **koch**(n) with recursion to plot the Koch curve after n iterations.

    koch(4);   % example: $n = 4$

The **kline** local function generates the curve using recursive calls, while the **plotline** local function is used to plot the curve.

```
function koch(n)
    global X1, Y1, X5, Y5, i
    if (n == 0)
    x = [0; 1];  y = [0; 0];
    line(x, y, 'Color', 'b');  axis equal;
    else
```

```
   lc = 10^n;  L = lc/(3^n);  l = ceil(L);
   kline(0, 0, lc, 0, l);
   axis equal;  set(gcf, 'Name', 'Koch Curve')
end
...........................................................
function kline(x1, y1, x5, y5, lm)
   global X1, Y1, X5, Y5, i
   length = sqrt((x5 - x1)^2 + (y5 - y1)^2);
   if (length > lm)
      x2 = (2*x1 + x5)/3;  y2 = (2*y1 + y5)/3;
      x3 = (x1 + x5)/2 - (y5 - y1)/(2.0*sqrt(3.0));
      y3 = (y1 + y5)/2 + (x5 - x1)/(2.0*sqrt(3.0));
      x4 = (2*x5 + x1)/3;  y4 = (2*y5 + y1)/3;
      kline(x1, y1, x2, y2, lm);  kline(x2, y2, x3, y3, lm);
      kline(x3, y3, x4, y4, lm);  kline(x4, y4, x5, y5, lm);   % recursion
   else  plotline(x1, y1, x5, y5);
   end
...........................................................
function plotline(a1,b1,a2,b2)
   global X1, Y1, X5, Y5, i
   x = [a1; a2];  y = [b1; b2];
   X1(i) = a1;  Y1(i) = b1;  X5(i) = a2;  Y5(i) = b2;
   i = i + 1;   line(x, y);
```

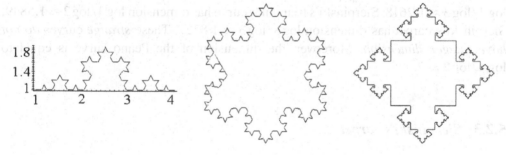

Fig. 5.14 Koch curves.

There exist several definitions of *dimension* leading to essentially different mathematical results. The **first definition** is related to the minimal number of coordinates necessary for uniquely specifying the location of points in a figure. In this case the dimension is an integer.

In the **second definition** of the (topological) dimension, a figure's dimension is taken to be one more than the dimension of a section dividing the figure into two separate parts. A set consisting of a finite (or countable) number of points is said to be null-dimensional. By this definition a smooth curve is one-dimensional, a plane (divided by a one-dimensional line) is two-dimensional, and a ball in space is three-dimensional. The topological dimension is also an integer.

The **third definition** of dimension is most closely related to our theme.

The *dimension of self-similarity* D is defined by the formula $D = \frac{\log N}{\log n}$, where N is the number of n-times-smaller equal parts into which a self-similar figure can be broken.

We plot sections dividing the square into $N = 4$ squares with edge $n = 2$ times smaller than the initial edge (Figure 5.13(c)), since the dimension of self-similarity of a square is $D = \frac{\log 4}{\log 2} = 2$. Analogously, for a segment $D = \frac{\log 2}{\log 2} = 1$, and for a cube (Figure 5.13(d)) $D = \frac{\log 8}{\log 2} = 3$, as desired.

Attempts at measuring the lengths of other self-similar curves lead to analogous results. When the scale of measuring is decreased, the length of the curve grows to infinity. This explains, for example, the difference of 20% in the length of the border between Portugal and Spain given (possibly) in reference books of these countries; different scales were probably used to measure the border. One may write down the formula for the length of the Koch curve in the form

$$L = A\lambda^{-\beta}, \quad \text{where } A = 3a^{\log 4/\log 3}, \quad \beta = \log 4/\log 3 - 1.$$

The coefficient β in the formula is related to the dimension of the object.

In deriving the dimensions of the polygon in Figures 5.14, 5.15, and 5.23, we find that the dimension of each part (and hence the whole) of the Koch curve is equal to $\log 4/\log 3 \approx 1.2618$; Sierpiński's triangular curve has dimension $\log 3/\log 2 \approx 1.5849$; Sierpiński's carpet has dimension $\log 8/\log 3 \approx 1.8727$. These *strange* curves *do not have integer dimension*. Moreover, the dimension of the Peano curve is equal to $\log 4/\log 2 = 2$.

5.2.3 *Sierpiński's carpet*

A square with edge a (the square of zero rank) is broken by four lines into nine equal squares with edge $\frac{1}{3}a$, and the interior of the middle square is removed. For each of the other eight closed squares (of first rank), which form the set C_1, the above process is repeated. We then obtain 64 squares of the second rank, whose union is C_2, and so on. Denote by C_n the union of all 8^n squares of nth rank with edges $a/3^n$. This set is connected and compact. Moreover, $C_{n+1} \subset C_n$ holds.

The intersection C of all the C_n is called *Sierpiński's carpet* or the Sierpiński (universal plane) curve. The curve is a two-dimensional generalization of the Cantor set.

Example. Plot Sierpiński's carpet, Figures 5.15(a,b). Derive the common area of all the removed squares.

Hint. Find the area of all removed squares as the following sum:

$$a^2 \left(\frac{1}{3^2} + \frac{8}{3^4} + \frac{8^2}{3^6} + \cdots \right) = \frac{a^2}{8} \sum_{i=1}^{\infty} \left(\frac{8}{9} \right)^i = \frac{a^2}{8} \left(\frac{1}{1 - 8/9} - 1 \right) = \frac{a^2}{8} \cdot 8 = a^2,$$

which coincides with the area of the given square. Hence, Sierpiński's carpet (the complement to all the removed squares) has zero area, which explains its status as a curve.

```
function sierpinski(n)
   S = 0;
   for k = 1 : n
   S = [S, S, S;  S, ones(3^(k-1)), S;  S, S, S];
   end
   imagesc(S);  colormap(gray),  axis equal,  axis off;
end
   sierpinski(3);                                          % example: n = 3
```

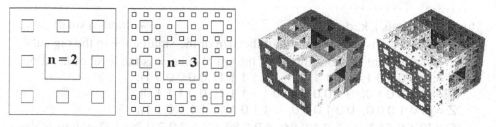

Fig. 5.15 Sierpiński's carpet.   Menger cube.

5.2.4 *Menger cube*

A 3-D analogue of Sierpiński's carpet is obtained as follows; Figures 5.15(c,d). Divide a cube (of zero rank) with edge a by six planes, parallel to its faces, into 27 equal cubes with edge $a/3$. Then we remove the central cube and all its neighbors along two-dimensional faces of this division. We obtain the set K_1 consisting of 20 cubes of the first rank. Repeating the same with each of the other closed cubes of first rank, obtain the set K_2 of 400 cubes of the second rank. Continuing this process we obtain the sequence of polyhedra $K_1 \supset K_2 \supset K_3 \supset \cdots$, whose intersection M is called the *Menger cube* (or *universal Menger curve*). Its universal property means that any space curve can be embedded into the set M.

Example. We plot the polyhedron K_n.

```
function menger(n);
    M = 0;
    for k = 1 : n
    A = zeros([3^k, 3^k, 3^k]);
    A(:, :, 1 : 3^(k-1)) = [M, M, M; M, ones(size(M)), M; M, M, M];
    A(:, :, 3^(k-v1) + 1 : 2*3^(k-1)) = ...
    [M, ones(size(M)), M; ones(size(M)), ones(size(M)), ones(size(M));
    M, ones(size(M)), M];
    A(:, :, 2*3^(k-1) + 1 : 3^k) = [M, M, M; M, ones(size(M)), M; M, M, M];
    M = A;
    end
    hold on;  d = 1;
    for i = 1 : 3^n;  for j = 1 : 3^n;  for k = 1 : 3^n
    if M(i, j, k) == 0;  cube(i, j, k, d, n);  end;        % sub-program cube
    end;  end;  end
    axis equal;  axis off;  view(3);
```

....

```
function cube(i, j, k, d, n);                             % draws a small cube;
            % d is the proportion of the size of the small cube to the big cube,
        % i, j, k are the coordinates of the small cube, n is the level of iteration.
    X = [0 0 0 0 0 1;  1 0 1 1 1 1;  1 0 1 1 1 1;  0 0 0 0 0 1];
    Y = [0 0 0 0 1 0;  0 1 0 0 1 1;  0 1 1 1 1 1;  0 0 1 1 1 0];
    Z = [0 0 1 0 0 0;  0 0 1 0 0 0;  1 1 1 0 1 1;  1 1 1 0 1 1];
    C = [0.1 0.84 1.1 1.1 0.1 0.84;  0.2 0.86 1.2 1.2 0.2 0.86;     % colour scales
    0.3 0.88 1.3 1.3 0.3 0.88;  0.4 0.90 1.4 1.4 0.4 0.90 ];        % for each cube
    X = (d*(X - 0.5) + i)/3^n;  Y = (d*(Y - 0.5) + j)/3^n;  Z = (d*(Z - 0.5) + k)/3^n;
    fill3(X, Y, Z, C);                                    % draw cube
end
```

menger(2);   % example: $n = 2$

5.3   Space Curves and Projections

5.3.1   *Intuitive projection*

How is a space curve $\mathbf{r}(t) = [x(t), y(t), z(t)]$ (or more generally, any space figure) plotted on a planar computer display? Sometimes one creates the image by intuition, as in the following *method of plotting the hyperboloid of one sheet*, using our knowledge of a *star-shaped polygon*.

Example. The vertices of the star-shaped (n,m) polygon are $P_k = [R\cos(2\pi km/n),$ $R\sin(2\pi km/n)]$, $1 \le k \le n$ (see Section 1.4). We transform the circumscribed circle of this polygon into an ellipse and raise and lower it (in the plane) a distance of ± 1 and to obtain two star-shaped polygons. Now we connect the ith vertex of one of them with the $(i+1)$st vertex of another, and obtain a part of the hyperboloid of one sheet. One may realize this plan by two methods. We plot the "intuitive projection" of the hyperboloid of one sheet by both methods. The first one is applicable when the greatest common divisor of m and n, gcd(m, n), is 1.

```
m = 11;   n = 50;                         % first method, Figure 5.16(b)
k = 1 : m*n;  T = k.*2*m*pi/n;
hold on;  axis equal;  plot(cos(T(k)), .5*sin(T(k)) + (-1).^k, 'r-');
t = 0 : .01 : 2*pi;   plot(cos(t), .5*sin(t) + 1,  cos(t), .5*sin(t) - 1);
```

The second method is applicable for all n, m.

```
m = 8;  n = 3*m;  T = @(j) j*2*pi/n;      % second method, Figure 5.16(c)
hold on;  axis equal;
t = 0 : .01 : 2*pi;   plot(cos(t), .5*sin(t) + 1,  cos(t), .5*sin(t) - 1);
for k = 1 : n;
  plot([cos(T(k)), cos(T(k+m))], [0.5*sin(T(k)) - 1, 0.5*sin(T(k+m)) + 1], 'r-');
end;
```

A two-dimensional (2-D) projection of a space curve $\mathbf{r}(t) = [x(t), y(t), z(t)]$ in Cartesian coordinates is plotted using either plot3 or ezplot3.

Examples. (2-D in 3-D.)

1. Show that the curve $\mathbf{r}(t) = [1 + 3t + 2t^2, 2 - 2t - 4t^2, 1 - t^2]$ lies in a plane and describe it.

Solution. Note that the curve $\mathbf{r}(t) = \mathbf{r}_1 + t\mathbf{r}_2 + t^2\mathbf{r}_3$ lies in a plane if the three vectors $\mathbf{r}_1$, $\mathbf{r}_2$, $\mathbf{r}_3$ are coplanar, i.e., the determinant consisting of their coordinates vanishes. Any two (linearly independent) vectors from among them define the plane. In our case:

```
r1 = [1, 2, 1];  r2 = [3, -2, 0];  r3 = [2, -4, -1];
det([r1; r2; r3])                              % Answer: 0
syms t U V;
r = r1 + r2*t + r3*t^2;    R = r1.*U + r2.*V;        % a curve and a plane
ezplot3(r(1), r(2), r(3), [-1, 2]); hold on; ezsurf(R(1), R(2), R(3))% Figure 5.16(a)
```

2. Plot closed curves in $\mathbb{R}^3$ using cycloidal curves (plane curves): astroid $(m = -1/4)$, deltoid $(m = -1/3)$, nephroid $(m = 1/2)$, and roses, etc.

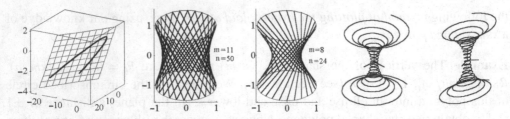

Fig. 5.16 (a) Plane curve in space. (b,c) Hyperboloid.   (d,e) Spiraling helix.

```
m = -1/4;  h = 1;                    % "3-dimensional" astroid (enter your data)
syms t;
r = [(1+m)*cos(m*t)-h*m*cos(t+m*t), (1+m)*sin(m*t)-h*m*sin(t+m*t), cos(t)];
ezplot3(r(1), r(2), r(3),  [0, 2*pi/abs(m)]);
```

3. Plot *two engaged circles* (tori).

```
hold on;  axis equal;  view(65, 25);
plot3(cos(t), sin(t), t - t, 'r', 'LineWidth', 5);
plot3(t - t,  cos(t) + .8,  sin(t),  'g', 'LineWidth', 5);          % Figure 5.17(a)
```

Fig. 5.17 Two engaged circles. Short chain. Chain armor.

4. Plot the perspective projection of a parallelepiped. The position of the eye is at $(0, 0, 10)$, so $d = 10$.

```
x = [-2, -2, 0, 2, 2, 0, 0, 0, -2, 0, 2];
y = [0, -2, -2, 0, 2, 0, -2, 0, 0, 2, 2];
z = [3, 3, 3, 1, 1, 3, 3, 3, 3, 1, 1];
d = 10;    s = d* ones(1, length(x));
x1 = d*x./(s - z);    y1 = d*y./(s - z);
w = zeros(1, length(x));
plot3(x1, y1, w);  view(0, 90);  axis equal;               % parallelepiped
```

To contrast, we simply project our box on the xy-plane. We use x, y, and w from the previous program, and add these statements:

```
plot3(x, y, w);  view(0, 90);  axis equal;  axis off                    %  parallelepiped
```

Other than Cartesian coordinates, there are two popular methods for fixing the location of a point P in space.

(a) Consider the *height z*, being the projection of the segment OP onto the axis OZ; the *distance* $|OP'| = \rho$, where P' is the projection of the point P onto the plane XY; and the *angle* $\varphi \in [-\pi, \pi]$ between the segment OP' and the axis OX. The triple (ρ, φ, z) is called the *cylindrical coordinates* of the point P. Their relation with the rectangular coordinates (x, y, z) is analogous to the case of polar coordinates:

$$x = \rho \cos \varphi, \qquad\qquad y = \rho \sin \varphi, \qquad\qquad z = z.$$

The coordinate surfaces are the following: cylinders $\rho = $ const, half-planes $\varphi = $ const through the axis OZ, and also horizontal planes $z = $ const.

(b) Consider the *distance* $|OP| = \rho$, the *(vertical) angle* $\phi \in [-\pi, \pi]$ between the segment OP and the axis OZ, and the *(horizontal) angle* $\theta \in [0, \pi]$ between the segment OP' and the axis OX. The triple (ρ, θ, ϕ) is called the *spherical coordinates* of the point P. Their relation with rectangular coordinates is given by

$$x = \rho \cos \theta \sin \phi, \qquad\qquad y = \rho \sin \theta \sin \phi, \qquad\qquad z = \rho \cos \phi.$$

The coordinate surfaces are spheres $\rho = $ const with centers at O, half-planes $\theta = $ const through the axis OZ, and circular cones $\phi = $ const with the common axis OZ. We plot figures in cylindrical/spherical coordinates by transforming the data,

```
[X,Y,Z] = pol2cart(rho, theta, z)
[X,Y,Z] = sph2cart(phi, theta, z)
```

and then applying the appropriate plot command (see examples in Section 5.4).

There is a variety of systems of curvilinear coordinates in $\mathbb{R}^3$ (in addition to the above three), and we can define a number of new ones in MATLAB.

In addition to the space *coordinates of an object*, an important role is also played by the *coordinates of the viewer*. Typing view(a, b) points the direction (in spherical coordinates) from which we look at the object with respect to the coordinate system of the object. The default is view(-37.5, 30), while view(2) is equivalent to view(0, 90) and gives a two-dimensional view of a surface looking down from above; view(3) sets the default three-dimensional view, $a = -37.5$, $b = 30$.

5.3.2 *Oblique and axonometric projections*

An *orthographic* projection of an object shows the details of only one of its main faces, which is why at least three projections are needed.

Axonometric projections show more of the object in each projection, at the price of having the wrong dimensions and angles.

Here is a summary of the properties of axonometric projections:

- Axonometric projections are parallel, so parallel lines on the object will appear parallel in the projection.
- There are no vanishing points. Thus, a wide image can be scrolled slowly while different parts of it are observed.
- Distant objects retain their size regardless of their distance from the observer. If the parameters of the projection are known, then the dimensions of any object can be computed from measurements taken on the projection.
- There are standards for axonometric projections. A standard may specify the orientation of the object relative to the observer, making it easy to compute distances directly from the projection.

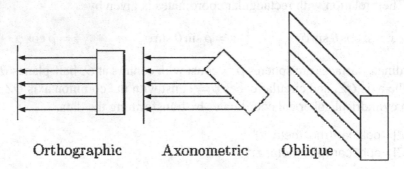

Orthographic   Axonometric   Oblique

Fig. 5.18 Comparing parallel projections.

For a clearer understanding of how to plot images using *parallel projection*, consider a simplified model that reduces the problem to the basic two-dimensional command plot. We also project the coordinate system $\{O; \mathbf{i}, \mathbf{j}, \mathbf{k}\}$ of space. Let O' be the origin of rectangular coordinates $\{O'; \mathbf{e}_1, \mathbf{e}_2\}$ of the two-dimensional window, and let $O'X'$, $O'Y'$ be the horizontal and vertical axes.

We consider two examples of projections $p\colon \mathbb{R}^3 \to \mathbb{R}^2$ that are similar to the standard *isometric* and *dimetric* axonometric projections (studied in greater detail in *descriptive geometry*). By the *Pohlke–Schwarz theorem*, the system $\{O; \mathbf{i}, \mathbf{j}, \mathbf{k}\}$ can be projected (within the transformation of similarity in the plane of the image) onto an arbitrary triple of vectors through the point O'.

Assume that the projections of the axes OX, OY form the angles

$$\pm(90° + \arctan(1/2)) \approx \pm 117°$$

with the axis $O'Y'$, and the projection of OZ coincides with the axis $O'X'$: $p(O) = O'$, $p(\mathbf{i}) = -\mathbf{e}_1 - \frac{1}{2}\mathbf{e}_2$, $p(\mathbf{j}) = \mathbf{e}_1 - \frac{1}{2}\mathbf{e}_2$, $p(\mathbf{k}) = \mathbf{e}_2$. In view of the *linearity* of projections, we obtain

$$x'\mathbf{e}_1 + y'\mathbf{e}_2 = p(x\mathbf{i} + y\mathbf{j} + z\mathbf{k}) = xp(\mathbf{i}) + yp(\mathbf{j}) + zp(\mathbf{k})$$
$$= x\left(-\mathbf{e}_1 - \frac{1}{2}\mathbf{e}_2\right) + y\left(\mathbf{e}_1 - \frac{1}{2}\mathbf{e}_2\right) + z\mathbf{e}_2 = (y - x)\mathbf{e}_1 + \left(z - \frac{x+y}{2}\right)\mathbf{e}_2.$$

Hence,

$$x' = y - x, \qquad\qquad y' = z - (x+y)/2.$$

The point $P = (x, y, z)$ projects onto the point $P' = (y - x, z - (x+y)/2)$ of the display. Let us call this projection *isometric*.

   Dimetric projection is defined analogously: the axes OY, OZ are projected onto the axes $O'X'$, $O'Z'$ of the two-dimensional window, and the projection of the axis OX is parallel to the bisector of the three-dimensional coordinate angle, namely,

$$x' = y - x/2, \quad y' = z - x/2.$$

Example. (Simplified axonometric projection.) Let us plot the *isometric projection* of a spiraling helix on the catenoid (the surface of revolution for the catenary); Figure 5.16(d).

```
s = 4;  v1 = [-s -s/2];  v2 = [s -s/2];  v3 = [0 s];
t = linspace(-3, 3, 600);  f = cosh(t);  g = 4*t;          % enter f, g
a = 20;  X = f.*cos(a*t);  Y = f.*sin(a*t);  Z = g;
hold on;  axis equal;  plot(Y - X, Z - (X + Y)/2, 'r-');    % isometric
arrow([0 0], v1);  arrow([0 0], v2);  arrow([0 0], v3);    % see p. 70
```

This program is illustrative only, but the basic approach can be implemented in various computer programs. For plotting images of space objects below, the appropriate commands in MATLAB are used.

5.4 More Examples of Space Curves

5.4.1 *Springs on surfaces*

There are well-known surfaces obtained by revolving a plane curve γ (the *meridian*) around a line (the *axis of revolution*) in space.

Trajectories of points of the curve γ (*parallels*) are circles contained in planes orthogonal to the axis of revolution. Let the curve γ lie in the plane XOZ and be given by parametric equations $\gamma\colon x = f(t),\, y = 0,\, z = g(t)$. Substituting $u = t$ and denoting by $v \in [0, 2\pi]$ the *angle of revolution* of the initial meridian, we obtain the following *equations of the surface of revolution:*

$$\begin{aligned}
\mathbf{r}(u,v) &= f(u)\cos(v)\mathbf{i} + f(u)\sin(v)\mathbf{j} + g(u)\mathbf{k} \\
&= [f(u)\cos(v),\, f(u)\sin(v),\, g(u)]; \qquad \text{see also Section 8.1.} \qquad (5.11)
\end{aligned}$$

The idea of *knitting* on the surface of revolution on the display is as follows. If one were to wind a thread on a transparent surface of revolution, then the shape of the surface would be visible. This effect exists for a dense winding of a surface of revolution, i.e., when the coils almost lie along parallels. To realize this idea, substitute t for u and at for v in (5.11), where the real parameter a is sufficiently large (say, $a = 27$):

$$\begin{aligned}
\mathbf{r}(t) &= [f(t)\cos(at),\, f(t)\sin(at),\, g(t)], \qquad \text{or} \\
\mathbf{r}(t) &= [f(t),\, at,\, g(t)] \qquad\qquad\qquad \text{in cylindrical coordinates.}
\end{aligned}$$

We plot springs on surfaces of revolution, taking

$f = R,\, g = t$ for the *helix* on a circular cylinder,
$f = t,\, g = bt$ for the circular conic helix,
$f = R\cos(t),\, g = R\sin(t)$ for the spring on a sphere,
$f = a + b\cos(t),\, g = b\sin(t)$ (with $a > b$) for the torus knot,
$f = a\sin(t),\, g = a\left(\cos(t) + \ln(\tan\frac{t}{2})\right)$ for the pseudosphere, etc.

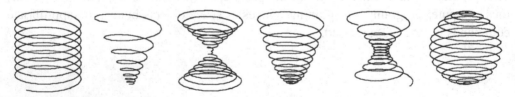

Fig. 5.19 Springs (curves) on algebraic surfaces of second order.

Examples.

1. Plot the following springs on surfaces of revolution using cylindrical and spherical coordinates:

(a) The curve $\mathbf{r}(t) = \left[at\cos t, at\sin t, \frac{a^2t^2}{2p}\right]$ lies on the paraboloid of revolution, and projects onto the plane XY as Archimedes' spiral.

```
a = 1;  t = 0:.01:24*pi;
[X, Y, Z] = pol2cart(t, a*t, t.^2/2);          % cylindrical → Cartesian
plot3(X, Y, Z);                                % Figure 5.19(d)
```

(b) The curve $\mathbf{r}(t) = [\cosh(at)\cos(t), \cosh(at)\sin(t), \sinh(at)]$ lies on a hyperboloid of revolution (Figure 5.19(e)).

```
a = .06;  t = -12*pi : .01 : 12*pi;
[X, Y, Z] = pol2cart(t, cosh(a*t), sinh(a*t));
plot3(X, Y, Z), axis equal;
```

(c) The curve $\mathbf{r}(t) = [a\cos t\cos(kt), b\cos t\sin(kt), c\sin t]$ lies on an ellipsoid (on a sphere if $a = b = c$, Figure 5.19(f)).

```
R = 1;  a = 28;  t = 0 : .01 : 2*pi;
[X, Y, Z] = sph2cart(a*t, t, R);               % spherical → Cartesian
plot3(X, Y, Z), axis equal;
```

2. Plot several complicated curves on a sphere (similar to *macrame*).

```
m = 7;  n = 10;  t = -pi:.01:pi;  [X, Y, Z] = sph2cart(m*t, n*t, 1);
plot3(X, Y, Z, 'LineWidth', 6), axis equal     % Figure 5.20(b)
```

For $m = 1$, $n = 17$ and $|t| \leq \frac{\pi}{2}$ the curve consists of 17 meridians, Figure 5.20(a).

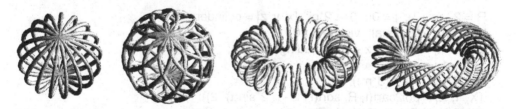

Fig. 5.20 Closed curves on a sphere and knots on a torus.

3. Tie a *torus knot* of arbitrary type on the torus of revolution and also plot a part of the *irrational winding*. For the torus knot $\mathbf{K}_{8,27}$, one may assume $R_1 = 3$, $R_2 = 1$ and $u = 27t$, $v = 8t$.

```
m = 27;  n = 8;  t = -pi : .01 : pi;              % take n = 1 to obtain Figure 5.20(c).
X = (3 + cos(m*t)).*cos(n*t);  Y = (3 + cos(m*t)).*sin(n*t);  Z = sin(m*t);
plot3(X, Y, Z, 'LineWidth', 5),  axis equal                    % Figure 5.20(d).
```

5.4.2 *Curves as intersections of surfaces*

We deduce explicit and parametric equations of several curves given as intersections of pairs of surfaces, and plot them.

Bi-cylindrical curve. The axes (OX and OZ) of two circular cylinders of radii $R \geq r > 0$ intersect at a right angle, and at the intersection of the cylinders we obtain the curve *bicylinder*, which for $R \neq r$ consists of two closed curves (Figure 5.21(d)), and for $R = r$ breaks up into two intersecting ellipses that lie in the planes $x = y$ and $x = -y$. The implicit and parametric equations of the curve are the following:

$$
\begin{cases} x^2 + z^2 - r^2 = 0, \\ y^2 + z^2 - R^2 = 0 \end{cases} \Longleftrightarrow \begin{cases} x = r\cos t, \\ y = \pm\sqrt{R^2 - r^2\sin^2 t}, \qquad t \in [0, 2\pi]. \\ z = r\sin t, \end{cases}
$$

First, plot two intersecting cylinders.

```
r = .5;  u = 0 : .15 : 2*pi + .15;  v = -2 : .15 : 2;
[U, V] = meshgrid(u, v);
surf(cos(U), sin(U), V);  hold on;  axis equal;         % bigger cylinder
surf(r*cos(U), V, r*sin(U))                             % smaller cylinder
```

Then plot the family of the bicylinders under a change in the radius of one of the cylinders.

```
R = 2;  m = 9;  t = 0 : .01 : 2*pi;  [x, y, z] = cylinder;
hold on;  axis equal;  view(40, 20);
surf(R*x, R*y, R*(2*z - 1));  shading('interp');
for i = 1 : m;
r = R*((2*m + i)/(3*m));
[X, Y, Z] = pol2cart(t, R, sqrt(R^2 - r.^2*sin(t).^2));
plot3(X, Y, Z);  plot3(X, Y, -Z);
end;                                                    % cylindrical → Cartesian
```

Viviani window. The sphere of radius R (with center O) intersects the cylinder of diameter R, one of whose rulings (parallel to the axis OZ) contains the center of the sphere. The intersection is the *Viviani curve* (Figures 5.21(b) and 8.9(a)). The equations of the curve are

$$\begin{cases} x^2 + y^2 + z^2 = R^2, \\ x^2 + y^2 - Rx = 0 \end{cases} \iff \mathbf{r}(t) = (R\cos^2 t, R\cos t \sin t, R\sin t),$$

where $t \in [0, 2\pi]$ is the longitude of points on a sphere. A point of self-intersection $(R, 0, 0)$ on a Viviani curve breaks it into two loops.

First, plot a sphere and a cylinder (Figure 8.9(b)).

```
r = .5;                                              % try r < 1
hold on;  axis equal;
[X, Y, Z] = sphere(20);  surf(X + r, Y, Z);          % a sphere
[x, y, z] = cylinder;  surf(r*x, r*y, 2*z - 1);      % a cylinder
```

Show that the projections of a Viviani curve onto the coordinate planes YZ, XZ, and XY look like a figure eight, a parabola, and a circle, respectively (Figure 5.21(c)).

```
syms t;
ezplot3(cos(t)^2, cos(t)*sin(t), sin(t),  [0, 2*pi]),  axis equal;
```

Sphere intersects ellipsoid. The sphere of radius R (with center O) intersects the standard ellipsoid with the axes $a > b > c$. The best method to parameterize these curves comes from mechanics, namely, from solutions of Euler's differential equations of the rotation of a rigid body about a fixed point in space:

$$l_1' + \left(\frac{1}{J_2} - \frac{1}{J_3}\right) l_2 l_3 = 0, \quad l_2' + \left(\frac{1}{J_3} - \frac{1}{J_1}\right) l_3 l_1 = 0, \quad l_3' + \left(\frac{1}{J_1} - \frac{1}{J_2}\right) l_1 l_2 = 0. \quad (5.12)$$

Here J_i are the moments of inertia of a rotating body about the principal axes at time t, ω_i are the angular velocities about these axes, and $l_i = J_i \omega_i$ are the corresponding components of the angular momentum. We assume that $J_1 < J_2 < J_3$ and $l^2 > 2EJ_2$. Euler's equations (5.12) have two integrals, and the (periodic) trajectory of the motion lies at the intersection of the sphere and the ellipsoid:

$$\begin{cases} l_1^2 + l_2^2 + l_3^2 = l^2, \\ \dfrac{l_1^2}{J_1} + \dfrac{l_2^2}{J_2} + \dfrac{l_3^2}{J_3} = 2E \end{cases} \iff \begin{cases} l_1 = J_1 \omega_1^0 \mathrm{cn}(z, k), \\ l_2 = J_2 \omega_2^0 \mathrm{sn}(z, k), \\ l_3 = J_3 \omega_3^0 \mathrm{dn}(z, k). \end{cases}$$

Here $\mathrm{cn}(z, k)$, $\mathrm{sn}(z, k)$, and $\mathrm{dn}(z, k)$ are the *Jacobi elliptic functions* (with period $4K$, where $K = $ Elliptic K(k), see Section 1.2); we call them CN(z,k), SN(z,k), and DN(z,k) in MATLAB, and

$$\omega_1^0 = \sqrt{\frac{2EJ_3 - l^2}{J_1(J_3 - J_1)}}, \qquad \omega_2^0 = \sqrt{\frac{2EJ_3 - l^2}{J_2(J_3 - J_2)}}, \qquad \omega_3^0 = \sqrt{\frac{l^2 - 2EJ_1}{J_3(J_3 - J_1)}},$$

$$z = \sqrt{\frac{(J_3 - J_2)(l^2 - 2EJ_1)}{J_1 J_2 J_3}}\, t, \qquad k = \sqrt{\frac{(J_2 - J_1)(2EJ_3 - l^2)}{(J_3 - J_2)(l^2 - 2EJ_1)}}.$$

Plot the curve $\mathbf{r}(t) = [X, Y, Z]$ (solution of Euler's equations (5.12) for $Y(0) = 0$) with the corresponding intersecting ellipsoid and sphere, varying the radius R of the sphere.

```
J1 = 1;  J2 = 2;  J3 = 16;  L = 2.1;  E = 1.1;                    % enter data
L^2 - 2*E*J2                                          % check that the expression > 0
omega01 = sqrt(( 2*E*J3 - L^2) / (J1*(J3 - J1)));
omega02 = sqrt(( 2*E*J3 - L^2) / (J2*(J3 - J2)));
omega03 = sqrt((-2*E*J1 + L^2) / (J3*(J3 - J1)));
k = sqrt((J2 - J1)*(2*E*J3 - L^2) / ((J3 - J2)*(L^2 - 2*E*J1)));
K = mfun('EllipticK', k);
t = 0 : .01 : 4*K + 1.7;   z = sqrt((J3 - J2)*(L^2 - 2*E*J1)/(J1*J2*J3))*t;
[SN, CN, DN] = ellipj(z, k);
X = J1*omega01*CN;  Y = J2*omega02*SN;  Z = J3*omega03*DN;
hold on;  axis equal;  view(0, 30);
plot3([X' X'], [Y' Y'], [Z' -Z'], 'LineWidth', 2);               % a curve
ellipsoid(0, 0, 0, L, L, L);                                     % a sphere
ellipsoid(0, 0, 0, sqrt(2*E*J1), sqrt(2*E*J2), sqrt(2*E*J3));    % an ellipsoid
```

Fig. 5.21 A circular helix and Viviani curve with shadows.   Bicylindric.

5.4.3 Curves with shadows

Unfortunately, in MATLAB figures of three-dimensional objects have no shadows. The *shadow of the point* $P = (x, y, z)$ in the plane ω under *parallel* lighting in the direction $\mathbf{d}$

is the point $Q = (x', y', z')$ of (3.5). Methods to obtain shadows on a cylinder or a sphere are also presented in what follows.

We plot the shadow (as a plane curve) of a space curve, the circular helix $X = R\cos t$, $Y = R\sin t$, $Z = Vt$, on a given plane.

(a) The *parallel projections* of the circular helix onto the plane orthogonal to its axis, depending on the correlation of angles among the axis, the direction of projection, and the tangent lines to the curve, are the usual curtate or prolate cycloids; see Figure 5.22(a). Let $\omega \colon Ax + By + Cz = 0$ be the plane of shadow and $\mathbf{d} = (a, b, c)$ be the (unit) direction of light, which is not parallel to the plane. The formula for the shadow transformation is shown in (3.5). If $\mathbf{d} \perp \omega$, i.e., $\mathbf{d} \parallel \mathbf{n} = (A, B, C)$, the projection associates to a point in space the closest point in ω, and so the projection is called *orthogonal*.

In the program we define $\mathbf{d}$ by two angles (ϕ, ψ) in spherical coordinates. Hence, $a = \sin\phi\cos\psi$, $b = \sin\phi\sin\psi$, $c = \cos\psi$.

```
A = 0;  B = 0;  C = 1;  R = 5;  V = .8;  phi = 45;  psi = 50;        % enter data
a = sind(phi)*cosd(psi);  b = sind(phi)*sind(psi);  c = cosd(psi);
t = 0: .01 : 7*pi;    X = R*cos(t);  Y = R*sin(t);  Z = V*t;
Xs = ((b*B + c*C)*X - a*B*Y - a*C*Z) / (a*A + b*B + c*C);
Ys = (-b*A*X + (a*A + c*C)*Y - b*C*Z) / (a*A + b*B + c*C);
Zs = (-c*A*X - c*B*Y + (a*A + b*B)*Z) / (a*A + b*B + c*C);
hold on;  axis equal;  view(40, 25);
plot3(X, Y, Z, 'LineWidth', 2);                              % a curve
plot3(Xs, Ys, Zs, 'LineWidth', 3);                           % a shadow
plane([1 2 -.1], [0 0 1], 20, 20);                           % a tray
```

(b) The *central projection* (shadow) of a circular helix onto the plane orthogonal to its axis from a point on the helix is a cochleoid; see Figure 5.21(a).

Let $S = (0, 0, h)$ be the center of projection, so the shadow $P' = (X_s, Y_s, Z_s)$ of a point $P = (X, Y, Z)$ is an intersection of $\omega \colon \{Z = 0\}$ with the ray SP. The formulae for the shadow transform $(X, Y, Z) \rightarrow (X_s, Y_s, Z_s)$ are easy:

$$X_s = \frac{h}{h - Z} X, \qquad\qquad Y_s = \frac{h}{h - Z} Y, \qquad\qquad Z_s = 0.$$

```
R = .8;  h = .2*8*pi;
t = 0 : .01 : 16*pi;    X = R*(cos(t) - 1);   Y = R*sin(t);   Z = .2*t;
Xs = h. / (h - Z).*X;    Ys = h. / (h - Z).*Y;    Zs = X - X;
hold on;  axis equal;  view(40, 25);
plot3(X, Y, Z, 'LineWidth', 2);                              % a helix
plot3(Xs, Ys, Zs, 'm', 'LineWidth', 2);            % a shadow (cochleoid)
plane([1 -10 -.1], [0 0 1], 20, 20);                         % a tray
```

Our next task is to plot the shadows (radial projections) of the circular helix (curve) onto the simple surfaces (a) cylinder and (b) sphere.

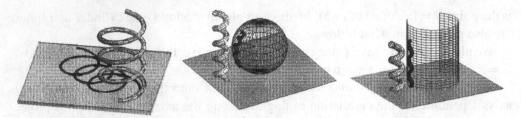

Fig. 5.22 Circular helix with a shadow in the plane, sphere, and cylinder.

(a) The standard helix is translated a distance of $8R$ in the x-direction; its radial projection (shadow when the curve is inside of the cylinder) lies on the cylinder $X^2 + Y^2 = R^2$ (Figure 5.22(c)). We suppose that the centers of projection lie on the axis, OZ, of the cylinder. The formulae for the radial projection (shadow) $(X, Y, Z) \rightarrow (X_s, Y_s, Z_s)$ are easy:

$$X_s = \frac{RX}{\sqrt{X^2 + Y^2}}, \qquad Y_s = \frac{RY}{\sqrt{X^2 + Y^2}}, \qquad Z_s = Z.$$

```
R = .2;
t = 0 : .01 : 6*pi;   X = R*cos(t) + 8*R;  Y = R*sin(t);  Z = R*t/2;
Xp = X./sqrt(X.^2+Y.^2);   Yp = Y./sqrt(X.^2+Y.^2);   Zp = Z;
hold on; axis equal; view(40, 25);
plot3(X, Y, Z, 'LineWidth', 2);                        % a curve (helix)
[x, y, z] = cylinder;  surf(x, y, 2.5*z);              % a cylinder (wall)
plot3(Xp, Yp, Zp, 'm', 'LineWidth', 2);                % a shadow
plane([0 0 -.1],  [0 0 1],  3, 2);                     % a tray
```

(b) The radial projection of the helix (shadow when the curve is inside of the sphere) lies on the sphere $X^2 + Y^2 + Z^2 = R^2$. Suppose that the origin is the center of the projection, Figure 5.22(b). The formulae for the radial projection (shadow) $(X, Y, Z) \rightarrow (X_s, Y_s, Z_s)$ are similar to the case of the cylinder:

$$X_s = \frac{RX}{\sqrt{X^2 + Y^2 + Z^2}}, \qquad Y_s = \frac{RY}{\sqrt{X^2 + Y^2 + Z^2}}, \qquad Z_s = \frac{RZ}{\sqrt{X^2 + Y^2 + Z^2}}.$$

```
R = .2; t = -2/R : .01 : 6*pi;
X = R*cos(t) + 8*R;   Y = R*sin(t);   Z = R*t/2;
M = sqrt(X.^2 + Y.^2 + Z.^2); Xp = X./M; Yp = Y./M; Zp = Z./M;
hold on; axis equal; view(40, 25);
plot3(X, Y, Z, 'LineWidth', 2);                        % a curve (helix)
[x, y, z] = sphere;  surf(x, y, z);                    % a sphere
plot3(Xp, Yp, Zp, 'm', 'LineWidth', 2);                % a shadow
```

```
plane([0 0 -1.1],  [0 0 1],  3, 2);                                    % a tray.
```

5.4.4 *Construction of curves using polynomials*

Parametric curves used in computer graphics are often based on polynomials. A polynomial of degree one has the form $\mathbf{r}_1(t) = \mathbf{a}_1 t + \mathbf{a}_0$ and is, therefore, a line in $\mathbb{R}^N$. A polynomial of degree 2 (quadratic) has the form $\mathbf{r}_2(t) = \mathbf{a}_2 t^2 + \mathbf{a}_1 t + \mathbf{a}_0$ and is always a parabola (see Example 1, p. 235). A polynomial of degree 3 (cubic) has the form $\mathbf{r}_3(t) = \mathbf{a}_3 t^3 + \mathbf{a}_2 t^2 + \mathbf{a}_1 t + \mathbf{a}_0$ and is the simplest curve that can have complex shapes and can also be a space curve (see exercises).

Indeed, one can try the interpolating polynomials for an arbitrary number of points. Namely, given the $n+1$ data points $\mathbf{P} = P_1, P_2, \ldots, P_{n+1}$, the problem is to find a vector function $\mathbf{L}(t)$ that will pass through all of them. We try an expression of the form of the Lagrange polynomial (see Section 1.5),

$$\mathbf{L}_n(t) = \sum_{i=1}^{n+1} L_{n,i}(t) P_i \quad \text{where} \quad L_{n,i}(t) = \prod_{j \neq i} \frac{t - t_j}{t_i - t_j}. \tag{5.13}$$

The following *barycentric form* of (5.13) is more attractive:

$$\mathbf{L}_n(t) = \varphi(t) \sum_{i=1}^{n+1} \frac{w_i}{t - t_i} P_i, \quad \text{where} \quad w_i = 1 \left/ \prod_{j \neq i}^{n+1} (t_i - t_j), \right.$$

The weights w_i no longer depend on t, and consequently have to be computed just once. The only quantity that depends on t is $\varphi(t) = \prod_{i=1}^{n+1} (t - t_i)$.

The MATLAB command polyfit does not work with several Lagrange polynomials simultaneously. The M-file **polyinterp.m** (Section A.1) can fit $\mathbf{L}_n(t)$; see exercises in what follows.

Example.

1. We are looking for a quadratic polynomial that passes through three points $\mathbf{P} = (P_1, P_2, P_3)$ of $\mathbb{R}^N$ and has the form

$$\mathbf{r}(t) = \mathbf{a}_2 t^2 + \mathbf{a}_1 t + \mathbf{a}_0 = (\mathbf{a}_2, \mathbf{a}_1, \mathbf{a}_0) \, T(t),$$

where $T(t) = [t^2; t; 1]$ is a column vector. Since the three points may be located anywhere, we make only the general assumption that $\mathbf{r}(t_0) = P_1, \mathbf{r}(t_2) = P_2$ (the endpoints), and $\mathbf{r}(t_1) = P_2$ (the interior point) for some $t_0 < t_1 < t_2$. We therefore write down three equations and solve them symbolically:

```
syms a0 a1 a2 t P1 P2 P3;
```

```
t0 = 0;   t1 = 1/2;   t2 = 1;                                      % enter data
r = @(t) a0 + a1*t + a2*t^2;
Eq1 = r(0) - P1;   Eq2 = r(t1) - P2;   Eq3 = r(t2) - P3;
[a0, a1, a2] = solve(Eq1, Eq2, Eq3)
```

$$a_0 = P_1 \quad a_1 = -3P_1 + 4P_2 - P_3 \quad a_2 = 2P_1 - 4P_2 + 2P_3 \qquad \text{\% Answer}$$

```
r2 = collect(a0 + a1*t + a2*t^2, [P1, P2, P3])
```

$$r_2 = (1 - 3t + 2t^2)P_1 + (4t - 4t^2)P_2 + (-t + 2t^2)P_3 \qquad \text{\% Answer}$$

```
g1 = subs(r2, [P1 P2 P3], [1 0 0]),
g2 = subs(r2, [P1 P2 P3], [0 1 0]),
g3 = subs(r2, [P1 P2 P3], [0 0 1])                       % coefficients of r₂(t)
```
% coefficients of $\mathbf{r}_2(t)$

Hence, for $t_0 = 0, t_1 = 1/2, t_2 = 1$, the solution is $\mathbf{r}(t) = g_1(t)P_1 + g_2(t)P_2 + g_3(t)P_3$, where the Lagrange polynomial weights are $g_1(t) = 2t^2 - 3t + 1$, $g_2(t) = -4t^2 + 4t$, $g_3(t) = 2t^2 - t$. Its matrix presentation is

$$\mathbf{r}(t) = (P_1\ P_2\ P_3) \begin{pmatrix} 2 & -3 & 1 \\ -4 & 4 & 0 \\ 2 & -1 & 0 \end{pmatrix} \begin{pmatrix} t^2 \\ t \\ 1 \end{pmatrix}. \tag{5.14}$$

2. The algebraic presentation of a cubic vector polynomial, in which the four coefficients (vectors $\mathbf{a}_0, \mathbf{a}_1, \mathbf{a}_2, \mathbf{a}_3 \in \mathbb{R}^N$) are unknown, is

$$\mathbf{r}(t) = \mathbf{a}_3 t^3 + \mathbf{a}_2 t^2 + \mathbf{a}_1 t + \mathbf{a}_0. \tag{5.15}$$

The velocity (tangent vector) to a curve $\mathbf{r}(t)$ is the derivative $\mathbf{r}'(t) = 3\mathbf{a}_3 t^2 + 2\mathbf{a}_2 t + \mathbf{a}_1$. The desired equations $\mathbf{r}(0) = P_1, \mathbf{r}(1) = P_2, \mathbf{r}'(0) = Q_1, \mathbf{r}'(1) = Q_2$ are

$$\mathbf{a}_0 = P_1, \qquad \mathbf{a}_3 + \mathbf{a}_2 + \mathbf{a}_1 + \mathbf{a}_0 = P_2, \qquad \mathbf{a}_1 = Q_1, \qquad 3\mathbf{a}_3 + 2\mathbf{a}_2 + \mathbf{a}_1 = Q_2.$$

They are easy to solve

```
[a3 a2 a1 a0] = solve('a0 = P1', 'a3 + a2 + a1 + a0 = P2', 'a1 = Q1',
    '3*a3 + 2*a2 + a1 = Q2')
```

and the solution is

$$\mathbf{a}_3 = 2(P_1 - P_2) + Q_1 + Q_2, \quad \mathbf{a}_2 = 3(P_2 - P_1) - 2Q_1 - Q_2, \quad \mathbf{a}_1 = Q_1, \quad \mathbf{a}_0 = P_1.$$

Substituting this into (5.15) gives the Hermite curve (note that $n_1 + n_2 = 1$)

$$\mathbf{r}(t) = n_1(t)P_1 + n_2(t)P_2 + n_3(t)Q_1 + n_4(t)Q_2, \tag{5.16}$$

$$n_1 = (1 - 3t^2 + 2t^3), \qquad n_2 = t^2(3 - 2t), \qquad n_3 = t(t-1)^2, \qquad n_4 = t^2(t-1).$$

```
r = collect(a0 + a1*t + a2*t^2 + a3*t^3, [P1  P2  Q1  Q2])
```

3. The *Newton polynomial* offers an alternative approach to the problem of polynomial interpolation: it is identical to the Lagrange polynomial, but the derivation is different. The Newton polynomial allows the user to easily add more points and thereby provide fine control over the shape of the curve.

Assume that $n+1$ data points $P_1, P_2, \ldots, P_{n+1}$ are given and are assigned knot values $t_1 = 0 < t_2 < \cdots < t_n < t_{n+1} = 1$. We are looking for a curve expressed by the degree-n parametric polynomial $\mathbf{r}(t) = \sum_{i=1}^{n+1} N_i(t) A_i$. This definition (originally proposed by Newton) is useful because each coefficient A_i depends only on points P_1 through P_i. If we decide to add a point P_{n+2}, then only one coefficient, A_{n+2}, and one basis function, $N_{n+1}(t)$, need to be recomputed. The definition of the basis functions is

$$N_1(t) = 1,$$
$$N_i(t) = (t - t_1)(t - t_2) \cdots (t - t_{i-1}), \qquad\qquad i > 1.$$

To calculate the unknown coefficients, we write the equations

$$
\begin{aligned}
P_1 &= P(t_1) = A_1, \\
P_2 &= P(t_2) = A_1 + A_2(t_2 - t_1), \\
P_3 &= P(t_3) = A_1 + A_2(t_2 - t_1) + A_3(t_3 - t_1)(t_3 - t_2), \\
&\cdots \\
P_{n+1} &= P(t_{n+1}) = A_1 + \cdots
\end{aligned}
\qquad (5.17)
$$

We use the method of divided differences to express all solutions in compact notation. The *divided difference of the knots* t_i, t_k is denoted by $[t_i t_k]$ and is defined as $[t_i t_k] = (P_i - P_k)/(t_i - t_k)$. The solutions of (5.17) are

$$
\begin{aligned}
A_1 &= P_1, \\
A_2 &= (P_2 - P_1)/(t_2 - t_1) = [t_2 t_1], \\
A_3 &= [t_3 t_2 t_1] = ([t_3 t_2] - [t_2 t_1])/(t_3 - t_1), \\
&\cdots \\
A_{n+1} &= [t_{n+1} \ldots t_2 t_1] = ([t_{n+1} \ldots t_2] - [t_n \ldots t_1])/(t_{n+1} - t_1).
\end{aligned}
$$

We prepare the function M-file **dd_newt** to compute A_i.

```
function A = dd_newt(T, P)
  N = length(T);
  for i = 1 : N;
   B(:, i, 1) = P(i, :);
  end;
  for k = 1 : N;  for i = 1 : N-k;
   B(:, i, k+1) = (B(:, i+1, k) - B(:, i, k))/(T(k+i) - T(i));
   end;  end
```

```
    A(:, :) = B(:, 1, :);
end
```

The program works for $\mathbb{R}^n$ $(n \geq 1)$.

Let $T = [1,2,3,4,5]$, $P_1 = (0,2)$, $P_2 = (1,0)$, $P_3 = (2,3)$, $P_4 = (3,5)$, $P_5 = (5,2)$. Then $A_1 = (0,2)$, $A_2 = (1,-2)$, $A_3 = (0,5/2)$, $A_4 = (0,-1)$, $A_5 = (1/24, 1/12)$.

```
    T = [1 2 3 4 5];    P = [0 2;  1 0;  2 3;  3 5;  5 2];
    A = dd_newt(T, P)                    % the columns of A are points A_i
```

4. We apply Lagrange polynomials to plane curves. For spirals we obtain a sufficiently good visualization! (Compare with Example 1, p. 57.)

```
    t = linspace(0, 3*pi, 12);
    x = sqrt(t).*cos(t);    y = sqrt(t).*sin(t);          % Fermat's spiral ρ = √t
    xy = [x; y];
    syms u;   L2 = polyinterp(t, xy, u);
```

Then plot:

```
    hold on;    plot(x, y, 'o - - r');   ezplot(L2(1, :), L2(2, :), [t(1), t(end)])
```

Lagrange interpolation for more complicated data is not quite as good:

```
    x = [.016, .287, .655, .716, .228, .269, .666, .979];   t = 0 : length(x) - 1;
    y = [.820, .202, .202, .521, .521, .820, .820, .227];
```

5. We also apply Lagrange polynomials to space curves. For the helix

$$\mathbf{r}(t) = [\cos t, \sin t, t], \qquad |t| \leq 2\pi$$

(see Figure 5.22), we obtain a good view!

```
    t = linspace(-2*pi, 2*pi, 12);
    x = cos(t);   y = sin(t);   z = t;                    % 12 points of helix
    xyz = [x; y; z];
    syms u;   L3 = polyinterp(t, xyz, u);
```

Then plot:

```
    ezplot3(L3(1), L3(2), L3(3), [t(1), t(end)]);   hold on;
    plot3(x, y, z, 'o - - r', 'Linewidth', 2);
```

5.5 Exercises

Section 5.1

1. For modeling *hypocycloids* (Figure 5.3), put $a = -3 : -1$ and $b = 6 : 8$ in the program in Example 2, p. 204 above. For modeling *prolate and curtate trochoids*, assume $h = 1.5$ and then $h = 0.5$.

2. Study each of the following interesting cycloidal curves.

(a) The *cardioid* (heart-shaped), the epicycloid with modulus m = 1 (Figures 5.2(a), 6.10(b)) and a curve of fourth degree with the equation $(x^2+y^2+2Rx)^2 = 4R^2(x^2+y^2)$. Let us substitute $x^2+y^2 = \rho^2$, $x = \rho\cos\varphi$ in the polynomial equation and reduce the fraction by ρ^2 to obtain the equation in polar coordinates $\rho = 2R(1-\cos\varphi)$. Slightly changing the equation of the cardioid, we obtain *Pascal's limaçon*, $\rho = l - 2R\cos\varphi$, with the loop and self-intersection for $l < 2R$; without a loop, but with the singular point O for $l > 2R$ (Figures 1.34(c), 6.10(a)).

(b) The *nephroid* (*nephros* means kidney-shaped; Proctor, 1878), the epicycloid with modulus m = 1/2. In Figure 5.2(a) of epicycloids above, it is the entry for a = 3, b = 6. This curve has two cuspidal points.

(c) The *astroid* (star-shaped), the hypocycloid with modulus m = -1/4. It is the sixth-degree curve $(x^2+y^2-R^2)^3 + 27R^2x^2y^2 = 0$. Its implicit (2nd method of plotting) and parameterized equations are

$$x^{\frac{2}{3}} + y^{\frac{2}{3}} = R^{\frac{2}{3}} \iff \mathbf{r}(t) = [R\cos^3(t/4), R\sin^3(t/4)] \quad \text{(see Figure 5.3(a))}.$$

(d) The *deltoid* (delta-shaped *curve of Steiner*), the hypocycloid with $m = -\frac{1}{3}$. It is the fourth-degree curve $(x^2+y^2)^2 + 8Rx(3y^2-x^2) + 18R^2(x^2+y^2) - 27R^4 = 0$.

For plotting individual *cycloidal curves*, fix h = 1 and, for example, R = 1. Plot by a dotted line the stationary circle. Then, changing the values of the modulus $m = 1, 1/2, 1/3$, etc., plot the *epicycloids:* the *cardioid,* the *nephroid,* etc.

For plotting *hypocycloids*, use negative values of the variable $m = -1/3, -1/4$, etc., and plot the *deltoid, astroid,* etc.

In both cases, do not forget, for m = 1/b, to extend the domain of the parameter t to the segment $[-b\pi, b\pi]$.

3. Continue the experiment (with polar coordinates) *up and down, around and near* for other values $k = 6, 7, \ldots$ or another relationship $\rho = f(t)$.

4. Plot Lissajous curves for several values of $n = a/b$ while φ changes. Deduce that the curve is closed and algebraic when n is a rational number. What figure is filled by the Lissajous curve with n irrational? For $n = 2$ find values of φ for which the Lissajous curve is the parabola and the figure eight (Figure 8.16(a)). Show that for $n = 1$ it is an ellipse (a line segment when $\varphi = 0$).

5. Plot the *SICI spiral*, $\mathbf{r}(t) = [\text{Ci}(t), \text{Si}(t)]$, where Ci is the cosine integral and Si is the sine integral. Check that the arc length of the curve from the point $t = 0$ to the point t is equal to $\ln(t)$, and that the curvature is $k(t) = t$ (see Section 6.4).

```
s = 1.5 : 0.05 : 30;
X = mfun('Ci', s);  Y = mfun('Si', s);
plot(X, Y);  axis equal;                                    % Figure 5.6(b)
```

6. Given a circle with center O and a point A inside it, find the point P on the circle such that the angle $\angle APO$ is maximal.

7. Plot the graph of an arbitrary 2-D vector field (similarly to the gradient of a function); see Figure 5.7(b).

```
v = -pi : .3 : pi;  [x, y] = meshgrid(v);
fx = y.*cos(x*y);  fy = x.*cos(x*y);  quiver(v, v, fx, fy)
```

8. Plot the space curve $\mathbf{r}(t) = (t, x_1(t), x_2(t))$ that is the solution of a system of two ODEs.

```
ODE2 = inline('[.1*x(1) + x(2); -x(1) + .1*x(2)]', 't', 'x');
[t, x2] = ode45(ODE2, [0 55], [.01, 0]);              % 2-D system
plot3(t, x2(:, 1), x2(:, 2))                          % Figure 5.9(b)
```

9. Carry out a similar derivation (see Example, p. 214) for the problem

$$J(y) = \int_{x_0}^{x_1} F(x, y, y', y'', \ldots, y^{(m)}) \, dx \to \min, \qquad (5.18)$$

where $m \geq 1$, and the admissible function $y(x)$ belongs to class C^{m+1} on the interval $[x_0, x_1]$ and satisfies the boundary conditions

$$y(x_0) = y_0, \qquad y'(x_0) = y_0', \qquad \ldots, \qquad y^{(m)}(x_0) = y_0, \qquad (5.19)$$
$$y(x_1) = y_1, \qquad y'(x_1) = y_1', \qquad \ldots, \qquad y^{(m)}(x_1) = y_1.$$

Hint. The minimization in this case leads to the following *Euler equation*:

$$F_{,y} - \frac{d}{dx}(F_{,y'}) + \frac{d^2}{dx^2}(F_{,y''}) - \cdots + (-1)^m \frac{d^m}{dx^m}(F_{,y^{(m)}}) = 0, \qquad (5.20)$$

while $\frac{d^i}{dx^i} F_{,y^{(i)}}$ are derived analogously to (5.7).

10. To generalize the previous case of (5.4), consider the problem defined in a different way by the functional

$$J(y) = \int_{x_0}^{x_1} F(x, y_1, \ldots, y_n, y_1', \ldots, y_n') \, dx \to \min, \qquad (5.21)$$

where $n \geq 1$, and F is a continuous function of $2n+1$ arguments. Suppose that the admissible functions $y_i(x)$ of one variable belong to class C^2 on the interval $[x_0, x_1]$, and that the boundary values are

$$y_i(x_0) = y_{i0}, \qquad y_i(x_1) = y_{i1}, \qquad 1 \leq i \leq n. \qquad (5.22)$$

Show that the minimization leads to a *system of Euler equations*

$$F_{,y_i} - \frac{d}{dx}(F_{,y_i'}) = 0, \qquad 1 \leq i \leq n. \qquad (5.23)$$

11. To generalize the case of (5.21) still further, show that similar calculations performed on the extreme problem

$$J(y) = \int_{x_0}^{x_1} F\left(x, y_1,\ldots,y_n, y_1',\ldots,y_n',\ldots,y_1^{(m)},\ldots,y_n^{(m)}\right) dx \to \min \qquad (5.24)$$

lead to the following system of Euler equations:

$$F_{,y_i} - \frac{d}{dx}\left(F_{,y_i'}\right) + \frac{d^2}{dx^2}\left(F_{,y_i''}\right) - \cdots + (-1)^m \frac{d^m}{dx^m}\left(F_{,y_i^{(m)}}\right) = 0, \quad 1 \le i \le n. \quad (5.25)$$

12. Write down the Euler equations (5.20), for $m = 2$, and (5.23), for $n = 2$, for given functions using MATLAB (see Example, p. 214). Verify for the functions:

$$F = (y'')^2 - 2(y')^2 + 4yy' + y^2 - 2y\sin(x), \quad F = (y')^2 + (z')^2 + 2yz, \quad \text{etc.}$$

13. Deduce Euler's equation (5.10) using MATLAB (see Example, p. 214).

| Section 5.2 |

1. Plot a Pythagorean triangle (Figure 5.12(d)), using transformations and the program **hilbert**(n).

2. Plot the Peano curve in the triangle (Figure 5.23).

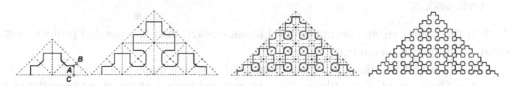

Fig. 5.23 Triangular Peano curve.

3. Plot Koch curves which are inside the triangle and the square. For obtaining new Koch curves, replace the equilateral triangle in the original Koch curve by an isosceles triangle, or try to find your own "generators."

4. Plot a Sierpiński triangular curve and its analogue in a pyramid.

5. Compute the dimension of the Koch curve, Sierpiński's triangular curve, Sierpiński's carpet, and the Peano curve. What is the length of a Koch curve? Find the area of the triangle- and square-shaped snowflakes and plot them. *Hint.* For the nth step of plotting, we obtain the polygon L_n consisting of $3 \cdot 4^n$ segments each of length $\frac{a}{3^n}$. Its total length is $L_n = 3(\frac{4}{3})^n a$. Fixing the scale $\lambda_n = \frac{a}{3^n}$, we find that the length of the Koch curve is equal to the length of the polygon corresponding to the nth stage of its plotting, $L_n = 3(\frac{4}{3})^n a$. For $n \to \infty$ we have $\lambda_n \to 0$, but the length of L_n runs to infinity. Meanwhile, the area bounded by the Koch curve is finite. Let the area of the initial

triangle be equal to S_0. Then the common area of the snowflake can be expressed by the geometrical series $S_0(1 + \frac{1}{3} + \frac{1}{3} \cdot \frac{4}{9} + \frac{1}{3}(\frac{4}{9})^2 + \cdots) = S_0(1 + \frac{1}{3} \cdot \frac{1}{1-4/9}) = 1.6 S_0$.

6. Plot the Menger polyhedron K_n, and find the volume of the deleted cubes.

Section 5.3   1. Plot the coordinate surfaces for cylindrical and spherical coordinates.

2. Generalize Example 3, p. 224, and plot the *chain* containing $2n$ circles.

```
n = 3;
hold on;  axis equal;  view(65, 25);
for i = 0 : n - 1;
 plot3(2*cos(t), 2*sin(t) + 6*i, t - t, 'r', 'LineWidth', 5);
 plot3(t - t, 2*cos(t) + 3 + 6*i, 2*sin(t), 'g', 'LineWidth', 5);
end;                                                        % Figure 5.17(b).
```

3. Plot the *chain armor* with m lines (Figure 5.17(c)).

```
n = 4;  m = 4;
hold on;  axis equal;  view(65, 45);
for j = 0 : m - 1;  for i = 0 : n - 1;
 plot3(2*cos(t) + 6*j, 2*sin(t) + 6*i, t - t, 'r', 'LineWidth', 5);
 plot3(t - t + 6*j, 2*cos(t) + 3 + 6*i, 2*sin(t), 'g', 'LineWidth', 5);
 plot3(2*cos(t) - 3+6*i, t - t + 6*j, 2*sin(t), 'b', 'LineWidth', 5);
end;  end;
```

4. If the planes α_1 and α_2 are parallel to each other, then any parallel projection π from α_1 onto α_2 is an isometry.

5. Given an ellipse, there is a parallel projection which maps it onto a circle.

6. Let l be a chord of an ellipse. Show that the midpoints of the chords parallel to l lie on a diameter of the ellipse.

Hint. Use Exercise 5 and the properties of a circle.

7. Let l be a diameter of an ellipse. Then there is another diameter m of the ellipse such that:

(a) the midpoints of all chords parallel to l lie on m;

(b) the midpoints of all chords parallel to m lie on l.

8. Plot the perspective projection of a tetrahedron, and an octahedron (use Example 4, p. 224).

9. Plot the *dimetric projection* of a spiraling helix on the catenoid.

```
s = 4;  a = 20;
v1 = [-s/2 -s/2];  v2 = [s 0];  v3 = [0 s]
t = linspace(-3, 3, 600);  f = cosh(t);  g = 4*t;              % enter f, g
X = f.*cos(a*t);  Y = f.*sin(a*t);  Z = g;
hold on;  axis equal;  plot(Y - X/2, Z - X/2, 'r-');     % dimetric, Figure 5.16(e)
arrow([0 0], v1);  arrow([0 0], v2);  arrow([0 0], v3);           % see p. 70
```

10. Plot the Viviani curve with shadow and revolution about the axis OZ (Figure 5.21(b)).

Section 5.4   1. Deduce the equations of a spring lying on a cylinder (*helix*) and then plot it: *The point P rotates uniformly about a line l (that is parallel to the axis OZ) and does so at a distance R from l. At the same time, P moves with constant velocity parallel to l.*

Hint. The *circular helix* $\mathbf{r}(t) = [R\cos t, R\sin t, bt]$ (Figure 5.19(a)). A generalization is a "spring" $\mathbf{r}(t) = [a\cos t, b\sin t, ct]$ on an elliptic cylinder.

    R = 1; a = 1;
    t = 0 : .01 : 16*pi;   plot3(R*cos(t), R*sin(t), a*t)

A circular helix can be obtained as the image of the line (oblique to the ruling) on the development of the circular cylinder.

2. Deduce equations of the "springs" on a cone and then plot them:

The line OL not orthogonal to OZ rotates uniformly around O with the angular velocity $\omega = 1$, and the point P moves along OL:

(a) with velocity proportional to the length OP, Figure 5.19(b).

Hint. $\mathbf{r}(t) = [Re^{at}\cos t, Re^{at}\sin t, be^{at}]$ is a *conic helix*.

    R = 1; a = .05;   t = 0:.01:16*pi; b = exp(a*t);
    plot3(R*b.*cos(t), R*b.*sin(t), b)

(b) with constant velocity; Figure 5.19(c).

Hint. $\mathbf{r}(t) = [at\cos t, at\sin t, bt]$ is the *circular conic helix*.

    R = 1; a = .05; t = 0 : .01 : 16*pi;
    plot3(R*a*t.*cos(t), R*a*t.*sin(t), a*t)

Plot a more general curve $\mathbf{r}(t) = [at\cos t, bt\sin t, ct]$ on the elliptic cone.

3. Show that a cubic polynomial passing through four points P_1, P_2, P_3, and P_4 in $\mathbb{R}^N$ so that $\mathbf{r}(0) = P_1$, $\mathbf{r}(1/3) = P_2$, $\mathbf{r}(2/3) = P_3$, $\mathbf{r}(1) = P_4$ has the form $\mathbf{r}(t) = g_1(t)P_1 + g_2(t)P_2 + g_3(t)P_3 + g_4(t)P_4$, where the Lagrange polynomial weights for the four points are

$$g_1(t) = -4.5t^3 + 9t^2 - 5.5t + 1, \qquad g_2(t) = 13.5t^3 - 22.5t^2 + 9t,$$
$$g_3(t) = -13.5t^3 + 18t^2 - 4.5t, \qquad g_4(t) = 4.5t^3 - 4.5t^2 + t.$$

(For computations modify the program of Example 1, p. 235). Its matrix presentation is $\mathbf{r}(t) = \mathbf{P}MT(t)$ $(0 \leq t \leq 1)$, where $T(t) = [t^3; t^2; t; 1]$ is a column vector and

$$\mathbf{P} = (P_1\ P_2\ P_3\ P_4), \qquad M = \begin{pmatrix} -9/2 & 9 & -11/2 & 1 \\ 27/2 & -45/2 & 9 & 0 \\ -27/2 & 18 & -9/2 & 0 \\ 9/2 & -9/2 & 1 & 0 \end{pmatrix}. \qquad (5.26)$$

4. The polynomial $\mathbf{L}_2(t)$ is easy calculate once t_0, t_1, t_2 have been determined (see Example 1, p. 235). Show that

(a) the Lagrange polynomial weights for three points P_1, P_2, and P_3 are

$$L_{2,0}(t) = \frac{(t-t_1)(t-t_2)}{(t_0-t_1)(t_0-t_2)}, \quad L_{2,1}(t) = \frac{(t-t_0)(t-t_2)}{(t_1-t_0)(t_1-t_2)}, \quad L_{2,2}(t) = \frac{(t-t_0)(t-t_1)}{(t_2-t_0)(t_2-t_1)};$$

(b) its matrix presentation (see (5.14)) is

$$\mathbf{L}_2(t) = (P_1 \; P_2 \; P_3) \begin{pmatrix} \frac{1}{(t_1-t_0)t_2} & \frac{-1}{t_2} - \frac{1}{t_1-t_0} & 1 \\ -\frac{1}{(t_2-t_1)(t_1-t_0)} & \frac{1}{t_1-t_0} + \frac{1}{t_2-t_1} & 0 \\ \frac{1}{(t_2-t_1)t_2} & -\frac{1}{t_2-t_1} + \frac{1}{t_2} & 0 \end{pmatrix} \begin{pmatrix} t^2 \\ t \\ 1 \end{pmatrix}. \tag{5.27}$$

5. Create the matrix M of (5.14) using the following program:

```
% enter polynomials g_i(t) of Example 1, p. 236.
M = [sym2poly(g1); sym2poly(g2); sym2poly(g3)]
```

6. Calculate $L_n(x)$, and then plot the data and the graph.

```
x = 2 : 8;  y = [32, 34, 36, 34, 39, 40, 37];                    % enter data
xi = linspace(2, 8, 100);   yi = polyinterp(x, y, xi);
plot(x, y, 'o - -',  xi, yi, '-')                                % see Figure 1.26(a).
```

The **polyinterp** function works correctly with symbolic variables:

```
symt = sym('t')
P = polyinterp(x, y, symt);   pretty(P)
```
$P = \frac{7}{72}t^6 - \frac{44}{15}t^5 + \frac{1277}{36}t^4 - \frac{2629}{12}t^3 + \frac{52231}{72}t^2 - \frac{72839}{60}t + 831$ % Answer.

This is a canonical form of the interpolating Lagrange polynomial $L_6(t)$.

7. Apply splines and Lagrange polynomials to interpolate the spring on the paraboloid $\mathbf{r}(t) = [t\cos t, t\sin t, t^2]$ for $0 \le t \le 4\pi$. Compare results.

6

Geometry of Curves

We will study some constructions and visualizations of the differential geometry of curves: the tangent line and Frenet frame (Section 6.1), the singular points (Section 6.2), the length and center of mass (Section 6.3), and the curvature and torsion (Section 6.4).

6.1 Tangent Lines

We will plot the (moving) tangent line and Frenet frame for various types of presentations of a curve. Then we will study some applications of tangent and normal lines to a curve. Curves can appear as envelopes of families of lines, circles, etc. Using MATLAB®, we will realize some methods of "mathematical embroidery" from the literature, and study evolutes, caustics, and parallel curves.

6.1.1 *Tangent lines, normals, and the Frenet frame*

The *tangent line* to the curve γ of class C^1 at the point $P \in \gamma$ is the limit position of the secant PQ as the second point of intersection Q approaches P along the curve. *Semi-tangent lines from the right* and *from the left* are defined analogously.

Since the secant through the points at t and $t + \Delta t$ is parallel to the vector $\Delta \mathbf{r} = \mathbf{r}(t + \Delta t) - \mathbf{r}(t)$, and by the definition of the derivative $\mathbf{r}'(t) = \lim_{\Delta t \to 0} \frac{\Delta \mathbf{r}}{\Delta t}$, the line through the point of the regular curve $\mathbf{r}(t) \in \gamma$ in the direction of the velocity vector $\mathbf{r}'(t) = [x'(t), y'(t), z'(t)]$ is the tangent line (Figure 6.1). The *unit tangent and normal vectors of the plane curve* $\mathbf{r}(t)$ are derived as follows:

V. Rovenski, *Modeling of Curves and Surfaces with MATLAB®*,
Springer Undergraduate Texts in Mathematics and Technology 7, DOI 10.1007/978-0-387-71278-9_6,
© Springer Science+Business Media, LLC 2010

$$\tau = \left(\frac{x'}{\sqrt{x'^2+y'^2}}, \frac{y'}{\sqrt{x'^2+y'^2}} \right), \quad v = \left(\frac{-y'}{\sqrt{x'^2+y'^2}}, \frac{x'}{\sqrt{x'^2+y'^2}} \right).$$

The tangent line at the point t of the parametric curve $\mathbf{r}(t) \subset \mathbb{R}^3$ has the following equation:

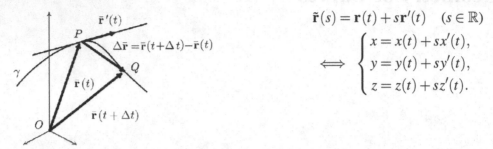

$$\tilde{\mathbf{r}}(s) = \mathbf{r}(t) + s\mathbf{r}'(t) \quad (s \in \mathbb{R})$$

$$\Longleftrightarrow \begin{cases} x = x(t) + sx'(t), \\ y = y(t) + sy'(t), \\ z = z(t) + sz'(t). \end{cases}$$

Fig. 6.1 Derivative of a vector-valued function and a tangent line.

Equations of the tangent line to a plane curve of class C^1 (for various types of equations of the curve) are the following (here X, Y are coordinates of points on the tangent line, and $s \in \mathbb{R}$):

graph	$Y = y + f'(x)(X - x),$
parametric	$X = x(t) + sx'(t),\ Y = y(t) + sy'(t),$
implicit	$\frac{\partial F}{\partial x}(x,y)(X - x) + \frac{\partial F}{\partial y}(x,y)(Y - y) = 0.$

Equations of the normal vector to the plane curve are the following:

graph	$X = x - sf'(x),\ Y = y + s,$
parametric	$X = x(t) - sy'(t),\ Y = y(t) + sx'(t),$
implicit	$X = x + s\frac{\partial F}{\partial x}(x,y),\ Y = y + s\frac{\partial F}{\partial y}(x,y).$

For an implicitly defined plane curve, $F(x,y) = 0$, the *vector gradient* $\nabla F = \left[\frac{\partial F}{\partial x}, \frac{\partial F}{\partial y} \right]$ is parallel to its normal vector.

Examples.

1. *A dog runs along the axis OY starting from $O = (0,0)$, and its owner (initially staying on the axis OX) follows the dog, pulling on the leash of a length a. What is the trajectory of the owner?*

The curve is called the *tractrix* (Figure 6.3(c)). We deduce its equation using the tangent line (see also Exercise 1, p. 279, with MATLAB).

Solution. From a right triangle with the tangent line as the hypotenuse, deduce $x = a\sin t$, where $t \in [0, \pi)$ is the angle between the tangent line and the axis OY. Thus

$dx = a\cos t\, dt$. In view of the equality $dx = \tan t\, dy$ (from the geometrical interpretation of the tangent line), one obtains $dy = a\frac{\cos^2 t}{\sin t}\, dt$. After integrating dx and dy, one deduces that $\mathbf{r}(t) = [a\sin t, a(\ln\tan(\frac{t}{2}) + \cos t)]$, which can be simplified to the form $y = a\ln(\frac{a - \sqrt{a^2 - x^2}}{x}) - \sqrt{a^2 - x^2}$. The following construction of the *tractrix of an arbitrary curve* generalizes the previous problem: *The dog runs along the curve* γ, *and his owner follows the dog along the tractrix of this curve.*

For example, the *circular tractrix* in polar coordinates has the equation

$$\varphi = -\arcsin(\rho/(2a)) + \sqrt{4a^2 - \rho^2}/\rho \quad (0 < \rho \le 2a).$$

2. Find the equation of a parabola $y = x^2 + ax + b$ that is tangent to the circle $x^2 + y^2 = 2$ at the point $P = (1,1)$, and plot the curves.

Solution.

```
syms x y a b;
f = y - (x^2 + a*x + b);   g = x^2 + y^2 - 2;
df = [diff(f, x), diff(f, y)];   dg = [diff(g, x), diff(g, y)];
t = pi/4;                               % enter your t for P(cos t, sin t)
P = [sqrt(2)*cos(t), sqrt(2)*sin(t)];
a0 = solve(subs(det([df; dg]), {x, y}, {P(1), P(2)}), a)
b0 = solve(subs(f, {x, y, a}, {P(1), P(2), a0}))
f0 = subs(f, {a, b}, {a0, b0})
```

$a_0 = -3 \qquad b_0 = 3 \qquad f_0 = y - x^2 + 3x - 3 \qquad$ % Answer (parabola).

Finally, plot the curves:

```
hold on;  axis equal;   ezplot(f0, [-2, 3, -2, 3]);
ezplot(g, [-2, 3, -2, 3]);
```

3. Let us plot the tangent lines and the normals to some curves. We consider three typical programs that are useful in solving exercises.

(a) The *graph* of a function $y = f(x)$. Start with the parabola. First prepare the data and calculate.

```
syms x t;  m = 20;
f = 'x^2/2';  x1 = -2;  x2 = 2;                    % enter your function
fx = diff(f, x);
N = [x, f] + t*[-fx, 1]/sqrt(1 + fx^2);
```

Then plot:

```
hold on;  h = ezplot(f, [x1, x2]);  set(h, 'Color', 'r', 'LineWidth', 3);
m = 20;    for i = 0 : m;
X = x1 + i*(x2 - x1)/m;
```

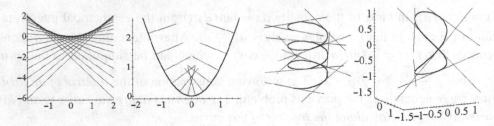

Fig. 6.2 Tangent lines and normals to a graph. Tangent lines to space curves.

```
p = subs(f + fx*(t - x), x, X);  ezplot(p, [x1, x2]);
end;                                              % Figure 6.2(a), tangent lines

hold on;  axis equal;    h = ezplot(f, [x1, x2]);  set(h, 'Color', 'r');
for i = 0 : m;
 Ni = subs(N, x,  x1 + i*(x2 - x1)/m);
 ezplot(Ni(1), Ni(2), [0, 1]);
end;                                              % Figure 6.2(b), normals
```

(b) A *parametric plane curve*. Start with a cycloid.

```
r = [t - sin(t), 1 - cos(t)];  t1 = 0;  t2 = 4*pi;              %  enter data
rt = diff(r, t);  N = [-rt(2), rt(1)];                   % velocity and normal
h = ezplot( r(1), r(2), [t1, t2]);  set(h, 'Color', 'r', 'LineWidth', 3);  hold on;
m = 15;  for i = 0 : m;
 T = t1 + i*(t2 - t1)/m;
 rT = subs(r + s*rt, t, T);
 rN = subs(r + s*N, t, T);
 ezplot(rT(1), rT(2), [-2, 2]);                            %  tangent line
 hn = ezplot(rN(1), rN(2), [-1, 1]);  set(hn, 'Color', 'g');        % normal
end;
```

(c) A *parametric space curve*. Start with a circular helix (Figure 6.2(c)). First prepare the data.

```
syms s t;
m = 19;  R = 1;  v = 1;   t1 = 0;  t2 = 6*pi;                     % enter data
r = [R*cos(t), R*sin(t), v*t];    rt = diff(r, t);
```

Then plot the helix with tangent lines:

```
ezplot3(r(1), r(2), r(3), [t1, t2]);  hold on;  axis equal;
S = linspace(-1, 1, 2);
for j = 0 : m;
 T = subs(r + s*rt,  t,  t1 + j*(t2 - t1)/m);
 plot3(subs(T(1), s, S) + S-S, subs(T(2), s, S) + S-S, subs(T(3), s, S) + S-S, 'r');
end;
```

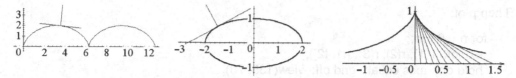

Fig. 6.3 Frenet frame field of cycloid and ellipse. Tangent lines to tractrix.

4. Plot the moving normal along the graph (Figure 6.2(b)).

```
for n = 0 : m;
 ezplot(f, [x1, x2]); hold on;  axis equal;
 X = x1 + n*(x2 - x1)/m;
 p = [X, subs(f, x, X)];
 d = 2*sqrt(1 + fx^2);
 p1 = subs([x - fx/d, f + 1/d], x, X);
 arrow(p, p1 - p, 'r');   pause(0.2);  hold off;
 end;
```

5. Plot the moving tangent and normal vectors along the cycloid.

```
syms s t;
r = [t - sin(t), 1 - cos(t)];    t1 = 0;  t2 = 4*pi;
rt = diff(r, t);
N = [-rt(2), rt(1)];
m = 30;  for n = 0 : m;
 ezplot(r(1), r(2), [t1, t2]); hold on;  axis equal;
 T = t1 + n*(t2 - t1)/m;
 rT = subs(r + s*rt, t, T);
 rN = subs(r + s*N, t, T);
 hn = ezplot(rN(1), rN(2), [0, 1]);  set(hn, 'Color', 'r');
 ht = ezplot(rT(1), rT(2), [0, 1]);  set(ht, 'Color', 'g');
 pause(0.1);  hold off;
 end;                                                  % Figure 6.3(a)
```

For a regular curve (say, an ellipse r = [a*sin(t), b*cos(t)] with t2 = 2*pi), use its unit tangent and normal vectors, i.e., the *Frenet frame* (Figure 6.3(b)).

```
d = 2*sqrt(rt*rt');    rt1 = rt/d;
N = [-rt1(1), rt1(2)];
```

6. Plot the moving tangent line to the Viviani window (Figure 6.2(d)).

```
syms s t;  m = 19;  R = 1;  v = 1;
r = [R*cos(t)^2, R*cos(t)*sin(t), R*sin(t)];    t1 = 0;  t2 = 2*pi;
rt = diff(r, t);                            % parametric equations and tangent line
```

Then plot:

```
for n = 0 : m;
  ezplot3(r(1), r(2), r(3), [t1, t2]);
  hold on;  axis equal;  grid off;  view(135, 70)
  T = subs(r + s*rt, t, t1 + n*(t2 - t1)/m);
  plot3(subs(T(1), s, S) + S-S, subs(T(2), s, S) + S-S, subs(T(3), s, S) + S-S, 'r');
  pause(0.2);  hold off;
end;
```

6.1.2 Asymptotes of curves

Asymptotes can be thought of as tangent lines at "infinite points" of a curve. One can distinguish *vertical, horizontal,* and *oblique* asymptotes of plane curves with respect to rectangular coordinates (see Figure 6.4(a)). An oblique asymptote to a graph can be *left* (if $x \to -\infty$), *right* (if $x \to \infty$), or *two-sided* (if the graph approaches the asymptote for both $x \to -\infty$ and $x \to \infty$). Some curves, such as the circular tractrix and various spirals (Section 1.4), etc., have *asymptotic points.*

We will consider four situations where asymptotes appear, and plot them: curves graphed in Cartesian and in polar coordinates, and parametric curves in the plane and in space.

Strategy.

1. We derive the coefficients in the equation $y = kx + b$ of the asymptote of the *graph* of $y = f(x)$ for $x \to \pm\infty$ as follows:

$$k_+ = \lim_{x \to \infty} \frac{f(x)}{x}, \qquad b_+ = \lim_{x \to \infty} [f(x) - k_+ x],$$

and analogously for the coefficients k_- and b_- when $x \to -\infty$.

Example. Calculate and plot the asymptotes of the graphs

$$\text{(a)} \quad y = \sqrt[3]{x^3 + 2}, \qquad\qquad \text{(b)} \quad y = \frac{x^3}{2(x+1)^2}.$$

(a) *Answer:* A two-sided asymptote $y = x$ (Figure 6.4(b)).

```
syms x;  f = (x^3 + 2).^(1/3);
k = limit(f/x, inf),  b = limit(f - k*x, inf)          % Answer: k = 1, b = 0.
f1 = 'sign(x.^3 + 2)*abs(x^3 + 2)^(1/3)';
hold on;  ezplot(f1, [-2 2]);  ezplot(k*x + b, [-2 2]);
```

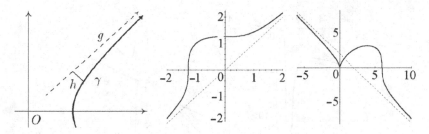

Fig. 6.4 Asymptotes of the graphs.

(b) *Answer:* Vertical and oblique asymptotes are the following:

$$x+1=0, \qquad\qquad x-2y-2=0.$$

```
syms x;  f = x^3/(2*(x+1)^2);
k = limit(f/x, inf),  b = limit(f - k*x, inf)          % Answer: k = 1/2, b = −1
hold on;  ezplot(f, [-12 2.8]);  ezplot(k*x + b, [-12 2.8])          % Figure 6.5(a)
```

2. We plot the asymptotes of the curve $\rho = \rho(\varphi)$ in *polar coordinates* (see Section 1.4) as follows:

(a) Derive the values φ_i for which $\lim_{\varphi \to \varphi_i} |\rho(\varphi)| = \infty$.

(b) Let $p_i = \lim_{\varphi \to \varphi_i} \rho^2(\varphi)/|\rho'(\varphi)|$. If this limit exists, then the infinite branch of the curve has the asymptote $\rho_i = p_i/\cos(\varphi - \varphi_i)$, and if the limit does not exist, then the infinite branch has no asymptotes.

If $p_i > 0$, then the asymptote lies on the right of the position vector of the curve as it runs to infinity. If $p_i < 0$, then the asymptote lies on the left side.

3. We calculate and plot the asymptotes of the *parametric plane curve* $[x(t), y(t)]$ by the following plan:

(a) Find points $\{a_i\}$ where the curve *runs to* ∞, i.e., $\lim_{t \to a_i}[x^2(t)+y^2(t)] = \infty$.

(b) For each $t = a_i$, derive the coefficient of inclination $k_i = \lim_{t \to a_i} y(t)/x(t)$.

(c) For each finite k_i, derive $b_i = \lim_{t \to a_i}[y(t) - k_i x(t)]$ and plot the oblique asymptote.

(d) For each infinite k_i, derive $x_i = \lim_{t \to a_i} x(t)$ and plot the vertical asymptote $x = x_i$.

Example. Calculate and plot the asymptotes of the curve $\left[\frac{t^2}{1-t}, \frac{t^3}{1-t^2}\right]$.

Answer: The curve has four points running to infinity. Its three asymptotes are $x = \frac{1}{2}$, $y = \frac{1}{2}x - \frac{1}{4}$, and $y = x + 1$ (Figure 6.5(b)).

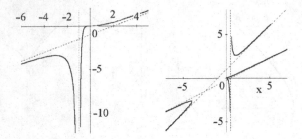

Fig. 6.5 Curves with asymptotes.

```
syms t;
X = t^2/(1 - t);  Y = t^3/(1 - t^2);  A = [-inf, -1, 1, inf];
for i = 1 : 4;
 k(i) = limit(Y/X, t, A(i));
 b(i) = limit(Y - k(i)*X, t, A(i));
end;       % Answer: k = [1  NaN  1/2  1],  b = [1  NaN  -1/4  1].

hold on;  ezplot(X, Y, [-6 -1.15]);  ezplot(X, Y, [-0.9 0.9]);
ezplot(X, Y, [1.15 5]);
ezplot(k(1)*'x' + b(1));  ezplot(k(3)*'x' + b(3));
```

4. We plot the asymptotes $y = B + Mx$, $z = C + Nx$ of the *parametric space curve* $\mathbf{r}(t) = [x(t), y(t), z(t)]$ as follows:

(a) Find points $\{a_i\}$ where the curve *runs to* ∞, i.e.,

$$\lim_{t \to a_i} [x^2(t) + y^2(t) + z^2(t)] = \infty.$$

(b) For each $t = a_i$, derive the coefficients of inclination

$$M_i = \lim_{t \to a_i} \frac{y(t)}{x(t)}, \qquad\qquad N_i = \lim_{t \to a_i} \frac{z(t)}{x(t)}.$$

(c) For each finite $\{M_i, N_i\}$, derive

$$B_i = \lim_{t \to a_i} [y(t) - M_i x(t)], \qquad\qquad C_i = \lim_{t \to a_i} [z(t) - N_i x(t)];$$

then plot the oblique asymptote.

(d) For each infinite M_i, derive $x_i = \lim_{t \to a_i} x(t)$; for each infinite N_i, derive $x_i = \lim_{t \to a_i} x(t)$; and then plot the vertical asymptotes.

6.1.3 *Envelope of a family of curves*

In Chapter 5, various curves appeared as sets of points satisfying certain geometrical conditions or equations. In this section, the curves appear in a different guise, as
envelopes of families of lines, circles, etc.

We assume the function $\varphi(x,y,t)$ belongs to class C^1.

The *envelope of a family of curves* $\{\gamma_t : \varphi(x,y,t) = 0\}$, where t is the parameter of
the family, is the curve γ: $\{x(t),y(t)\}$ that is tangent at each of its points $\gamma(t)$ to the
corresponding curve γ_t.

One derives the envelope of a family $\{\gamma_t : \varphi(x,y,t) = 0\}$ of curves from the system
of equations (for instance, by elimination of t)

$$\varphi(x,y,t) = 0, \qquad \varphi'_t(x,y,t) = 0. \tag{6.1}$$

Notice that this system (6.1) defines the *discriminant set*, sometimes containing extraneous solutions (just as points of inflection appear in the problem of deriving extrema
of the function $f(x)$ using the condition $f'(x) = 0$).

Examples. Find the envelope of the family of segments that cut out:

(a) triangles with area S from the coordinate angle XOY.

Solution. The equation of the family is γ_t: $\frac{x}{t} + \frac{yt}{2S} - 1 = 0$ or $\varphi = yt^2 - 2St + 2Sx = 0$.
From $\varphi'_t = 2yt - 2S = 0$ we obtain $t = S/y$. Then, substituting in the first equation, we
obtain $xy = S/2$: a *hyperbola*.

```
N = 10; t = linspace(0, 1, 20);   hold on;
for n = 1 : N;
x = t*n/N;  y = (1 - t)*N/n;
plot(x, y); plot(x, -y); plot(-x, -y); plot(-x, y);
end                                              % Figure 6.6(a)
```

(b) triangles with hypotenuse a from the coordinate angle XY.

Solution. The equation of the family is γ_t: $\frac{x}{a\cos t} + \frac{y}{a\sin t} = 1$, the *astroid*. Recall that
the line segment PQ between the points $P = (x_1, y_1)$ and $Q = (x_2, y_2)$ has equations
$x = tx_2 + (1-t)x_1$, $y = ty_2 + (1-t)y_1$, where $t \in [0,1]$.

```
N = 19; t = linspace(0, 1, 20); hold on;
for n = -N : N;
s = pi*n/N;
plot(t*cos(s), (1 - t)*sin(s));   plot(t*cos(s), -(1 - t)*sin(s));
end                                              % Figure 6.6(b)
```

(c) segments with sum of lengths from the coordinate axes OX and OY equal to a.

Answer: The equation of the family is $\gamma_t: \frac{x}{t} + \frac{y}{a-t} = 1$.

```
N = 9;  t = linspace(0, 1, 20);  hold on;
for n = -N : N;
 x = t*n/10;  y = (1 - t)*(1 - abs(n)/10);
 plot(x, y);  plot(x, -y);
end                                              % Figure 6.6(c)
```

Figure 6.6 contains only families of curves, but each family creates the illusion that the figure also contains its envelope.

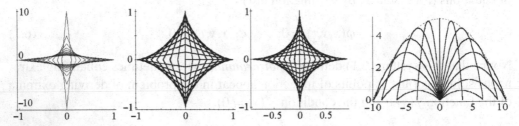

Fig. 6.6 Embroidery of the hyperbola $xy = 1$, the astroid, and similarly the curve $x^{3/2} + y^{3/2} = 1$. Parabola of safety.

Parabola of safety. The following problem is useful in ballistics.

Prove that trajectories of points (shells), leaving the origin (a high-angle gun) with constant velocity v at different angles t $(0 < t < \pi)$ to the axis OX (the horizon) have a parabola in the role of envelope.

Solution. We have the equation of a trajectory (parabola): $x = vs\cos t$, $y = vs\sin t - \frac{1}{2}gs^2$, where s is the time. Eliminating s, we derive the equation of the family $\varphi(x,y,t) = y - x\tan t + \frac{gx^2}{2v^2\cos^2 t}$. From the second equation $\varphi'_t(x,y,t) = 0$ we obtain $\tan t = \frac{v^2}{gx}$. Then $\frac{1}{\cos^2 t} = 1 + \tan^2 t = 1 + v^4/(g^2 x^2)$. Substituting in the first equation $\varphi(x,y,t) = 0$ with the aim of eliminating t, after simple transformations we obtain the equation $y = \frac{v^2}{2g} - \frac{g}{2v^2}x^2$ of the *parabola of safety* (Figure 6.6(d)).

```
syms x t;
g = 9.8;  v = 10;
X = @(s) v*t*cos(pi/18*s);  Y = @(s) v*t*sin(pi/18*s)-g*t^2/2;
```

Then we plot (using animation):

```
for s = 0 : 17;
 h = ezplot(v^2/(2*g) - g/(2*v^2)*x^2, [-v^2/g, v^2/g]);
```

```
set(h, 'Color', 'r', 'LineWidth', 2);  hold on;
T = 2*v*sin(pi/18*s)/g;
ezplot(X(s), Y(s), [0, T]);
pause(0.2);  hold off;
end;
```

6.1.4 *Mathematical embroidery*

Mathematical embroidery is a method of plotting curves as envelopes of certain families of lines (segments) or circles. MATLAB allows us to realize on the display the algorithms of mathematical embroidery, and to find new examples.

Mathematical embroidery using line segments.

1. For the embroidery of the *deltoid*, divide the circle of radius R with center O, starting with the point $A = (R, 0°)$, into arcs of $5°$ and number the points of partition counterclockwise. Then, starting from the point $A' = (R, 180°)$, divide the circle into arcs of $10°$ and number the points of partition clockwise. Join by line segments the points of the partition with matching numbers for A and A'. Next, continue the segments to their intersection with the larger circle of radius $3R$ (Figure 6.7(a)).

Hint. First, plot the base circle of radius $R = 1$ and the boundary circle of radius $R = 3$. Then use integers $-36 \leq n \leq 35$ and plot lines from the family $y = \frac{y_2 - y_1}{x_2 - x_1}(x - x_1) + y_1$, where $x_1 = \cos(T)$, $y_1 = \sin(T)$, $x_2 = \cos(\pi - Tm)$, and $y_2 = \sin(\pi - Tm)$.

```
m = 2;                                              %  deltoid, Figure 6.7(a)
syms x;
t = linspace(-pi, pi, 40);  hold on;  axis equal
for i = -35 : 35;
 if mod(i, 6)  = 0;  T = pi/18*i;
  ezplot((sin(pi-T*m) - sin(T))/(cos(pi-T*m) - cos(T))*(x-cos(T)) + sin(T));
  axis([-3 3 -3 3]);
  plot(cos(t), sin(t), '–g');                           % the base circle
  plot(3*cos(t), 3*sin(t), 'r', 'LineWidth', 2);        % the boundary circle
 end;
end
```

2. The embroidery of the *astroid* is similar to that of the deltoid. The only difference is that arcs are taken from the point $A' = (R, 180°)$ by $15°$, and segments are continued till their intersection with the larger circle of radius $2R$ (Figure 6.7(b)).

```
m = 3;                                              %  astroid, Figure 6.7(b)
syms x;
```

```
t = linspace(-pi, pi, 40);  hold on;  axis equal
for i = -36 : 35;
 if mod(i + 27, 18)  = 0;  T = pi/18*i;
 ...                                          % [ as several lines of the program above ]
 end
```

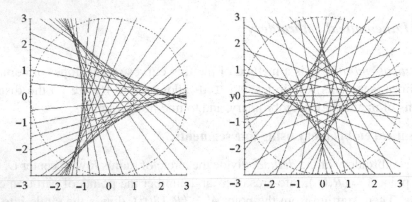

Fig. 6.7 Embroidery of the astroid and the deltoid.

3. The embroidery of the *cardioid* needs only one (base) circle of radius R: the arcs from points $A = (R, 0°)$ and $A' = (R, 180°)$ are equal, for example, $10°$, and are fixed in the same direction (say, counterclockwise). The segments are chords of the base circle; they join points from A with points from A' having index *twice* as large (Figure 6.8).

The embroidery of the *nephroid* is similar to that of the cardioid. The only difference is that the chords of the base circle join points from A with points from A' having index *three times* as large (Figure 6.9).

```
m = 2;              % m = 2 cardioid; m = 3 nephroid, Figures 6.8(a,b), 6.9(a,b)
syms t;
hold on;  axis equal;
for i = -35 : 35;
 T = pi/36*(18 + i*m);    S = pi/36*(-18 + i);
 ezplot(t*cos(T) + (1 - t)*cos(S), t*sin(T) + (1 - t)*sin(S));  axis([-1 1 -1 1]);
end;
h = ezplot(cos(t), sin(t));  set(h, 'Color', 'r', 'LineWidth', 2);
```
```
syms t;
M = 1;                       % envelope: M = 1 – cardioid, M = 0.5 – nephroid.
ezplot((1 + M)*sin(M*t) - M*sin(t + M*t), -(1 + M)*cos(M*t) + M*cos(t + M*t),
[0, 2*pi/M]);                            % Figures 6.8(c), 6.9(c)
```

One may plot the cardioid by using its complex equation $z(t) = 2e^{it} + e^{2it}$:

```
t = 0 : .1 : 2*pi;   plot(2*exp(i*t) + exp(2*i*t))
```

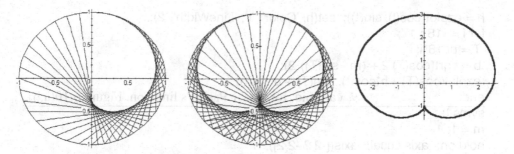

Fig. 6.8 Embroidery of the cardioid.

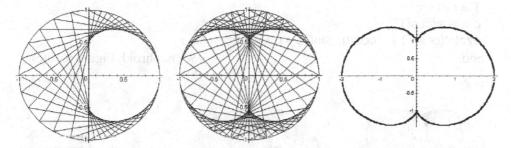

Fig. 6.9 Embroidery of the nephroid.

Mathematical embroidery using circles.

1. Starting with the point $A = (R, 0°)$, break the basic circle of radius R with center O into arcs (of $10°$, for example). Then plot a circle of radius $R_i = |P_iA|$ about each division point P_i. The heart-shaped envelope of all these circles is the *cardioid*. In other words, the union of circles through the point $A \in \omega$ whose centers lie on the given circle ω is the plane domain bounded by the cardioid (Figure 6.10(b)).

2. The embroidery of *Pascal's limaçon* by circles generalizes that of the cardioid. The only difference is that the fixed point $A = (kR, 0°)$ does not belong to the base circle. For $k > 1$, Pascal's limaçon has a loop, and for $0 < k < 1$ it does not (Figure 6.10(a)). The complex equation of Pascal's limaçon is $z(t) = e^{it}(a + b\cos t)$ $(a, b > 0)$ (Verify!).

```
plot(exp(i*t).*(3 + 4.*cos(t)))
```

3. The embroidery of the *nephroid* by circles is similar to that of the cardioid. The only difference is that the radii of the circles R_i are assumed equal to the distances from the points P_i to the diameter through the point A (the axis OX), i.e., $R_i = d(P_i, |OA|)$ (Figure 6.10(c)).

```
syms t;
m = 1;
hold on; axis equal; axis([-3 3 -3.5 2]);
```

```
h = ezplot(cos(t), sin(t));  set(h, 'Color', 'r', 'LineWidth', 2);
for i = -18 : 17;
 T = pi/18*i;
 b = sqrt(cos(T)^2 + (m - sin(T))^2);
 ezplot(cos(T) + b*cos(t),  sin(T) + b*sin(t));
end;                    % cardioid, for m ≠ 1 Pascal's limaçon, Figure 6.10(a,b)
syms t;
m = 1;
hold on;  axis equal;  axis([-2 2 -2 2]);
h = ezplot(cos(t), sin(t));  set(h, 'Color', 'r', 'LineWidth', 2);
for i = -18 : 17;
 T = pi/18*i;
 c = abs(cos(T));
 ezplot(cos(T) + c*cos(t),  sin(T) + c*sin(t));
end;                                        %  nephroid, Figure 6.10(c).
```

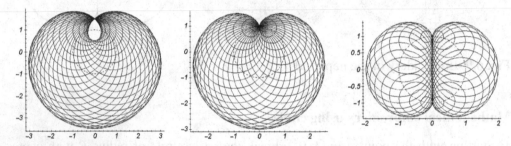

Fig. 6.10 Embroidery using circles: Pascal's limaçon, cardioid, and nephroid.

6.1.5 *Evolutes and evolvents, and parallel curves*

The normal vectors at points of the given curve of class C^1 can be chosen as the family of lines for plotting the envelope.

The envelope γ_2 of a family of normals to the given curve γ_1 is called the *evolute*. The curve γ_1 is called the *evolent* (*involute*) of the curve γ_2, if γ_2 is the evolute of γ_1.

In fact, the evolute is related to the evolvent in the same way that differentiation is related to indefinite integration. The evolute of the plane curve $\mathbf{r}(t)$ coincides with the *curve of centers of curvature* (see Section 6.4). In particular, the evolute of a circle is a single point. The evolute is defined by the equation $\mathbf{r}_{ev}(t) = \mathbf{r}(t) + \frac{1}{k(t)}\mathbf{v}(t)$, or in coordinate form

$$\begin{cases} x_{ev}(t) = x(t) - \dfrac{y'(t)}{k(t)\sqrt{x'^2+y'^2}} = x(t) - y'(t)\dfrac{x'^2+y'^2}{x'y''-x''y'}, \\[3mm] y_{ev}(t) = y(t) + \dfrac{x'(t)}{k(t)\sqrt{x'^2+y'^2}} = y(t) + x'(t)\dfrac{x'^2+y'^2}{x'y''-x''y'}. \end{cases}$$

Various evolvents of the unit-speed plane curve $\mathbf{r}(s)$ are given by

$$\mathbf{r}_{EV}(s) = \mathbf{r}(s) - (s - s_0)\,\boldsymbol{\tau}(s),$$

where $\mathbf{r}$, τ, s are related to the given curve, but the constant s_0 depends on the choice of the evolvent.

Examples.

1. The evolute of the parabola $y = x^2$ (with parametric equations $x = t$, $y = t^2$) is the *half-cubic parabola*

$$x_{ev} = t - \frac{1+(2t)^2}{1\cdot 2 - (2t)\cdot 0}2t = -4t^3, \qquad y_{ev} = t^2 - \frac{1+(2t)^2}{(2t)\cdot 0 - 1\cdot 2} = 3t^2 + \frac{1}{2}.$$

To show this we develop a program that can be used for various curves.

```
syms t;
x = t;  y = t^2;  t1 = -2;  t2 = 2;  t3 = t2 - 1;          % enter data
xt = diff(x, t);  yt = diff(y, t);
xtt = diff(xt, t);  ytt = diff(yt, t);
x1 = simplify(x - yt*(xt^2 + yt^2)/(xt*ytt - xtt*yt))
y1 = simplify(y + xt*(xt^2 + yt^2)/(xt*ytt - xtt*yt))
```

$$x_1 := -4t^3 \qquad y_1 := 3t^2 + 0.5 \qquad\qquad \text{\% Answer}$$

Then obtain the curves with their normals (Figure 6.11(a)):

```
m = 16;
hold on;  axis equal;
h = ezplot(x, y, [t1, t2]);  set(h, 'Color', 'r', 'LineWidth', 2);
g = ezplot(x1, y1, [t1 + 1, t3]);  set(g, 'Color', 'g', 'LineWidth', 2);
for i = m/4 : 3*m/4;
 T = t1 + i*(t2 - t1)/m;
 N = [x, y] + 's'*[x1 - x, y1 - y];
 ezplot(subs(N, t, T),  [0, 1]);
end
```

2. The *evolvent of a circle* belongs, as does the cycloid, to the category of cycloidal curves. Namely, it is the trajectory of a point on a line rolling without slipping along the circle.

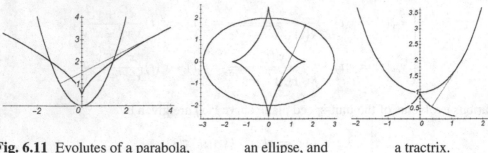

Fig. 6.11 Evolutes of a parabola,   an ellipse, and   a tractrix.

Show that the evolvents of the unit-speed circle $\mathbf{r}(s) = [R\cos(s/R), R\sin(s/R)]$ are the curves

$$\mathbf{r}_{EV}(s) = [R\cos(s/R) + (s - s_0)\sin(s/R),\ R\sin(s/R) - (s - s_0)\cos(s/R)]$$

(Figure 6.12(a)). Plot them with the following program:

```
syms s;
R = 1;   x = R*cos(s/R);  y = R*sin(s/R);                % enter data (equations)
hold on;  axis equal;
h = ezplot(x, y);   set(h, 'Color', 'g', 'LineWidth', 2);
for c = 0 : 5;
 X = x - (s - c)*diff(x, s);   Y = y - (s - c)*diff(y, s);
 ezplot(X, Y, [c, c + 1.5*pi]);
end
```

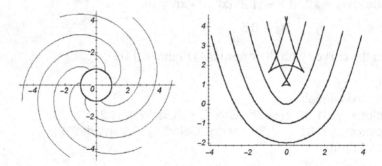

Fig. 6.12 Evolvents of a circle. Parallel curves of a parabola.

The envelope of the pencil of rays emanating from a given point or direction and (a) reflected or (b) refracted by the given curve is called (a) the *catacaustic* or (b) the *diacaustic* (or *caustic*, for short) of this curve with respect to the given point.

Let γ be the graph $y = y(x)$, and consider a pencil of rays parallel to the unit vector $\mathbf{e} = [a,b]$. The caustic $[X,Y]$ has the following equations:

$$X(x) = x - \frac{(b-ay')^2}{2y''}y' - \frac{(b-ay')(a+by')}{2y''},$$

$$Y(x) = y + \frac{(b-ay')^2}{2y''} - \frac{(b-ay')(a+by')}{2y''}y'.$$

Let $\mathbf{r} = \mathbf{r}(s)$ be the natural parameterization of γ. The equation of the caustic $\tilde{\mathbf{r}}$ is

$$\tilde{\mathbf{r}} = \begin{cases} \mathbf{r} + \dfrac{(\mathbf{e},\mathbf{n})}{2k(s)}[(\mathbf{e},\mathbf{n})\,\mathbf{n} - (\mathbf{e},\tau)\,\tau] & \text{the rays are parallel to } \mathbf{e}, \\[3mm] \mathbf{r} + \dfrac{(\mathbf{r},\mathbf{n})}{2k(s)\mathbf{r}^2 + (\mathbf{r},\mathbf{n})}[(\mathbf{r},\mathbf{n})\,\mathbf{n} - (\mathbf{r},\tau)\,\tau] & \text{the rays run from the origin } O. \end{cases}$$

Using normal segments to the curve of constant length d (on one side or the other), we obtain *parallel curves* (*equidistants*). Equations of equidistants γ_d of the curve γ: $x = x(t)$, $y = y(t)$ are the following:

$$\mathbf{r}_d(t) = \mathbf{r}(t) + d \cdot \mathbf{v}(t) \iff \begin{cases} x_d(t) = x(t) - dy' \,/\, \sqrt{x'^2 + y'^2}, \\[2mm] y_d(t) = y(t) + dx' \,/\, \sqrt{x'^2 + y'^2}. \end{cases}$$

If $|d|$ is less than the radius of curvature of the smooth curve γ, the equidistant γ_d is smooth; otherwise, γ_d may contain singular points. The difference between the perimeters of two closed smooth parallel curves is $2\pi d$.

A parallel curve of an ellipse (a *toroid*) appears as the boundary curve of a parallel projection of the torus of revolution. The circular axis of the torus projects into the ellipse.

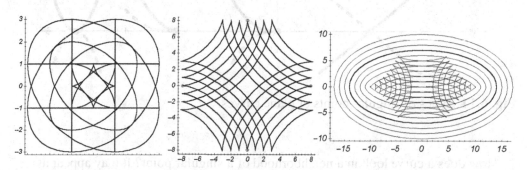

Fig. 6.13 Parallel curves to astroid and ellipse.

Example. Plot equidistants for several curves using the program below.

```
xt = diff(x, t);  yt = diff(y, t);
S = sqrt(xt^2 + yt^2);
xd = @(d) x - d*yt/S;    yd = @(d) y + d*xt/S;
hold on;  axis equal;
h = ezplot(x, y, [t1, t2]);  set(h, 'Color', 'g', 'LineWidth', 2);
for i = -3 : 3;
 ezplot(xd(i), yd(i),  [t1, t2]);
end
```

First prepare the data:

```
syms t;
a = 8;  x = a*cos(t)^3;  y = a*sin(t)^3;  t1 = -pi;  t2 = pi;        % Figure 6.13(b).
a = 8;  b = 7;  x = a*cos(t);  y = b*sin(t);  t1 = 0;  t2 = 2*pi;    % Figure 6.13(c).
x = t;  y = t^2;  t1 = -2;  t2 = 2;                        %  parabola, Figure 6.12(b).
```

6.2 Singular Points

In this section we study singular points for different types of equations of a curve. The last subsection contains more advanced examples.

The definitions of singular and regular points of curves (see Section 5.1.1) depend on the class of parameterization. Figure 6.14 presents several types of singular points of plane curves.

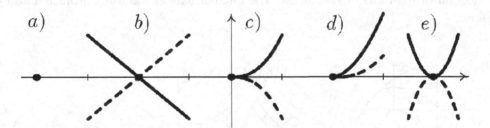

Fig. 6.14 Types of singular points of plane curves.

How does a curve look in a neighborhood of a singular point? It may appear as

(a) an *isolated* point, if a neighborhood of the singular point does not contain other points of a curve;

(b) a *double* point (or *node*), if two branches of the curve intersect at the point but are not tangent;

(c) a *cuspidal* point of the *first kind*, if two branches of the curve lie on different sides of the common semi-tangent line at the point;

(d) a *cuspidal* point of the *second kind*, if two branches of the curve lie on the same side of the common semi-tangent line at the point;

(e) a point of *self-tangency*, if two branches of the curve are tangent at the singular point.

6.2.1 Singular points of parametric curves

There is a useful *sufficient* condition for the point t_0 of the plane curve $\mathbf{r}(t)$ of class C^∞ with the property $\mathbf{r}'(t_0) = 0$ to be singular.

Theorem 6.1. *Let a curve γ of class C^∞ be given by the equations $x = x(t)$ and $y = y(t)$. Then the point $(x_0, y_0) \in \gamma$ (for $t = t_0$) is*

(a) *a regular point if the first nonzero derivative of the functions $x(t)$ and $y(t)$ at this point is odd,*

(b) *a singular (cuspidal) point if for $t = t_0$ the first nonzero derivative (in series) of the functions $x(t)$ and $y(t)$ is even.*

Examples.

1. Let $\mathbf{r}(t) = [t^2, t^3]$. Show that $(0,0)$ is a cuspidal point of the 1st kind.

   ezplot('t^2', 't^3', [-1, 1]), grid on;   % Figure 6.15(a)

2. Show that $(0,0)$ is a cuspidal point of the 2nd kind of $\mathbf{r}(t) = [t^2, t^4 + t^5]$.

   ezplot('t^2', 't^4 + t^5', [-1.2, 0.8]), grid on;   % Figure 6.15(b)

3. Plot the curve $\mathbf{r}(t) = [2t - t^2, 3t - t^3]$ and *semi-tangent lines* to it at its singular points (Figure 6.15(c)).

Hint. Show that $t_0 = 1$ is a singular point. Find the coefficient of inclination of a semi-tangent line by the formula $\lim_{t \to -t_0} x'(t)/y'(t)$ (or $t \to t_0$).

```
syms t;
r = [2*t - t^2, 3*t - t^3];
rt = diff(r, t)                          % Answer: r_t = [2 - 2t, 3 - 3t^2]

t1 = solve(rt(1), rt(2), t),   pp = subs(r, t1)      % Answer: t_1 = 1, pp = [1, 2].
k = limit(rt(2)/rt(1), t1), Tp = pp + [t, k*t]    % Answer: k = 3, T_p = [t + 1, 3t + 2]
ezplot(rt(1), rt(2), [-1.9, 2.2]); hold on;
h = ezplot(Tp, [-2.3, 0]); set(h, 'Color', 'r');             % Figure 6.15(c).
```

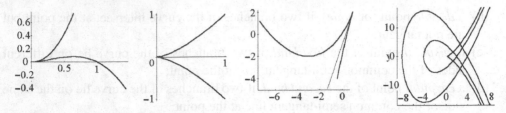

Fig. 6.15 Cuspidal points. Semi-tangent line. Family of curves.

6.2.2 Singular points of implicitly defined curves

What are sufficient conditions for the point (x_0, y_0) of the implicitly defined plane curve $F(x,y) = 0$ to be singular?

Denote by $F_x, F_y, F_{xx}, F_{xy}, F_{yy}, F_{xxx}, F_{xxy}, F_{xyy}, F_{yyy}$ the partial derivatives at the given point. Assume that not all partial derivatives of second order at this point (x_0, y_0) are zero, and let $\Delta = F_{xx}F_{yy} - F_{xy}^2$.

Let $\mathrm{Res}(f, g)$ be the *resultant* (see Section 1.5) of the following two polynomials: $f(t) = F_{xx}t^2 + F_{xy}t + F_{yy}$ and $g(t) = F_{xxx}t^3 + F_{xxy}t^2 + F_{xyy}t + F_{yyy}$. In other words,

$$\mathrm{Res}(f, g) \text{ is the determinant of } R = \begin{vmatrix} F_{xx} & F_{xy} & F_{yy} & 0 & 0 \\ 0 & F_{xx} & F_{xy} & F_{yy} & 0 \\ 0 & 0 & F_{xx} & F_{xy} & F_{yy} \\ F_{xxx} & F_{xxy} & F_{xyy} & F_{yyy} & 0 \\ 0 & F_{xxx} & F_{xxy} & F_{xyy} & F_{yyy} \end{vmatrix}.$$

One may test the program **Resfg** (see Section 1.5):

```
syms Fxx Fxy Fyy Fxxx Fxxy Fxyy Fyyy;
a = [Fxx, Fxy, Fyy];
b = [Fxxx, Fxxy, Fxyy, Fyyy];
Resfg(a, b)                          % obtain the structure of the matrix R.
```

The cochleoid, a transcendental curve (see Figure 1.19(b)), has an infinite number of branches through the pole and tangent to the polar axis; hence, the pole is a singular point of *infinite order*. If the Taylor series of the function $F(x,y)$ at the point (x_0, y_0) starts from some number $k > 2$, then we obtain by definition a *singular point of order* k. The case $k = 2$ is treated in the following theorem.

Theorem 6.2. *Three cases are possible for the point* (x_0, y_0) *on the curve* γ: $F(x,y) = 0$ *with conditions* $F_x(x_0, y_0) = 0$, $F_y(x_0, y_0) = 0$:

$$\Delta \begin{cases} > 0, & \text{an isolated point of the level set } F = 0, \text{ type (a),} \\ < 0, & \text{the intersection of two branches of the curve } F = 0, \text{ type (b),} \\ = 0, & \begin{cases} \det R(x_0, y_0) \neq 0, & \text{cuspidal point of the first kind, type (c),} \\ \det R(x_0, y_0) = 0, & \text{further investigation is required.} \end{cases} \end{cases}$$

Now (using **Resfg**) we prepare M-file **singDR** to detect singular points.

```
function DR = singDR(F, x0, y0)                    % based on Theorem 6.2
  syms x y;
  a = [diff(F, x, 2),  diff(diff(F, x), y),  diff(F, y, 2)];
  b = [diff(F, x, 3),  diff(diff(F, x, x), y),  diff(diff(F, x), y, 2),  diff(F, y, 3)];
  dF = subs([diff(F, x),  diff(F, y)], {x, y}, {x0, y0});
  if dF == 0;   dF,                             disp( 'the point is singular' );
  if a == [0 0 0];          warning('ddF = 0; further investigation is required');
  D = subs(a(1)*a(3) - a(2)^2, {x, y}, {x0, y0});
  if D > 0;   D,          disp( 'D > 0; an isolated point of the level set F = 0' );
  elseif D < 0;   D,              disp( 'D < 0; intersection of two branches' );
  else
  detR = det(subs(Resfg(a, b), {x, y}, {x0, y0}));
  if detR == 0;   D, detR,
                  warning( 'D = 0, detR = 0; further investigation is required' );
  else   D, detR,      disp( 'D = 0, detR ≠ 0; cuspidal point of the first kind' );
  end end end
  else   dF,                           warning('dF ≠ 0; the point is regular');
  end
end
```

Examples.

1. We detect the type of singular point $(0,0)$ on $F = 0$ using **singDR**.

(a) The equation $(x^2 + y^2)(x - 1) = 0$ defines the line $x = 1$ together with an isolated point $(0,0)$ at which $F_{xx} = F_{yy} = -1$, $F_{xy} = 0$, and $\Delta > 0$.

```
syms x y;   F = (x^2 + y^2)*(x - 1);            % obtain D = 4 > 0;
singDR(F, 0, 0)                      % isolated point of the level set F = 0.
```

(b) The equation $(x^2 + y^2)^2 - 2a^2(x^2 - y^2) = 0$ defines the lemniscate of Bernoulli (Figure 1.21(c)), with an additional node singular point $(0,0)$ at which $F_{xx} = -2a^2$, $F_{yy} = 2a^2$, $F_{xy} = 0$, and $\Delta < 0$. In a neighborhood of the point $(0,0)$ this curve consists of two elementary curves.

```
syms x y;   F = (x^2 + y^2 - 2*x)^2 - (x^2 + y^2);        % obtain D = -12;
singDR(F, 0, 0)                       % intersection of two branches.
```

(c) The equation $y^2 - x^3 = 0$ defines the half-cubic parabola (Figure 6.11(a)), with a cuspidal point of the first kind $(0,0)$ at which $F_{xx} = F_{xy} = 0$, $F_{yy} = 1$, and $\Delta = 0$.

```
syms x y;   F = y^2 - x^3;                    % obtain D = 0; detR = 288;
singDR(F, 0, 0)                               % cuspidal point of the first kind.
```

(d) The equation $y(y - x^2) = 0$ defines the parabola $y = x^2$, with the line $y = 0$ osculating at the singular point $(0,0)$, where $F_{xx} = F_{xy} = 0$, $F_{yy} = 1$, and $\Delta = 0$.

```
syms x y;   F = y*(y - x^2);                  % obtain D = -12;
singDR(F, 0, 0)                               % intersection of two branches.
```

2. What are the relations between a and b when the curve $y^2 = x^3 + ax + b$ has a double point? Plot some of these curves. *Answer*: $16a^3 = -27b^2$.

```
syms a b x y t;
F = y^2 - x^3 - a*x - b;
Fx = diff(F, x);     Fy = diff(F, y);
s = solve(F, Fx, Fy, y, a, b)          % Answer: s = {b = 2x^3, a = -3x^2, y = 0}
F1 = subs(F, {a, b}, {-3*t^2, 2*t^3})  % Answer: F1 = y^2 - x^3 + 3t^2 x - 2t^3
S = solve(F1, y)          % Answer: S = (2t + x)^{1/2}(t - x), -(2t + x)^{1/2}(t - x)
hold on;  for i = 0 : 2;
Y = subs(S, t, 2*i);
ezplot(Y(1), [-9, 9]);  ezplot(Y(2), [-9, 9]);
end                                            % Figure 6.15(d).
```

6.2.3 *Unusual singular points of plane curves*

Some transcendental curves have singular points of more complicated types, for example, when the derivative of a function (in the equation of the curve) is not continuous (see Figure 6.16).

Examples.

1. The right derivative of the function $y = e^{1/x}$ has a discontinuity at $x = 0$. At the origin, the graph of the function is interrupted. Such points are called *stopping points*.

```
f = @ (x) exp(1).^(1./x);
x1 = -5 : .01 : 0;  y1 = f(x1);
x2 = .5 : .01 : 5;  y2 = f(x2);
plot(x1, y1, 'r',  x2, y2),  grid on;          % Figure 6.16(a)
```

2. The left derivative of the function $y = \frac{x}{1 + e^{1/x}}$ has a discontinuity at the point $x = 0$. At the origin, the graph of the function is broken: two branches of the curve form the angle $\varphi \in (0, \pi)$ at this point. Such points are called *angular points*.

```
f = @ (x) x./(1 + exp(1).^(1./x));
X = -1 : .01 : 2;
plot(f(X), 'r', 'LineWidth', 3);  hold on;  grid on;
plot(X);  plot(X - X);                                    % Figure 6.16(b)
```

3. Why is $(0,0)$ a singular point of order 3 of the implicitly given curve $x^6 - 2a^2x^3y - b^3y^3 = 0$ $(a, b \neq 0)$?

Solution. Three branches of the curve are tangent to the line $y = 0$ at $(0,0)$. For $x \approx 0$ neglect the term x^6 in the equation and then divide it by y: the curve is similar to the *half-cubic parabola* $2a^2x^3 + b^3y^2 = 0$. For $y \approx 0$ neglect the term b^3y^3 and divide the equation by x^3: the curve is similar to the *cubic parabola* $x^3 - 2a^2y = 0$.

```
v = -0.5 : .002 : 0.5;   [x, y] = meshgrid(v);
f = 'x.^6 - 2*x.^3.*y - y.^3';
contour(x, y, f, [0 0]);                                   % Figure 6.16(c)
```

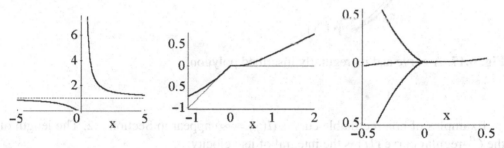

Fig. 6.16 Unusual singular points of plane curves.

6.3 Length and Center of Mass

We start from basic formulae for calculating the length and center of mass of a curve. Then we find these characteristics for a regularly inscribed polygon (that approximates the curve) and compare them with the results obtained by integrating velocity along the curve.

A polygon σ with vertices $P_1, P_2, \ldots, P_n$, is called *regularly inscribed in the curve* $\gamma = \mathbf{r}(t)$ $(a \leq t \leq b)$ if there exists a partition $a = t_1 < \cdots < t_n = b$ of the segment $[a,b]$ such that $P_i = \mathbf{r}(t_i)$.

The *length of the polygon* is equal to $l(\sigma) = \sum_{i=1}^{n-1} |P_{i+1}P_i|$.

The length of the polygon $P_1(\rho_1, t_1), \ldots, P_n(\rho_n, t_n)$ in polar coordinates is given by the formula

$$l(\sigma) = \sum_{i=1}^{n-1} \sqrt{\rho_{i+1}^2 + \rho_i^2 - 2\rho_{i+1}\rho_i \cos(t_{i+1} - t_i)},$$

derived with the help of the cosine theorem.

The *length* $l(\gamma) < \infty$ *of the rectifiable curve* γ is the least upper bound of the lengths of all regularly inscribed polygons.

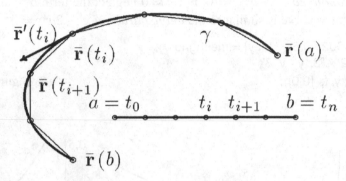

Fig. 6.17 A curve and its regularly inscribed polygon.

Examples of non-rectifiable curves ($l(\gamma) = \infty$) appear in Section 5.2. The length of the C^1-regular curve $\mathbf{r}(t)$ is the integral of its velocity:

$$l(\gamma) = \int_a^b |\mathbf{r}'(t)| \, dt. \tag{6.2}$$

The *natural parameterization* of the regular curve $\mathbf{r}(t)$ is the vector function $\tilde{\mathbf{r}}(s) = \mathbf{r}(t(s))$, where $t(s)$ is the inverse function to $s(t) = \int_{t_0}^{t} |\mathbf{r}'(t)| \, dt + s_0$.

The formulae for calculating the length of a curve γ:

1. γ: $\mathbf{r}(t) = [x(t), y(t), z(t)] \subset \mathbb{R}^3$: $l(\gamma) = \int_a^b \sqrt{x'^2 + y'^2 + z'^2} \, dt$,
2. γ: $y = f_1(x), z = f_2(x)$ in $\mathbb{R}^3$: $l(\gamma) = \int_a^b \sqrt{1 + f_1'^2 + f_2'^2} \, dx$,
3. γ: $\mathbf{r}(t) = [x(t), y(t)] \subset \mathbb{R}^2$: $l(\gamma) = \int_a^b \sqrt{x'^2 + y'^2} \, dt$,
4. γ: $y = f(x)$ in $\mathbb{R}^2$: $l(\gamma) = \int_a^b \sqrt{1 + f'^2} \, dx$,
5. γ: $\rho = \rho(\varphi)$ in polar coordinates: $l(\gamma) = \int_a^b \sqrt{\rho^2 + \rho'^2} \, d\varphi$.

The formula $m(\gamma) = \int_a^b \mu(t) |\mathbf{r}'(t)| \, dt$ for the *mass of the curve* $\mathbf{r}(t)$ constructed from thin wire with density $\mu(t)$ generalizes (6.2). The *center of mass* of the plane curve

$\mathbf{r}(t) = [x(t), y(t)]$ $(a \leq t \leq b)$ has the coordinates $x_c(\gamma) = M_y/m(\gamma)$, $y_c(\gamma) = M_x/m(\gamma)$, where

$$M_x = \int_a^b \mu(t) y(t) \sqrt{x'^2 + y'^2}\, dt, \qquad M_y = \int_a^b \mu(t) x(t) \sqrt{x'^2 + y'^2}\, dt$$

are *moments of the first order* of the curve with respect to OX and OY.

Example. The *metric form* of curvilinear coordinates in the plane $\mathbf{r} \colon \mathbb{R}^2 \to \mathbb{R}^2$ (i.e., $\mathbf{r}(u, v) = [x(u, v), y(u, v)]$) is the quadratic form

$$ds^2 = E(u, v)\, du^2 + 2F(u, v)\, du\, dv + G(u, v)\, dv^2,$$

where $E = \mathbf{r}_u^2 = x_u^2 + y_u^2$, $F = \mathbf{r}_u \cdot \mathbf{r}_v = x_u x_v + y_u y_v$, and $G = \mathbf{r}_v^2 = x_v^2 + y_v^2$; see Section 4.3 and the M-file **EFG.m** in Section 7.5. For *affine* coordinates $x = a_1 u + a_2 v + a_0$, $y = b_1 u + b_2 v + b_0$ the functions E, F, and G are constant.

The *length of the curve* $\gamma \colon u = u(t)$, $v = v(t)$ $(t_1 \leq t \leq t_2)$ in curvilinear coordinates in $\mathbb{R}^2$ is given by the formula

$$l(\gamma) = \int_{t_1}^{t_2} \sqrt{E(u(t), v(t)) u'(t)^2 + 2F(u(t), v(t)) u'(t) v'(t) + G(u(t), v(t)) v'(t)^2}\, dt.$$

The notation

$$g_{11} = E, \qquad g_{12} = F, \qquad g_{22} = G$$

is often used. There are analogous formulae ds^2, $g_{ij}(u_1, u_2, u_3)$, and $l(\gamma)$ for curvilinear coordinates $\mathbf{r}(u_1, u_2, u_3) = [x(u_1, u_2, u_3), y(u_1, u_2, u_3), z(u_1, u_2, u_3)]$ in space:

$$ds^2 = \sum_{i,j=1}^{3} g_{ij}(u_1, u_2, u_3)\, du_i\, du_j \quad \text{where } g_{ij} = \mathbf{r}_{u_i} \mathbf{r}_{u_j},$$

$$l(\gamma) = \int_{t_1}^{t_2} \sqrt{\sum_{i,j=1}^{3} g_{ij}(u_1, u_2, u_3)\, du_i\, du_j}\, dt.$$

We use two methods (integrals, and a regularly inscribed polygon) and compare results, to solve the problem: *find the length and center of mass of a given curve* (see examples and exercises below).

Strategy:

   (1) Enter the data for the curve and plot it.
   (2) Derive the length and center of mass of the curve, using an integral.
   (3) Enter the number of edges n of the inscribed polygon and plot it.
   (4) Derive the length and the center of mass of the polygon.
   (5) Compare numeric results of steps (2) and (4) for some values of n.

Examples. Consider four typical situations and programs.

(a) Curves plotted as *graphs* of functions $y = f(x)$. See Exercise 2(a)–(c), p. 282.

```
syms x;
a = -2;  b = 2;  f = x^2:                          %   define your function
fx = diff(f, x);  nf = sqrt(1 + fx^2);
L1 = double(int(nf, x, a, b))              % Answer: L1 = 9.2936 (the length)
My = double(int(x*nf, x,  a, b));  Mx = double(int(f*nf, x,  a, b));
rc1 = [My, Mx]/L1                                  % Answer: rc1 = [0, 1.823]
n = 6;  X = @(i) a + i*(b - a)/n;
F = @(i) subs(f, x, X(i));
for i = 0 : (n - 1);
 D(i + 1) = sqrt((F(i + 1) - F(i))^2 + (X(i + 1) - X(i))^2);
 Cx(i + 1) = (X(i + 1) + X(i))/2*D(i + 1);
 Cy(i + 1) = (F(i + 1) + F(i))/2*D(i + 1);         % the center of mass
end
L2 = sum(D),  d = abs(L1 - L2)/L1       % Answer: L2 = 9.2936,  d = 0.0075
rc2 = [sum(Cx) sum(Cy)]/L2,  rc2 - rc1     % Answer: rc2 = [0, 1.85], [0, 0.03]
```

Then plot a graph and the regularly inscribed polygon.

```
hold on;  axis equal;  ezplot(f, [a, b]);          % a graph, Figure 6.18(a)
plot( X(0 : n),  F(0 : n),  '--or' );                        % a polygon
```

(b) *Parametric plane curves.* See Exercise 2(d)–(f), p. 282.

```
syms t;
a = 0;  b = 2*pi;  x = t - sin(t);  y = 1 - cos(t);        % cycloid, Figure 6.18(b)
xt = diff(x, t);  yt = diff(y, t);
vt = simplify(sqrt(xt^2 + yt^2));
L1 = double(int(vt, t, a, b));                            % Answer: L1 = 8.0000
Mx = double(int(y*vt, t, a, b));   My = double(int(x*vt, t, a, b));
rc1 = [My/L1, Mx/L1]                                % rc1 = [3.14, 1.33]
n = 10;  T = @(i) a + i*(b - a)/n;
X = @(i) subs(x, t, T(i));
Y = @(i) subs(y, t, T(i));
for i = 0 : (n - 1);
 D(i + 1) = sqrt((Y(i + 1) - Y(i))^2 + (X(i + 1) - X(i))^2);
 Cx(i + 1) = (X(i + 1) + X(i))/2*D(i + 1);
 Cy(i + 1) = (Y(i + 1) + Y(i))/2*D(i + 1);
end:
L2 = sum(D),  d = abs(L1 - L2)/L1       % Answer: L2 := 7.9680,  d := 0.0040
rc2 = [sum(Cx) sum(Cy)]/L2,  rc2 - rc1;        % rc2 = [3.14, 1.3], [0, 0.034]
```

Then plot a curve with the regularly inscribed polygon.

```
hold on;  axis equal;
ezplot(X, Y, [a, b]);                    % parameterized plane curve, Figure 6.18(b)
plot(X(0 : n),  Y(0 : n),  '- - or');                              % a polygon
```

(c) *Curves in polar coordinates*. See Exercise 2(g)–(h), p. 282.

```
syms t;  a = 0; b = 2*pi;  rho = .2*t;      % Archimedes' spiral, Figure 6.18(c)
rt = diff(rho, t);  nr = sqrt(rho^2 + rt^2);
L1 = double(int(nr, t, a, b))                      % Answer: L1 = 4.2513
n = 12;
T = @(i) a + i*(b - a)/n;  R = @(i) subs(rho, t, T(i));
for i = 0 : (n - 1);
 D(i + 1) = sqrt(R(i + 1)^2 + R(i)^2 - 2*R(i + 1)*R(i)*cos(T(i + 1) - T(i)));
end
L2 = sum(D),  d = abs(L2 - L1)/L1        % Answer: L2 = 4.1892,   d = 0.0146
```

Then plot a curve with the regularly inscribed polygon.

```
ezpolar(rho, [a, b]);  hold on;    % a curve in polar coordinates, Figure 6.18(c)
polar(T(0 : n),  R(0 : n),  '- - ro');                             % a polygon
```

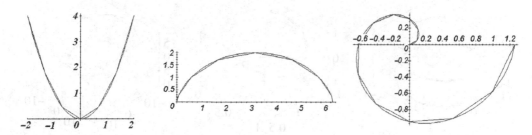

Fig. 6.18 Length of a plane curve.

(d) *Parametric space curves*. See Exercise 2(i)–(k), p. 282, and Figure 6.19.

```
syms t;  a = 0;  b = 2*pi;                              % enter data
x = cos(t)^2;  y = cos(t)*sin(t);  z = t - sin(t);      % Viviani window
xt = diff(x, t);  yt = diff(y, t);  zt = diff(z, t);
vt = sqrt(xt^2 + yt^2 + zt^2);                                % velocity
L1 = double(int(vt, t,  a, b))              % Answer: L1 = 9.5076 (length)
Myz = double(int(x*vt, t, a, b));
Mxz = double(int(y*vt, t, a, b));
Mxy = double(int(z*vt, t, a, b));
rc1 = [Myz, Mxz, Mxy]/L1;              % Answer: rc1 = [0.5169, 0, 3.1416]
```

```
n = 8;  T = @(i) a + i*(b - a)/n
X = @(i) subs(x, t, T(i));
Y = @(i) subs(y, t, T(i));
Z = @(i) subs(z, t, T(i));
for i = 0 : (n-1);
 D(i + 1) = sqrt((Z(i + 1)-Z(i))^2 + (Y(i + 1) - Y(i))^2 + (X(i + 1) - X(i))^2);
 Cx(i + 1) = (X(i+1) + X(i))/2*D(i + 1);
 Cy(i + 1) = (Y(i+1) + Y(i))/2*D(i + 1);
 Cz(i + 1) = (Z(i+1) + Z(i))/2*D(i + 1);
end
L2 = sum(D);  d = abs(L2 - L1)/L1          % Answer: L_2 = 9.0283, d = 0.0504
rc2 = [sum(Cx) sum(Cy) sum(Cz)]/L2                % rc_2 = [0.51, 0, 3.14]
rc1 - rc2                                    % Answer: [0.0051, 0, 0]
```

Finally, plot a curve with the regularly inscribed polygon.

```
ezplot3(x, y, z, [a, b]);                        % a space curve
hold on;  axis equal;  view(3);
plot3(X(0 : n), Y(0 : n), Z(0 : n), '--or');             % a polygon
```

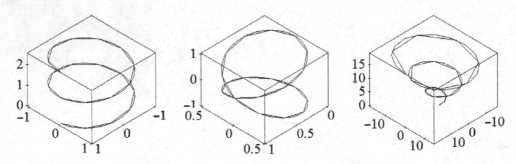

Fig. 6.19 Length of a space curve.

6.4 Curvature and Torsion

Using the formulae for curvature and torsion, we will calculate and plot the moving Frenet frame and the osculating circle of plane and space curves, and illustrate the fundamental theorem of algebra. Then we will discuss the natural equations (the main theorem in the theory) of curves.

6.4.1 *Basic formulae*

The *curvature* of a curve γ at a point P is the real number $k(P) = \lim_{P' \to P} \Delta\varphi / \Delta s = \lim_{\Delta s \to 0} \Delta\varphi / \Delta s$ (if the limit exists), where Δs is arc length of γ between points P and P', and $\Delta\varphi$ is the angle formed by tangent lines at these points. The *radius of curvature* of γ at the point P is the real number $R(P) = 1/k(P)$.

In other words, the curvature of a curve is the measure of its deviation from a straight line in a neighborhood of a given point, and the curvature becomes greater as this deviation becomes greater.

If $\mathbf{r}(s)$ is a natural parameterization of a C^2-regular curve in $\mathbb{R}^3$, then the curvature is $k(s) = |\mathbf{r}''(s)|$ (acceleration).

For an arbitrary C^2-regular parameterization $\mathbf{r}(t)$ (of this curve), the curvature is given by the formula $k(t) = \frac{|\mathbf{r}' \times \mathbf{r}''|}{|\mathbf{r}'|^3}$. Hence,

1. γ: $\mathbf{r} = [x(t), y(t), z(t)] \subset \mathbb{R}^3$: $\qquad k(t) = \frac{\sqrt{(x'y''-x''y')^2+(x'z''-x''z')^2+(y'z''-y''z')^2}}{(x'^2+y'^2+z'^2)^{3/2}}$.

2. γ: $\mathbf{r} = [x(t), y(t)]$ (i.e., $z(t) = 0$): $\qquad k(t) = \frac{|x'y''-x''y'|}{(x'^2+y'^2)^{3/2}}$.

3. γ: $y = y(x)$, a graph in $\mathbb{R}^2$: $\qquad k(x) = \frac{|y''|}{(1+y'^2)^{3/2}}$.

4. γ: $F(x,y) = 0$ at a point $(x,y) \in \gamma$: $\qquad k(x,y) = \frac{|F_{xx}F_y^2 - 2F_{xy}F_xF_y + F_{yy}F_x^2|}{(F_x^2+F_y^2)^{3/2}}$.

5. γ: $\rho = \rho(\varphi)$, a graph in polar coordinates: $\qquad k(\varphi) = \frac{|\rho^2+2\rho'^2-\rho\rho''|}{(\rho^2+\rho'^2)^{3/2}}$.

6. γ: $z = x(t) + iy(t)$, in complex plane $\mathbb{C}$: $\qquad k(t) = \frac{-\operatorname{Im}(z'\,\overline{z''})}{|z'|^3}$.

The *curvature* $\kappa(s)$ of a plane curve is given as $\frac{d\theta}{ds}$, where $\theta(s)$ is the angle between the local tangent line and the x-axis. Here $s \in [0, l]$ is the natural length parameter of the curve (see Section 6.3).

A plane curve $\mathbf{r}(s) = [x(s), y(s)]$ with a given curvature function $\kappa(s) = \theta'$ may be reconstructed, in view of $x'(s) = \cos\theta$, $y'(s) = \sin\theta$, by the formulae

$$x(s) = x_0 + \int_0^s \cos\theta \, ds, \qquad y(s) = y_0 + \int_0^s \sin\theta \, ds, \qquad \theta = \int_0^s \kappa \, ds. \qquad (6.3)$$

The *osculating circle* at the point t of a curve $\mathbf{r}(t)$ lies in the *osculating plane* (which by definition contains the vectors $\mathbf{r}'(t)$ and $\mathbf{r}''(t)$ at the point $\mathbf{r}(t)$); its center coincides with the *curvature center* of the curve at the given point, i.e., the endpoint of the vector $\frac{1}{k(t)}\mathbf{v}(t)$ whose initial point is $\mathbf{r}(t)$; and its radius is equal to $R(t) = 1/k(t)$. This circle is characterized as the "nearest" to the given curve among all circles through the point on the curve. The set $\tilde{\gamma}$ of all curvature centers of the plane curve γ is its *evolute*.

The *torsion* τ of a C^3-regular curve in $\mathbb{R}^3$ is the rotating velocity of the osculating plane (or the binormal vector) along the curve with natural parameterization. Hence, the torsion of a plane curve is identically zero. If $\mathbf{r}(s)$ is a C^3-regular curve in $\mathbb{R}^3$ with a natural parameter s, the torsion at points with $k \neq 0$ is given by the formula $\tau(s) = \frac{(\mathbf{r}', \mathbf{r}'', \mathbf{r}''')}{k^2}$.

For arbitrary C^3-regular parameterization $\mathbf{r}(t)$ of this curve, the torsion at points with $k \neq 0$ is given by the formula $\tau(t) = \frac{(\mathbf{r}', \mathbf{r}'', \mathbf{r}''')}{(\mathbf{r}' \times \mathbf{r}'')^2}$.

We will illustrate some global theorems regarding the curvature of curves. A property of a curve (or of a geometrical object) is called *global* if it describes the object as a whole (for example, length, area, closing of the curve, connectedness, or boundedness). A property is called *local* if it can be checked in an arbitrarily small part of the object or neighborhoods of its points (for example, the value of the curvature or torsion, or the presence of singular points). In a *global theorem*, a global property is essential either in the conditions or in its claim. In a *local theorem*, all conditions and claims are local.

6.4.2 *Curvature of a plane curve*

For the closed curve $\mathbf{r}(t)$ one defines the *coefficient of engagement with the point Q*, which shows how many times a point on $\mathbf{r}(t)$ under monotone change of $t \in I$ rotates about Q. Check that the coefficient of engagement keeps its (integer) value when the curve is continuously deformed in the plane without intersecting the point Q.

The *winding number m* of a closed curve $\mathbf{r}(t)$ (i.e., the number of its loops taking into account their orientation $\pm$) is equal to the coefficient of engagement of its unit tangent vector $\tau(t)$ with respect to the origin O.

The surprising fact is that the *integral curvature* IC$= \int_\gamma k$ of the plane curve equals $2\pi(m+1)$. In particular, the IC of a simple closed plane curve equals 2π. The coefficient of engagement of a plane curve (used in the proof of the *fundamental theorem of algebra*) asserts the existence of a complex root of any polynomial $w = f_n(z)$ of degree $n > 0$.

Examples.

1. **Integral curvature.** The IC of an ellipse (a simple closed curve) is equal to 2π. Show that for Pascal's limaçon (it has one loop), IC $= 4\pi$.

```
              % derive curvature of the limaçon using Program 6.1, p. 276
  int(k, t, 0, 2*pi)/(2*pi);                              % Answer: 2
  tt = linspace(0, 2*pi, 41);
  plot(tt, subs(k, t, tt),  tt, subs(diff(k, t), t, tt)),  grid on;      % Figure 6.20(b)
```

2. The curvature of the equiangular spiral $z(t) = e^{(a+i)t}$ is $k = e^{-at}/\sqrt{a^2+1}$.

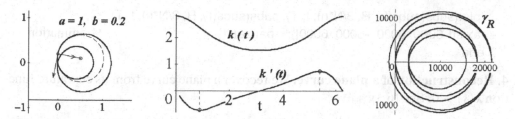

Fig. 6.20 Vertices of Pascal's limaçon.   Existence of complex roots.

```
syms a t real;
z = exp((a + i)*t);
zt = diff(z, t),  ztt = diff(zt, t)        % z' = e^((a+i)t)(a+i), z'' = e^((a+i)t)(a+i)^2
P = simplify(zt*conj(ztt))                 % e^(2at)(a^2+1)(a-i)
imP = factor(imag(expand(P)))              % -e^(2at)(a^2+1)
k = simplify(-imP/abs(zt)^3)               % (a^2+1)/(e^(at)|a+i|^3) = e^(-at)/√(a^2+1).
```

3. **Fundamental theorem of algebra.** Experiment using the next program, which, along with Figure 6.20(c), illustrates this theorem. The idea of the program is as follows. Let $f_n(0) \neq 0$. The values of f_n along the circle $|z| = R$ belong to the curve γ_R: $f_n(R(\cos(t) + i\sin(t)))$ $(t \in [0, 2\pi])$, which for "small" R lies near the point $f_n(0)$; i.e., *its coefficient of engagement with O is zero*. For "large" R the curve γ_R almost coincides with the circle taken n times, $\omega = S^1(R, n)$: $R^n(\cos(nt) + i\sin(nt))$; i.e., *its coefficient of engagement with O is equal to n*. (In the program, the curve γ_R under increasing R will pass through the origin $n = 5$ times.) Hence, there exists $R_0 > 0$ such that the curve γ_{R_0} passes through the origin. In other words, the polynomial has a root of the form $z_0 = R_0(\cos(t_0) + i\sin(t_0))$.

```
syms z R t y;
n = 5;
p = round(100*(rand(1, n + 1) - .5));
Z = 1;
for j = 1 : n;   Z = [ Z z^j ];    end;
pz = sum(Z*p') + 10000          % Answer: p = [-37, -35, 0, 0, 97, 98]
```

Hence, the polynomial p gives us $p(z) = -37z^5 - 35z^4 + 97z + 10098$. Then we restrict $p(z)$ to a circle of radius R:

```
f = subs(pz, z, 'R*(cos(t) + i*sin(t))')
fy = simplify( real(f) ),    fx = simplify( imag(f) )
```
$f_y := -592R^5 \cos(t)^4 \sin(t) + \ldots + 140R^4 \cos(t) \sin(t)$
$f_x := -592R^5 \cos(t)^5 + \ldots - 35R^4$ % Answer.
```
N = 20;   T = linspace(0, 2*pi, 51);
for n = 0 : N;
```

```
   plot(subs(subs(fx, R, 3/N*n), t, T), subs(subs(fy, R, 3/N*n), t, T), 'r');
   axis([-6000 6000 -6000 6000]);   pause(.1)                    % animation
end
```

4. **Reconstruction of a plane curve.** We recover a plane curve from its curvature function $k(s) > 0$. See (6.3).

```
a = 0;  b = pi;                                          % enter data
sys2 = inline('[2;  cos(X(1)); sin(X(1))]', 's', 'X');      % replace 2 by your k(s)
                  % For k(s) = s you obtain a clothoid, Figure 5.6(a).
[s, Xs] = ode45(sys2, [a b], [0,0,0]);      % Answer: θ = Xs₁, x = Xs₂, y = Xs₃
plot(Xs(:, 2), Xs(:, 3));  axis equal      % a circle for k = const (a plane curve)
```

In the code comments: $\theta = Xs_1, x = Xs_2, y = Xs_3$

5. **Osculating circle.** We plot the *moving* osculating circle of a plane curve (ellipse, logarithmic spiral, etc.).

```
a = 2;  b = 1;  t1 = 0;  t2 = 2*pi;
r = [a*cos(t), b*sin(t)];                                 % ellipse
r1 = diff(r, t);  r2 = diff(r1, t);  d = sqrt(r1*r1');
T = simplify(r1/d);  N = [-T(2), T(1)];      % unit tangent and normal vectors
k = simplify((r1(1)*r2(2) - r1(2)*r2(1))/d^2)    % the curvature k = 2/(3 sin²t+1)
Circ = r + N/k + (N*cos(2*pi*s) + T*sin(2*pi*s))/k;    % the osculating circle
m = 18;  for n = 0 : m;                        % prepare animation
  ezplot(r, [t1, t2]);  hold on;  axis equal;
  tn = t1 + n*(t2 - t1)/m;   C = subs(Circ, t, tn);
  rT = subs(r + s*T, t, tn);   rN = subs(r + s*N, t, tn);
  hn = ezplot(rN(1), rN(2), [0, 1]);  set(hn, 'Color', 'r');
  ht = ezplot(rT(1), rT(2), [0, 1]);  set(ht, 'Color', 'r');
  hc = ezplot(C(1), C(2), [0, 1]);  set(hc, 'Color', 'g');
  pause(0.1);  hold off;
end;                                                   % Figure 6.21(a).
```

In the comments: $k = \dfrac{2}{3\sin^2 t+1}$

6.4.3 *Curvature and torsion of a space curve*

We compute the characteristics of a C^3-regular space curve at an arbitrary and given point (starting with the circular helix).

Program 6.1 (Investigation of a space curve).

```
syms t x1 y1 z1 real;
syms a R real;  x = R*cos(t);  y = R*sin(t);  z = a*t;  t0 = 1;       % enter data
% x = cos(t);  y = sin(t);  z = t^2;  t0 = pi/4;            % another example
r = [x, y, z];
```

```
r1 = diff(r, t);   r2 = diff(r1, t);   r3 = diff(r2, t);            % derivatives of r(t)
Tp = [(x1 - x)/r1(1),  (y1 - y)/r1(2),  (z1 - z)/r1(3)];            % tangent line
p1 = simplify(det([[ x1 - x, y1 - y, z1 - z]', r1', r2' ]));        % osculating plane (T,N)
b1 = cross(r1, r2);    B = simplify(b1/sqrt(b1*b1'));               % binormal
d = sqrt(r1*r1');   T = simplify(r1/d);                             % unit tangent vector
N = cross(B, T);                                                    % the main normal
p2 = collect(T*[x1 - x, y1 - y, z1 - z]', [x1, y1, z1]);           % normal plane (N,B)
p3 = collect(N*[x1 - x, y1 - y, z1 - z]', [x1, y1, z1]);           % rectifying plane (T,B)
k = simplify((r1(1)*r2(2) - r1(2)*r2(1))/d^3);
tau = -simplify(det([r1', r2', r3'])/(b1*b1'));                     % Ans.: k = R/(a²+R²), τ = a/(a²+R²)
```

Here the k and τ formula: $k = \dfrac{R}{a^2+R^2}$, $\tau = \dfrac{a}{a^2+R^2}$

For economy of space only the two results $k(t)$ and $\tau(t)$ are given.

```
x0 = limit(x, t, t0);   z0 = limit(z, t, t0);            % Compute the point t0.
Tp0 = simplify(subs(Tp, t, t0))
T0 = subs(T, t, t0),   B0 = subs(B, t, t0),   N0 = subs(N, t, t0)
p10 = subs(p1, t, t0),   p20 = subs(p2, t, t0),   p30 = subs(p3, t, t0)
k0 = subs(k, t, t0)
tau0 = subs(tau, t, t0)
```

The *absolute integral curvature* AIC $= \int_\gamma |k|$ of a closed space curve γ is not less than 2π, and equality holds exactly for plane convex curves. If for a closed space curve AIC $< 4\pi$ holds, then the curve *is not engaged*.

Examples.

1. One may use Program 6.1 to find the AIC of the Viviani window.

```
R = 1;   r = [R*cos(t)^2, R*cos(t)*sin(t), R*sin(t)];   t1 = 0;   t2 = 2*pi;
.........................                                % compute k(t)
AIC = int(k,  t1, t2)
```

2. Plot the moving Frenet frame and osculating circle of a space curve, using the test example of the Viviani curve.

```
syms t s;
R = 2;   t1 = 0;   t2 = 2*pi;
r = [R*cos(t).^2, R*cos(t).*sin(t), R*sin(t)];
r1 = diff(r, t);   r2 = diff(r1, t);   r3 = diff(r2, t);
d = simplify(sqrt(r1*r1'));
T = simplify(r1/d);
b1 = cross(r1, r2);   B = simplify(b1/sqrt(b1*b1'));
N = cross(B, T);
k = simplify(sqrt(b1*b1')/d^3);
C = r + N/k + (N*cos(2*pi*s) + T*sin(2*pi*s))/k;
S = linspace(0, 1, 2);   K = linspace(0, 1, 20);
```

Then plot the animation:

```
m = 32;
for n = 0 : m;
tn = t1 + n*(t2 - t1)/m;
Tn = subs(r + s*T, t, tn);
Nn = subs(r + s*N, t, tn);
Bn = subs(r + s*B, t, tn);
Cn = subs(C, t, tn);
ezplot3(r(1), r(2), r(3), [t1, t2]);  hold on;  axis equal;
plot3(subs(Tn(1), s, S)+S-S, subs(Tn(2), s, S)+S-S, subs(Tn(3), s, S)+S-S);
plot3(subs(Nn(1), s, S)+S-S, subs(Nn(2), s, S)+S-S, subs(Nn(3), s, S)+S-S);
plot3(subs(Bn(1), s, S)+S-S, subs(Bn(2), s, S)+S-S, subs(Bn(3), s,vS)+S-S);
plot3(subs(Cn(1), s, K)+K-K, subs(Cn(2), s, K)+K-K, subs(Cn(3), s, K)+K-K);
 pause(0.1);  hold off;
end;                                              % Figures 6.21(b,c).
```

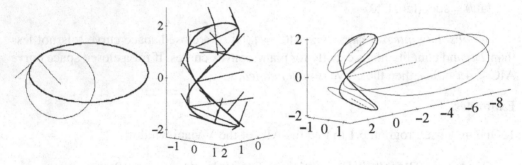

Fig. 6.21 Frenet frame field and osculating circles.

Natural equations of a curve. The theorem on the existence and uniqueness up to congruence of naturally parameterized curves in space with given *continuous* curvature and torsion (i.e., rotating velocity of the binormal) is the culmination of the *classical theory of curves*. The condition of continuity imposed on the curvature and the torsion allows us to apply theorems of ordinary differential equations to the Frenet formulae, but this a priori condition narrows the class of curves for which the theorem holds. The following result, where continuity of functions is replaced by the rather weak condition (b), removes this gap in the classical theory of curves, and completes it.

Theorem 6.3. *Let $f_i \colon I \to \mathbb{R}$ $(i = 1, 2)$ be two functions.*

1. If the following conditions are satisfied,

(a) $f_1(t) > 0$, $f_2(t) \geq 0$ *on all of I, and*

(b) *each of the functions f_i has a primitive function,*

then there exists a unit-speed curve $\gamma: I \to \mathbb{R}^3$ with curvature $k = f_1$ and torsion $\tau = f_2$ that has a differentiable Frenet frame.

2. *For the existence of the unit-speed curve $\gamma: I \to \mathbb{R}^3$ with curvature $k = f_1$ and torsion $\tau = f_2$, up to congruence, it is necessary and sufficient that conditions (a), (b) be satisfied together with the following property:*

(c) *The set $\{t \in I: f_2(t) > 0\}$ is connected.*

Example. Plot a space curve $\mathbf{r}(t) \subset \mathbb{R}^3$ of class C^3 composed of two copies of the plane curve $y = x^4$ for $x \geq 0$ (Figure 6.22(c)), and find its curvature and torsion.

Solution. Obviously, $y'' = 12x^2$ and $k(0) = 0$. Rotate the branch for $x \leq 0$ about OY through the angle $90°$ and obtain the required curve $\mathbf{r}(t)$. The curvature $k(t)$ of $\mathbf{r}(t)$ vanishes at only one point, but with torsion $\tau(t) \equiv 0$ except one point,

$$x(t) = t, \qquad y(t) = \{0 \text{ if } t < 0 \text{ else } t^4\}, \qquad z(t) = \{t^4 \text{ if } t < 0 \text{ else } 0\}.$$

Finally we check that the torsion $\tau \equiv 0$ (both branches are plane curves) and the curvature vanish only at the point $(0,0,0)$.

```
t1 = linspace(-1.3, 0, 21);  t2 = linspace( 0, 1.3, 21);
z1 = t1.^4;
y2 = t2.^4;
r1 = [t1, t1 - t1, z1];
r2 = [t2, y2, t2 - t2];
hold on;  view(3);  grid on;
plot3(t1, t1-t1, z1, 'g', 'LineWidth', 2);
plot3(t2, y2, t1-t1, 'r', 'LineWidth', 2);          % Figure 6.22(c)
```

6.5 Exercises

| Section 6.1 | 1. Using symbolic manipulations in MATLAB, deduce the equation of the tractrix and then plot it. |

Hint:

```
syms t a Dx Dy Dt;
X = a*sin(t);
q1 = diff(X, t)*Dt;  q2 = tan(t)*Dy;
eq1 = Dx - q1,  eq2 = Dx - q2        % eq1 = Dx - a cos t Dt, eq2 = Dx - tan t Dy
```

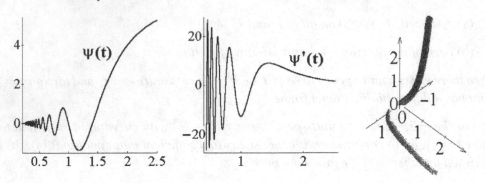

Fig. 6.22 No regular curve with $k = |\psi'|$.   Curve with $\tau(t) \equiv 0$.

```
eq3 = subs(eq2, Dx, q1)
D = simple(solve(eq3, Dy)/Dt)
Y = simplify(int(D, t))
ezplot([X/a, Y/a], [.2, pi - .1]);
```
% Answer: $eq_3 = a\cos t\, Dt - \tan t\, Dy$
% Answer: $D = a\cos t / \tan t$
% Answer: $Y = a\cos t + a\ln\left(\frac{\sin t}{\cos t + 1}\right)$

Continuation: Now derive and plot the tangent line to the tractrix. Try to animate the tangent line.

```
y0 = simplify(Y - X*diff(Y, t)/diff(X, t));
h = ezplot([X/a, Y/a], [.17, pi-.17]);  hold on;  set(h, 'Color', 'r');        % a curve
for i = 1 : 10;
  tt = pi/2 + i/9;
  Y0 = subs(y0, t, tt)/a;
  X1 = subs(X, t, tt)/a;
  Y1 = subs(Y, t, tt)/a;
  ezplot((1 - t)*[X1, Y1] + t*[0, Y0], [0, 1]);          % tangent lines (segments)
end;                                                      % Figure 6.3(c)
```

2. Calculate and plot the asymptotes of the graph $y = \sqrt[3]{6x^2 - x^3}$.
Answer: the two-sided asymptote $y = -x + 2$ (Figure 6.4(c)).

```
syms x;
f = (6*x^2 - x^3).^(1/3);
k = limit(f/x, -inf);  b = limit(f - k*x, -inf);           % Answer: $k = -1\ b = 2$
f1 = 'sign(6*x^2 - x^3)*abs(6*x^2 - x^3)^(1/3)';
hold on;  ezplot(f1, [-6 10]);
ezplot(k*x + b, [-6 10]);
```

3. Plot the following space curve and find its three asymptotes.
Hint: coordinate axes.

```
f = exp(-t^2);
r = [f*(t - 2)*(t - 1)/t, f*(t -v2)*t/(t - 1), f*t*(t - 1)/(t - 2)];
ezplot3(r(1), r(2), r(3), [-.5, .5])
```

4. Check that the discriminant $y = 0$ of the family $\varphi = y^3 - (x - t)^2$ consists of singular points of members of the family.

```
u = linspace(-1.1, 1.1, 41); hold on;
  for t = 1 : 7;
    plot(u.^3 + 2*t, u.^2/10);
  end
```

5. Plot a parabola, an ellipse, and a hyperbola, using the following algorithms:

(a) Fix a line l and a point S that does not belong to this line. Start to plot from points $P_t \in l$ lines l_t that are orthogonal to the segment $P_t S$. After plotting sufficiently many lines l_t, note that they envelop a curve γ, which is symmetric with respect to the perpendicular SA from the point S onto the line l. The line l itself also is tangent to this curve (when the point P_t coincides with A). Prove that γ is a parabola.

(b) Fix a circle with center at O and a point S inside of the circle. Join S by a line with any point P_t of the circle and push through P_t the line orthogonal to SP_t. Show that all these lines are tangent to an ellipse.

(c) Fix a circle with center C and a point S outside the circle. Join S with points P_t on the circle, and plot perpendiculars to the segments SP_t from the points P_t. Prove that the two branches of the envelope form a hyperbola.

6. Show that the evolute of an ellipse $\mathbf{r}(t) = [a\cos(t), b\sin(t)]$ is the *prolate astroid*

$$x_{ev} = \frac{a^2 - b^2}{a} \cos^3(t), \qquad y_{ev} = \frac{b^2 - a^2}{b} \sin^3(t) \qquad \text{(Figure 6.11(b))}.$$

Assume (in the program of Example 1, p. 259):

```
a = 3;  b = 2;
x = t;  y = t^2;
t1 = -2;  t2 = 2;  t3 = t2 - 1;
```

7. Prove that the catenary is the evolute of the tractrix (Figure 6.11(c)). In the program of Example 1, p. 259, assume a = 1; t1 = .25; t2 = pi - .25; etc.

8. Check that the evolute of the *epicycloid* (the *hypocycloid*) is again the epicycloid (resp., the hypocycloid), similar to the given curve with coefficient of similarity $\frac{1}{2m+1}$, and rotated with respect to it through the angle $m\pi$.

9. Plot the evolvents of the *figure eight* curve $\mathbf{r}(t) = [\sin(t), \sin(2t)]$.

10. Write programs to illustrate the following problems:

(a) A pencil of parallel rays is reflected by the inner side of a half-circle; plot the caustic of reflecting rays. *Answer:* a nephroid.

(b) The catacaustic of the logarithmic curve $y = a\log(x/a)$ for rays parallel to OX is the catenary $x = a\cosh((y + a)/2)$.

(c) The catacaustic of a deltoid is an astroid.

Section 6.2 1. Show that the point $t = 0$ of the curve $x(t) = t^3$, $y(t) = t^5$ is regular with respect to parameterizations of class C^1 (the curve is the graph of the function $y = x^{5/3}$ of class C^1), but $t = 0$ is singular with respect to analytic parameterizations.

2. Find the connected curve whose singular points belong to k branches of the curve. *Hint.* Consider the roses $\rho = \sin(n\varphi)$ in rectangular coordinates, where $n = k$ for odd k, and $n = \frac{1}{2}k$ for even k.

3. Show that

(a) Pascal's limaçon $(x^2 + y^2 - 2ax)^2 - b^2(x^2 + y^2) = 0$ for $b > 2a > 0$ has a singularity at $(0,0)$ with analogous properties to that of the curve $a^2 y^2 + x^4 - y^4 = 0$;

(b) an analogous situation to (b) of Example 1, p. 265, appears for Pascal's limaçon for $0 < b < 2a$ (Figure 1.34(c)), and for the folium of Descartes (Figure 5.6(c));

(c) an analogous situation to (c) of Example 1, p. 265, appears for the cycloid and the cycloidal curves (astroid, deltoid, etc.);

(d) the point $(0,0)$ on the curve $4x^2(x^2 - a^2) + m^2 y^4 = 0$ has analogous properties to (d) of Example 1, p. 265.

4. Find the singular points of the following curves, and explain their type:

(a) the cycloid $Rt - R\sin(t)$, $y = R - R\cos(t)$,

(b) the cissoid $y^2 = x^3/(2a - x) \iff \mathbf{r}(t) = \left[2a/(1+t^2), 2a/(t(1+t^2))\right]$,

(c) the trisectrix of Maclaurin $x(x^2 + y^2) = a(3x^2 - y^2)$,

(d) the astroid $x = R\cos^3(t/4)$, $y = R\sin^3(t/4)$,

(e) $y^3 = ax^2 + x^3$,

(f) the trefoil or cloverleaf $\frac{27}{4}a^2(x^2 + y^2 - a^2) + ax^3 - 3axy^2 - (x^2 + y^2)^2 = 0$,

(g) the rosette, given by the polar equation $\rho = a\sin\left(\frac{\varphi}{2}\right)$, or by the equation in rectangular coordinates $4(x^2 + y^2)^3 - 4a^2(x^2 + y^2)^2 + a^4 y^2 = 0$.

Answer for (g): the curve has a double point of self-tangency at the origin and two nodes at the points $(0, \pm a/\sqrt{2})$.

Section 6.3 1. Show that the natural parameterization has unit speed: $|\tilde{\mathbf{r}}'(s)| = 1$.

2. Find the lengths and centers of mass (for $\mu \equiv 1$) of the following curves by two methods (integrals, and a inscribed polygon), and compare the results:

(a) parabola $y = x^2$ and (b) half-cubic parabola $y = x^{3/2}$, $0 \le x \le 4$,

(c) catenary $y = a\cosh(x/a)$, $0 \le x \le x_0$,

(d) arc of a cycloid $\mathbf{r}(t) = [a(t - \sin t), a(1 - \cos t)]$, $0 \le t \le 2\pi$,

(e) ellipse $\mathbf{r}(t) = [a\cos t, b\sin t]$, $0 \le t \le 2\pi$,

(f) astroid $\mathbf{r}(t) = [a\cos^3 t, a\sin^3 t]$, $0 \le t \le \frac{\pi}{2}$,

(g) Archimedes' spiral $\rho(t) = at$, $0 \le t \le 2\pi$,

(h) cardioid $\rho(t) = a(1 + \cos t)$, $0 \le t \le 2\pi$,

(i) circular helix $\mathbf{r}(t) = [a\cos t, a\sin t, vt]$, $0 \le t \le 4\pi$,

(j) conic circular helix $\mathbf{r}(t) = [t\cos t, t\sin t, vt]$, $0 \le t \le 6\pi$,

(k) Viviani curve $\mathbf{r}(t) = [\cos^2 t, \cos t \sin t, \sin t]$, $0 \le t \le 2\pi$.

3. Find the center of mass of the arc of a circle with angle φ and density $\mu(x) = 1$.

4. The arc length of the ellipse $\mathbf{r} = [a\cos t, b\sin t]$ is expressed by the *elliptic integral of the second kind* $l(t) = a\int_0^t \sqrt{1 - \varepsilon^2 \sin^2 t}\, dt = aE(\varepsilon, t)$ (see Section 1.2), where $\varepsilon = \sqrt{1 - (b/a)^2}$ is its eccentricity. The integral $E(\varepsilon) = E(\varepsilon, \pi/2)$ related to one-fourth of the length of the ellipse is called the *full elliptic integral.* Show that the length of the Viviani curve $\mathbf{r} = [\cos^2 t, \cos t \sin t, \sin t]$ is equal to $4E(\sqrt{2}/2)$.

5. Find the natural parameterization of

(a) a circle, (b) a circular helix, (c) a catenary, (d) an ellipse, (e) a conic helix.

Hint. (a) For the circle $\mathbf{r}(t) = [R\cos t, R\sin t]$, we have $s' = R \Rightarrow s = Rt \Rightarrow t = s/R$, where t is the angle of the vector $\mathbf{r}(t)$ with the axis OX. Its natural parameterization is $\tilde{\mathbf{r}}(s) = \mathbf{r}(s/R) = [R\cos(s/R), R\sin(s/R)]$.

6. Find the length of the torus knot $\mathbf{K}(m, n)$; see Section 5.4.

7. Derive the lengths of the edges of a triangle and its area through the coordinates of its vertices.

| **Section 6.4** | 1. Check that Pascal's limaçon, $\rho = a\cos(\varphi) - b$, has positive curvature $k(t)$ (Figure 6.20(a,b)), but it is not a convex curve. |

```
a = 1; b = .2;
t1 = 0;  t2 = 2*pi;
r = [(a*cos(t) - b)*cos(t), (a*cos(t) - b)*sin(t)];
ezplot(r),  grid on;                              %  see Figure 6.20(a)
```

The function $k'(t)$ has two extrema (Figure 6.20(b)), and hence the curve has two *vertices*; see the program. Also the curvature of the lemniscate has only one maximum and one minimum.

This means that the "convex" property of the curve in the *theorem on the existence of at least four vertices on an oval* is not superfluous.

2. Plot any oval (convex closed curve) with $2n > 4$ vertices.
Hint. Such an oval intersects some circle in $2n$ points.

3. Calculate the IC for the lemniscate of Bernoulli and the roses $\rho = \sin(5\varphi)$. Check that the IC of a prolate epicycloid with modulus $m = \frac{a}{b}$, where a, b are relatively prime, equals $2\pi b$.

4. At the point t_0 of the curve $\mathbf{r} = \mathbf{r}(t) \subset \mathbb{R}^3$ derive the *tangent line, binormal vector, main normal vector, osculating plane, normal plane, rectifying plane, Frenet frame, curvature,* and *torsion.*

(1) $\mathbf{r}(t) = [e^t, e^{-t}, t\sqrt{2}], t_0 = 0,$ (2) $\mathbf{r}(t) = [\frac{t^2}{2}, \frac{2t^3}{3}, \frac{t^4}{2}], t_0 = 1,$

(3) $\mathbf{r}(t) = [t, t^2/2, t^3/6], t_0 = 1,$ (4) $\mathbf{r}(t) = [t, \frac{t^2}{3}, \frac{1}{2t}], t_0 = 1,$

(5) $\mathbf{r}(t) = [\sin(t), \cos(t), t^2], t_0 = \frac{\pi}{4},$ (6) $\mathbf{r}(t) = [2t, \ln t, t^2], t_0 = 1,$

(7) $\mathbf{r}(t) = [\cos(t), \sin(t), t^3 - 9t], t_0 = 0,$ (8) $\mathbf{r}(t) = [t, t^2, e^t], t_0 = 0,$

(9) $\mathbf{r}(t) = [t, t^3, t^2 + 4], t_0 = 1,$ (10) $\mathbf{r}(t) = [t, t^2 + 2, t^3 + 3], t_0 = 1.$

5. Let a sphere have contact of order ≥ 3 with a curve γ at p, and let $\mathbf{r}(t)$ be a C^3-regular parameterization of γ, with $\mathbf{r}(t_0) = p$. Then the center of the sphere is $C = \mathbf{r}(t_0) + \frac{1}{k}N + \frac{k'}{k^2\tau}B$, where N is the main normal and B the binormal at p. Such a sphere is called the *osculating sphere* of γ at p.

(a) Compute and plot the osculating spheres of the curves $\mathbf{r} = \mathbf{r}(t) \subset \mathbb{R}^3$ given in Exercise 4 at the point t_0.

(b) Show that the osculating sphere is the limit of a variable sphere passing through four points of γ as these points approach p.

6. Calculate the *absolute integral curvature* AIC $= \int_\gamma |k|$ of the torus knot $\mathbf{K}(m,n)$: $\mathbf{r}(t) = [(3+\cos(mt))\cos(nt), (3+\cos(mt))\sin(nt), \sin(mt)]$ for small m, n (see Figure 5.20, Section 5.4).

7. Plot a unit-speed curve γ in $\mathbb{R}^2$ with curvature k that *has no primitive function*. In view of the theorem above, there *does not exist a naturally parametric curve with curvature k*.

Hint. Define the function $\psi(t)$ by the formulae $\psi(0) = 0$ and $\psi(t) = t^2 \sin(2\pi/t^2)$ $(t \neq 0)$, and set $\mathbf{e}_1(t) = \cos(\psi)\mathbf{i} + \sin(\psi)\mathbf{j}$. Obviously, the derivative $\psi'(t)$ exists, and $\mathbf{e}_1' = \psi'\mathbf{e}_2$ holds, where $\mathbf{e}_2(t) = -\sin(\psi)\mathbf{i} + \cos(\psi)\mathbf{j}$. Deduce that there exists a curve $\mathbf{r}(t) \subset \mathbb{R}^2$ such that $\mathbf{r}' = \mathbf{e}_1$ is naturally parametric and $k(t) = |\psi'(t)|$ is its curvature.

Assume that $\varphi: \mathbb{R} \to \mathbb{R}$ is a primitive function of $k(t)$. Then φ is non-decreasing, but since the segments $I_n = [\frac{1}{\sqrt{n+.5}}, \frac{1}{\sqrt{n+.25}}]$ do not intersect, the sequence $a_n = \varphi(\frac{1}{\sqrt{n+.5}} - \frac{1}{\sqrt{n+.25}})$ converges. At the same time, $\varphi' \geq 0$ on each segment I_n (check this). Thus for I_n we have $\varphi' = k = |\psi'| = \psi$, and $a_n = \psi(\frac{1}{\sqrt{n+0.5}} - \frac{1}{\sqrt{n+0.25}}) = \frac{1}{n+0.25}$ holds, contrary to the convergence of the sequence $\{a_n\}$. Hence, by Theorem 6.3 the curve $\mathbf{r}(t)$ is not regular.

```
syms t;
f = t^2*sin(2*pi/t^2);
grid on;  ezplot( f, [.2, 2.5]),
ezplot(diff(f, t), [.45, 2.5]);                    % Figures 6.22(a,b).
```

8. Write a program to solve the Frenet natural equations for space curves. Then use it for drawing a curve with given curvature and torsion.

9. Write a program showing how a circle through three points on a curve tends to the osculating circle at P as the three points approach P.

7

Geometry of Surfaces

In Section 7.1 we consider the basic notions of a parametric surface and a regular surface, and use a number of MATLAB® commands to produce surfaces by various methods. (Similar definitions for curves were studied in Section 5.1). In Section 7.2 we calculate and plot the tangent planes and normal vectors of a surface. As an application we solve conditional extremum problems in space. In Section 7.3 we consider parametric and implicitly defined surfaces with singularities. In Section 7.4 we use changes in coordinates and linear transformations in space to calculate and plot the osculating paraboloid at a point of a surface. This elementary approach is given only for methodical reasons. In Section 7.5 we calculate characteristics related to the first and second fundamental forms, the Gaussian and mean curvatures, write down the equations of geodesics using the M-file (program) of Section A.8, and plot geodesics on surfaces.

7.1 Regular Surface

7.1.1 *What is a surface?*

Euclid defines a *surface* intuitively, as a two-dimensional figure either swept by the path of a moving curve, or bounding a solid body. Planes, some polyhedra, and curved surfaces, each defined in some practical way, are studied in school. The mathematically correct definition of a surface is based on notions from *topology*, but it starts from the key notion of an *elementary surface*, which can be imagined as an *elementary domain* of the plane after continuous deformation that is stretched in and out in space.

We call $G \subset \mathbb{R}^n$ $(n = 2, 3)$ an *open set* if for each point $P \in G$ there exists $\varepsilon > 0$ such that the ball $B(P, \varepsilon)$ of radius ε with center P (the disk, when $n = 2$) lies inside of G. The complement $F = \mathbb{R}^3 \setminus G$ of an open set G is called a *closed set*. Open sets

V. Rovenski, *Modeling of Curves and Surfaces with MATLAB®*,
Springer Undergraduate Texts in Mathematics and Technology 7, DOI 10.1007/978-0-387-71278-9_7,
© Springer Science+Business Media, LLC 2010

(or domains) are convenient for working with continuous and differentiable functions defined on such sets. An arbitrary open set is the union of a finite or countable number of disjoint domains.

Recall that a *homeomorphism* is a one-to-one correspondence between points in two geometric figures that is continuous in both directions.

As in the definition of an elementary curve, we use the notion of a *homeomorphism* from one geometric figure to another, i.e., a one-to-one map that is continuous and has a continuous inverse.

A set M in space is called an *elementary surface* if it is the image of a planar elementary domain G under the homeomorphism $\mathbf{r}\colon G \to \mathbb{R}^3$. An open set in the plane (in space) that is homeomorphic to a disk (respectively, a ball) is called an *elementary domain*.

The following two elementary domains in the plane with coordinates u, v are often used: the disc $u^2 + v^2 < R^2$ of radius R and the rectangle with sides a, b parallel to the coordinate axes.

Since we will also be interested in *self-intersecting* surfaces, we give the following definition:

A set M in $\mathbb{R}^3$ is called a *surface* if it can be covered by a finite or countable number of elementary surfaces. A surface M is called a *simple surface* if it has no points of self-intersection.

The notion of an elementary surface is sufficient for *local study* of surfaces (i.e., in a neighborhood of some point). An elementary surface is a simple one. The following surfaces are simple but not elementary: a *sphere*, Figure 8.1(b) (which can be covered by two disks); a *circular cylinder*, Figure 7.3(a) (covered by two cylindrical strips, each homeomorphic to a rectangle); a *torus*, Figure 7.1(b) (covered by four rectangles).

The following surfaces are not simple: the *Klein bottle*, Figure 7.3(c) (whose points of self-intersection form a circle); a *cone*, Figure 7.16(a) (whose vertex is a *singular point*); a *pair of intersecting planes* (whose self-intersection points form a line); the *pinched torus*, Figure 7.1(c) (whose self-intersection points form two line segments). A more complicated example is the *union of a countable number of cylinders with the common generatrix OZ*: $M = \bigcup_{n \in \mathbb{N}}\{x^2 + (y - n)^2 = n^2\}$. It cannot be covered by a finite number of elementary surfaces.

If we fix rectangular coordinates (with the orthonormal basis $\{\mathbf{i}, \mathbf{j}, \mathbf{k}\}$) with origin O in $\mathbb{R}^3$, then the coordinates x, y, z of a point on an elementary surface given by the map $\mathbf{r}$ are scalar functions of the coordinates u, v of a point in the elementary domain G:

$$x = x(u,v), \qquad y = y(u,v), \qquad z = z(u,v). \tag{7.1}$$

So, $\mathbf{r} = \overrightarrow{OP}$ (the position vector of $P = (x,y,z) \in M$) is the vector function

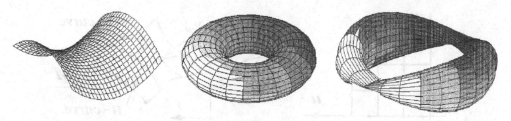

Fig. 7.1 An elementary surface, the torus, and the pinched torus.

$$\mathbf{r} = \mathbf{r}(u,v) = x(u,v)\,\mathbf{i} + y(u,v)\,\mathbf{j} + z(u,v)\,\mathbf{k}, \qquad (u,v) \in G. \tag{7.2}$$

If the starting point of a continuous vector $\mathbf{r}(u,v)$ coincides with the origin O, then its endpoint, as a rule, moves along a *surface* in $\mathbb{R}^3$ called the *hodograph* of the vector-valued function. For example, a plane in space is the hodograph of a linear vector-valued function

$$\mathbf{r}(u,v) = [a_1 u + b_1 v + c_1, \ a_2 u + b_2 v + c_2, \ a_3 u + b_3 v + c_3].$$

Example. We can derive some calculus-type operations on vector-valued functions $\mathbf{r}(u,v) = [x(u,v), y(u,v), z(u,v)]$ of class C^k.

```
syms u v;
x = u^2;  y = v^2;  z = u*v;          % define your functions
r = [x, y, z];
```

The *limit* $\lim_{(u,v) \to (u_1,v_1)} \mathbf{r}(u,v)$, and the *partial derivatives* $\frac{\partial}{\partial u}\mathbf{r} = \mathbf{r}_u$, $\frac{\partial}{\partial v}\mathbf{r} = \mathbf{r}_v$:

```
u1 = 2;  v1 = 1;                                   % define values
r1 = subs([x, y, z], {u, v}, {u1, v1})     % Answer: r1 = [4,1,2]

ru = diff(r, u), rv = diff(r, v)      % Answer: ru = [2u,0,v],  rv = [0,2v,u]
```

Similarly, we calculate the second partial derivatives $\mathbf{r}_{uu}$, $\mathbf{r}_{vv}$, $\mathbf{r}_{uv}$, the *Taylor expansion* (see Section 7.4), etc.

```
taylor(sin(u*v^2), 11)              % Answer: uv^2 − (u^3 v^6)/6 + (u^5 v^{10})/120.
```

We call (7.1) or (7.2), representing a map $\mathbf{r} \colon G \to \mathbb{R}^3$ from the planar open set G into $\mathbb{R}^3$, the *parametric equations of the surface* $M = \mathbf{r}(G)$: the pair of real numbers (u,v) constitutes the *curvilinear coordinates* of the point $P = (x,y,z)$ on the surface. Taking $v = v_0$ as a constant and $u = t$ as varying, we obtain the coordinate *u-curve* on the surface M, which also is the space curve $\mathbf{r}(t,v_0)$. Similarly, for a constant $u = u_0$ and $v = t$ varying, we obtain the *v-curve* on the surface M, that is, the space curve

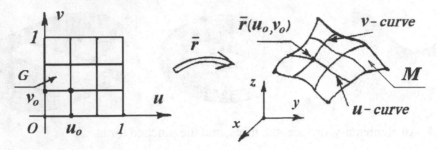

Fig. 7.2 Parameterized (elementary) surface.

$\mathbf{r}(u_0, t)$. These two families of curves (v-curves and u-curves) form the *coordinate net* of the surface, Figure 7.2.

A *curve* γ *on the surface M* is defined by equations in the curvilinear coordinates $u = u(t), v = v(t)$ ($t \in I$). The equations of $\gamma(t)$ in space are

$$\tilde{x}(t) = x(u(t), v(t)), \qquad \tilde{y}(t) = y(u(t), v(t)), \qquad \tilde{z}(t) = z(u(t), v(t)).$$

```
subs(r, {u, v}, {u(t), v(t)});
```

Sometimes we consider surfaces with a *boundary:* the boundary of an elementary surface is homeomorphic to a circle (the image of the boundary of an elementary domain); the boundary of the *half-plane* is a line; the boundary of a *cylinder* (between its parallels) is a pair of circles, Figure 7.3(a); the boundary of the *Möbius band*, Figure 7.3(b), is a circle.

The Möbius band is obtained when we twist a thin strip (of paper) through a 180° angle and then glue its lateral sides. A remarkable property of the Möbius band is that it has only *one side*. Moreover, a surface that "contains" a Möbius band is called *one-sided* (or *non-orientable*). Another example of a one-sided surface is the *Klein bottle*, Figure 7.3(c).

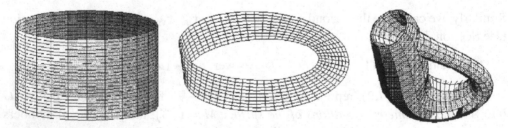

Fig. 7.3 Circular cylinder, Möbius band, and Klein bottle.

A simple surface M is called *complete* if the limit point of any sequence of its points which converges in $\mathbb{R}^3$ belongs to M. If a complete surface does not have a *boundary*, it is called *closed*. A closed and bounded surface is called *compact*.

Paraboloids, hyperboloids and cylinders are examples of closed surfaces. The sphere, the torus, and Klein bottle are compact surfaces. Omitting a closed set from a complete surface, we obtain a non-complete surface.

7.1.2 *Regular surfaces*

By definition, a simple surface can be defined (locally) by parametric equations: see (7.1), (7.2). Suppose that the coordinate functions $x(u,v)$, $y(u,v)$, $z(u,v)$ are defined on some planar elementary domain G and are of class C^k. In this case the u- and v-curves (space curves) are of class C^k, and we find their velocity vectors

$$\mathbf{r}_u = x'_u(u,v)\mathbf{i} + y'_u(u,v)\mathbf{j} + z'_u(u,v)\mathbf{k} \qquad \text{(of the } u\text{-curve)},$$
$$\mathbf{r}_v = x'_v(u,v)\mathbf{i} + y'_v(u,v)\mathbf{j} + z'_v(u,v)\mathbf{k} \qquad \text{(of the } v\text{-curve)}.$$

Definition 7.1. A surface M is called *regular of class* C^k (resp. C^∞ or C^ω) if each of its points has a neighborhood with the parametric equations

$$\mathbf{r}(u,v) = [x(u,v), y(u,v), z(u,v)],$$

where the functions $x(u,v)$, $y(u,v)$, $z(u,v)$, defined on the elementary domain $G \in \mathbb{R}^2$, are of class C^k (resp. C^∞ or C^ω); moreover, the vectors $\mathbf{r}_u(u,v)$ and $\mathbf{r}_v(u,v)$ should be linearly independent at every point $(u,v) \in G$. A surface is *smooth* when $k = 1$.

A point P on a simple surface M is called a *regular (smooth) point* if some neighborhood of P admits a C^k-parametric equations $\mathbf{r}(u,v)$ such that the vectors $\mathbf{r}_u$ and $\mathbf{r}_v$ are linearly independent at P; otherwise, P is called a *singular point* (for example, the vertex on the top of a circular cone or the bottom fold of such a cone). A curve on a surface all of whose points are singular is called a *singular curve* (for example, two line segments on the pinched torus, Figure 7.1(c)).

A surface without singular points is smooth. Its small open sets look like plane disks slightly deformed in space.

Although a regular surface admits a number of parametric equations locally, there does not exist a "natural" way to parameterize it. We consider the transition from one set of parametric equations to another.

Example. Show that $\mathbf{r}(u,v) = \left[\frac{a(uv+1)}{u+v}, \frac{b(u-v)}{u+v}, \frac{uv-1}{u+v}\right]$ is a C^∞-regular parameterization of part of a hyperboloid of one sheet.

What are the coordinate curves of these parameterizations?

Solution. (i) First check the regularity condition (b) of Exercise 1, p. 314.

```
syms u v;  r = [u + v, u - v, u^2 + v^2];
ru = diff(r, u);  rv = diff(r, v);  cross(ru, rv)
```

$$[2u + 2v, \ 2u - 2v, \ -2] \qquad\qquad \text{\% obtain a nonzero vector.}$$

Then substitute the vector-valued function into the explicit equation:

```
syms x y z u v;
simplify(subs(z - (x^2 + y^2)/2, {x, y, z}, {u + v, u - v, u^2 + v^2}))        % obtain 0
```

We plot graphs by two methods:

```
ezsurf(r);    ezsurf((x^2 + y^2)/2);
```

The command ezmesh returns the same surface

```
ezmesh(u + v, u - v, u^2 + v^2);                                    .
```

Lemma 7.1. *Let* $\mathbf{f}\colon G \to M$ *be* C^k-*regular parametric equations of a surface* M *and let* $\mathbf{h}\colon G_1 \to G$ *be an injective map of class* C^k *of the elementary domain* G_1 *into the elementary domain* G *with nonzero jacobian* J_h. *Then the composition* $\mathbf{f}_1 = \mathbf{f}\circ\mathbf{h}\colon G_1 \to M$ *is also a* C^k-*regular parametric patch on* M *(Figure 7.4).*

Proof. Define $\mathbf{f}$ by the vector-valued function $\mathbf{r} = [x(u,v),\ y(u,v),\ z(u,v)]$. Let $\mathbf{h}$ be given by $u = \varphi(\alpha,\beta)$, $v = \psi(\alpha,\beta)$. The composition $\mathbf{f}_1$ is

$$\tilde{\mathbf{r}}(\alpha,\beta) = [x(u(\alpha,\beta),v(\alpha,\beta)),\ y(u(\alpha,\beta),v(\alpha,\beta)),\ z(u(\alpha,\beta),v(\alpha,\beta))].$$

By standard theorems from calculus, the map $\mathbf{f}_1\colon \tilde{\mathbf{r}} = \mathbf{r}(\varphi(\alpha,\beta),\ \psi(\alpha,\beta))$ is of class C^k. To prove the regularity of the map $\mathbf{f}_1$ we start with $\tilde{\mathbf{r}}_\alpha = \mathbf{r}_u\varphi_\alpha + \mathbf{r}_v\psi_\alpha$, $\tilde{\mathbf{r}}_\beta = \mathbf{r}_u\varphi_\beta + \mathbf{r}_v\psi_\beta$. Moreover, $|\mathbf{r}_u \times \mathbf{r}_v| \neq 0$, in view of the regularity of the parameterization $\mathbf{f}$. Hence

$$|\tilde{\mathbf{r}}_\alpha \times \tilde{\mathbf{r}}_\beta| = |\mathbf{r}_u \times \mathbf{r}_v| \cdot |\varphi_\alpha\psi_\beta - \varphi_\beta\psi_\alpha| = |\mathbf{r}_u \times \mathbf{r}_v| \cdot J_h \neq 0. \qquad \square$$

7.1.3 *Methods of generating surfaces*

The *graph* of a function $f\colon G \subset \mathbb{R}^2 \to \mathbb{R}$ in two real variables (x,y) is the set $\Gamma_f = \{(x,y,z) \in \mathbb{R}^3 : z = f(x,y), (x,y) \in G\}$. The *level line* with height c of f is the set $f^{-1}(c) = \{f(x,y) = c\}$; for example, isobars and isotherms on geographical maps.

For a function $F(x,y,z)$ defined on the open set $G \subset \mathbb{R}^3$, the similar set $F^{-1}(c) = \{F(x,y,z) = c\}$ is called the *level surface* of height c of F. In this notation, the real number c is called the *height* of the level set.

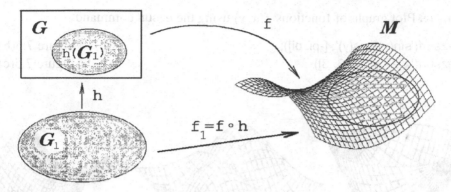

Fig. 7.4 A change of variables on a surface.

Among surfaces of the second order, the paraboloids $z = x^2/p \pm y^2/q$, and the parabolic cylinders $z = x^2$, can be viewed as graphs.

A function in several variables has associated to it a number of *level sets* (forming the *contours of the function*), which are useful for studying the qualitative properties of the functions.

Intersecting the graph Γ_f with the horizontal plane $z = c$ and projecting the intersection onto the plane XY, we obtain the level line $f^{-1}(c)$; see Figure 7.5. Instead of graphs of complicated functions, it is often more convenient to draw their contours $\{f^{-1}(c)\}$ ($c \in \mathbb{R}$). Fixing level lines through equal intervals of values $0, \pm d, \pm 2d, \ldots,$ we can judge the steepness of the graph by the density of level lines. Level lines lie closer together where the slope of the graph to the horizontal plane is larger. *Projections with real marks* are used in the construction of earthworks and in drawing. Level sets are also useful in problems with extrema; see Section 7.2.

In addition to *parametric equations* of the surface $M \subset \mathbb{R}^3$, there are other analytic methods to represent it by formulae:

(1) as the *graph* of a function in two variables: $z = f(x,y)$,
(2) in *implicit* form, as the level set of a function in three variables: $F(x,y,z) = c$.

Similar formulae are used for surfaces in the cases of *cylindrical, spherical*, and other *curvilinear coordinates in space*. These three analytical methods of generating simple surfaces are equivalent to one another in a *local* sense (i.e., considering neighborhoods of points), but they are not equivalent for surfaces in the *large*. For example, the sphere $x^2 + y^2 + z^2 = R^2$ intersects a number of lines of any direction at two points and hence is not a graph.

Example. Plot graphs of functions $f(x, y)$ using the ezsurf command.

```
ezsurf('sin(x)*cos(y)', [-pi, pi]);                    % Figure 7.5(b)
ezsurf('x^2 - y^2', [-3, 3])                           % Figure 7.5(c)
```

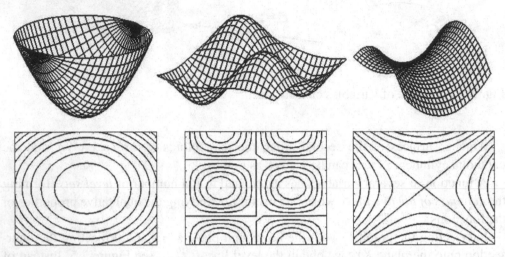

Fig. 7.5 Graphs of $x^2 + y^2$, $\sin(x)\cos(y)$, $x^2 - y^2$ and their contours.

A graph is a particular case of a parametric surface, because the equation $z = f(x, y)$ is equal to $\mathbf{r}(u, v) = u\mathbf{i} + v\mathbf{j} + f(u, v)\mathbf{k}$:

```
r = [u, v, f(u,v)];
```

Note that graf3d demonstrates various surface plots in MATLAB: one can adjust the type of plot, the type of shading, the color map, and so on.

Sufficient conditions for the smoothness of a surface graph are analogous to those for a smooth plane curve.

The notion of an implicitly defined surface also has its difficulty. The equation for a continuous function in (x, y, z) without additional requirements can define an arbitrary closed set in space (for example, a ball). However, there are obvious sufficient conditions for the implicit equation $F(x, y, z) = c$ to define a surface, analogous to the case of an implicitly defined planar curve. Recall that the *gradient* of a function $F(x, y, z)$ is the vector field $\nabla F = \left[\frac{\partial F}{\partial x}, \frac{\partial F}{\partial y}, \frac{\partial F}{\partial z} \right]$.

Using the implicit function theorem, one can prove the following result.

Theorem 7.1. *Let $F(x,y,z)$ be a function of class C^k on the open set $G \subset \mathbb{R}^3$, and suppose that the gradient ∇F is nonzero on the level set $M_c = \{(x,y,z) \in G : F(x,y,z) = c\}$ of height c. Then M_c is a C^k-regular surface.*

In addition to analytical methods, there are other ways to represent a surface: for example, using a *table with coordinates of an array of points*. Also, one may use *spline surfaces* (see Section 9.6) that are continuously glued from patches parameterized by vector polynomials.

Examples.

1. For the function $F = x^2 + y^2 - z^2$, the gradient $\nabla F = [2x, 2y, -2z]$ is zero only at the point $O = (0,0,0)$,

```
f = x^2 + y^2 + z^2;
df = [diff(f, x), diff(f, y), diff(f, z)]
```

which is a singular point (the vertex) on the cone M_0. Other level sets M_c ($c \neq 0$) are smooth surfaces: hyperboloids (of one sheet when $c > 0$ and of two sheets when $c < 0$). See Figure 7.16. In a similar way, we see that the ellipsoid is a smooth surface.

2. Let $f(x,y)$ be a function of class C^k in the open set G. Then the equation $z = f(x,y)$ defines a C^k-regular surface graph. To show this, take $x = u$, $y = v$ as the parameters of the vector-valued function $\mathbf{r}(u,v) = [u, v, f(u,v)]$. We obtain linearly independent vectors $\mathbf{r}_u = [1, 0, f_u'(u,v)]$ and $\mathbf{r}_v = [0, 1, f_v'(u,v)]$.

3. One may use surf to plot a surface by a table of points.

$w \backslash^t$	35	30	25	20	15	10	5	0	-5	-10
5	32	27	22	16	11	6	0	-5	-10	-15
10	22	16	10	3	-3	-9	-15	-22	-27	-34
15	16	9	2	-5	-11	-18	-25	-31	-38	-45
20	12	4	-3	-10	-17	-24	-31	-39	-46	-53
25	8	1	-7	-15	-22	-29	-36	-44	-51	-59
30	6	-2	-10	-18	-25	-33	-41	-49	-56	-64
35	4	-4	-12	-20	-27	-35	-43	-52	-58	-67
40	3	-5	-13	-21	-29	-37	-45	-53	-60	-69
45	2	-6	-14	-22	-30	-38	-46	-54	-62	-70

The function *wind chill factor* is defined as *how cold it feels (for example, $WC = 32°$ F) for a given wind speed W (5 miles per hour) and a given air temperature ($T = 35°$ F).*

```
L = [ 32 22 16 12 8 6 4 3 2;  27 16 9 4 1 -2 -4 -5 -6;  22 10 2 -3 -7 -10 -12 -13
-14;  16 3 -5 -10 -15 -18 -20 -21 -22;  11 -3 -11 -17 -22 -25 -27 -29 -30;  6 -9
-18 -24 -29 -33 -35 -37 -38;  0 -15 -25 -31 -36 -41 -43 -45 -46;  -5 -22 -31 -39
-44 -49 -52 -53 -54;  -10 -27 -38 -46 -51 -56 -58 -60 -62;  -15 -34 -45 -53 -59
-64 -67 -69 -70];
surf(L);
```

Here L is a 9×10 matrix. The intersection of the surface with the horizontal plane $WC = -20$ (Figure 7.6(a)) is the level line where one feels the coldness at $-20°$ F.

We plot a surface and a plane:

```
i = 1:10;  j = 1:9;
[x, y] = meshgrid(-10 + 5*(j-1), 5 + 5*(i-1));
z = L(i, j);
hold on; view(3); xlabel('W'); ylabel('T'); zlabel('WC');
surf(x, y, z);  surf(x, y, -20 + 0*z);
```

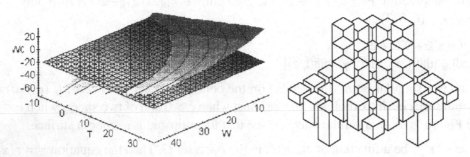

Fig. 7.6 The *wind chill factor* and the level $WC = -20°$ F. Command bar3.

7.2 Tangent Planes and Normal Vectors

7.2.1 *Basic formulae and properties*

A *tangent line of a smooth surface M* is defined as a line that is tangent to some smooth curve on M. The plane containing all tangent lines to the smooth surface M at the point P is called the *tangent plane* to M at P. It is denoted by T_PM.

Let $\mathbf{r}(u, v)$ be a smooth parameterization of a surface M and let $\gamma: u = u(t), v = v(t)$, $t \in I$, be a smooth curve on M through the point $P = \gamma(t_0)$. The equation of γ as a space curve is $\tilde{\mathbf{r}}(t) = \mathbf{r}(u(t), v(t))$ $(t \in I)$. Hence, the velocity vector of γ is a linear combination of the vectors $\mathbf{r}_u$, $\mathbf{r}_v$:

$$\tilde{\mathbf{r}}'(t) = \mathbf{r}_u \cdot u'(t) + \mathbf{r}_v \cdot v'(t).$$

In fact, the linearly independent vectors $\mathbf{r}_u$, $\mathbf{r}_v$ define the plane that is tangent to the surface M at the point P.

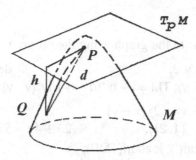

Fig. 7.7 Tangent plane at a point of a surface.

Let $M \subset \mathbb{R}^3$ be a surface, $P \in M$ a point, σ any plane through P, $Q \in M$ any point near P, $d = |PQ|$ the length of the segment between P and Q, and $h = \rho(Q, \sigma)$ the distance between Q and the plane σ.

Notice that *the tangent plane T_pM of a smooth surface M is characterized among all planes containing P by the condition $\lim_{Q \to P} h/d = 0$.*

Let $\mathbf{r}(u,v) = [x(u,v), y(u,v), z(u,v)]$ be a smooth parameterization of the surface, where $(u,v) \in G \subset \mathbb{R}^2$. Consider points $P = (u,v)$ and $A = (X,Y,Z)$ on the tangent plane σ. The vector equation of σ is

$$\mathbf{r} = \mathbf{r}(u,v) + a\mathbf{r}_u(u,v) + b\mathbf{r}_v(u,v) \qquad (a, b \in \mathbb{R}). \tag{7.3}$$

We rewrite the equation of σ using a point and two vectors:

$$(\overrightarrow{AP}, \mathbf{r}_u, \mathbf{r}_v) = \begin{vmatrix} X - x(u,v) & Y - y(u,v) & Z - z(u,v) \\ x_u(u,v) & y_u(u,v) & z_u(u,v) \\ x_v(u,v) & y_v(u,v) & z_v(u,v) \end{vmatrix} = 0. \tag{7.4}$$

In particular, when M is the graph $z = f(x,y)$, we obtain

$$\begin{vmatrix} x - x_0 & y - y_0 & z - f(x_0, y_0) \\ 1 & 0 & -f'_x \\ 0 & 1 & -f'_y \end{vmatrix} = z - f(x_0, y_0) f'_x(x_0, y_0)(x - x_0) - f'_y(x_0, y_0)(y - y_0) = 0,$$

i.e., the tangent plane at $P = (x_0, y_0, f(x_0, y_0))$ is given by

$$z = f(x_0, y_0) + f'_x(x_0, y_0)(x - x_0) - f'_y(x_0, y_0)(y - y_0). \tag{7.5}$$

The vector $\mathbf{n} = \frac{\mathbf{r}_u \times \mathbf{r}_v}{|\mathbf{r}_u \times \mathbf{r}_v|}$ is called the *unit normal* to the surface M^2 at $P = (u,v)$. A line through a point of the surface orthogonal to the tangent plane is called a *normal line*.

Examples.

1. We plot the tangent plane to the graph using the program below.

```
syms x y u v;  f = x^2 + y^2;                          % define a function f(x,y)
fx = diff(f, x);  fy = diff(f, y);  TM = f + fx*(u - x) + fy*(v - y);      % tangent plane
```
$$TM = x^2 + y^2 + 2x(u - x) + 2y(v - y) \qquad \text{% Answer}$$
```
TMP = subs(TM, {x, y}, {1,2});          % 2u + 4v − 5 = 0 tangent plane at P
hold on;  view(3);  ezsurf(f);  ezsurf(TMP);               % Figure 7.8(a)
```

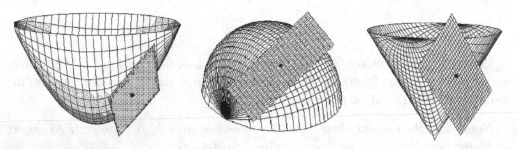

Fig. 7.8 Tangent planes to a paraboloid, a sphere, and a cone.

2. Check that the unit normal $\mathbf{n} = \frac{\mathbf{r}_u \times \mathbf{r}_v}{|\mathbf{r}_u \times \mathbf{r}_v|}$ does not depend on the parameterization of the surface, and that $\mathbf{n}$ changes direction only when the jacobian changes sign under coordinate displacement (see Lemma 7.1).

For the implicitly defined smooth surface M^2: $F(x,y,z) = 0$, the gradient $\nabla F = (F_x', F_y', F_z')$ is orthogonal to the surface. Thus, the tangent plane at $P = (x_0, y_0, z_0) \in M$ has the following equation:

$$F_x'(x_0, y_0, z_0)(x - x_0) + F_y'(x_0, y_0, z_0)(y - y_0) + F_z(x_0, y_0, z_0)(z - z_0) = 0. \qquad (7.6)$$

If a surface bounds a body in space, then it has *exterior* and *interior* sides. If a surface is the graph of a function $z = f(x,y)$, one can define its *top* and *bottom* sides. Moreover, a surface has two sides if and only if there is a continuous field of unit normals on the whole surface. A surface that does not admit a continuous field of unit normals is one-sided.

Equations of the normal to the surface:

(a) $\mathbf{r}(u,v) + \frac{\mathbf{r}_u \times \mathbf{r}_v}{|\mathbf{r}_u \times \mathbf{r}_v|} t$ at the point (u,v) of the parametric surface,

(b) $\frac{x - x_0}{F_x'(x_0, y_0, z_0)} = \frac{y - y_0}{F_y'(x_0, y_0, z_0)} = \frac{z - z_0}{F_z'(x_0, y_0, z_0)}$ at the point (x_0, y_0, z_0) of the
implicitly given surface $F(x,y,z) = 0$,

(c) $\frac{x - x_0}{f_x'(x_0, y_0)} = \frac{y - y_0}{f_y'(x_0, y_0)} = \frac{z - f(x_0, y_0)}{-1}$ at the point (x_0, y_0) of the graph $z = f(x,y)$.

7.2.2   *Extrema of functions defined on surfaces*

An *extremum* is a value of a continuous function that is a local maximum or local minimum. Extrema of a function defined on a surface are related by notions of the tangent plane and the normal vector. It is known from calculus that a necessary condition for an extremum of a smooth function $z = f(x,y)$ in the open set G is the vanishing of all partial derivatives of f at the point or, equivalently, that the gradient ∇f is zero at this point:

$$\{f'_x(x_0,y_0) = 0, \ f'_y(x_0,y_0) = 0\} \iff \nabla f(x_0,y_0) = 0. \tag{7.7}$$

The meaning of (7.7) is that *the tangent plane* (7.4) *of the graph of f at* $(x_0,y_0,f(x_0,y_0))$ *is parallel to the coordinate plane XY.*

For the function $f(x,y)$ of class C^2, we use the notations

$$a_{11} = f''_{xx}(x_0,y_0), \quad a_{12} = f''_{xy}(x_0,y_0), \quad a_{22} = f''_{yy}(x_0,y_0), \quad D = a_{11}a_{22} - a_{12}^2.$$

We have the following:

- *If $D > 0$, then at the point (x_0,y_0) the function f has a local extremum (maximum when $a_{11} < 0$ and minimum when $a_{11} > 0$).*
- *If $D < 0$, then at the* saddle *point (x_0,y_0) the function f does not have a local extremum.*
- *If $D = 0$, then the point (x_0,y_0) might be an extremum of f.*

The general form of the *conditional extremum problem* is the following: *what is the maximal or minimal value of the continuous function $g\colon \mathbb{R}^3 \to \mathbb{R}$ on a given surface M? As a rule, the maximum (minimum) is reached at points where the surface M is tangent to some level surface of g.*

Theorem 7.2. *Let $F(x,y,z)$ be a smooth function on an open set $G \subset \mathbb{R}^3$, and let $M_c = F^{-1}(c)$ be the level surface (of height c). Assume that M_c is smooth, i.e., $\nabla F \neq 0$ holds along M_c. Suppose that $g\colon G \to \mathbb{R}$ is a smooth function and $P \in M_c$ is an extremal point for g on M_c. Then the tangent plane of M_c at P is orthogonal to the gradient $\nabla g(P)$, i.e., there is $\lambda \in \mathbb{R}$ such that $\nabla g(P) = \lambda \nabla F(P)$.*

The real λ in Theorem 7.2 is called the *Lagrange multiplier*. The condition in Theorem 7.2 is equivalent to the system

$$\begin{cases} F(x,y,z) = c, \\ F'_x(x,y,z) = \lambda g'_x(x,y,z), \\ F'_y(x,y,z) = \lambda g'_y(x,y,z), \\ F'_z(x,y,z) = \lambda g'_z(x,y,z). \end{cases}$$

If the surface M_c is compact, then any continuous function g takes its maximum and minimum on M_c. Hence, Theorem 7.2 can be used for the selection of possible candidates (among all critical points of g) for these extremal points. There is an analogous theorem for extrema of the function $g(x, y)$ along a plane curve γ; see Section 5.1.

Examples.

1. The *height function* $g(x, y, z) = z$ on the torus defined by revolution of the circle γ: $\{y = 0, x^2 + (z - 2)^2 = 1\}$ about the x-axis has four critical points (Figure 7.9): in

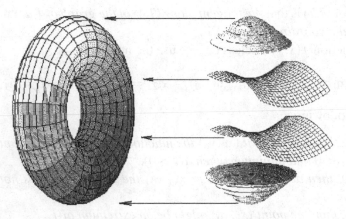

Fig. 7.9 Critical points of the height function on the torus.

addition to two extrema (maximum at $P_1 = (0, 0, 3)$ and minimum at $P_2 = (0, 0, -3)$) there are also two saddle points $P_3 = (0, 0, 1)$ and $P_4 = (0, 0, -1)$. If M_c is not compact, there may be no extrema (for instance, the height function $g(x, y, z) = z$ on the cylinder M_1: $x^2 + y^2 = 1$).

2. Find the local extrema of the function $z = 3x^2 y - x^3 - y^4$.

Solution. First plot the graph and the level lines of f:

```
syms x y;  f = 3*x^2*y - x^3 - y^4;
ezsurf(f, [-6 10 -4 5])
```

Then calculate the partial derivatives f'_x, f'_y,

```
fx = diff(f, x),  fy = diff(f, y)        % Answer: fx = -3x^2 + 6xy, fy = 3x^2 - 4y^3
```

and finally solve the system $\{f'_x = 0,\ f'_y = 0\}$:

```
S = solve(fx, fy, x, y);   S.x(2),  S.y(2)                    % Answer: 6, 3.
```

Two points are possibly extrema: $P_1 = (0,0)$ and $P_2 = (6,3)$. Calculate the second partial derivatives of f and the term D:

> fxx = diff(fx, x), fxy = diff(fx, y), fyy = diff(fy, y)
> $f_{xx} = -6x - 6y$, $f_{xy} = 6x$, $f_{yy} = -12y^2$ % Answer
> Delta = fxx*fyy - fxy^2;   % Answer: $\Delta = 72xy^2 + 36y^3 - 36x^2$

At the point P_1 we have:

> subs([fxx, fxy, fyy, Delta], {x, y}, {0, 0})
> f0 = subs(f, {x, y}, {S.x(2), S.y(2)})   % Answer: $[0,0,0,0]$, 27.

Since $a_{11} = 0$, $a_{12} = 0$, $a_{22} = 0$ hold, then $D = 0$ and the point P_1 requires additional investigation. The value of f at this point is zero: $f(0,0) = 0$. Further, for $x < 0, y = 0$, we have $f(x,y) = -x^3 > 0$, and for $x = 0$, $y \neq 0$, we have $f(x,y) = -y^4 < 0$. Consequently, in any neighborhood of P_1 the function $f(x,y)$ has values larger than $f(P_1)$ and smaller than $f(P_1)$. Hence, $f(x,y)$ does not have an extremum at P_1. Analogous calculations at P_2 yield $a_{11} = -18$, $a_{12} = 36$, $a_{22} = -108$, and hence $D = 648 > 0$. Since $a_{11} < 0$, the function has a local maximum at P_2.   □

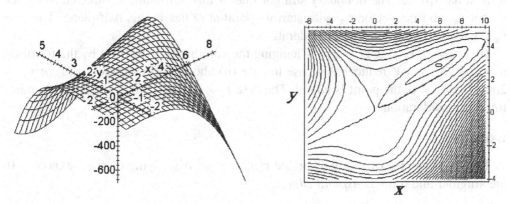

Fig. 7.10 Extrema of $z = 3x^2y - x^3 - y^4$: the graph and the level lines.

7.3 Singular Points on Surfaces

7.3.1 *Parametric surfaces*

We give a test for singular points on a parametric surface.

A point (u_0, v_0) on a surface M is singular if the vector function $\mathbf{n} = \frac{\mathbf{r}_u \times \mathbf{r}_v}{|\mathbf{r}_u \times \mathbf{r}_v|}$ does not tend to some limit value as $(u, v) \to (u_0, v_0)$.

The structure of a parametric surface $\mathbf{r}(u, v)$ of class C^k, $k \ge 2$, near a singular point $P = (u_0, v_0)$ is usually investigated using a Taylor expansion of second or third order. Starting with zero values of parameters u_0, v_0 from the given point P, we have

$$\mathbf{r}(u,v) = \mathbf{r}(p) + \mathbf{r}_u(p)u + \mathbf{r}_v(p)v + \frac{1}{2}\left(\mathbf{r}_{uu}(p)u^2 + \mathbf{r}_{vv}(p)v^2 + 2\mathbf{r}_{uv}(p)uv\right)$$

$$+ \frac{1}{6}\left(\mathbf{r}_{uuu}(p)u^3 + 3\mathbf{r}_{uuv}(p)u^2v + 3\mathbf{r}_{uvv}(p)uv^2 + \mathbf{r}_{vvv}(p)v^3\right) + \cdots + \bar{\boldsymbol{\varepsilon}}_k(u^2 + v^2)^{\frac{k}{2}}.$$

From the definition of a singular point it follows that $\mathbf{r}_u \times \mathbf{r}_v = 0$.

The following simple cases are possible:

(i) $\mathbf{r}_u = 0$, $\mathbf{r}_v = 0$. If, in addition, $\mathbf{r}_{uu}\mathbf{r}_{uv}\mathbf{r}_{vv} \ne 0$, the point is *edged*. The case $\mathbf{r}_{uu}\mathbf{r}_{uv}\mathbf{r}_{vv} = 0$ requires further consideration. Tangent rays at a singular point form a dihedral angle. The edge of this angle is directed along the vector $\mathbf{r}_{uv}$, and its half-planes contain the vectors $\mathbf{r}_{uu}$ and $\mathbf{r}_{vv}$ whose starting points coincide with O.

(ii) $\mathbf{r}_u \ne 0$, $\mathbf{r}_v = 0$. If, in addition, $\mathbf{r}_u \times \mathbf{r}_{vv} \ne 0$, the tangent rays at the singular point form a half-plane. The boundary straight line of this half-plane is directed along the vector $\mathbf{r}_u$, and the vector $\mathbf{r}_{vv}$ with starting point at O lies in this half-plane. The case $\mathbf{r}_u \times \mathbf{r}_{vv} = 0$ requires further consideration.

(iii) $\mathbf{r}_u \ne 0$, $\mathbf{r}_v \ne 0$, $\mathbf{r}_v = a\mathbf{r}_u$. Changing the variables u, v to U, V by the formulae $u = U - aV$, $v = V$ reduces this case to case (ii) above. If, in addition, $\mathbf{r}_u \times (a^2\mathbf{r}_{uu} - 2a\mathbf{r}_{uv} + \mathbf{r}_{vv}) \ne 0$, the point is *edged*. The case $\mathbf{r}_u \times (a^2\mathbf{r}_{uu} - 2a\mathbf{r}_{uv} + \mathbf{r}_{vv}) = 0$ requires further consideration.

Examples.

1. The points on the cylindrical surface $\mathbf{r}(u, v) = [u^2, u^3, v]$ lying on the z-axis ($u = 0$) are singular and form a *cuspidal edge*.

```
syms u v;
r = [u^2, u^3, v];
ru = diff(r,u);  rv = diff(r,v);
subs(cross(ru, rv), {u, v}, {0, 0})          % Answer: 0, 0, 0.
ruu = diff(r, u, 2);  ruv = diff(r,u, v);  rvv = diff(r, v, 2);
det([ ruu;  ruv;  rvv ])          % Answer: 0 (requires further consideration).
ezsurf( r(1), r(2), r(3),  [-1, 1, 0, 2] );          % Figure 7.11(b)
```

Each plane orthogonal to the cuspidal edge intersects the surface along a curve for which the point of the cuspidal edge is a singular point of the same type.

2. Check that the point $u = v = 0$ of each of the following parametric surfaces is a singular point of a certain type:

(a) The surface $\mathbf{r}(u,v) = [u^2, uv, v^2]$; Figure 7.11(c).

```
syms u v;    r = [u^2, u*v, v^2];
ru = diff(r, u),  rv = diff(r,v)         % Answer: ru = [2u, v, 0], rv = [0, u, 2v]
subs(cross(ru, rv), {u, v}, {0, 0});                              % Answer: [0,0,0]

ruu = diff(r, u, 2),  ruv = diff(r,u, v),  rvv = diff(r, v, 2)
                        % Answer: ruu = [2,0,0], ruv = [0,1,0], rvv = [0,0,2]
det([ruu; ruv; rvv])                                             % Answer: 4
ezsurf( r(1), r(2), r(3),  [-1, 1, 0, 2] );                      % Figure 7.11(c)
```

(b) The *Whitney umbrella* $\mathbf{r}(u,v) = [uv, u, v^2]$; Figure 8.26(b).

```
r = [u*v, u, v^2];  ezsurf(r, [-1, 1, -1, 1]);
ru = diff(r, u);  rv = diff(r, v);  rvv = diff(r, v, 2);
                        % Answer: ru = [v,1,0], rv = [u,0,2v], rvv = [0,0,2]
subs(cross(ru, rv), {u, v}, {0, 0});                    % Answer: [2,0,0]
```

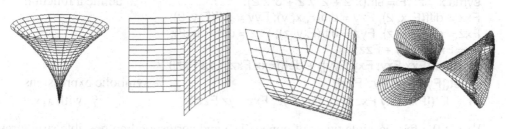

Fig. 7.11 Singular points and curves on surfaces.

7.3.2 *Implicitly given surfaces*

In studying an implicitly given surface $F(x,y,z) = 0$, where $F \in C^k$, $k \le 2$, the notion of a singular point must be modified. In particular, self-intersections also must be kept in mind. If the necessary conditions $F'_x = F'_y = F'_z = 0$ at a given point are satisfied, one applies Taylor's expansion to obtain information. It is convenient to assume our point is at the origin. Then

$$F(x,y,z) = \frac{1}{2}(F_{xx}x^2 + F_{yy}y^2 + F_{zz}z^2 + 2F_{xy}xy + 2F_{yz}yz + 2F_{xz}xz)$$
$$+ \frac{1}{6}(F_{xxx}x^3 + F_{yyy}y^3 + F_{zzz}z^3 + 3F_{xxy}x^2y + 3F_{yyx}y^2x + 3F_{yyz}y^2z$$
$$+ 3F_{zzy}z^2y + 3F_{zzx}z^2x + 3F_{xxz}x^2z + 6F_{xyz}xyz) + \cdots + \varepsilon_k(x^2 + y^2 + z^2)^{k/2}.$$

For a clearer picture of the shape of the surface $F = 0$ in a neighborhood of a singular point, one first determines the type of the surface

$$F_{xx}x^2 + F_{yy}y^2 + F_{zz}z^2 + 2F_{xy}xy + 2F_{yz}yz + 2F_{xz}xz = 0. \qquad (7.8)$$

This surface is a cone and is one of of five types. To calculate the type of the cone (7.8) we need to consider the real *invariants*

$$I_1 = F_{xx} + F_{yy} + F_{zz}, \quad I_2 = \begin{vmatrix} F_{xx} & F_{xy} \\ F_{xy} & F_{yy} \end{vmatrix} + \begin{vmatrix} F_{xx} & F_{xz} \\ F_{xz} & F_{zz} \end{vmatrix} + \begin{vmatrix} F_{yy} & F_{yz} \\ F_{yz} & F_{zz} \end{vmatrix}, \quad I_3 = \begin{vmatrix} F_{xx} & F_{xy} & F_{xz} \\ F_{xy} & F_{yy} & F_{yz} \\ F_{xz} & F_{yz} & F_{zz} \end{vmatrix}.$$

```
syms x y z;  F = sin(x^2 + 2*y^2 + 3*z^2);          % define a function
Fxx = diff(F, x, 2); Fxy = diff(F, x, y); Fyy = diff(F, y, 2);
Fxz = diff(F, x, z); Fyz = diff(F, y, z); Fzz = diff(F, z, 2);
I1 = Fxx + Fyy + Fzz;
I2 = det([Fxx Fxy; Fxy Fyy]) + det([Fxx Fxz; Fxz Fzz])
+ det([Fyy Fyz; Fyz Fzz])                            % symbolic expressions
I3 = det([Fxx Fxy Fxz; Fxy Fyy Fyz; Fxz Fyz Fzz])    % with x, y, z
```

We list the five possible types of cones (7.8) and corresponding possible structures of the implicitly given surface $F = 0$ near a singular point.

(a) $I_3 \neq 0$, $I_2 > 0$, $I_1 I_3 > 0$. The cone consists of one singular point. This singular point of the surface $F = 0$ is *isolated*.

(b) $I_3 \neq 0$, $I_2 \leq 0$ or $I_1 I_3 < 0$. The cone has the shape of an (oblique) circular cone. The surface near the singular point looks like a deformed neighborhood of the vertex of the above cone.

(c) $I_3 = 0$, $I_2 > 0$. The cone degenerates into a line. Rotating the coordinate system, we obtain the case where the z-axis coincides with the above line. The equation of the surface in the new coordinates takes the form $\tilde{F}(x,y,z) = 0$. Moreover,

$$\tilde{F}(x,y,z) = a^2 x^2 + b^2 y^2 + \frac{1}{6}\left(\tilde{F}_{xxx}x^3 + \cdots\right) + \cdots + \varepsilon_k\left(x^2 + y^2 + z^2\right)^{k/2}.$$

In this case if $\tilde{F}_{zzz} \neq 0$ holds, then a neighborhood of the singular point has the shape of a *spike*. All tangent rays of the surface at the singular point coincide. The case $\tilde{F}_{zzz} = 0$ requires further consideration.

(d) $I_3 = 0, I_2 < 0$. The cone consists of two intersecting planes, and the surface $F = 0$ consists of two intersecting sheets (their intersection being along a smooth curve).

(e) $I_3 = 0$, $I_2 = 0$. The cone consists of the doubly covered plane. Rotating the coordinate system, we obtain the case where the z-axis is orthogonal to the above plane. The equation of the surface in the new coordinates takes the form $\tilde{F}(x,y,z) = 0$ for

$$\tilde{F}(x,y,z) = az^2 + \frac{1}{6}\left(\tilde{F}_{xxx}x^3 + \cdots + \tilde{F}_{yyy}\right) + \cdots + \varepsilon_k\left(x^2 + y^2 + z^2\right)^{k/2}.$$

If the following equation of third order in u,

$$\tilde{F}_{xxx} + 3\tilde{F}_{xxy}u + 3\tilde{F}_{xyy}u^2 + \tilde{F}_{yyy}u^3 = 0, \tag{7.9}$$

has a *unique real root*, then the tangent rays at a singular point form a half-plane, which is one half of the cone. If (7.9) has *three different real roots*, then tangent rays at the singular point form three separate sectors in the plane xy, and the surface can be imagined as three funnels converging to the vertex, flattening to the plane (x,y) as they come near to the vertex. The case where (7.9) has one real root with multiplicity 3 requires further consideration.

Example. Check that the point $x = y = z = 0$ of each of the following implicitly given surfaces is a singular point of a certain type:

(a) $F(x,y,z) = (x^2 + y^2 + z^2)(z-1)$ is the union of a plane and a point.

(b) The surface of revolution of the lemniscate of Bernoulli, $(x^2 + y^2 + z^2)^2 - 2a^2(z^2 - x^2 - y^2)$, Figure 8.25(b), and "a figure eight," Figure 8.16(a,b).

```
f = inline('(x.^2 + y.^2 + z.^2).^2 - 2.*(z.^2 - x.^2 - y.^2)', 'x', 'y', 'z');
impl(f, [-1, 1, -1, 1, -2, 2], 0),  axis equal
```

(c) The surface $x^2 + y^2 - z^3 = 0$ from Exercise 2, p. 319.

```
f = inline('x.^2 + y.^2 - z.^3', 'x', 'y', 'z');
impl(f, [-2, 2, -2, 2, -.01, 2], 0),  axis equal        % Figure 7.11(a).
```

(d) The pair of intersecting planes $x^2 - y^2 = 0$.

(e) Consider the surface $z^2 - x^2y + y^3 = 0$ (see Figure 7.11(d)). We plot it by gluing together six pieces (graphs):

```
N = 41;   X = meshgrid(-2 : 4/N : 2);
[X2, Y2] = meshgrid(0 : 2/N : 2);
Z = sqrt(Y2.^3 - X.^2.*Y2);  Z2 = sqrt(-Y2.^3 + X2.^2.*Y2);
hold on; axis equal; view(3);
surf(-X2, -Y2, (Y2 < X2).*Z2);  surf(-X2, -Y2, -(Y2 < X2).*Z2);
surf( X2, -Y2, (Y2 < X2).*Z2);  surf( X2, -Y2, -(Y2 < X2).*Z2);
surf(X, Y2, (Y2 > abs(X)).*Z);  surf(X, Y2, -(Y2 > abs(X)).*Z);
```

The resulting graph contains an additional part of the plane $z = 0$. To remove this part, we replace zeros by NaNs in arrays Z, Z_2 (see the program), inserting the following commands (before the line starting with hold on):

```
for i = 1 : N;
    for j = 1 : abs((N + 1) - 2*i)-2;  Z(j, i) = NaN;  end;
    for j = 2 : i - 2;  Z2(i, j) = NaN;  end;
end
```

If (7.9) has three real roots, and one of them is of multiplicity two, then the surface near a singular point looks like the Whitney umbrella, Figure 8.26(b).

7.4 Osculating Paraboloid

An osculating paraboloid helps us visualize the shape of a surface near an arbitrary point. We derive and plot it by the "elementary" method. An application of parallel transport and rotation (composition of two rotations about coordinate axes) leads to the simple standard situation where the tangent plane at the given point, the origin, is horizontal; using the second-order Taylor formula, we then derive the osculating paraboloid, and finally return to the initial coordinates and plot its image.

7.4.1 *Properties of the osculating paraboloid*

Let M be a regular surface and P a point of M. Let F be a paraboloid with vertex at P and axis parallel to the normal of the surface at P. Take a point Q on the surface close to P, and denote by $d = |PQ|$ the distance between them; let $h = \text{dist}(Q, F)$ denote the (vertical) distance from the point Q to the paraboloid F.

A paraboloid F through a point P on a surface M tangent to the plane $T_P M$ and lying "closest" to the surface in a small neighborhood of P (i.e., $\lim_{Q \to P} \frac{h}{d^2} = 0$) is called an *osculating paraboloid* at P.

An osculating paraboloid gives the best second-order approximation of the shape of the surface near the point P, with respect to the distance to P. Its shape helps us to classify points on the C^2-regular surface.

We fix rectangular coordinates in $\mathbb{R}^3$ such that the origin coincides with P, the z-axis is parallel to the normal $\mathbf{n}$ to the surface at P, and the coordinate plane XY co-incides with the tangent plane of the surface at P. Now, the surface near $O = P$ is the graph of $z = f(x, y)$. Moreover, $f(0,0) = 0$ because O belongs to the surface, and $f_x(0,0) = 0$ and $f_y(0,0) = 0$ hold because XY is the tangent plane to the surface at O. Using the Taylor expansion of $f(x, y)$ at $(0,0)$, we obtain the equation of the surface

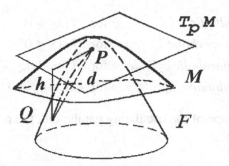

Fig. 7.12 Osculating paraboloid at a point on a surface.

as $z = \frac{1}{2}(rx^2 + 2sxy + ty^2) + o(x^2 + y^2)$, where $r = f_{xx}(0,0)$, $s = f_{xy}(0,0)$, $t = f_{yy}(0,0)$. The osculating paraboloid F at P is defined by the equation $z = \frac{1}{2}(rx^2 + 2sxy + ty^2)$, or in canonical form (with coordinate axes directed along the symmetry axes of the paraboloid) by $z = k_1 x^2 + k_2 y^2$, where k_1, k_2 are eigenvalues of the quadratic form $rx^2 + 2sxy + ty^2$ at P. This differs from the surface near P by an infinitesimal magnitude of the second order. The square of the distance of the point Q from P is $d^2 = x^2 + y^2 + \varphi^2(x,y)$ with $\varphi^2 = (x^2 + y^2)o(x^2 + y^2)$. Hence, we have $\lim_{Q \to P} \frac{h}{d^2} = \lim_{(x,y) \to 0} \frac{|o(x^2+y^2)|}{x^2+y^2+\varphi^2(x,y)} = 0$.

Thus the following claim has been proved: *For any point of a C^2-regular surface there is a unique osculating paraboloid that can degenerate in special cases to a parabolic cylinder or a plane.*

Example. The second-order equation

$$z = a_{11}x^2 + 2a_{12}xy + a_{22}y^2 \tag{7.10}$$

defines a *paraboloid* in $\mathbb{R}^3(x,y,z)$, excluding the degenerate cases of the parabolic cylinder and the plane. Rotation of the coordinate system in the plane XY by a suitable angle leads to the canonical form of (7.10)

$$z = k_1 x^2 + k_2 y^2. \tag{7.11}$$

We can find k_1 and k_2 as roots of the characteristic polynomial:

```
syms a11 a12 a22;  k = solve((a11 - t)*(a22 - t) - a12*a12, t)
```

Depending on the coefficients k_1, k_2, the *paraboloid* (7.11) is of one of the following types:

$$
\begin{cases}
\textit{elliptic} & \text{for} \quad k_1 k_2 > 0, \\
\textit{hyperbolic} & \text{for} \quad k_1 k_2 < 0, \\
\textit{parabolic cylinder} & \text{for} \quad k_1 = 0,\, k_2 \neq 0, \\
\textit{planar} & \text{for} \quad k_1 = k_2 = 0.
\end{cases}
$$

Depending on the shape of the osculating paraboloid, the point P of the surface has one of the following types:

$$
\begin{cases}
\textit{elliptic} & \text{for} \quad k_1 k_2 > 0, \\
\textit{hyperbolic} & \text{for} \quad k_1 k_2 < 0, \\
\textit{parabolic} & \text{for} \quad k_1 = 0,\, k_2 \neq 0 \text{ or } k_1 \neq 0,\, k_2 = 0, \\
\textit{spherical} & \text{for} \quad k_1 = k_2 \neq 0, \\
\textit{planar} & \text{for} \quad k_1 = k_2 = 0.
\end{cases}
$$

The first three cases are general. In the last two cases F is a paraboloid of revolution or a plane. The elliptical and the hyperbolic points for an arbitrary surface are separated by curves consisting of parabolic points.

Example. Surfaces with $P = O$ of 5 types (see definition):

```
syms x y;
ezsurf((x^2 + 9*y^2)/2,  [-1 1 -1 1])          % Figure 7.13(a)
ezsurf((x^2 - y^2)/2,  [-1 1 -1 1])            % Figure 7.13(b)
ezsurf((-x^2 + y^7)/2,  [-1 1 -1 1])           % Figure 7.13(c)
ezsurf((x^2 + y^2)/2,  [-1 1 -1 1])            % Figure 7.13(d)
ezsurf((-x^7 - y^6)/2,  [-1 1 -1 1])           % Figure 7.13(e)
```

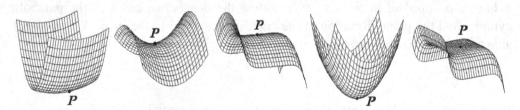

Fig. 7.13 The five types of smooth points on surfaces.

One can define *asymptotic* and *principal directions* at a point of a surface coinciding respectively with the asymptotic directions of the osculating paraboloid and with its symmetry axes. At each non-spherical or non-planar point there are exactly two mutually orthogonal principal directions. Based on the osculating paraboloid, one can also define the *normal curvature at a point on a surface in a given direction*, and the Gaussian and mean curvatures of a surface. We note only the basic properties:

(a) *There exist two asymptotic directions at a hyperbolic point, one at a parabolic point, and none at an elliptical point.*

(b) *All directions at a planar point are asymptotic and also principal.*

7.4.2 *Plotting an osculating paraboloid*

We will write a program for deriving the coefficients k_1 and k_2 and the equations of an osculating paraboloid. Recall that parallel translations and rotations about axes, and also their compositions, are rigid motions of space, i.e., they preserve the shape and the relative position of figures. We use 3×3 matrices to represent rotations about the axes OX and OY:

$$R_x(\psi) = \begin{vmatrix} 1 & 0 & 0 \\ 0 & \cos\psi & \sin\psi \\ 0 & -\sin\psi & \cos\psi \end{vmatrix}, \qquad R_y(\varphi) = \begin{vmatrix} \cos\varphi & 0 & -\sin\varphi \\ 0 & 1 & 0 \\ \sin\varphi & 0 & \cos\varphi \end{vmatrix} \quad \text{(see p. 90)}.$$

Let $z = f(x, y)$ be the equation of the surface M, and S the osculating paraboloid for M at $P = (x_0, y_0, z_0)$. The program is divided into four steps and is tested for the following data: $z = ax^2 + by^2$ $(a, b > 0)$ with $a = 1$, $b = 2$ at the point $x_0 = 1$, $y_0 = 0$.

(1) Parallel translation along the vector $\overrightarrow{PO}$ congruently moves the surface M and the paraboloid S into a surface M_1, with the osculating paraboloid S_1 at the origin.

```
syms x y z x1 y1 z1 t real;
a = 1;  b = 2;    f = a*x^2 + b*y^2;
x0 = 1;  y0 = 0;  z0 = subs(f, {x, y}, {x0, y0})          % Answer: z0 = 1
f1 = simplify(subs(f, {x, y}, {x + x0, y + y0}) - z0)     % f1 = x² + 2x + 2y²
```

(2) Assume that the common normal $\mathbf{n} = [a_1, b_1, c_1]$ to M_1 and S_1 at O is not parallel to the plane XY. Denote by $\mathbf{n}_1 = [0, b_1, c_1]$ the projection of this vector onto the coordinate plane YZ, and by ψ the angle between $\mathbf{n}_1$ and the z-axis. Obviously, $\cos\psi = c_1 / \sqrt{b_1^2 + c_1^2}$, $\sin\psi = b_1 / \sqrt{b_1^2 + c_1^2}$. Below we set $d = |\mathbf{n}|$, $d_1 = |\mathbf{n}_1|$.

```
n = subs([-diff(f1, x), -diff(f1,y), 1], {x, y}, {0, 0});   % Answer: n = [−2,0,1]
d1 = sqrt(n(2)^2 + n(3)^2);                                 % Answer: d1 = 1
d = sqrt(n*n');                                             % Answer: d = 2.2361.
```

Denote by R the composition of the following two rotations R_x and R_y in space:

(a) The rotation R_x through the angle ψ about the axis OX transposes M_1 and S_1 into the tangent pair M_2 and S_2 with common normal $\mathbf{n}_2$ at O; note that $\mathbf{n}_2 = \left[\frac{a_1}{d}, 0, \frac{d_1}{d}\right]$ lies in the plane XZ, but in view of our assumption it is not parallel to the axis OX.

Rx = [1, 0, 0; 0, n(3)/d1, n(2)/d1; 0, -n(2)/d1, n(3)/d1]

(b) The rotation R_y through the angle $\varphi = \angle(\mathbf{n}_2, OZ)$ around the axis OY translates M_2 and S_2 into the tangent pair M_3 and S_3 with common normal $\mathbf{n}_3$ at O, directed along OZ.

Ry = [d1/d, 0, n(1)/d; 0, 1, 0; -n(1)/d, 0, d1/d]

Find the rotation R and formulae for change of the coordinates:

R = Rx*Ry;   n3 = n*R   % checking $n_3 \parallel OZ$. Answer: $n_3 = [0, 0, 2.24]$.

coor = [x1, y1, z1]*R'   % Answer: $[\frac{1}{5}\sqrt{5}x_1 - \frac{2}{5}\sqrt{5}z_1, y_1, \frac{2}{5}\sqrt{5}x_1 + \frac{1}{5}\sqrt{5}z_1]$

If the common normal $\mathbf{n} = [a_1, b_1, c_1]$ to M_1 and S_1 at the point O is parallel to the plane XY, use the rotation in the plane XY about the point O through the angle $\psi' = \angle(\mathbf{n}_1, OY)$. The pair M and S are then translated into the tangent pair M_2' and S_2', and their common normal $\mathbf{n}_2'$ at O is directed along the axis OY. Finally, change the names of the variables: $y \to z$ and $z \to y$.

(3) The surface M_3 and its osculating paraboloid S_3 at the point O are in an optimal position, and we can start the calculation of the equation of S_3. Let $g(x,y,z) = 0$ be the equation of M_3.

g = collect(coor(3) - subs(f1, {x y}, {coor(1), coor(2)}), [x1, y1, z1])

gg = [diff(g, x1), diff(g, y1), diff(g, z1)]

$g = -\frac{1}{5}x_1^2 + \frac{4}{5}z_1 x_1 + z_1 5^{1/2} - \frac{4}{5}z_1^2 - 2y_1^2$ % Answer

subs(gg, {x1, y1, z1}, {0, 0, 0})   % checking $n_3 \parallel OZ$. Answer: $[0, 0, 2.24]$.

Although the explicit equation $z = z_3(x,y)$ of the surface M_3 in a neighborhood of O is unknown, techniques of calculus allow us to find all derivatives of the implicitly given function $z_3(x,y)$ and then derive the coefficients a_{11}, a_{12}, a_{22}. Obviously, $g_x'(0,0) = g_y'(0,0) = z_{3|x}'(0,0) = z_{3|y}'(0,0) = 0$,

$$a_{11} = \frac{1}{2}z_{3|xx}''(0,0) = \frac{g_{xx}''}{-2g_z'}(0,0), \qquad a_{12} = \frac{1}{2}z_{3|xy}''(0,0) = \frac{g_{xy}''}{-2g_z'}(0,0),$$

$$a_{22} = \frac{1}{2}z_{3|yy}''(0,0) = \frac{g_{yy}''}{-2g_z'}(0,0).$$

gz = subs(diff(g, z1), {x1, y1, z1}, {0, 0, 0})   % Answer: $g_z = \sqrt{5}$

gxx = subs(diff(g, x1, 2), {x1, y1, z1}, {0, 0, 0})/2

gyy = subs(diff(g, y1, 2), {x1, y1, z1}, {0, 0, 0})/2

gxy = subs(diff(diff(g, x1), y1), {x1, y1, z1}, {0, 0, 0})

$g_{xx} = -\frac{1}{5}$ $g_{yy} = -2$ $g_{xy} = 0$ % Answer

a11 = -gxx./(2*gz),   a12 = -gxy./(2*gz),   a22 = -gyy./(2*gz)

$a_{11} = .04472$ $a_{12} = 0$ $a_{22} = .4472$ % Answer

Write the final equation (7.10) of S_3:

S3 = a11*x1^2 + 2*a12*x1*y1 + a22*y1^2   % Answer: $S_3 = .045\,x_1^2 + .447\,y_1^2$

Solve the characteristic (quadratic) equation $(a_{11} - k)(a_{22} - k) - a_{12}^2 = 0$ and find the coefficients k_1 and k_2

k = solve((a11 - t)*(a22 - t) - a12*a12, t)   % Answer: $k = [0.0447, 0.4472]$
Scan = k(1)*x^2 + k(2)*y^2   % canonical equation $S_{can} = .0447\,x^2 + .447\,y^2$

We plot M_3 and S_3 (in the new coordinates).

hold on; axis equal; view(3);
ezsurf(S3, [-2, 1.4, -1, 1]);
gg = inline('-1/5*x1.^2 + 4/5*z1.*x1 + z1.*5^(1/2) - 4/5*z1.^2 - 2*y1.^2',
'x1', 'y1', 'z1');
impl(gg, [-2, 1.4, -1, 1, -.2, 1], 0);   % Figure 7.14(a)

(4) The rotation R^{-1} is given by the inverse matrix, which in the case of an orthogonal matrix coincides with its transpose R^*. It transforms M_3 and S_3 into the pair M_1 and S_1 and helps us to find the equation of S_1.

co1 = [x, y, z]*R   % Answer: $co_1 = [\frac{1}{5}5^{1/2}x + \frac{2}{5}5^{1/2}z, \ y, \ -\frac{2}{5}5^{1/2}x + \frac{1}{5}5^{1/2}z]$
S1 = simplify(subs(z1 - S3, {x1,y1,z1}, {co1(1),co1(2),co1(3)}))   % Answer

$S_1 = -0.894\,x + 0.447\,z - 0.00894\,x^2 - 0.0358\,xz - 0.0358\,z^2 - 0.447\,y^2$

We plot M_1 and S_1 (in the old coordinates).

hold on; axis equal; view(3);
ezsurf(f1, [-.6, .6, -1, 1]);
ff1 = inline('-2/5*x*5^(1/2)+1/5*z*5^(1/2)-1/250*x.^2*5^(1/2)
-2/125*x.*z*5^(1/2)-2/125*z.^2*5^(1/2)-1/5*5^(1/2)*y.^2', 'x', 'y', 'z');
impl(ff1, [-2, 1, -1, 1, -1, 2], 0);   % Figure 7.14(b),

Using the formulae for parallel translation along the vector $\overrightarrow{OP}$, deduce from the equation for S_1 the equation of the osculating paraboloid S. The figure with M and S is analogous to Figure 7.14(b).

S = simplify(subs(S1 + z0, {x, y, z}, {x - x0, y - y0, z - z0}))

$S = -0.8408\,x + 0.3667 + 0.5545\,z - 0.008944\,x^2$
$-0.03578\,xz - 0.03578\,z^2 - 0.4472\,y^2$ % Answer.

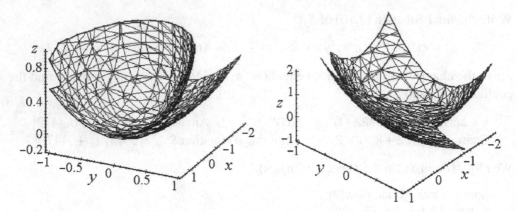

Fig. 7.14 Osculating paraboloids at O: (a) S_3 for M_3, (b) S_1 for M_1.

7.5 Curvature and Geodesics

7.5.1 *Gaussian and mean curvatures*

Let $\mathbf{r}(u,v)$ ($|u| \leq a$, $|v| \leq b$) be a (surface) patch M^2 in $\mathbb{R}^3$.

The expression $I = E\,du^2 + 2F\,du\,dv + G\,dv^2$, where

$$E = \langle \mathbf{r}_u, \mathbf{r}_u \rangle, \qquad\qquad F = \langle \mathbf{r}_u, \mathbf{r}_v \rangle, \qquad\qquad G = \langle \mathbf{r}_v, \mathbf{r}_v \rangle,$$

is called the *first fundamental form* of M^2.

The expression $II = L\,du^2 + 2M\,du\,dv + N\,dv^2$, where

$$L = \frac{(\mathbf{r}_{,uu}, \mathbf{r}_{,u}, \mathbf{r}_{,v})}{|\mathbf{r}_{,u} \times \mathbf{r}_{,v}|}, \qquad M = \frac{(\mathbf{r}_{,uv}, \mathbf{r}_{,u}, \mathbf{r}_{,v})}{|\mathbf{r}_{,u} \times \mathbf{r}_{,v}|}, \qquad N = \frac{(\mathbf{r}_{,vv}, \mathbf{r}_{,u}, \mathbf{r}_{,v})}{|\mathbf{r}_{,u} \times \mathbf{r}_{,v}|},$$

is called the *second fundamental form* of M^2.

For the plane $\mathbb{R}^2(x,y) \subset \mathbb{R}^3$, $I = dx^2 + dy^2$ and $II = 0$.

We create an M-file **EFG.m** to compute the first fundamental form:

```
function f1 = EFG(r)
    syms u v real;
    ru = diff(r, u);  rv = diff(r, v);
    E = ru*ru';  F = ru*rv';  G = rv*rv';
    f1 = simplify([E, F, G]);
end
```

For example:

```
syms u v R  real;
Rsphere = [R*cos(u)*cos(v)  R*sin(u)*cos(v)  R*sin(v)];
```

EFG(Rsphere)   % Answer: $[R^2\cos^2 v,\ 0,\ R^2]$

The length of a curve γ: $u = \phi(t)$, $v = \psi(t)$ ($t \in [t_1, t_2]$) on a surface $\mathbf{r}(u,v)$ is given by the integral $l(\gamma) = \int \sqrt{I}$ (see Example, p. 269),

$$l(\gamma) = \int_{t_1}^{t_2} \sqrt{E(u(t),v(t))u'^2 + 2F(u(t),v(t))u'v' + G(u(t),v(t))v'^2}\ dt.$$

For example, the length of a great circle ω: $u = t$, $v = 0$ on a unit sphere is

$$l(\omega) = \int_0^{2\pi} \sqrt{\cos^2(v(t))u'^2(t) + 1}\ dt = \int_0^{2\pi} \sqrt{\cos^2(0)\cdot 1 + 0}\ dt = 2\pi.$$

The area of a domain $D \subset M^2$ is $\text{Area}(D) = \iint_D \sqrt{EG - F^2}\ du\,dv$. Hence, the area of a unit sphere is $\text{Area}(S^2(1)) = \int_0^{2\pi}\int_0^\pi \sqrt{\cos^2(v)}\ du\,dv = 4\pi$.

Next we create an M-file **LMN.m** to compute the second fundamental form:

```
function f2 = LMN(X)
  syms u v real;
  ru = diff(r, u);  rv = diff(r, v);
  ruu = diff(ru, u);  ruv = diff(ru, v);  rvv = diff(rv, v);
  n = cross(ru, rv);  UN = n/simple(sqrt(n*n'));
  L = UN*ruu';  M = UN*ruv';  N = UN*rvv';
  f2 = simplify([L, M, N]);
end
```

For example:

```
catenoid = [u,  cosh(u)*cos(v), cosh(u)*sin(v)];
LMN(catenoid)                                    % Answer: [−1,0,1]
```

The *normal curvature* of M^2 at P in the direction $\mathbf{v} = \mathbf{r}_u du + \mathbf{r}_v dv$ is $k_n(P,\mathbf{v}) = II/I(P,\mathbf{v})$. Let $k_1 \le k_2$ be the *principal curvatures* of a surface,

$$k_1(P) = \min\{II/I(P,\mathbf{v}) : \mathbf{v} \neq 0\} \qquad \text{and} \qquad k_2(P) = \max\{II/I(P,\mathbf{v}) : \mathbf{v} \neq 0\}.$$

Then, the *Gaussian curvature* of the surface patch is $K = k_1 k_2$ and its *mean curvature* is $H = \frac{1}{2}(k_1 + k_2)$. The explicit formulae for H and K are:

$$K = \frac{LM - N^2}{EG - F^2}, \qquad H = \frac{LG - 2MF + NE}{2(EG - F^2)}. \tag{7.12}$$

The principal curvatures of M^2 are $k_{1,2} = H \pm \sqrt{H^2 - K}$.

Finally, we create an M-file **GK.m** to calculate the Gaussian curvature K. (A function **HK.m** for computing the mean curvature H differs from **GK.m** by one line, with f = simplify((G*L + E*N - 2*F*M)/(2*E*G - 2*F^2)).)

```
function f = GK(r)
   syms u v real;
   S = EFG(r);    T = LMN(r);
   E = S(1);  F = S(2);  G = S(3);
   L = T(1);  M = T(2);  N = T(3);
   f = simplify((L*N - M^2)/(E*G - F^2));
end
```

For example:

helicoid = [u*cos(v), u*sin(v), v];
GK(helicoid)   % Answer: $-1/\cosh^4 u$
enneper = [u - u^3/3 + u*v^2, -v + v^3/3 - v*u^2, u^2-v^2];
GK(enneper)   % Answer: $-4/(u^2+v^2+1)^4$

7.5.2 Geodesics

A curve γ on a surface $M^2 \subset \mathbb{R}^3$ is called a *geodesic* if $\gamma''(t)$ is zero or perpendicular to the surface at the point $\gamma(t)$, i.e., parallel to its unit normal, for all values of the parameter t. There is an interesting mechanical interpretation of geodesics: *a particle moving on the surface, and subject to no forces except a force acting perpendicular to the surface that keeps the particle on the surface, would move along a geodesic.*

Any geodesic has constant speed: $\frac{d}{dt}\|\gamma'\|^2 = 2\langle \gamma', \gamma'' \rangle = 0$. The simplest examples of geodesics are lines on the plane and the following:

(1) If a (segment of a) line belongs to M^2, then it is a geodesic.
    (In fact, a line $\gamma(t) = \mathbf{a} + t\mathbf{b}$ has zero acceleration $\gamma''(t)$ everywhere.)
(2) Meridians on a surface of revolution. Great circles on a sphere.

A *normal section* of a surface M^2 is the intersection $C = M^2 \cap \alpha$ (with a plane α) such that α is perpendicular to the surface at each point of C. One may show that *any normal section of a surface is a geodesic.*

The *geodesic equations* (see, for example, [Pressley])

$$(Eu' + Fv')' = \frac{1}{2}(E_u(u')^2 + 2F_u u'v' + G_u(v')^2),$$

$$(Fu' + Gv')' = \frac{1}{2}(E_v(u')^2 + 2F_v u'v' + G_v(v')^2) \tag{7.13}$$

are nonlinear differential equations, and are usually difficult or impossible to solve explicitly. The geodesic equations can be reduced to the form

$$u'' = f_1(u,v,u',v'), \qquad v'' = f_2(u,v,u',v'), \tag{7.14}$$

where f_i are smooth functions of the four variables u, v, u', and v' (see the formulae for x_3' and x_4' in the program **sgeod.m**).

From the theory of ordinary differential equations, *for any value of $t = t_0$ and given initial values $u(t_0)$, $v(t_0)$, $u'(t_0)$, $v'(t_0)$, there is a unique solution $u(t)$ and $v(t)$ to (7.14), defined and smooth for all $t \in [t_0 - \varepsilon, t_0 + \varepsilon]$, where ε is some positive number.* Hence, there is a unique (unit-speed) geodesic through any given point of a surface in any given direction. An isometry between two surfaces takes the geodesics of one surface to the geodesics of the other.

The following well-known theorem allows us to describe the qualitative behavior of the geodesics on M^2.

Clairaut's theorem . *Let γ be a geodesic on a surface of revolution M^2, let ρ be the distance of a point of M^2 from the axis of rotation, and let ψ be the angle between γ and the meridians of M^2. Then, $\rho \sin \psi$ is constant along γ.*

We can change (7.14) into a system of first-order equations. First we set $x_1 = u$, $x_2 = v$, $x_3 = u'$, and $x_4 = v'$. Hence

$$x_1' = x_3, \quad x_2' = x_4, \quad x_3' = f_1(x_1, x_2, x_3, x_4), \quad x_4' = f_2(x_1, x_2, x_3, x_4). \quad (7.15)$$

Now we create a function to implement this system:

```
function f = sgeod(r)
    syms u v;
    ru = diff(r, u);  rv = diff(r, v);                        % these 2 lines
    E = ru*ru';  F = ru*rv';  G = rv*rv';                     % can be replaced by
    % S = EFG(r);  E = S(1);  F = S(2);  G = S(3);            % the following line:
    g = E*G - F*F;
    Eu = diff(E, u);  Ev = diff(E, v);
    Fu = diff(F, u);  Fv = diff(F, v);
    Gu = diff(G, u);  Gv = diff(G, v);
    f(1)= simplify(-subs((G*Eu - F*(2*Fu - Ev))/(2*g)*'x(3)*x(3)'+ (G*Ev - F*Gu)/
        g*'x(3)*x(4)' +G*(2*Fv - Gu) - F*Gv)/(2*g)*'x(4)*x(4)', {u v}, {'x(1)', 'x(2)'}));
    f(2) = simplify(-subs((E*(2*Fu - Ev) - F*Eu)/(2*g)*'x(3)*x(3)'+ (E*Gu - F*Ev)/
        g*'x(3)*x(4)'+(E*Gv - F*(2*Fv - Gu))/(2*g)*'x(4)*x(4)', {u v}, {'x(1)', 'x(2)'}));
end
```

For example:

```
syms u v real;
r = [u*cos(v), u*sin(v), u];                                 % cone
r = [(5 + cos(u))*cos(v), (5 + cos(u))*sin(v), sin(u)];      % torus
```

We run the M-file **sgeod.m** to obtain the equations of the geodesics on a surface.

sgeod(r)   % Answer: $[f_1(x_1,x_2,x_3,x_4),\ f_2(x_1,x_2,x_3,x_4)]$ of (7.15)

$[2x_3x_4\tan x_2,\ -\frac{1}{2}\sin(2x_2)x_3^2]$ % f_1, f_2 for a sphere

$[0, 0]$ % f_1, f_2 for a cylinder

$[-x_4^2\sin x_1(\cos x_1+5),\ 2x_3x_4\sin x_1/(\cos x_1+5)]$ % f_1, f_2 for a torus

$[x_1x_4^2,\ -2x_1x_3x_4/(x_1^2+1)]$ % f_1, f_2 for a helicoid

$[\frac{1}{2}x_1x_4^2,\ -2x_3x_4/x_1]$ % f_1, f_2 for a cone

Then we copy–paste f_1, f_2 from the above answer (replace "," by ";"), create an inline function with $[x_3; x_4; f_1; f_2]$, and call ode45 to get a solution (see also examples of numerical solutions in Section 5.1.3).

```
Eqs = inline('[x(3); x(4); x(1)*x(4)^2/2; -2*x(3)*x(4)/x(1)]', 't', 'x')    % for a cone
x0 = [1, 10, 1, 3];  t1 = 6;                    % enter initial data and t1
[t, X] = ode45(Eqs, [0, t1], x0);        % numeric solution of Cauchy's problem
```

Finally, we find a curve $\tilde{\gamma}$: $\mathbf{r}(u(t),v(t)) \subset \mathbb{R}^3$ and plot it on the surface:

```
r1 = subs(r, {u v}, {X(:,1), X(:,2)});
hold on;  plot3(r1(:,1), r1(:,2), r1(:,3));
ezmesh(r(1), r(2), r(3), [0, 1.4, 0, 2*pi]);  view(3);           % Figure 7.15(c)
```

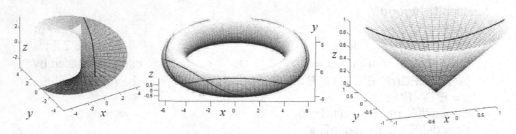

Fig. 7.15 Geodesics on surfaces: (a) helicoid, (b) torus, (c) cone.

7.6 Exercises

Section 7.1

---

1. Prove that each of the following conditions is equivalent to the property of linear independence of the vectors $\mathbf{r}_u(u,v)$ and $\mathbf{r}_v(u,v)$:

(a) The matrix $\begin{pmatrix} x_u & y_u & z_u \\ x_v & y_v & z_v \end{pmatrix}$ at the point (u,v) has rank 2.

(b) The cross product $\mathbf{n} = \mathbf{r}_u \times \mathbf{r}_v$ is nonzero at the point (u,v).

2. Prove that the vector-valued function

(i) $\mathbf{r}(u,v) = [u+v, u-v, u^2+v^2]$ is a C^∞-regular parameterization of the paraboloid $z = \frac{1}{2}(x^2+y^2)$,

(ii) $\mathbf{r}(u,v) = [u^2, uv, v^2]$, where $u > 0$, $v > 0$, defines a C^∞-regular coordinate system on part of the hyperbolic paraboloid $y^2 = xz$.

3. Find which of the following surfaces are compact, and plot them using the command **impl** of Section A.8.

$$\text{(a) } x^2 - y^4 + z^6 = 1, \qquad\qquad \text{(b) } x^2 - 2x + y^2 + z^4 = 1,$$
$$\text{(c) } x^2 + y^2 z^2 = 1, \qquad\qquad \text{(d) } x^2 + y^4 + z^6 = 1.$$

```
f = inline('x.^2 - y.^4 + z.^6 - 1', 'x', 'y', 'z');              % (a)
f = inline('x.^2 - 2*x + y.^2 + z.^4 - 1', 'x', 'y', 'z');        % (b)
f = inline('x.^2 + y.^2.*z.^2 - 1', 'x', 'y', 'z');              % (c)
f = inline('x.^2 + y.^4 + z.^6' - 1, 'x', 'y', 'z');             % (d)
impl(f, [-2, 2, -2, 2, -2, 2], 0)
```

4. Find a singular curve (generatrix) on a cylinder over a plane curve with one singular point, Figure 7.11(b).

5. Show that the level set of height c for the function $f(x)$ consists of roots of the equation $f(x) = c$. In such a way, $\mathbb{R}^3$ (or the plane) is fibered by level surfaces (or level lines) of F. The graph of $f(x,y)$ is the level surface of height 0 for the function $z - f(x,y)$ in three real variables.

6. One component of the *hyperboloid of two sheets* $\frac{x^2}{a^2} + \frac{y^2}{b^2} - \frac{z^2}{c^2} = -1$ is defined as the graph $z = c\sqrt{1 + \frac{x^2}{a^2} + \frac{y^2}{b^2}}$, and hence admits a parameterized form. On the whole, the *hyperboloid of two sheets* intersects "almost every" line at two points and hence cannot be a graph. Also, the following simple surfaces are not graphs: the circular cylinder, the hyperboloid of one sheet, the Möbius band.

7. Plot graphs of functions over the disk using the command cylinder:

```
z = 0 : 1/21 : 1;
[X, Y, Z] = cylinder(sqrt(z));  surf(X, Y, Z);                    % Figure 7.5(a)
```

Plot level lines of these functions using the command contour:

```
v = linspace(-pi, pi, 21);
[X, Y] = meshgrid(v);
contour(X.^2 + Y.^2)
contour(sin(X).*cos(Y))
contour(X.^2 + Y.^2)                                             % Figures 7.5(d–f)
```

8. Plot an implicitly given surface using **impl**. The method is coarse, as can be seen from the example of the cone $x^2 + y^2 - z^2 = 0$.

```
f = inline('x.^2 + y.^2 - z.^2', 'x', 'y', 'z');
impl(f, [-2, 2, -2, 2, -2, 2], 0),  axis equal
```

9. Plot both (up and down) sheets of a cone using the command ezsurf, starting with parametric equations.

```
x = 'u*cos(v)';  y = 'u*sin(v)';  z = 'u';
ezsurf(x, y, z);                              % Figure 7.16(a)
```

Plot the hyperboloid of one sheet:

```
ezsurf('sqrt(u.^2+1)*cos(v)', 'sqrt(u.^2+1)*sin(v)', 'u')    % Figure 7.16(c)
```

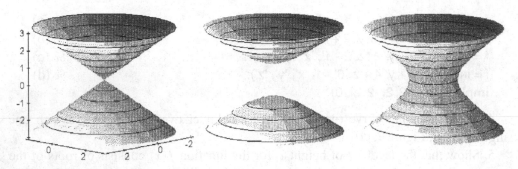

Fig. 7.16 The cone and hyperboloids.

10. Plot the graphs of the following complex expressions using the cplxmap command, namely, plot the magnitude of the function while coloring the resulting surface using the argument of the function. For example:

```
cplxroot(3), title('z^(1/3)');              % the Riemann surface for the cube root
cplxmap(z, mfun('gamma', z)), title('gamma(z)')       % the gamma function
```

11. Verify that the command bar3 produces diagrams or surfaces defined by finite lists or matrices.

```
A = hilb(6)
B = toeplitz([1, 2, 3, -3, -2, -1])
bar3(A + B)                                  % Figure 7.6(b).
```

Section 7.2   1. Write the equations of the tangent plane and the normal vector of the ellipsoid $\frac{x^2}{a^2} + \frac{y^2}{b^2} + \frac{z^2}{c^2} = 1$ at the point $P = (x_0, y_0, z_0)$, and plot the surface with the tangent plane and the normal vector at some P.

2. Prove that if the smooth surface M and the plane α have only one common point P, then α is the tangent plane to the surface at P.

3. Write equations of the tangent plane and the normal vector of surfaces satisfying
   (a) $\mathbf{r}(u,v) = [u+v, u-v, uv]$ at the point $M = (2,1)$,

(b) $\mathbf{r}(u,v) = [u\cos v, u\sin v, av]$ at an arbitrary point,

(c) $x^2 + 2y^2 - 3z^2 - 4 = 0$ at the point $M = (3,1,-1)$.

4. Prove that all tangent planes of the surface $z = x \cdot f(y/x)$ pass through the coordinate center. Plot an example.

5. Prove (using symbolic calculations) that the tangent planes of the surface $xyz = a^3$ form tetrahedra of constant volume with the three coordinate planes. Which property of tangent lines of the hyperbola does this problem generalize?

Hint. Denote by TM the tangent plane at the point (x,y,z) on the surface. Let x_1, y_1, z_1 be the segments of the intersection of TM with the coordinate axes:

```
syms x y z a X Y Z real;  F = x*y*z - a^3;
gr = [diff(F, x), diff(F, y), diff(F, z)];
TM = gr*[X - x, Y - y, Z - z]';
x1 = solve(subs(TM, {Y, Z}, {0, 0}), X);          % Answer: x1 = 3x
y1 = solve(subs(TM, {X, Z}, {0, 0}), Y);          % Answer: y1 = 3y
z1 = solve(subs(TM, {Y, X}, {0, 0}), Z);          % Answer: z1 = 3z
V = subs(x1*y1*z1, z, a^3/(x*y))                  % Answer: V = 27a^3.
```

The volume $V = 27a^3$ does not depend on the point on the surface.

6. Prove that if all normals on a surface pass through the same point, then this surface is a region on the sphere.

7. Prove that a one-sided surface cannot be represented as a graph $z = f(x,y)$ or in implicit form by $F(x,y,z) = 0$.

8. A curve on a surface $z = f(x,y)$ that forms the maximal angle with the plane XY at an arbitrary point is called a *maximal sloping curve*. Prove that a tangent line to such curve and a normal vector of the surface belong to the same plane that passes through the given point and is parallel to the z-axis.

9. Let $f(x,y) \in C^1$, $\nabla f \neq 0$, be a function. Prove that:

(a) maximal sloping curves of the surface $z = f(x,y)$ (see Exercise 8) and the curve of the intersection of this surface with the planes $\{z = \text{const}\}$ form an orthogonal net;

(b) these curves project onto an orthogonal net of the plane XY.

10. Solve the following conditional extremum problem: Find the dimensions of the cuboid of largest volume that can be fitted inside the ellipsoid $\frac{x^2}{a^2} + \frac{y^2}{b^2} + \frac{z^2}{c^2} = 1$, assuming that each edge is parallel to a coordinate axis.

Hint. Let $P = (x,y,z)$ be the vertex of the cuboid in the first octant, so the edges of the rectangular parallelepiped are $2x, 2y, 2z$. We therefore wish to optimize the volume function $V = 8xyz$ under the constraint $F(x,y,z) = 1$, where $F(x,y,z) = \frac{x^2}{a^2} + \frac{y^2}{b^2} + \frac{z^2}{c^2} = 1$. It is sufficient to solve the system $\text{grad}(V) = \lambda \cdot \text{grad}(F)$ in the variables x, y, z, and λ under the above constraint.

```
syms x y z a b c lambda;
V = 8*x*y*z;
F = x^2/a^2 + y^2/b^2 + z^2/c^2;
```

```
GradV = [diff(V, x), diff(V, y), diff(V, z)];
GradF = [diff(F, x), diff(F, y), diff(F, z)];
i = 1 : 3;  Eq = GradV(i) - lambda*GradF(i);
S = solve(F - 1, Eq(1), Eq(2), Eq(3), x, y, z, lambda);
S.x(4),  S.y(4),  S.z(4),  S.lambda(4)
```

$x_0 = \frac{1}{3}3^{1/2}a$, $y_0 = -\frac{1}{3}3^{1/2}b$, $z_0 = -\frac{1}{3}3^{1/2}c$, $\lambda_0 = \frac{4}{3}3^{1/2}abc$ % Answer

```
V0 = subs(V, {x, y, z}, {x0, y0, z0})
```
          % Answer: $V_0 = \frac{8}{9}3^{1/2}abc$

```
subs(F, {x, y, z}, {x0, y0, z0})
```
          % Answer: 1.

Conclusion: Taking the positive roots, find that the maximal cuboid has edges $x_0 = y_0 = z_0 = \frac{1}{\sqrt{3}}$ and volume $V_0 = \frac{8\sqrt{3}}{9}abc$.

11. Find extrema of the function $z = \frac{1}{3}x^3 + 9y^3 - 4xy$.

Conclusion: The function has the minimum $f_{min} = -\frac{64}{81}$ at $P_1 = (\frac{4}{3}, \frac{4}{9})$; the point $P_2 = (0,0)$ is a saddle.

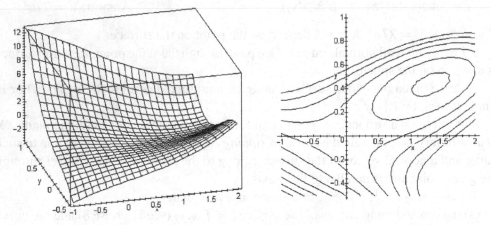

Fig. 7.17 Extrema of $z = \frac{1}{3}x^3 + 9y^3 - 4xy$: graph and the level lines.

12. Find extrema of the functions:

(a) $xy + 2x - \log(x^2y)$,   (b) $xy + \frac{1}{2x+2y}$,

(c) $\exp x + 2y(x^2 - y^2)$,   (d) $xy\log(x^2 + y^2)$,

(e) $\frac{x}{y} + \frac{1}{x} + y$.

13. Let $M \subset \mathbb{R}^3$ be a smooth surface and P_0 a point that does not belong to M. Prove that the shortest segment from P_0 to M (if it exists) is orthogonal to M. *Hint.* Use Theorem 7.2 about Lagrange multipliers for the *distance function* $g_{p_0}: Q \to \rho(P_0, Q)$.

| **Section 7.3** | 1. Check that $u = v = 0$ is a singular point of a certain type of the transformed Whitney umbrella $\mathbf{r}(u,v) = [(u+av)v, u+av, v^2]$. |

2. Show that the point $(0,0)$ on the surface $\mathbf{r}(u,v) = [u, v, (u^2 + v^2)^{1/3}]$, Figure 7.11(a), is a singular point:

```
r = 0 : 1/21 : 1;  t = -pi : 2*pi/41 : pi;
[R, T] = meshgrid(r, t);
X = R.*cos(T);  Y = R.*sin(T);  Z = R.^(2./3);
surf(X, Y, Z);
```

One may also obtain X, Y, Z by the command

```
[X, Y, Z] = pol2cart(T, R, R.^(2./3));
```

Hint. Use the program :

```
syms u v real;
r = [u, v, (u^2 + v^2)^(1/3)];
ru = diff(r, u);  rv = diff(r, v);
ruv = cross(ru, rv);
n1 = ruv/sqrt(ruv*ruv');
limit(limit(n1, u, 0), v, 0)                    % Answer:  [0 NaN 0]
```

3. Write two programs (M-files similar to **singDR**, Section 6.2) to detect singular points on parametric and implicitly given surfaces.

4. Plot the surface $z^2 - (x-y)^2(x+y) = 0$ and determine the type of the singular point $x = y = z = 0$.

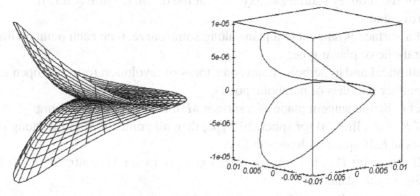

Fig. 7.18 *Wings* and the curve $\tilde{\gamma}$.

5. Prove that $O = (0,0)$ is a singular point of the surface (Figure 7.18(a))

$$\mathbf{r}(u,v) = [u^2 - v^2, 2uv, u^5].$$

```
r = linspace(0, 1, 41);
t = linspace(0, 2*pi, 61);
```

```
[R, T] = meshgrid(r, t);
[X, Y, Z] = pol2cart(R.^2.*cos(2.*T),  R.^2.*sin(2.*T),  (R.*sin(T)).^5);
surf(X, Y, Z);
```

Prove that while moving along the circle γ: $u = \rho \cos t$, $v = \rho \sin t$, the projection of a point of the curve (see Figure 7.18(b))

$$\tilde{\gamma}: \mathbf{r}(\rho \cos t, \rho \sin t) = [\rho^2 \cos(2t), \rho^2 \sin(2t), (\rho \sin t)^5]$$

onto the plane XY wraps twice about the origin O.

Section 7.4

1. Show that all points (a) in the plane are of planar type; (b) on the cylinder or cone are of parabolic type; (c) on the ellipsoid (sphere) are of elliptical (spherical) type. This explains the terminology.

2. Show that a torus of revolution (with the z-axis) has points of the three main types. The parabolic points fill two parallels (with maximal and minimal height z) and divide the torus into two open sets: one with elliptical and another with hyperbolic points.

3. Study the types of points on a second-order surface.

4. Deduce the equation of the osculating paraboloid for the following surfaces at given points and plot them:

(a) For the ellipsoid $\frac{x^2}{a^2} + \frac{y^2}{b^2} + \frac{z^2}{c^2} = 1$ at the point $P = (0,0,c)$.
Answer: $\frac{z}{c} = \left(1 - \frac{1}{2}\left(x^2/a^2 + y^2/b^2\right)\right)$.
(b) For the hyperboloid $x^2 + y^2 - z^2 = 1$ at the point $P = (0,1,0)$.
(c) For the monkey saddle $z = 3xy^2 - x^3$ at the origin (Figure 8.26(a)).

5. Prove that

(a) if a surface is tangent to a plane along some curve, then each point on this curve is of parabolic or planar type;

(b) elliptical and hyperbolic points on a torus of revolution form two open sets with two boundary parallels of parabolic points.

6. Let α be the tangent plane of a surface M at P. Prove the following:

(a) If P is of elliptical (or spherical) type, then all points of M close enough to P lie on the same half-space with respect to α.

(b) If the point P is hyperbolic, there exist points on M arbitrarily close to P on either side of the plane α.

(c) If the point P is parabolic or planar, both of the above cases can occur (find examples).

7. Show that

(a) if the boundary of a surface belongs to some plane, then either this surface belongs to a plane or the surface has elliptical points.

(b) every compact surface has elliptical points.

8. Plot the results of the program (Section A.8) for several surfaces.

Section 7.5   1. Find the first/second fundamental forms of each surface (using **EFG.m, LMN.m**): sphere, cone, helicoid, catenoid, pseudosphere, torus.

2. Calculate the Gaussian, the mean, and the principal curvatures of

(a) a surface-graph $z = f(x,y)$, where $f \in C^2$,

(b) a surface of revolution of γ: $x = f(u)$, $y = 0$, $z = g(u)$ about the z-axis,

(c) a ruled surface $\mathbf{r}(u,v) = \mathbf{a}(u) + v\mathbf{b}(u)$.

3. For the pseudosphere:

(i) calculate the length of a parallel;

(ii) calculate its total area;

(iii) calculate the principal curvatures;

(iv) show that the Gaussian curvature is a negative constant.

4. Surfaces with $H = 0$ are called *minimal*. Show that

(a) the only minimal surfaces of revolution are a plane and a catenoid;

(b) among all ruled surfaces, only a plane and a helicoid are minimal.

5. Plot geodesics on a sphere, a torus, a helicoid, and a Möbius strip using **sgeod.m** (see Figure 7.15(a,b)).

6. Describe and plot four different geodesics on the hyperboloid of one sheet $x^2 + y^2 - z^2 = 1$ passing through the point $(1,0,0)$.

7. Show that geodesics on a circular cylinder are

(a) parallels (circles),

(b) meridians (straight lines),

(c) helices.

Plot geodesics of each type. If P and Q are distinct points of a circular cylinder, there are either two or infinitely many geodesics joining P and Q. Which pairs P, Q have the former property?

8. Describe the geodesics on a circular cone.

9. What kind of a surface of revolution has the property that every parallel is a geodesic?

8

Examples of Surfaces

In this chapter we will study three very important and commonly occurring classes of surfaces (algebraic surfaces, surfaces of revolution, and ruled surfaces) and envelopes of surfaces. In Section 8.1 we will study surfaces of revolutions and plot the Möbius band and a map on a torus that cannot be colored with six colors. In Section 8.2 we will develop an M-file (the *resultant*) to deduce the polynomial equations of surfaces. In Section 8.3 we will study ruled surfaces of various types and calculate their striction curves and the distribution parameter. In Section 8.4, using the notion of a tangent plane (see Section 7.2), we will plot some envelopes of families of surfaces (see also Section 6.1 for curves).

8.1 Surfaces of Revolution

The *cylinder, cone, sphere, torus, paraboloid, hyperboloids*, and *ellipsoid with two equal axes* are simple examples of surfaces of revolution. Many examples occur in the real world.

Let α be a plane in $\mathbb{R}^3$, g a line in α, and γ a curve in α. If α rotates in $\mathbb{R}^3$ about g, the set M obtained from γ is called the *surface of revolution* with *axis* g generated from the *profile* curve γ. The curve of intersection of M with any plane through the axis of revolution (in particular, the profile curve) is called a *meridian*, and the curve of intersection of M with any plane orthogonal to the axis of revolution (a circle, a point, or the empty set) is called a *parallel*.

Note that all normals to a surface of revolution intersect the same line (the axis of revolution); in fact, surfaces of revolution (and their parts) are characterized by this property.

V. Rovenski, *Modeling of Curves and Surfaces with MATLAB®*,
Springer Undergraduate Texts in Mathematics and Technology 7, DOI 10.1007/978-0-387-71278-9_8,
© Springer Science+Business Media, LLC 2010

Deleting any meridian from a surface of revolution of an elementary curve that does not intersect the axis yields an elementary surface. For convenience we choose rectangular coordinates such that α coincides with the plane XZ and such that the line g coincides with the z-axis, and we denote the angle of rotation by $u \in [0, 2\pi)$. Suppose that the curve γ is given in the parametric form $x = f(u) \geq 0$, $z = g(u)$, where $u \in I$. If the point $P = (x_0, 0, z_0)$ belongs to the profile curve, a parallel through it is given by the equations $[x_0 \cos v, x_0 \sin v, z_0]$. Thus the surface of revolution M can be defined by the equations

$$\mathbf{r}(u, v) = [f(u) \cos v, f(u) \sin v, g(u)], \qquad 0 \leq v < 2\pi, \qquad (8.1)$$

which are called the *standard parametric equations* of a surface of revolution. The coordinate net consists of two families of curves: parallels and meridians. Substituting functions $f(u)$, $g(u)$ into (8.1) yields various examples of surfaces of revolution about the z-axis.

Example. The surface of revolution of the curve γ: $\mathbf{r}(u) = [u, 0, 4 + \sin(2u)]$ about the z-axis is given in Figure 8.1(a) together with its parallels and meridians.

```
syms u v;
ezsurf((4 + sin(2*u))*cos(v), (4 + sin(2*u))*sin(v), u, [0, 4*pi, 0, 2*pi])
```

A single command cylinder graphs a surface of revolution of a curve $x = f(z), 0 \leq z \leq 1$, in the xz-plane.

```
Z = linspace(0, 1, 41);
cylinder(4 + sin((4*pi)*2*Z))                        % Figure 8.1(a)
```

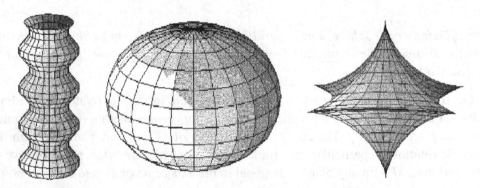

Fig. 8.1 (a) Revolving $[t, 0, 4 + \sin(2t)]$, (b) sphere, (c) astroidal sphere.

The reader can use the following program to plot various surfaces of revolution and their profile curves in rotation.

```
f = @(t) cos(t);  g = @(t) sin(t);  u0 = -pi/2;  u1 = pi/2;        % data (half-circle)
u = linspace(u0, u1, 41);  v = linspace(0, 2*pi, 41);
plot(f(u), g(u)),  axis equal                                      % plane curve
[U, V] = meshgrid(u, v);
X = f(U).*cos(V);    Y = f(U).*sin(V);    Z = g(U);
N = 24;  for n = 0 : N;                                            % preparing animation
mesh(X, Y, Z);  hold on;  axis equal;
plot3(f(u)*cos(n*2*pi/N), f(u)*sin(n*2*pi/N),  g(u));
hold off;  pause(0.2);
end                                                                % surface with rotating profile curve
```

	M	$f(u)$	$g(u)$	Domain $G(u,v)$
1	cylinder	R	u	$v \in [0,2\pi),\, u \in \mathbb{R}$
2	cone	u	$ku \ (k \neq 0)$	$v \in [0,2\pi),\, u \in \mathbb{R}$
3	sphere	$R\sin u$	$R\cos u$	$v \in [0,2\pi),\, u \in [0,\pi]$
4	2-hyperboloid	$R\sinh u$	$R\cosh u$	$v \in [0,2\pi),\, u \in \mathbb{R}$
5	1-hyperboloid	$R\cosh u$	$R\sinh u$	$v \in [0,2\pi),\, u \in \mathbb{R}$
6	catenoid	$a\cosh(\frac{u}{a})$	u	$v \in [0,2\pi),\, u \in \mathbb{R}$
7	pseudospere	$a\sin u$	$a(\cos u + \log\tan(\frac{u}{2}))$	$v \in [0,2\pi),\, u \in (0,\pi)$
8	torus	$R+r\cos u$	$r\sin u \ (R > r)$	$u,v \in [0,2\pi)$

1. Revolving a line which is parallel to the axis yields the *circular cylinder:*

   f = @(t) 1+t-t; g = @ (t) t; u0=0; u1=2; % $\mathbf{r} = [R\cos v, R\sin v, u]$, Figure 8.1(b)

2. Revolving a line which intersects the axis yields the *cone:*

   f = @(t) t; g = @ (t) t; u0=-1; u1=1;   % $\mathbf{r} = [u\cos v, u\sin v, ku]$, Figure 7.16(a)

3. Revolving a half-circle about its diameter yields the *sphere:*

$$\mathbf{r} = [R\sin u\cos v, R\sin u\sin v, R\cos u] \qquad \text{(Figure 8.1(b))}.$$

The coordinates u and v on this sphere are the *geographic coordinates* longitude and latitude. By deleting one meridian (say, the *"date line"*), we obtain an elementary surface.

4. Revolving a hyperbola yields a *hyperboloid of two sheets* (considered as a *sphere of imaginary radius* in a pseudo-Euclidean space $\mathbb{R}_1^3$):

$$\mathbf{r} = [R\sinh u\cos v, R\sinh u\sin v, \pm R\cosh u] \qquad \text{(Figure 7.16(b))}.$$

5. Revolving one branch of a hyperbola yields a *hyperboloid of one sheet:*

$$\mathbf{r} = [R\cosh u\cos v, R\cosh u\sin v, R\sinh u] \qquad \text{(Figure 7.16(c))}.$$

6. Revolving a catenary yields a *catenoid:*

$$\mathbf{r} = [a\cosh(u/a)\cos v,\, a\cosh(u/a)\sin v,\, u] \qquad \text{(Figure 8.2(b))}.$$

7. Revolving a tractrix about its asymptote yields a *pseudosphere:*

$$\mathbf{r} = [a\sin u\cos v,\, a\sin u\sin v,\, a(\cos u + \log(\tan(u/2)))] \qquad \text{(Figure 8.2(a))}.$$

8. Revolving a circle yields a *torus:*

$$\mathbf{r} = [(R+r\cos u)\cos v,\, (R+r\cos u)\sin v,\, r\sin u] \qquad \text{(Figure 7.1(b))}.$$

f = @(t) 2 + cos(t); g = @(t) sin(t); u0 = -pi; u1 = pi;   % torus

This table of surfaces of revolution (above) can be extended.

If the profile γ is given implicitly by $F(x,z) = 0$, then the surface of revolution about the z-axis has the equation, in rectangular or cylindrical coordinates,

$$F\left(\sqrt{x^2+y^2},z\right) = 0, \qquad\qquad F(\rho,z) = 0. \qquad (8.2)$$

In particular, if the profile curve is the graph $z = h(x)$, then the surface of revolution is also the graph in rectangular or cylindrical coordinates:

$$z = h\left(\sqrt{x^2+y^2}\right), \qquad\qquad z = h(\rho). \qquad (8.3)$$

In the case of parametric equations $\mathbf{r} = [u\cos v,\, u\sin v,\, h(\sqrt{u^2+v^2})]$, the coordinate net consists of parallels and meridians. The whole parallel through a singular point of γ consists of singular points (see Figure 8.2(a,c)). If the profile curve γ has no singular points and does not intersect the axis, then the surface of revolution also has no singular points (see Figure 8.2(b)).

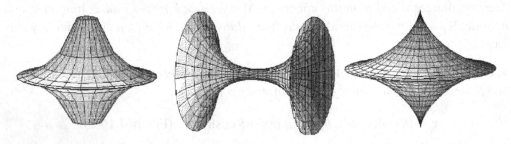

Fig. 8.2 Pseudosphere,   catenoid,   and   revolving astroid.

A *screw motion* of space is the composition of a revolution about some axis and a parallel displacement along the same axis. Using this map, we obtain the following generalization of a surface of revolution.

If a plane curve $\gamma(u)$ rotates about some axis in its plane and at the same time moves uniformly in the direction of this axis, then we obtain the *screw surface* with generatrix $\gamma(u)$. If the generatrix is a straight line, then the screw surface is called a *helicoid*, which is also a particular case of a ruled surface (see Section 8.3). A screw surface with generatrix $\gamma \colon [f(u), g(u)]$ is defined by the equation (where $h = \text{const}$)

$$\mathbf{r}(u,v) = [f(u)\cos v, \, f(u)\sin v, \, g(u) + hv] \qquad (0 \le v < 2\pi, \, u \in I).$$

Examples.

1. An example of a screw surface is the *twisted sphere*, Figure 8.25(c),

$$\mathbf{r}(u,v) = [R\sin u \cos v, \, R\sin u \sin v, \, R\cos u + hv] \qquad (|v| \le \pi, \, |u| \le \pi).$$

2. **Möbius band.** Turn the line segment $I_0 = [R, 0, u]$, where $|u| \le 1$, $R = \text{const} > 1$, about its center $(R, 0, 0)$ through the angle $\frac{v}{2} \in [0, \pi]$ in the plane XZ: $I_{v/2} = [R + u\sin(v/2), 0, u\cos(v/2)]$, and then rotate the line segment $I_{v/2}$ about the z-axis through the angle $v \in [0, 2\pi]$. We obtain a wonderful surface, Figure 7.3(b):

$$\mathbf{r}(u,v) = [(R + u\sin(v/2))\cos v, \, (R + u\sin(v/2))\sin v, \, u\cos(v/2)].$$

```
u = linspace(-1, 1, 11);  v = linspace(0, 2*pi, 31);
[U, V] = meshgrid(u, v);
X = (5 + sin(V/2).*U).*cos(V);  Y = (5 + sin(V/2).*U).*sin(V);  Z = cos(V/2).*U;
surf(X, Y, Z),  axis equal;
```

The n-times-turned (through $180°$) twisted strip has the similar equations

$$\mathbf{r}(u,v) = [(R + u\sin(nv/2))\cos v, \, (R + u\sin(nv/2))\sin v, \, u\cos(nv/2)].$$

```
n = 3;  u = linspace(-1, 1, 11);  v = linspace(0, 2*pi, 31);
[U, V] = meshgrid(u, v);
X = (5 + sin(n*V/2).*U).*cos(V);  Y = (5 + sin(n*V/2).*U).*sin(V);
Z = cos(n*V/2).*U;
surf(X, Y, Z),  axis equal;
```

The surface is homeomorphic to the cylinder for even n, and it is homeomorphic to the Möbius band and has one side for odd n.

3. The *four color map theorem* states that, given a plane separated into regions, the regions may be colored using no more than four colors in such a way that no two

adjacent regions receive the same color. One can also consider the coloring problem on surfaces other than the plane. One may show that

(a) the problem for a sphere is equivalent to that of the plane;
(b) seven colors are sufficient to color a map on a torus.

Let us plot the torus divided into the seven regions, each one of which touches every other (Figure 8.3).

```
r1 = 3;  r2 = 1;  a = 1;
syms u v uk vk;
x = (r1 + r2*cos(u))*cos(v);  y = (r1 + r2*cos(u))*sin(v);  z = r2*sin(u);
xk = (r1 + r2*cos(uk))*cos(vk+a*uk);  yk = (r1 + r2*cos(uk))*sin(vk+a*uk);
zk = r2*sin(uk);
r = subs([xk, yk, zk], vk, 2*pi/7*(7-1));
ezplot3(r(1), r(2), r(3));                  % boundary between 6 and 7 regions
hold on;  axis equal;  view(3);  ezmesh(x, y, z)              % torus
for k = [1 3 5];  ezsurf(xk, yk, zk, [0, 2*pi 2*pi/7*(k-1), 2*pi/7*k])  end;
```

The two-dimensional map (scheme) is the following:

```
u = linspace(0, 5, 41);
hold on;  axis equal;  axis([0 14 0 5]);
plot([0, 14, 14, 0, 0],  [0, 0, 5, 5, 0], 'r', 'LineWidth', 3);          % rectangle
for i = 1 : 9;  plot(u + 2*i - 6, u);  end;
```

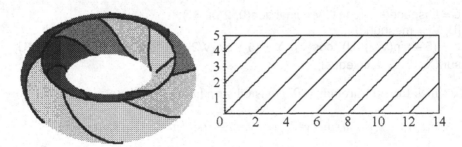

Fig. 8.3 Seven countries on the torus   and   the two-dimensional scheme.

4. Slowly changing the equations of a torus (using periodic flattening of the circle to a line segment) leads to a *pinched torus* (Figure 7.1(c)):

```
X = '(1 + .2*sin(u))*cos(v)';  Y = '(1 + .2*sin(u))*sin(v)';  Z = '.2*sin(v)*cos(u)';
ezsurf(X, Y, Z,  [0, 2*pi, 0, 2*pi]),  axis equal;
```

8.2   Algebraic Surfaces

A large class of surfaces is obtained by using polynomials in three variables.

A surface is called *algebraic* if it can be given by an equation $F(x,y,z) = 0$, where F is a polynomial in variables x, y, z. The degree of F is called the *order* of the surface. If F is the product of nonconstant polynomials, then the surface is called *decomposable*. A surface that is not algebraic is called a *transcendental surface*.

We will show that *if a profile curve is algebraic, then the surface of revolution is also algebraic.*

Indeed, if $f_n(x,z) = 0$, where f_n is a polynomial of order n, is the equation of the profile curve γ, then the surface of revolution (about the z-axis) has the equation $f_n(r,z) = 0$, where $r^2 - x^2 - y^2 = 0$.

Using the M-file **Resfg** (the *resultant* — see Sections 1.5, 6.2), we eliminate r from the system $f_n(r,z) = 0$, $r^2 - x^2 - y^2 = 0$ and obtain the polynomial equation of the surface $F_m(x,y,z) = 0$, where the degree $m \leq n + 2$.

Examples.

1. Any line intersects an algebraic surface of order n at no more than n distinct points, unless it is completely contained in the surface.

Hint. Substituting parametric equations of an arbitrary line $x = a_1 + tb_1$, $y = a_2 + tb_2$, $z = a_3 + tb_3$ in the equation of the surface, we obtain the polynomial $F(a_1 + tb_1, a_2 + tb_2, a_3 + tb_3) = \sum_{0 \leq i \leq n} d_i t^i$ in t of order not more than n. If all coefficients of this polynomial are zero, then the line belongs to the surface. Otherwise, the polynomial has at most n real roots, which correspond to the points of intersection of the line with the surface. (Multiple roots correspond to points of tangency or singular points.)

2. The *cylinder* over (or *revolving* of) a sine curve is a transcendental surface. A *plane* in space is defined by the linear equation $ax + by + cz + d = 0$, hence it is an algebraic surface of the first order. *Surfaces of the second order* (the sphere and the ellipsoid, elliptic and hyperbolic paraboloids, hyperboloids of one and two sheets, the elliptic cone, cylinders over curves of second order) are studied in *analytical geometry*.

3. The *swallow-tail* surface $\mathbf{r} = [u, -4v^3 - 2uv, 3v^4 + uv^2]$, Figure 8.4, is a fifth-order algebraic surface.

```
syms u v;
ezsurf(u, -4*v^3 - 2*u*v,  3*v^4 + u*v^2,  [-4, 4, -1, 1]),  axis equal;
```

Using **Resfg** (the *resultant* of two polynomials — see Section 6.2) and setting $x = u$, eliminate the variable v from the system of equations $4v^3 + 2xv + y = 0$, $3v^4 + xv^2 - z = 0$ and obtain a polynomial in x, y, z of the fifth order:

```
syms u v x y z;
u = x;
f = 4*v^3+2*u*v+y,  g = 3*v^4+u*v^2-z    % f = 4v^3 + 2xv + y, g = 3v^4 + xv^2 − z
a = [4, 0, 2*x, y];  b = [3, 0, x, 0, -z];    % coefficients of polynomials in v
F = det(Resfg(a, b))                          % implicit equation is F = 0
```

$$F = 4x^3y^2 + 27y^4 - 16x^4z + 128x^2z^2 - 144xzy^2 - 256z^3 \qquad \text{\% Answer.}$$

Next, show that the swallow-tail surface divides the space of polynomials $\mathbb{R}^3(x,y,z) = \{t^4 + xt^2 + yt + z\}$ into three domains, in one of which the polynomials (looking like a pyramid) have four real roots, in the next, two roots, and in the last domain, non-real roots.

Solution. If the polynomial $F(t;x,y,z)$ has a multiple root t, then the system $F = 0$, $F'_t = 0$ has a solution. Computing $F'_t = 4t^3 + 2xt + y$ and eliminating t from the system is equivalent to applying the command **Resfg** (*resultant*) for F and F'_t as polynomials in t.

```
syms x t;
at = [1, 0, x, y, z];              % coefficients of f_t = t^4 + xt^2 + yt + z
bt = [4, 0, 2*x, y];               % coefficients of g_t = 4t^3 + 2xt + y
d = det(Resfg(a1, b1))             % We obtain that d = −F
```

$$d = -4x^3y^2 - 27y^4 + 16x^4z - 128x^2z^2 + 144xzy^2 + 256z^3 \qquad \text{\% Answer}$$

```
simplify(subs(d, {'y', 'z'},  {'-4*t^3 - 2*x*t',  '3*t^4 + x*t^2'}))    % obtain 0.
```

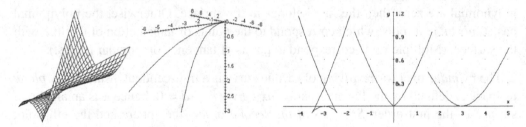

Fig. 8.4 Swallow-tail surface, its cuspidal edge,   sections by the planes $x = c$.

4. In the *revolving of an astroid* (an algebraic curve of the sixth order), eliminate r from the system of equations $(r^2 + z^2 - a^2)^3 + 27a^2z^2r^2 = 0$, $r^2 - x^2 - y^2 = 0$ and obtain $((x^2 + y^2) + z^2 - a^2)^3 + 27a^2z^2(x^2 + y^2) = 0$, i.e., a surface of the sixth order.

```
syms x y z r a;
fr = r^2 - x^2 - y^2;    ar = [1, - x^2 - y^2];    % polynomial in r^2
gr1 = collect(expand(gr), r)
```

$$gr_1 = r^6 - r^4(3a^2 - 3z^2) + r^2(3a^4 + 21a^2z^2 + 3z^4) - 3a^2z^4 + 3a^4z^2 - a^6 + z^6$$

```
br = [1, -3*a^2 + 3*z^2, 3*a^4 + 21*a^2*z^2 + 3*z^4,
    -3*a^2*z^4 + 3*a^4*z^2 - a^6 + z^6];
d1 = simple(det(Resfg(ar, br)))
```

$$d_1 = 3y^2z^4 + 3y^4z^2 + x^4(3y^2 - 3a^2 + 3z^2) + x^2(3a^4 - 6a^2y^2 + 21a^2z^2 + 3y^4 + 6y^2z^2 + 3z^4) - a^6 + x^6 + y^6 + z^6 - 3a^2y^4 + 3a^4y^2 - 3a^2z^4 + 3a^4z^2 + 21a^2y^2z^2$$

```
d2 = ((x^2 + y^2) + z^2 - a^2)^3 + 27*a^2*z^2*(x^2 + y^2);
simplify(d1 - d2)                                          % obtain 0.
```

$$((x^2 + y^2) + z^2 - a^2)^3 + 27a^2z^2(x^2 + y^2) = 0 \qquad\text{% Answer}$$

5. In the *revolving of a circle*, eliminate r from the system $(r-a)^2 + z^2 - b^2 = 0$, $r^2 - z^2 - y^2 = 0$ (with $a > b > 0$) and obtain a polynomial that defines a torus of revolution:

```
syms r a b x y z;
fr = (r - a)^2 + z^2 - b^2;    gr = r^2 - x^2 - y^2;
fr1 = collect(expand(fr), r)              % fr1 = a^2 - 2ar - b^2 + r^2 + z^2
```
$$\text{% } fr_1 = a^2 - 2ar - b^2 + r^2 + z^2$$
```
ar = [1, -2*a, a^2 - b^2 + z^2];  br = [1, 0, -x^2 - y^2];   % polynomials in r
F = simple(det(Resfg(ar, br)))
```

$$F = 2y^2z^2 - x^2(2a^2 + 2b^2 - 2y^2 - 2z^2) + a^4 + b^4 + x^4 + y^4 + z^4 \\ -2a^2b^2 - 2a^2y^2 + 2a^2z^2 - 2b^2y^2 - 2b^2z^2$$

It has the short form $F_1 = (a^2 - b^2 + x^2 + y^2 + z^2)^2 - 4a^2(x^2 + y^2)$.

```
F1 = (a^2 - b^2 + x^2 + y^2 + z^2)^2 - 4*a^2*(x^2 + y^2);
simplify(F - F1)                                          % obtain 0
```

8.3 Ruled Surfaces

We consider some important classes of surfaces that can be constructed by moving one curve in space, the *ruling*, along another curve, the *directrix*.

A surface M is called *ruled* if it is generated by a one-parameter family of lines (*rulings*). A space curve on a ruled surface that intersects each ruling at one point is called a *directrix curve*. For each point of some directrix $\rho(u)$ ($u \in I$), we define the unit vector $\mathbf{a}(u)$ that is parallel to the ruling g_u through this point. As a result, we obtain the *standard parametric equations* of a ruled surface

$$\mathbf{r}(u,v) = \rho(u) + v\mathbf{a}(u) \qquad (u \in I, \ v \in \mathbb{R}), \tag{8.4}$$

where the v-curves coincide with the rulings, and the parameter v is equal (up to sign) to the segment of the ruling between the point $\mathbf{r}(u,v)$ and the directrix $\rho(u)$; see Figure 8.5(a).

Particular cases of ruled surfaces are the following:

- the *cylindrical surface* $\mathbf{r}(u,v) = \rho(u) + v\mathbf{a}_0$: the vector-valued function $\mathbf{a}(u) = \mathbf{a}_0$ is constant and represents the direction of the *axis*;
- the *conic surface* $\mathbf{r}(u,v) = \rho_0 + v\mathbf{a}(u)$: the vector-valued function $\rho(u) = \rho_0$ is constant and represents the *vertex* (the singular point);
- the *tangent developable surface* $\mathbf{r}(u,v) = \rho(u) + v\rho'(u)$: all rulings are tangent to the directrix curve, i.e., $\mathbf{a}(u) = \rho'(u)$.

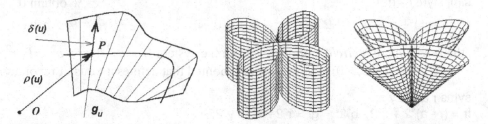

Fig. 8.5   Ruled surface.   Cylinder and cone over a four-leafed rose.

A ruled surface with the condition $\mathbf{a}'(u) \neq 0$ is called *non-cylindrical*. A non-cylindrical ruled surface whose rulings are parallel to some fixed *directrix plane* is a *Catalan surface*. A Catalan surface is called a *conoid* if all its rulings intersect a constant line (the *axis of the conoid*). A *conoid* is *right* if its axis is orthogonal to the directix plane.

A conoid is generated by a line that moves along a fixed line orthogonal to it (the axis of the conoid), and at the same time rotates about this line. If the velocity of rotation is proportional to the lifting velocity of the ruling, then the conoid is called a *helicoid*. One can distinguish a *right-side* and a *left-side* helicoid. The simplest conoid is the hyperbolic paraboloid. It is defined by moving a line that is parallel to a fixed plane and is guided by two fixed helices (has two axes!).

Examples.

1. The *swallow-tail* algebraic surface, Figure 8.4, is at the same time a ruled surface: $\mathbf{r} = [0, -4v^3, 3v^4] + u[1, -2v, v^2]$ (replace $u \leftrightarrow v$ to obtain the standard representation).

2. The Catalan surfaces are characterized among all ruled surfaces by the conditions $(\mathbf{a}, \mathbf{a}', \mathbf{a}'') = 0$, $\mathbf{a}'' \neq 0$.

Hint. Observe that the three vectors $\mathbf{a}, \mathbf{a}', \mathbf{a}''$ are parallel to the directrix plane.

3. One can construct a right helicoid as a particular case of the following surface. A *helicoid* is a surface obtained from a line that rotates about a fixed axis with constant angular velocity, intersects the axis at the constant angle θ, and at the same time has

translational motion (with constant velocity) along the axis. If $\theta = 90°$, then the helicoid is *right* (Figure 8.8(a)), and if $\theta \neq 90°$, then it is *skew:*

$$\mathbf{r}(u,v) = [0,0,cu] + av[\cos u, \sin u, 0] = [av\cos u, av\sin u, cu].$$

A visual demonstration of separate rays on this surface gives the steps of a spiral staircase. A generalization of the right helicoid is the *elliptical helicoid* $\mathbf{r}(u,v) = [av\cos u, bv\sin u, cu]$. The following surface (*the conoid*), of Figure 8.10(c), looks like the helicoid when $a = 1, c = 0.1$, but the velocity of rotation of the ruling is exponential:

$$\mathbf{r}(u,v) = [0,0,\exp(cu)] + av[\cos u, \sin u, 0] = [av\cos u, av\sin u, \exp(cu)].$$

```
ezsurf('v*cos(u)', 'v*sin(u)', '2*u')                    % Figure 8.8(a)
```

4. The *conoid of Wallis* (for example, with $u \in [0,1]$, $v \in [0,\pi]$, shown in Figure 8.6(b)) is given by $\mathbf{r}(u,v) = [v\cos u, v\sin u, c\sqrt{a^2 - b^2\cos^2(u)}]$.

```
syms u v;   r = [v*cos(u),  v*sin(u),  sqrt(1 - 3*cos(u)^2)];
ezsurf(r(1), r(2), r(3), [-.99 .99], [0, pi])    % a = c = 1, b = √3, Figure 8.6(b)
```

5. *Plücker's conoid* has the simple equation $z = \frac{2xy}{x^2+y^2} \iff \mathbf{r}(x,y) = [x,y,\frac{2xy}{x^2+y^2}]$, but in this case we do not see the rulings in the picture.

```
syms x y;   r = [x, y, 2*x*y/(x^2+y^2)];   ezsurf(r(1), r(2), r(3),  [-1, 1],  [-1 1])
```

Using *cylindrical coordinates* $x = u\cos v$, $y = u\sin v$, $z = z$ allows us to write other parametric equations of this surface

$$\mathbf{r}(u,v) = [0,0,\sin(2u)] + v[\cos u, \sin u, 0] = [v\cos u, v\sin u, \sin(2u)],$$

where the rulings are the v-curves,

```
syms u v;   n = 2;   r = [v*cos(u/n), v*sin(u/n), sin(u) ];      % define n = 2,3,...
ezsurf(r(1), r(2), r(3),  [0 1],  [0, 2*pi*n])          % n = 2, Figure 8.6(a)
```

We see that the surface is given by the rotation of the ray about the z-axis with simultaneous 2π-periodic oscillatory motion along the segment $[-1,1]$ of the z-axis. *Generalized Plücker's conoids* (having n folds instead of two) are obtained by the rotation of a ray about the z-axis, with simultaneous oscillatory motion (with period $2\pi n$) along the segment $[-1,1]$ of the z-axis (see Figure 8.7(a,b) for the cases of $n = 5, 8$):

$$\mathbf{r}(u,v) = [0,0,\sin(nu)] + v[\cos u, \sin u, 0] = [v\cos u, v\sin u, \sin(nu)].$$

A ruled surface is called *developable* if its tangent plane is the same at all points of any ruling, or equivalently, if the normal $\mathbf{n}$ preserves its direction (is stationary) along each ruling.

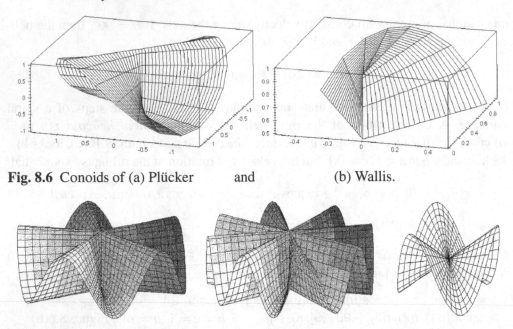

Fig. 8.6 Conoids of (a) Plücker   and   (b) Wallis.

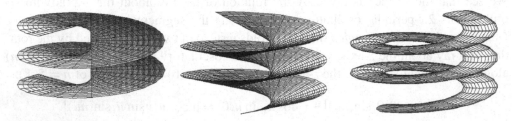

Fig. 8.7 Generalized Plücker's conoids for $n = 5, 8$ and   saddle-3.

Simple examples of developable ruled surfaces are cylinders and cones. A more complicated example is a tangent developable surface. To show this, find $\mathbf{r}_u = \rho'(u) + v\rho''(u)$ and $\mathbf{r}_v = \rho'(u)$. Hence, the normal $\mathbf{n}_1 \parallel \mathbf{r}_u \times \mathbf{r}_v = v\rho''(u) \times \rho'(u) \parallel \rho''(u) \times \rho'(u)$ preserves its direction as v varies.

For example, the tangent developable surface to the helix:

```
syms u v;   rho = [cos(u) sin(u) u];   r = rho + v*diff(rho, u);
n = simple(cross(diff(r, u), diff(r, v)))/v      % obtain n = [− sin(u), cos(u), −1]
```

The inverse statement also holds: *A developable ruled surface consists of cylindrical, conical, and tangent developable surfaces.*

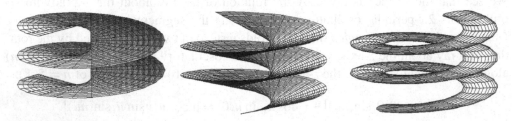

Fig. 8.8 Helicoid, tangent developable surface, a tube.

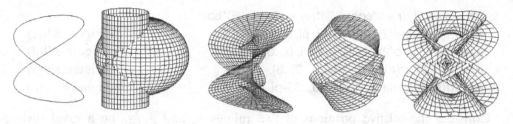

Fig. 8.9 Viviani curve and its normal, binormal, and tangent developable ruled surfaces.

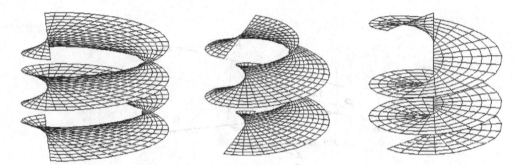

Fig. 8.10 Binormal ruled surface and $\mathbf{r}_{\pi/6}$ for a helicoid and exponential helicoid.

Example. Plot the *normal* and the *binormal surfaces*

$$\mathbf{r}_1(u,v) = \rho(u) + v\,v(u), \qquad \mathbf{r}_2(u,v) = \rho(u) + v\,\beta(u)$$

generated by the main normals $v(u)$ and binormals $\beta(u)$, respectively, to the given space curve $\rho(u)$. Plot the surfaces $\mathbf{r}_1(u,v)$ and $\mathbf{r}_2(u,v)$ for the helicoid (see the right conoid and Figure 8.10(a)) and for the Viviani curve (see Figure 8.9(c,d)). Plot the intermediate surfaces for some values of θ (see Figure 8.10(b)):

$$\mathbf{r}_\theta(u,v) = \rho(u) + v\cos(\theta)v(u) + v\sin(\theta)\beta(u).$$

Hint. Use the program

```
syms u v;  r = [cos(u)^2, cos(u)*sin(u), sin(u)];
ezplot3(r(1), r(2), r(3), [-pi, pi]);            % Viviani curve, Figure 8.9(a)
ru = diff(r, u);  ruu = diff(ru, u);
tt = ru./sqrt(ru*ru');
b = cross(ru, ruu);  bb = b./sqrt(b*b');
nn = cross(bb, tt);
rn = r + v*nn;  rt = r + v*tt;  rb = r + v*bb;
```

```
rth = simplify(r + v*cos(pi/6)*nn + v*sin(pi/6)*bb);
ezsurf(rn(1), rn(2), rn(3), [-pi, pi, 0, .3], 31);          % Figure 8.9(c)
ezsurf(rb(1), rb(2), rb(3), [-pi, pi, 0, .3], 31);          % Figure 8.9(d)
ezsurf(rt(1), rt(2), rt(3), [-pi, pi, 0, .3], 31);          % Figure 8.9(e)
ezsurf(rth(1), rth(2), rth(3), [0, .3, -pi, pi], 31);       % Figure 8.10(c)
```

Consider the relative positions of two rulings g_u and $g_{u+\Delta u}$ on a ruled surface $\mathbf{r}(u,v) = \rho(u) + v\mathbf{a}(u)$, where $|\mathbf{a}(u)| = 1$. If the surface is non-cylindrical, these rulings are skew lines in space. Let PP' be the *shortest segment* between rulings, where $P = (u,v)$, $P' = (u+\Delta u, v+\Delta v)$, and u, v are unknown (Figure 8.11). The bottom P of

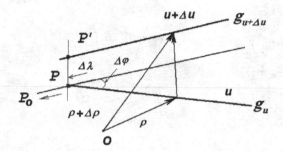

Fig. 8.11 Striction point on a ruling.

the perpendicular PP' tends along the ruling g_u to the limit value P_0 as $\Delta u \to 0$; P_0 is called the *point of striction* on the ruling g_u. It shows the narrowest (striction) place on the ruled surface near this ruling. The position vector of the striction point is

$$\mathbf{r}(u) = \rho(u) - \frac{\rho'(u) \cdot \mathbf{a}'(u)}{(\mathbf{a}'(u))^2} \mathbf{a}(u),$$

and so we have a *striction curve. It encircles the ruled surface at its narrowest, and is independent of the choice of the directrix curve* $\rho(u)$.

The ruling g_u comes into the new position $g_{u+\Delta u}$, with rotation through some angle $\Delta\varphi$ and with a deviation from the initial position at the distance $\Delta\lambda = |PP'|$. For a non-cylindrical ruled surface we have

$$p(u) = \lim_{\Delta u \to 0} (\Delta\lambda)/(\Delta\varphi) = (\mathbf{a}, \mathbf{a}', \rho')/|\mathbf{a}'|^2. \qquad (8.5)$$

The function (8.5) is called the *distribution parameter* $p(u)$ of the non-cylindrical ruled surface (8.4).

Examples.

1. The hyperboloid of revolution of one sheet $\frac{x^2+y^2}{a^2} - \frac{z^2}{c^2} = 1$ is a doubly ruled non-cylindrical surface. The striction curve of each of its families is the circle of its intersection with the plane XY. From the example we see that the striction curve does not intersect the rulings at right angles.

2. Find the striction curve and the distribution parameter of the hyperbolic paraboloid.

Solution. The parametric equations of the hyperbolic paraboloid $z = xy$ are $\mathbf{r}_u = [u,0,0] + v[0,1,u]$. We calculate the striction curve $\rho(u) = [u,0,0]$ and the function $\mathbf{a}(u) = [0,1,bu]/\sqrt{b^2+u^2}$.

```
syms b u v real;
d = [0, 1, b*u]/sqrt(1+(b*u)^2);
r = [u, 0, 0];
ru = diff(r, u);  du = diff(d, u);
A = [d; du; ru];
p = factor(det(A)/(du*du'))          % obtain p = (1+b²u²)/b
```

8.4 Envelope of a One-Parameter Family of Surfaces

Previously, surfaces have been defined as figures satisfying certain geometrical conditions or equations. In this section, surfaces appear as envelopes of one-parameter families of spheres, planes, and other surfaces.

A set of surfaces $\{M_t\}$ that depend on the parameter t is called a *one-parameter family of surfaces*. A smooth surface M is called an *envelope* of the family $\{M_t\}$ if it is tangent at each of its points to at least one surface from the family, and in each of its domains is tangent to an infinite number of surfaces from the family. Let $F(x,y,z;t)$ be a function of class C^1. One derives the envelope of a family of surfaces $\{M_t : F(x,y,z;t) = 0\}$ from the system of equations

$$F(x,y,z;t) = 0, \qquad\qquad F_t'(x,y,z;t) = 0 \qquad\qquad (8.6)$$

by elimination of t, similarly to (6.1). In fact, (8.6) defines the *discriminant set*, which includes extraneous solutions (analogous to the inflection points in a solution of the problem of extrema of a function $f(x)$ using the condition $f'(x) = 0$).

Examples.

1. It is known that the envelope of a one-parameter family of planes has the structure of a ruled developable surface.

2. The envelope of the family of spheres M_t: $(x-t)^2 + y^2 + z^2 - R^2 = 0$ (Figure 8.14(a)) is the circular cylinder with axis OX (Figure 8.14(b)).

3. It is easy to check that the envelope of the family of spheres of radius R with centers on the circle $\mathbf{r}(t) = [a\cos t, a\sin t, 0]$, where $a > R$, is a torus of revolution (Figure 8.12). This family M_t is defined by the equation $(x - a\cos t)^2 + (y - a\sin t)^2 + z^2 - R^2 = 0$.

4. Find the envelope of spheres $\varphi(x, y, t) = (x-t)^2 + y^2 + z^2 - R^2 = 0$ from Example 2, p. 338. First calculate $F_t' = 2(x-t) = 0$. From this obtain $t = x$, and then substitute into the equation $F = 0$ and obtain $y^2 + z^2 = R^2$. Hence, the envelope is the cylinder of radius R with axis OX (Figure 8.14(b)).

```
syms u v;
[X, Y, Z] = sphere;
hold on;  axis equal;  view(3);
ezmesh(v, cos(u), sin(u),  [-pi - 1, 0, -.5, 6]);
for i = 1 : 6;
  surf( .8*i + X,  Y,  Z);
end;
```

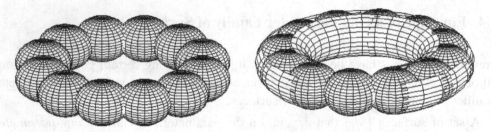

Fig. 8.12 The torus is the envelope of a family of spheres.

5. Plot the envelope from Example 3, p. 338 (Figure 8.12(b)):

```
syms u v;
[X, Y, Z] = sphere;
hold on;  axis equal;  view(3)
ezmesh((3 + cos(u))*cos(v), (3 + cos(u))*sin(v), sin(u), [-pi - 1, .7, pi/3, 2*pi])
for i = 1 : 12;
  surf(3*cosd(i*30) + X,  3*sind(i*30) + Y,  Z)
end;
```

6. The discriminant set of the family $F(x, y, z; t) = y^3 - (x-t)^2$ of cylindrical surfaces with the z-axis coincides with a plane $\{y = 0\}$ that is not tangent to the given surfaces but is the union of their singular points; see Figure 8.13(b) (and Figure 8.13(a) with a similar 2-D case).

```
syms u v;
hold on; view(3);
for i = 0 : 7;
 ezsurf([u^3 + 2*i, u^2/10, v], [-1.6, 1.6], 20);
end;                                                    % surface
hold on; for i = 1 : 7;
 ezplot(u^3 + 2*i,  u^2/.5, [-1.1, 1.1]);
end;                                                    % curves
```

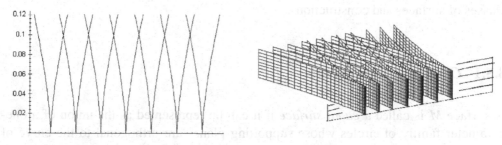

Fig. 8.13 Discriminant set (envelope is empty).

A curve along which the envelope is tangent to some surface from a given family is called a *characteristic*. Equations (8.6) for the envelope are at the same time the equations of a characteristic for fixed t. For an envelope of planes, the characteristics are lines; for spheres, they are circles.

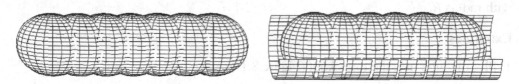

Fig. 8.14 The cylinder is the envelope of a family of spheres.

The envelope of the family of spheres $M_t : (x - a(t))^2 + (y - b(t))^2 + (z - c(t))^2 - R(t)^2 = 0$ of radii $R(t)$, with a curve of centers $\mathbf{r}(t) = [a(t), b(t), c(t)]$ with nonzero curvature, is a *canal surface* (see Section 8.5)

$$\mathbf{r}(u,v) = \mathbf{r}(t) + R(t)(-R'(t)\,\tau(t) \pm \sqrt{1 - (R'(t))^2}\,(\cos(\theta)\,\nu(t) + \sin(\theta)\,\beta(t))),$$

where τ, v, β is a Frenet frame field of the curve $\mathbf{r}(t)$. If the radius is constant, $R(t) = R > 0$, this surface is a *tube* (see Section 8.5). Moreover, the characteristics will be the generators (circles).

8.5   More Examples of Surfaces

Any surface gives rise to other surfaces through a variety of general constructions. In addition to the surfaces of Sections 8.2–8.4, we now study some other interesting classes of surfaces and constructions.

8.5.1   *Canal surfaces and tubes*

A surface M is called a *canal surface* if it can be represented as the union of a one-parameter family of circles whose supporting planes are orthogonal to the curve of centers of these circles. By definition, a canal surface with a C^3-regular curve of centers $\alpha(u)$ and a function of radius $R(u)$ admits regular parametric equations of the form

$$\mathbf{r}(u,v) = \alpha(u) + R(u)(\cos v\, v(u) + \sin v\, \beta(u)),$$

where the (unit) main normal $v(u)$ and the binormal $\beta(u)$ are orthogonal to the tangent vector $\boldsymbol{\tau}(u)$ of the curve $\alpha(u)$. For every u the function $R(u)(\cos v\, v(u) + \sin v\, \beta(u))$, where $v \in [0, 2\pi)$, defines a circle (the *generator*) of radius $R(u)$ that belongs to M. If the radius is constant, $R(u) = R$, such a surface is called a *tube over the curve* $\alpha(u)$ with radius R.

Examples.

1. Plot the tube over the torus knot (Figure 8.15(c)).

```
syms t s real;
x = cos(t)*(10 + 4*sin(9*t));  y = sin(t)*(10 + 4*sin(9*t));  z = 4*cos(9*t);
r = [x, y, z];  dr = diff(r, t);  ddr = diff(r, t, 2);
T = dr/sqrt(dr*dr');
b = cross(dr, ddr);  B = b/sqrt(b*b');
N = cross(B, T);
rs = r + N*cos(s) + B*sin(s);
hold on;  axis equal;  view(3);
ezmesh((10 + 4*cos(t))*cos(s), (10 + 4*cos(t))*sin(s), sin(t),  [0, 2*pi]);
ezsurf(rs(1), rs(2), rs(3),  [0, 2*pi], 121);    two surfaces
```

2. A surface of revolution (see Section 8.1) is a canal surface with the curve of centers on the axis of revolution. The circular cylinder is a tube over a line; the torus is a tube over a circle. A tube generated by a circular helix can be seen in Figure 8.8(c).

The program for a canal surface over an arbitrary curve is:

```
r = [x, y, z];  dr = diff(r, u);  ddr = diff(r, u, 2);
T = dr/sqrt(dr*dr');
b = cross(dr, ddr);  B = b/sqrt(b*b');
N = cross(B, T);
rs = r + N*R*cos(v) + B*R*sin(v);
ezsurf(rs(1), rs(2), rs(3),  [u0, u1, 0, 2*pi]),  axis equal;
```

Before using this program one should define the following functions:

```
syms u v real;
x = u;  y = u^2;  z = u/2;          % define x(u), y(u), z(u)
R = u/5;  u0 = 0;  u1 = 2;          % define R(u), u0, u1
```

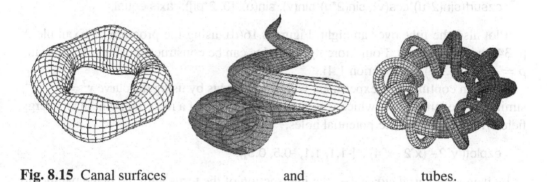

Fig. 8.15 Canal surfaces   and   tubes.

8.5.2 *Translation surfaces*

A surface M that admits parametric equations of the form $\mathbf{r}(u,v) = \mathbf{r}_1(u) + \mathbf{r}_2(v)$ is called a *translation surface*. In other words, a translation surface is obtained by moving some curve (*generator*) that is always parallel to its initial position and is guided with one of its points on another curve, the *directrix*. The simplest translation surface is a cylindrical surface. It can be obtained by parallel displacement of the directrix, which intersects every generator.

Examples.

1. The *Bohemian dome* is obtained by translation of an ellipse along a circle: $\mathbf{r}(u,v) = [a\cos u, a\sin u + b\cos v, c\sin v]$ (Figure 8.17(a)).

```
syms u v;
ezsurf(8*cos(u), 8*sin(u) + 3*cos(v), -3*sin(v),  [0 2*pi 0 pi]);  axis equal;
```

2. Plot eight different *figure eights*.

Solution. We plot the surface defined by moving the figure eight along a circle $[R\cos(v), 0, R\sin(v)]$ (Figure 8.16(c)):

```
syms u v;  R = .5;
ezsurf(sin(2*u) + R*cos(v),  sin(u),  R*sin(v),  [0, 2*pi]);
```

The *Lissajous curve* $\mathbf{r}(u) = [\sin(nu), \sin u]$ for $n = 2$ has the shape of the number 8 (Figure 8.16(a)). We revolve the figure eight about the z-axis (Figure 8.16(b)):

```
syms u v;
ezsurf(sin(2*u)*cos(v), sin(2*u)*sin(v), sin(u), [0, 2*pi]);  axis equal;
```

Plot also the tube over an eight, Figure 8.16(d), using the program of Example 2, p. 341 with $R = $ const. Four more figure eights can be constructed using a lemniscate, $\rho = \sqrt{2\cos(2\varphi)}$ (see Section 1.4).

One can continue this experiment with figure eights by using the curve $y^2 = x^2 - x^4$, similar to the lemniscate, which is related to the motion of a material point in an energy field with two symmetric potential holes,

```
ezplot('y^2 - (x^2 - x^4)',  [-1.1, 1.1, -0.5, 0.5]);
```

or by using the *spiral curve,* i.e., the intersection of the torus with the plane parallel to its axis. (The points P on a spiral curve satisfy the equality $|PA|^2 \cdot |PB|^2 = c \cdot |PO|^2 + c_1$, where A, B are fixed points, O is the midpoint of AB, and c, c_1 are real numbers. Hence, these curves generalize the ovals of Cassini. See Section 1.4.)

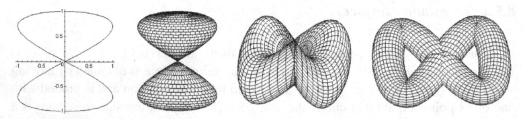

Fig. 8.16 Figure eights.

8.5.3   *Twisted surfaces*

Consider a generalization of the Möbius band and the Klein bottle: let γ: $\mathbf{r}(t) = [\varphi(t), \psi(t)]$ be a plane curve with the condition $\mathbf{r}(-t) = -\mathbf{r}(t)$. Then the *twisted surface* with profile curve γ is defined by the equations

$$\mathbf{r}_1(u,v) = (a + \cos(bu)\,\varphi(v) - \sin(bu)\,\psi(v))\,[\cos u, \sin u, 0]$$
$$+ (\sin(bu)\,\varphi(v) + \cos(bu)\,\psi(v))\,[0, 0, 1].$$

Example. Plot the Lissajous curve $\mathbf{r}(u) = [\sin(nu), \sin u]$, $|u| \leq \pi$, with $n = 4$, Figure 8.17(b), and its twisted surface, Figure 8.17(c).

```
ezplot('sin(4*t)', 'sin(t)', [0, 2*pi]);                    % a curve
syms u v;
ezsurf((2 + cos(u/2)*sin(4*v) - sin(u/2)*sin(v))*cos(u),
(2 + cos(u/2)*sin(4*v) - sin(u/2)*sin(v))*sin(u),
sin(u/2)*sin(4*v) + cos(u/2)*sin(v), [-pi/8, 3*pi/2, 0, pi]);   % surface
```

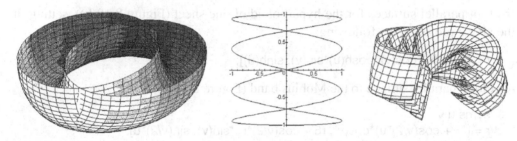

Fig. 8.17 Bohemian dome.   The Lissajous curve and its twisted surface.

8.5.4   *Parallel surfaces*

Let M: $\mathbf{r}(u,v)$ be a smooth surface of class C^1 and $\mathbf{n}(p)$ its normal vector field. For an arbitrary real h plot the surface $M(h)$, placing from each point $P \in M$ the line segment of length $|h|$ in the direction of the normal $\mathbf{n}(P)$ when $h > 0$, and the line segment of length $|h|$ in the direction $-\mathbf{n}(P)$ when $h < 0$. The surface $M(h)$ is called a *parallel (or equidistant) surface* for M and is given by the equations $\mathbf{r}(u,v;h) = \mathbf{r}(u,v) + h\,\mathbf{n}(u,v)$.

A parallel surface for a plane is again a plane. The parallel surfaces for the sphere and the cylinder of radius R, for $|h| < R$, are spheres and cylinders of radius $R \pm h$.

Obviously, the properties of the parallel surface $M(h)$ are defined by the properties of M and the value of h. For small h the surface $M(h)$ is regular; its normal $\mathbf{n}(P, h)$ coincides with $\mathbf{n}(P)$, and its tangent planes at corresponding points are parallel to one another. Parallelness of two surfaces is a symmetric property: M is also a parallel surface for $M(h)$.

Examples. Plot parallel surfaces to several surfaces.

(a) Plot some parallel surfaces to (part of) an ellipsoid (Figure 8.18(a)):

```
syms u v;
r = [cos(u)*cos(v), cos(u)*sin(v), 2*sin(u)];
ru = diff(r, u);  rv = diff(r, v);
nn = cross(ru, rv);
n = nn/sqrt(nn*nn');
hold on;  axis auto;  view(40, 50);
for i = [-1.5, -1, -.6, .2];
 ri = r + i*n;
 ezsurf(ri(1), ri(2), ri(3), [-pi/3, pi/3, pi/8, 2*pi - pi/8]);
end;
```

(b) Plot parallel surfaces for the hyperboloid of one sheet (Figure 8.18(b)), setting, in the above program, the following:

```
r = [cosh(u)*cos(v), cosh(u)*sin(v), sinh(u)];
```

(c) Plot a parallel surface to the Möbius band (Figure 8.18(c)):

```
syms u v;
r = [(5 + cos(v/2)*u)*cos(v), (5 + cos(v/2)*u)*sin(v), sin(v/2)*u];
ru = diff(r, u);  rv = diff(r, v);
nn = cross(ru, rv);
n = nn/sqrt(nn*nn');
hold on;  axis auto;  view(40, 50);
ezmesh(r(1), r(2), r(3),  [-1 1 0 2*pi]);
h = 1;
rh = r + h*n;
ezsurf(rh(1), rh(2), rh(3),  [-1 1 0 4*pi]);
```

Since the Möbius band is one-sided, one cannot select a continuous unit normal field along the whole surface. Thus we move twice (in the parameter u) along the Möbius band; see the formulae in the program. Moreover, the unit normal takes both possible values at each point of the surface. Note that a parallel surface for the Möbius band has two sides (is orientable).

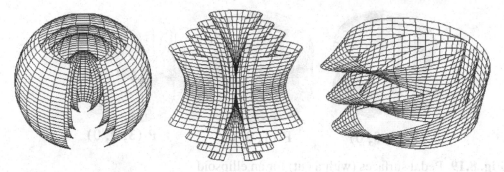

Fig. 8.18 Parallel surfaces for (a) ellipsoid, (b) hyperboloid of 1 sheet, (c) Möbius band.

8.5.5   *Pedal and podoid surfaces*

Consider the construction of a surface analogous to pedal curves.

A *pedal surface* (French *padaire*) $M(P)$ consists of the bottoms of perpendiculars from the fixed point $P \in \mathbb{R}^3$ onto various tangent planes of the given surface M. If $\mathbf{r}(u,v)$ are the parametric equations of M, then the pedal surface with respect to P is given by

$$\mathbf{r}_1(u,v) = \mathbf{r}_p + \frac{((\mathbf{r}(u,v) - \mathbf{r}_p), \mathbf{r}_u, \mathbf{r}_v)}{|\mathbf{r}_u \times \mathbf{r}_v|^2}(\mathbf{r}_u \times \mathbf{r}_v).$$

The pedal surface for a plane, obviously, coincides with a point. The following construction of surfaces is in a simple relation to the pedal surface. The locus of points $M(P)$ that are symmetric to P with respect to all possible tangent planes of the surface M is called a *podoid surface*.

Example. Plot the pedal surfaces of the ellipsoid with respect to its center, one of its vertices, and any external point.

```
syms u v;    p = [3 0 0];
r = ([cos(u)*cos(v), 2*cos(u)*sin(v), 3*sin(u)]);
ru = diff(r, u);  rv = diff(r, v);
nn = cross(ru, rv);
coef = (r - p)*nn'/(nn*nn');
rr = p + nn*coef;
ezsurf(rr(1), rr(2), rr(3), [-pi/2 pi/2 pi/2 2*pi]);  view(120, 20);
```

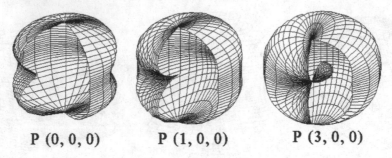

P (0, 0, 0)   P (1, 0, 0)   P (3, 0, 0)

Fig. 8.19 Pedal surfaces (with a cut) for an ellipsoid.

8.5.6 *Cissoidal and conchoidal maps*

The following construction of surfaces is analogous to the *difference of graphs* $f(x,y) = f_1(x,y) - f_2(x,y)$ of two functions in rectangular coordinates. The *cissoidal map* is applied to any two surfaces (for curves see Section 1.4), whose equations in spherical coordinates are $\rho = \rho_1(\varphi, \theta)$ and $\rho = \rho_2(\varphi, \theta)$, with the aim of constructing the surface $\rho = \rho_1(\varphi, \theta) - \rho_2(\varphi, \theta)$. In other words, on each ray from the point O we set the line segment OM equal to the segment on the ray between the two given surfaces.

The following construction of surfaces is analogous to the known conchoidal transformation on plane curves. A *conchoidal surface* for a given surface (for a curve see Section 1.4) is obtained by increasing (or decreasing) the position vector of each point of the surface by a fixed line segment l. If the equation of a surface is $\rho = \rho_1(\varphi, \theta)$ in spherical coordinates, then the equations of its conchoidal surface are $\rho = \rho_1(\varphi, \theta) \pm l$.

Examples.

1. Write down the equations and plot the cissoidal surface of a sphere with diameter $|OA| = 2a$ and tangent plane α at the point $A = (2a, 0, 0)$.

Solution. The *cissoid* of a line $x = 2a$ in the plane XZ, Figure 8.20(a), has the equation

$$\rho = 2a \frac{\sin^2 \varphi}{\cos \varphi} \iff z^2 = x^3/(2a - x).$$

    x = linspace(0, 1.8, 41); f = sqrt(x.^3./(2 - x)); plot(x, f, x, -f);   % a=1

We need the surface of revolution about the axis OX of the part of the cissoid with $z \geq 0$, which has the equation $z^2 + y^2 = x^3/(2a - x)$. Since **impl.m** gives only an approximate image, we use the parametric equations of part of the cissoid $x = t$, $z = \sqrt{t^3/(2a - t)}$ $(0 \leq t < 2a)$ and plot the surface of revolution, Figure 8.20(c), by the formula (8.7):

$$\mathbf{r}(u,v) = \left[u, \sqrt{\frac{u^3}{2a-u}} \cos v, \sqrt{\frac{u^3}{2a-u}} \sin v \right] \qquad (0 \le u < 2a,\ 0 \le v < 2\pi).$$

```
u = linspace(0, 1.9, 41);  v = linspace(0, 2*pi);
[U, V] = meshgrid(u, v);
f = sqrt(U.^3./(2 - U));
surf(U, f.*cos(V), f.*sin(V));
```

2. Write down the equation and plot the *conchoidal surface of the plane* $\{z = a\}$ with $l = 1$.

Hint. The conchoid of the line $z = a$ in XZ,

$$\rho = \frac{a}{\sin \varphi} \pm l \iff (z^2 + x^2)(z-a)^2 - l^2 z^2 = 0 \qquad \text{(Figure 8.20(b))},$$

for $l > a$ has a loop, for $l = a$ has a cuspidal point, and for $0 < l < a$ has an isolated point. Using rotation about the z-axis, we obtain the conchoid of the plane $z = a$: $(z^2 + x^2 + y^2)(z-a)^2 - l^2 z^2 = 0$. Since **impl.m** gives a coarse image, we use the parametric equations of part of the conchoid $x = \frac{a}{\sin t} \pm l \cos t,\ z = \frac{a}{\sin t} \pm l \sin t$ $(0 \le t < \pi/2)$ and plot, using (8.1), the surface of revolution that we need with $a = 2,\ l = 1$ (Figure 8.20(d)),

$$\mathbf{r}(u,v) = [(2/\sin u \pm 1) \cos u \cos v, (2/\sin u \pm 1) \cos u \sin v, (2/\sin u \pm 1) \sin u].$$

```
ezpolar('2/sin(t)-1',  [.2, pi - .2]);                          % profile curve
t = linspace(.1, pi - .1, 41);  s = linspace(-pi, pi, 41);
[U, V] = meshgrid(t, s);
h = 2./sin(U) - 1;
surf(h.*cos(U).*cos(V), h.*cos(U).*sin(V), h.*sin(U));          % surface
```

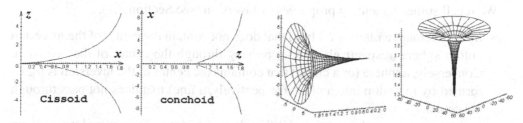

Fig. 8.20 Cissoid of the sphere, conchoid of the plane, and their profile curves.

8.5.7 *Inversion of a surface*

Inversion is the simplest (nonlinear) transformation of space after rigid motions and
affine maps. Inversion with respect to a sphere is a generalization of symmetry with
respect to a plane, but in our case the points near the center (lying inside the sphere)
are mapped far from the center (outside the sphere) and conversely (see Section 4.1;
inversion of the plane is treated in Section 4.1.2).

Inversion is an involutive map, i.e., $(i_{O,R})^2 = \mathrm{Id}$ (identity map). The sphere $S(O,R)$
is the set of fixed points of the inversion.

Example. Inversion with center O and radius 1 of the circular helix, Figure 8.21(a),
$\mathbf{r}(t) = [3 + \cos t, \sin t, 0.1t]$ is the curve in Figure 8.21(b).

> r = [3 + cos(u), sin(u), .1*u]; ezplot3(r(1), r(2), r(3), [0, 8*pi]);   % curve
> ri = r/(r*r'); ezplot3(ri(1), ri(2), ri(3), [-25*pi, 25*pi]);   % inversion of a curve

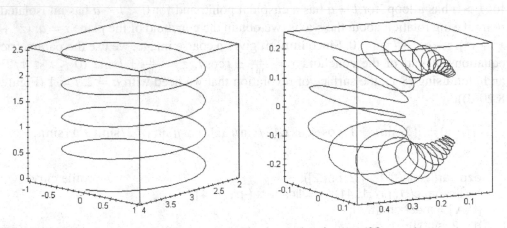

Fig. 8.21 The circular helix (4 turns) and its inversion image (30 turns).

We recall some elementary properties of inversion (see Section 4.1).

- Inversion maps a plane (or a line) that does not contain the center of the inversion
  onto a sphere (respectively, circle) passing through the center of the inversion.
  Conversely, a sphere (or a circle) that contains the center of the inversion is transformed by inversion into a plane (respectively, a line) that does not pass through
  the center.
- Inversion maps a sphere (or circle) that does not contain the center into a sphere
  (respectively, circle) that also does not contain the center of the inversion.
- Inversion preserves angles between vectors, and hence preserves angles between
  intersecting curves.

Using inversion, one can compare the behavior of very similar non-compact surfaces (and curves) at infinity.

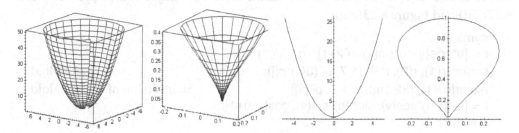

Fig. 8.22 Paraboloid of revolution and its inversion (outside $x^2 + y^2 \leq 1$).

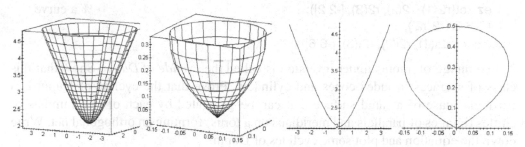

Fig. 8.23 Hyperboloid of one sheet and its inversion (outside $x^2 + y^2 \leq 1$).

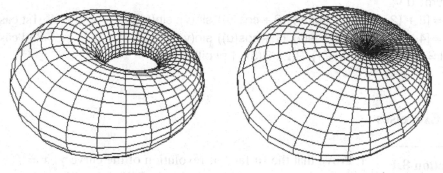

Fig. 8.24 Cyclides of Dupin.

Examples.

1. Plot an inversion image of a paraboloid of revolution (Figure 8.22(a,b)) a hyperboloid of one sheet (Figure 8.23(a,b)), and the inversion images of their profiles (Figure 8.22(c,d) and Figure 8.23(c,d)):

```
syms u v;
r = [u*cos(v), u*sin(v), u^2+1];  ri = r/(r*r');
ezsurf(r(1), r(2), r(3), [1 7 -pi (3/4)*pi]);                    %  paraboloid
ezsurf(ri(1), ri(2), ri(3), [1 7 -pi pi]);            % inversion of a paraboloid
r = [sinh(u)*cos(v), sinh(u)*sin(v), cosh(u)+1];
ri = r/(r*r');
ezsurf(r(1), r(2), r(3), [1 7 -pi (3/4)*pi]);                   %  hyperboloid
ezsurf(ri(1), ri(2), ri(3), [1 7 -pi pi]);            % inversion of hyperboloid
r2 = [u, u^2 + 1];                                               % parabola
r2 = [sinh(u), cosh(u)+1];                                        % hyperbola
ezplot(r2(1), r2(2), r2(3), [-2 2]);                              % a curve
r2i = 2/(r2*r2');
ezplot(r2i(1), r2i(2), r2i(3), [-6 6]);
```

2. The image of a torus under inversion is called the *cyclide of Dupin*. Show that this class of surfaces includes cones and cylinders. Prove that the cyclide of Dupin is a particular case of a canal surface that can be generated by each of two families of circles (images of parallels and meridians on a torus) forming an orthogonal net. Write down the equation and plot some cyclides of Dupin.

Hint. Plot two images, Figure 8.24, of the torus $\mathbf{r}(u,v) = [b + (a + \cos u)\cos v, (a + \cos u)\sin v, \sin u]$ under inversion with center O and radius 1: $a = 2$, $b = 8$, and $a = 1$, $b = 4$.

```
syms u v;
r = [8 + (2 + cos(u))*cos(v), (2 + cos(u))*sin(v), sin(u)];       % 1st case
r = [4 + (1 + cos(u))*cos(v), (1 + cos(u))*sin(v), sin(u)];       % 2nd case
R = r/(r*r');  ezsurf(R(1), R(2), R(3),  [-pi pi]);
```

8.6 Exercises

Section 8.1

1. Prove that the surface of revolution of the curve γ: $x = f(u)$, $z = g(u) \geq 0$, where $u \in I$, about the axis OX is given by

$$\mathbf{r}(u,v) = [f(u), g(u)\cos v, g(u)\sin v] \qquad (0 \leq v < 2\pi). \qquad (8.7)$$

2. Write down the equation and plot the *surface of revolution of the astroid* about one of its axes of symmetry (Figure 8.2(c), which looks like a top), the revolution of

the lemniscate about each of its axes of symmetry, and the revolution of the four-leafed rose about an axis of symmetry.

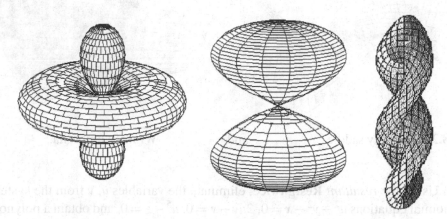

Fig. 8.25 Revolution of a four-leafed rose and lemniscate. Twisted sphere.

3. The construction of a surface of revolution can be generalized if the points of a curve $\gamma(u)$ are moving not along circles (parallels), but along plane curves that are homothetic to a given curve $\gamma_1(v)$. Write down the equations of generalized surfaces of revolution in the cases where $\gamma_1(v)$ is an ellipse, a parabola, a four-leafed rose, etc.

4. Complete the program of Example 3, p. 327 to obtain each region colored differently.

5. Find the singular points on the surface of revolution of the astroid about its axis of symmetry (Figure 8.2(c)).

6. The cubes of the coordinate functions in the equations of the sphere define parametric equations of the *astroidal sphere* $\mathbf{r} = [(R\sin u\cos v)^3, (R\sin u\sin v)^3, (R\cos u)^3]$. It looks like an octahedron with concave faces and edges, Figure 8.1(c). Check that the equation $x^{2/3} + y^{2/3} + z^{2/3} = R^{2/3}$ defines an astroidal sphere.

Section 8.2

1. Show that the following surfaces $\mathbf{r}(u,v)$ are algebraic:

(a) *The monkey saddle* $\mathbf{r} = [u, v, u^3 - 3uv^2]$, Figure 8.26(a) (third order).

(b) *The Whitney umbrella* $\mathbf{r} = [uv, u, v^2]$, Figure 8.26(b) (third order).

(c) *The wings* $\mathbf{r}(u,v) = [u^2 - v^2, 2uv, u^5]$, Figure 7.18(a) (tenth order).

Then explain why the surface $\mathbf{r}(u,v) = [f_1(u,v), f_2(u,v), f_3(u,v)]$, where $f_i(u,v)$ are polynomials, is an algebraic surface.

Hint. (a) Consider an explicit equation of this surface.

(b) Consider the function $x^2 - y^2z$.

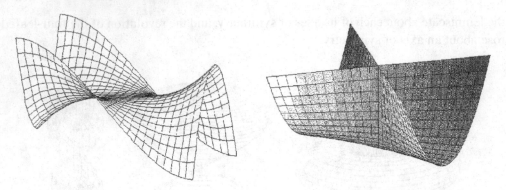

Fig. 8.26 Monkey saddle   and   Whitney umbrella.

(c) Using the *resultant* **Resfg** twice, eliminate the variables u, v from the system of polynomial equations $u^2 - v^2 - x = 0$, $2uv - y = 0$, $u^5 - z = 0$, and obtain a polynomial F in x, y, z of the tenth order:

```
syms x y z u v;
a = [-1,  0,  u^2 - x];  b = [2*u, -y];          % polynomials in v
r1 = det(Resfg(a, b))                            % Answer: r₁ = 4u⁴ − 4u²x − y²
a1 = [4, 0, -4*x, 0, -y^2]; b1 = [1, 0, 0, 0, 0, -z]   % polynomials in u
F = det(Resfg(a1, b1))
```

$$ F = -y^{10} - 320y^4 z^2 x - 1280 z^2 x^3 y^2 + 1024 z^4 - 1024 x^5 z^2 \qquad \text{\% Answer.} $$

2. The *swallow-tail* surface has a cuspidal edge (Figures 8.4(a,b)) and a curve of self-intersection.

(a) To imagine this surface better, consider its intersections with parallel planes $x = c$, Figure 8.4(c).

(b) Show that the points (x, y, z), where the polynomial $F(t; x, y, z) = t^4 + xt^2 + yt + z$ has the multiple root t, form the swallow-tail surface (see Example 3, p. 329) given by equations $y = -4t^3 - 2xt$, $z = 3t^4 + xt^2$.

(c) The cuspidal edge on the swallow-tail surface is given by the condition that t is a triple root of the equation $F = 0$, i.e., $F''_{tt} = 0$. Show that this singular curve is defined by the parametric equations $\mathbf{r}(t) = [-6t^2, 8t^3, -3t^4]$, Figure 8.4(b).

Hint. Eliminate x, y, z from the system of 3 equations, then plot:

```
syms t x y z;
F = t^4 + x*t^2 + y*t + z;
r = solve(F, diff(F, t), diff(F, t, 2),  x, y, z);
[r.x  r.y  r.z]                          % [−6t², 8t³, −3t⁴]
ezplot3(r.x, r.y, r.z,  [-1, 1])
```

(d) Show that the swallow-tail surface is the union of all tangent lines to the space curve $\mathbf{r}(t) = [t^2, t^3, t^4]$.

Section 8.3

1. Plot the following surfaces and verify their ruled structure:

(a) The *hyperboloid of one sheet* $x^2/a^2 + y^2/b^2 - z^2/c^2 = 1$:

$$\mathbf{r}(u,v) = \rho(u) \pm v(\rho'(u) + [0,0,c]),$$

where $\rho(u) = [a\cos u, b\sin u, 0]$ is an ellipse, and $\pm$ are two ways to fix the rulings.

(b) The *hyperbolic paraboloid* $z = x^2/a^2 - y^2/b^2$:

$$\mathbf{r}(u,v) = [au, 0, u^2] + v[a, \pm b, 2u] = [a(u+v), \pm bv, u^2 + 2uv],$$

where $\pm$ are two ways to fix the rulings. There are other convenient parametric equations when this surface (after rotation) is given by $z = axy$,

$$\mathbf{r}(u,v) = [u,0,0] + v[0,1,au] = [u,v,avu].$$

(c) The *elliptical cone* $x^2/a^2 + y^2/b^2 - z^2/c^2 = 0$. It can be parameterized with rulings in the role of the v-curves,

$$\mathbf{r}(u,v) = v[a\cos u, b\sin u, c] \quad (v \in \mathbb{R},\ u \in [0,2\pi)).$$

(d) The *conical surface* with vertex $S = (a,b,0)$ and a four-leafed rose $\rho = \cos(2\varphi)$ as its section by the plane $\alpha\colon z = 1$ (see Figure 8.5(c)):

$$\mathbf{r}(u,v) = [a,b,0] + v[\cos(2u)\cos u, \cos(2u)\sin u, 1] \quad (0 \le v \le 2,\ 0 \le u \le 2\pi).$$

(e) The *cylinder* with the z-axis and a four-leafed rose as its directrix:

$$\mathbf{r}(u,v) = c[\cos(2u)\cos u, \cos(2u)\sin u, 0] + v[0,0,1], \quad \text{Figure 8.5(b)}.$$

(f) The *tangent developable surface* with the helix as the directrix curve:

$$\mathbf{r}(u,v) = [\cos u, \sin u, au] + v[-\sin u, \cos u, a], \quad \text{Figure 8.8(b)}.$$

2. Prove that:

(a) A cylinder over a smooth plane curve is also a smooth surface.

(b) The unique singular point on a cone over a smooth curve that differs from a part of the plane is its vertex.

(c) The singular points of a tangent developable surface form its directrix curve, which is called the *cuspidal edge*.

3. Plot the tangent developable surface over the Viviani curve:

$$\mathbf{r}(u) = [R\cos u^2, R\cos u\sin u, R\sin u] \quad (u \in [-\pi,\pi]), \quad \text{Figure 8.9(b)}.$$

4. Find conditions for when the surface $z = f(x,y)$ is a ruled developable surface. Compute using MATLAB® and give examples.

5. Show that the saddle type surface, Figure 8.7(c), generalizes both the Plücker conoid and the monkey saddle: $\mathbf{r}(u,v) = [v\cos u, v\sin u, v^m \sin(vu)]$.

6. Show that:

(a) The striction curve of a tangent developable surface coincides with its cuspidal edge.

(b) For the hyperboloid of one sheet, $\frac{x^2}{a^2} + \frac{y^2}{b^2} - \frac{z^2}{c^2} = 1$, the striction curve is a space curve of the fourth order, singular for each system of rulings.

(c) The axis of a right conoid is its striction curve.

7. Show that the striction curve of the helicoid coincides with its axis and that its distribution parameter is a constant function.

8. Show that the distribution parameter of a developable (non-cylindrical) ruled surface is identically zero.

Section 8.4   1. Find a family of surfaces whose discriminant is (a) a line, (b) a point.

Answer: (a) $(x - t)^2 + y^2 = t^2, t \neq 0$, (b) $(x - t)^2 + y^2 + z^2 = t^2, t \neq 0$.

2. Find the envelope of the family of planes that cut from the coordinate octant $x, y, z \geq 0$ tetrahedra of constant volume V.

Answer: $xyz = \frac{2}{9}V$; see also Exercise 5, p. 317.

3. Write down the equation of the family of spheres for which the envelope is the cone $x^2 + y^2 = a^2 z^2, (z \neq 0)$ without its vertex.

Answer: $x^2 + y^2 + (z - t)^2 = a^2 t^2 / (a^2 + 1)$.

4. Let $\mathbf{r}(s)$ be a space curve with the natural parameter s and with nonzero curvature, let $\{\tau, \nu, \beta\}$ be its Frenet frame field, and let $\tilde{\mathbf{r}}$ be a point in space. Prove that:

(a) The envelope of the *family of osculating planes* $(\tilde{\mathbf{r}} - \mathbf{r}(s)) \cdot \beta(s) = 0$ is a tangent developable surface for which the given curve is the cuspidal edge; for each t the characteristic coincides with the tangent line.

(b) The envelope of the *family of normal planes* $(\tilde{\mathbf{r}} - \mathbf{r}(s)) \cdot \tau(s) = 0$ is a tangent developable surface that is the locus of centers of osculating spheres (for a given curve).

(c) The envelope of the *family of rectifying planes* $(\tilde{\mathbf{r}} - \mathbf{r}(s)) \cdot \nu(s) = 0$ contains the given curve (as its *geodesic*); the characteristics coincide with the *momentary axis of revolution* of the Frenet frame field $\{\tau, \nu, \beta\}$. (This developable ruled surface is called *rectifying*, because after unrolling onto the plane, the given curve becomes a *straight line*.)

5. Prove that the family of surfaces given by $F(x,y,z) = t$, where F is an arbitrary regular function in three variables x, y, z, has no envelope.

Section 8.5   1. Write down equations of the canal surface with $R(u) = ku$ and the tube over the circular/conic helix. Plot the surfaces (*sea shell*, Figure 8.15(b)).

2. Show that the following surfaces are translation surfaces:

(a) The elliptical and hyperbolic paraboloids.

(b) The part of the helicoid $\mathbf{r}(u,v) = [u\cos v, u\sin v, av]$ with $u \leq c$.

Hint. Replace the coordinate system by $u = c\cos\frac{\varphi - \psi}{2}$, $v = \frac{\varphi + \psi}{2}$, where $0 \leq \varphi - \psi < \frac{\pi}{2}$; assume $\mathbf{r}_1(t) = 2[c\cos t, c\sin t, at]$; and deduce the equation of the helicoid $\mathbf{r}(\varphi, \psi) = \mathbf{r}_1(\varphi) + \mathbf{r}_1(\psi)$.

(c) The surface that is the locus of centers of segments whose endpoints belong to two given space curves.

3. Show that the *Möbius band* (see Example 2, p. 327) is twisted for $b = \frac{1}{2}$, using the line segment γ: $\mathbf{r}(t) = [t, 0]$ ($|t| \leq 1$).

4. Plot the canal surface defined by moving the circle of radius $4 + \sin(kt)$ ($k = 3, 4, \dots$) along the circle $\mathbf{r}(t) = [10\cos t, 10\sin t, 0]$ (Figure 8.15(a)):

```
syms u v;
k = 4;  f = 4 + sin(k*v);                          % enter your data
ezsurf((12 + f*sin(u))*cos(v), (12 + f*sin(u))*sin(v), f*cos(u),  [0, 2*pi]);
```

5. Show that the Figure Eight immersion of the *Klein bottle*, Figure 7.3(c), has simple parametric equations: it is twisted for $b = \frac{1}{2}$ using the Lissajous curve γ: $\mathbf{r}(t) = [\sin t, \sin(2t)]$ ($|t| \leq \pi$), and is given by the equations

$$\begin{aligned}
\mathbf{r}_1(u,v) &= (a + \cos(u/2)\sin v - \sin(u/2)\sin(2v))[\cos u, \sin u, 0] \\
&\quad + (\sin(u/2)\sin v + \cos(u/2)\sin(2v))[0, 0, 1].
\end{aligned} \tag{8.8}$$

In this immersion, the self-intersection circle is a geometric circle in the XY-plane. The positive constant $a = 3$ is the radius of this circle. The parameter u gives the angle in the XY-plane, and v specifies the position around the 8-shaped cross section.

6. Show that the parallel surface for a tube of radius R over a space curve with small h is again a tube of radius $R \pm h$ over the same curve.

7. Show that a parallel surface for a developable ruled surface is again a developable ruled surface.

8. Prove that a pedal surface of the sphere is the surface of revolution of a cardioid if the point P belongs to the sphere. Plot this surface.

9. Plot the pedal surfaces of

(a) a torus with respect to its center;

(b) a parabola with respect to its vertex (a cissoid).

Then plot the analogous pedal surface of a paraboloid.

10. Write down the equations for the podoid surface of a given surface.

Show that the podoid surface

(a) of a developable ruled surface degenerates to a curve;

(b) of a surface can be obtained from a pedal surface using similarity with center at P and with coefficient 2;

(c) of a sphere is again a sphere.

11. Show that the conchoidal surface of the sphere is the surface of revolution of Pascal's limaçon. Plot these surfaces.

12. Study the analogous *cissoidal and conchoidal mappings of surfaces* using a line (the z-axis) instead of the point O, and cylindrical coordinates. Plot examples of such surfaces.

9

Piecewise Curves and Surfaces

After an introductory section, in Section 9.2 we will consider Bézier curves. Then we will discuss the important notions of parametric C^k-continuity and geometric C^k-continuity of piecewise curves. Section 9.3 is devoted to (parametrically C^k-continuous) Hermite interpolation and its applications to spline curves. In Section 9.4 we will study β-splines, which are geometrically C^k-continuous. We will also briefly survey their particular case, B-splines, which are parametrically C^k-continuous. In Sections 9.5 and 9.6 we will study similar problems for piecewise surfaces. Along the way we will create and apply several M-files that are given in Section A.4.

9.1 Introduction

The main problem discussed in this chapter is this: given points (*control vertices*) $\mathbf{P} = \{P_1, P_2, \ldots, P_n\}$ arbitrarily placed in the plane or in space, construct a smooth curve passing near or through these points and satisfying some additional conditions. A polygon that joins neighboring points from $\mathbf{P}$ is called a *control polygon*; P_1 and P_n are called *boundary* points, and $P_2, \ldots, P_{n-1}$ are called *inner* points. The equation of the curve will be written in the form of weighted sum $\mathbf{r}(t) = \sum_{i=1}^{n} a_i(t) P_i$ ($t \in [a, b]$), where the functions $a_i(t)$ must be derived.

Spline curves are obtained by using a similar scheme to that in Section 1.6. We fix the net $a = t_1 < t_2 < \cdots < t_m = b$ and build the vector function $\mathbf{r}(t)$ with the following two conditions:

(i) $\mathbf{r}$ restricted to $[t_i, t_{i+1}]$ is a polynomial of degree at most d,

(ii) $\mathbf{r} \in C^{d-1}[a, b]$.

We call $\mathbf{r}$ a *spline of degree* $d \geq 1$ *with respect to the* $\{t_i\}$, the *knots*. For *interpolating* splines the *nodes* $\{\mathbf{r}(t_i)\}$ coincide with the control vertices. For *smoothing* splines the

V. Rovenski, *Modeling of Curves and Surfaces with MATLAB®*,
Springer Undergraduate Texts in Mathematics and Technology 7, DOI 10.1007/978-0-387-71278-9_9,
© Springer Science+Business Media, LLC 2010

nodes need not coincide with the control vertices. Each control vertex participates in at most $d + 1$ curve segments, so the curve features *local control:* moving one control vertex affects only those segments and not the entire curve.

A spline may also be defined on an infinite interval. From the above definition, we see that a linear spline $\mathbf{r}(t)$ (with $d = 1$) is just a continuous polygonal arc. The most common case is that of *cubic splines* $(d = 3)$.

Similarly to curves, a piecewise (spline) surface can be obtained by constructing a number of *patches* and connecting them. The method used to construct the patch should allow for smooth connection of the patches. A surface patch can be displayed either as a wire frame or as a solid surface.

Recall that a *regular* curve γ can be generated by a *regular parameterization* $\mathbf{r}(t)$; i.e., $\mathbf{r}'(t) \neq 0$ for all $t \in I$. A curve γ is called *parametrically C^k-continuous* if it can be generated by a regular parameterization $\mathbf{r}(t)$ of class C^k (i.e., $\mathbf{r}^{(k)}(t)$ is a continuous vector function).

One may consider a C^k-regular change of parameter $t' = \varphi(t)$.

If $\mathbf{r}(t)$ is a C^k-regular parameterization of γ, then the natural parameterization $\mathbf{r}(s)$ of the curve is also regular C^k.

The examples of piecewise plane curves illustrate the fact that the continuity of the vector $\mathbf{r}'$ (depending on the choice of parameterization of the curve) does not imply the continuity of the tangent line.

Example.

Verify the *condition of continuity* $\mathbf{r}_1(1) = \mathbf{r}_2(0)$ for the curves composed of two segments: $\mathbf{r}_1(t)$ (called *"arriving"*) and $\mathbf{r}_2(t)$ (called *"departing"*), $0 \leq t \leq 1$. Find characteristics of the curve at its node, and then plot it using the statements:

```
hold on;    ezplot(r1, [t1, t2]);
h = ezplot(r2, [t3, t4]);  set(h, 'Color', 'r');
plot(subs(r1(1), t, t2), subs(r1(2), t, t2), 'o');
```

(a) $\mathbf{r}_1(t) = [4t, 4t]$, $\mathbf{r}_2(t) = [t + 1, t + 1]$ are two segments (Figure 9.1(a)). The first derivatives at the node $(1, 1)$ are not equal: $\mathbf{r}_1'(1) = [4, 4]$, $\mathbf{r}_2'(0) = [1, 1]$, but the curve has continuous tangent line.

```
r1 = [4*t, 4*t];    t1 = 0;  t2 = 1/4;
r2 = [1+t, 1+t];    t3 = 0;  t4 = 1;                                    % data
```

(b) $\mathbf{r}_1(t) = [-(1 - t)^2, (1 - t)^2]$, $\mathbf{r}_2(t) = [t^2, t^2]$ are two segments (Figure 9.1(b)). Although the first derivatives at $(0, 0)$ are equal, $\mathbf{r}_1'(1) = \mathbf{r}_2'(0) = [0, 0]$, and the curve $\mathbf{r}'(t)$ is continuous, it is not C^1-continuous.

```
r1 = [-(1 - t)^2, (1 - t)^2]; t1 = 0;  t2 = 1;
r2 = [t^2, t^2]; t3 = 0;  t4 = 1;
```

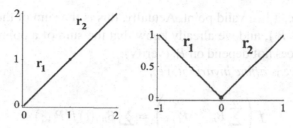

Fig. 9.1 Two-component curves.

9.2 Bézier Curves

An alternative for least-squares smoothing is provided by Bézier curves. An interesting application of Bézier-curves is to font design. The boundary of each character in a font is specified by Bézier curves. Different font sizes are obtained by applying scaling transformations to the control points, and making use of the property of invariance under affine transformations. Italic fonts may be obtained by applying a shear transformation or a projection.

9.2.1 *Bézier curves of degree n*

The *Bézier curve* defined by control points $\mathbf{P} = P_1, \ldots, P_{n+1}$ of $\mathbb{R}^N$ is the nth-degree polynomial given by $\mathbf{r}(t) = \sum_{i=0}^n B_{n,i}(t) P_{i+1}$ for $t \in [0,1]$. The multipliers at the points in the above equation are called *Bernstein polynomials* and are defined as $B_{n,i}(t) = \binom{n}{i} t^i (1-t)^{n-i}$. When $P_i \in \mathbb{R}^2$ ($i = 1, \ldots, n+1$), then the curve is planar, and when $P_i \in \mathbb{R}^3$, the curve is spatial. Equivalently, the Bézier curve is represented in matrix form as

$$\mathbf{r}(t) = \mathbf{G}[B_{n,n}(t); B_{n,n-1}(t); \ldots; B_{n,1}(t); B_{n,0}(t)], \tag{9.1}$$

where $\mathbf{G} = (P_1, P_2, \ldots, P_{n+1})$ is the *geometric matrix*. If $N = 3$, the geometric matrix has the form $\mathbf{G} = \begin{pmatrix} x_1 & x_2 & \cdots & x_{n+1} \\ y_1 & y_2 & \cdots & y_{n+1} \\ z_1 & z_2 & \cdots & z_{n+1} \end{pmatrix}$.

The difference of two points, $P_2 - P_1 = (\Delta x_1, \ldots, \Delta x_N)$, is the vector from P_1 to P_2. The sum of a point P_0 and a vector $\mathbf{v}$ is well defined and is a point $P_0 + \mathbf{v}$. The sum of points, in general, is not well defined. There is, however, one important special case where the sum of points is well defined, the so-called *barycentric sum*. If we multiply each point P_i by a weight w_i and if $\sum_{i=0}^n w_i = 1$, then the sum $\sum_{i=0}^n w_i P_i$ is affine invariant, i.e., it is a valid point. Actually, this is the sum of the point P_0 and the vector $\sum_{i=0}^{n} w_i(P_i - P_0)$, and we already know that the sum of a point and a vector is a point. The result does not depend on P_0 (verify).

The Bézier curve is *affine invariant*, i.e.,

$$f\left(\sum_{i=0}^{n} B_{n,i}(t)P_{i+1}\right) = \sum_{i=0}^{n} B_{n,i}(t)f(P_{i+1})$$

for any affine transformation f, because $\sum_{i=0}^{n} B_{n,i}(t) = 1$. It has the form $\sum_{i=0}^{n} w_i(t)P_i$, where the weights $w_i(t)$ are barycentric. The Bézier curve is a result of *blending* several points with the Bernstein polynomials.

Example. For $n = 2$, the Bernstein polynomials (weights) are

$$B_{2,0}(t) = (1-t)^2, \qquad B_{2,1}(t) = 2t(1-t), \qquad B_{2,2}(t) = t^2,$$

and the *quadratic Bézier curve* is $\mathbf{r}(t) = (1-t)^2 P_1 + 2t(1-t)P_2 + t^2 P_3$. The matrix presentation of the curve (in $\mathbb{R}^3$) takes the form $\mathbf{r}(t) = \mathbf{G}MT(t)$ $(0 \le t \le 1)$, where $\mathbf{r}(t) = [x(t); y(t); z(t)]$ and $T(t) = [t^2; t; 1]$ are column vectors, and

$$\mathbf{G} = (P_1\, P_2\, P_3), \qquad M = \begin{pmatrix} 1 & -2 & 1 \\ -2 & 2 & 0 \\ 1 & 0 & 0 \end{pmatrix}. \tag{9.2}$$

Given P_1, P_2, P_3, one can evaluate and plot the quadratic Bézier curve.

```
P = [0  0;  2 -1;  3  3];
syms t;   r = (1 - t)^2*P(1,:) + 2*t*(1 - t)*P(2,:) + t^2*P(3,:)
hold on;   ezplot(r(1), r(2), [0, 1]);              % Bézier curve (parabola)
plot(P(:, 1), P(:, 2), 'o- -r');                     % 3 points
```

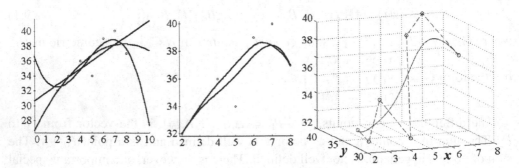

Fig. 9.2 (a) Least-squares method. (b, c) Bézier curves.

One may prove by induction that the $(n+1)$-point Bézier curve is represented by $\mathbf{r}(t) = \mathbf{G}MT(t)$ $(0 \le t \le 1)$, where $T(t) = [t^n; t^{n-1}; \ldots; t; 1]$ is a column vector and M is a symmetric matrix with zero elements below the second diagonal. The ith row of M consists of the coefficients of $B_{n,i}(t)$:

$$[B_{n,n}(t); B_{n,n-1}(t); \ldots; B_{n,1}(t); B_{n,0}(t)] = M[t^n; t^{n-1}; \ldots; t; 1].$$

Its determinant therefore equals (up to sign) the product of the diagonal elements, which are all nonzero. Hence, M is nonsingular. M can also be written as the product $\bar{M}D$, where $D = \text{diag}(\binom{n}{0}, \binom{n}{1}, \ldots, \binom{n}{n})$ is the diagonal matrix, and

$$\bar{M} = \begin{pmatrix} \binom{n}{n}(-1)^n & \binom{n-1}{n-1}(-1)^{n-1} & \cdots & \binom{n-n}{n-n}(-1)^0 \\ \binom{n}{n-1}(-1)^{n-1} & \binom{n-1}{n-2}(-1)^{n-2} & \cdots & 0 \\ \vdots & \vdots & \cdots & 0 \\ \binom{n}{0}(-1)^0 & 0 & 0 & 0 \end{pmatrix} \tag{9.3}$$

A Bézier curve goes through initial/end control points. Its nice approximation properties are explained by the Weierstrass theorem. Note that the addition of a point increases the degree of a Bézier curve. Another defect is that the whole Bézier curve changes when one of the control points moves.

Examples.

1. Compute the symmetric matrix (9.3) for arbitrary n.

```
n = 4;                                              % enter data
for i = 1 : n + 1; for j = 1 : n + 1;
   M(i, j) = gamma(n + 1)./gamma(j)./gamma(n - j + 2).*
      gamma(n + 2 - j)./gamma(n + 3 - i - j)./gamma(i).*(-1).^(n + 2 - i - j);
end; end
M                                                   % matrix (9.3) for a given n.
```

2. Calculate and plot the plane Bézier curve using the M-file **bezier**.

```
x = 2 : 8;  y = [32, 34, 36, 34, 39, 40, 37];       % enter data
r = bezier(x, y)
hold on;   ezplot(r(1), r(2), [0, 1]);
plot(x, y, 'o - - r');                              % Figure 9.2(b)
```

The first component of $\mathbf{r}(t)$ is linear when the step of the x-net is constant:

```
r = bezier(x, y);   r(1);                           % Answer: 2 + 6t
```

Figures 9.2(b,c) illustrate some standard properties of Bézier curves. For instance, these curves always lie in the convex hulls of their control points, since base Bernstein functions are nonnegative with sum equal to 1. Thus, the curve is said to have the *convex hull property*.

simplify(sum(B))   % obtain 1 (B contains $B_{n,i}(t)$, see M-file **bezier.m**)

3. Calculate and plot the Bézier curve in $\mathbb{R}^3$ using the M-file **bezier**.

x = 2 : 8;
yz = [32, 34, 36, 34, 39, 40, 37; 33, 32, 34, 32, 37, 38, 36];   % enter data
r = **bezier**(x, yz)
$r = [6t + 2, \; 70t^6 - 222t^5 + 225t^4 - 80t^3 + 12t + 32,$
$\qquad 74t^6 - 240t^5 + 270t^4 - 140t^3 + 45t^2 - 6t + 33]$ % Answer $\mathbf{r}(t) \subset \mathbb{R}^3$
ezplot3(f(1), f(2), f(3), [0, 1]);   hold on;
plot3(x, yz(1, :), yz(2, :), 'o - - r');   % Figure 9.2(c).

9.2.2 Piecewise cubic Bézier curves

The *cubic* (*elementary*) *Bézier curve* is determined by four control points P_1, P_2, P_3, P_4
and the base Bernstein polynomials

$$B_{3,0}(t) = (1-t)^3, \quad B_{3,1}(t) = 3t(1-t)^2, \quad B_{3,2}(t) = 3t^2(1-t), \quad B_{3,3}(t) = t^3.$$

Its equation is $\mathbf{r}(t) = B_{3,0}(t)P_1 + B_{3,1}(t)P_2 + B_{3,2}(t)P_3 + B_{3,3}(t)P_4$.

Obviously, $\mathbf{r}'(t) = -3(1-t)^2 P_1 + 3(1-t)(1-7t)P_2 + 3t(2-3t)P_3 + 3t^2 P_4$; hence,
$\mathbf{r}(0) = P_1$, $\mathbf{r}(1) = P_4$, $\mathbf{r}'(0) = 3(P_2 - P_1)$, $\mathbf{r}'(1) = 3(P_4 - P_3)$.

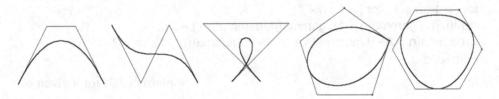

Fig. 9.3 Elementary and closed Bézier curves.

The definition of elementary Bézier curve is equivalent to

$$\mathbf{r}(t) = (((1-t)P_1 + 3tP_2)(1-t) + 3t^2 P_3)(1-t) + t^3 P_4, \qquad 0 \le t \le 1. \tag{9.4}$$

The matrix presentation of (9.4) has the form $\mathbf{r}(t) = \mathbf{G}MT(t)$ ($0 \le t \le 1$), where $\mathbf{r}(t) = [x(t); y(t); z(t)]$ and $T(t) = [t^3; t^2; t; 1]$ are column vectors and

$$\mathbf{G} = (P_1\,P_2\,P_3\,P_4), \qquad M = \begin{pmatrix} -1 & 3 & -3 & 1 \\ 3 & -6 & 3 & 0 \\ -3 & 3 & 0 & 0 \\ 1 & 0 & 0 & 0 \end{pmatrix}. \qquad (9.5)$$

Here M is called the *basic matrix* of the cubic Bézier curve, and $\mathbf{P}$ is its *geometric matrix*. The derivatives of the Bézier curve are

$$\mathbf{r}'(t) = \mathbf{G}MT'(t), \qquad \mathbf{r}''(t) = \mathbf{G}MT''(t), \qquad \mathbf{r}'''(t) = \mathbf{G}MT'''(t), \qquad (9.6)$$

where $T'(t) = [0;\ 1;\ 2t;\ 3t^2]$, $T''(t) = [0;\ 0;\ 2;\ 6t]$, $T'''(t) = [0;\ 0;\ 0;\ 6]$.

The gluing of many elementary Bézier segments leads to a *piecewise* (long) *Bézier curve*.

The *piecewise cubic Bézier curve* defined by the array $\mathbf{P} = P_1, \ldots, P_{3m+1}$ ($m \geq 1$) is a curve $\gamma(t)$ ($0 \leq t \leq m$) represented as the union of m elementary cubic Bézier curves $\gamma^{(1)} \cup \cdots \cup \gamma^{(m)}$; the section $\gamma^{(i)}(t)$ corresponds to the points $P_{3i-2}, P_{3i-1}, P_{3i}, P_{3i+1}$ for $i-1 \leq t \leq i$ and $i = 1, \ldots, m$.

For simplicity, assume the condition $P_{3i} = \frac{1}{2}(P_{3i-1} + P_{3i+1})$ and do not include points P_{3i} in the array $\mathbf{P}$. Thus in the procedure **bezier2d** (for the plane curve), $\mathbf{P}$ contains $2m+2$ points, and $\gamma(t)$ ($0 \leq t \leq m$) consists of m segments, defined as follows: $\gamma^{(1)}$ by the points P_1, P_2, P_3; $\gamma^{(i)}$ ($2 \leq i \leq m-1$) by the points P_{2i-1}, P_{2i}; and $\gamma^{(m)}$ by the points $P_{2m-1}, P_{2m}, P_{2m+1}$.

The result of $\mathbf{B} = \textbf{bezier2d}(x, y)$ is the matrix with parametric equations of the curve components $[x_i(t), y_i(t)]$, where $i = 1, \ldots, m-1$.

Examples.

1. Calculate and plot three cubic (elementary) Bézier curves of different shapes using the program **bezier**(x, y).

```
x = [0, 1, 2, 3];  y = [0, 2, 2, 0];                    % Figure 9.3(a)
x = [0, 1, 2, 3];  y = [0, 2, 0, 2];            %  zigzag, Figure 9.3(b)
x = [2, 4, 1, 3];  y = [0, 2, 2, 0];            %  loop, Figure 9.3(c)
r = bezier(x, y)                                % use one of x, y
hold on;  ezplot(r(1), r(2), [0, 1]);  plot(x, y, 'o--r');
```

2. Plot the plane piecewise cubic Bézier curve, Figure 9.4(a), using the command **bezier 2d**(x, y).

```
x = [.056, .287, .655, .716, .228, .269, .666, .929];
y = [.820, .202, .202, .521, .521, .820, .820, .227];
r = bezier2d(x, y)         % matrix (the equations of the curve components)
hold on;  plot(x, y, 'o--r');
for i = 1 : (length(x)/2 - 1);
 ezplot(r(1, i), r(2, i), [i - 1, i]);
end
```

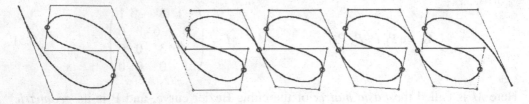

Fig. 9.4 Woven (piecewise) Bézier curve.

3. *Continuation* of 2. Plot the self-repeating piecewise cubic Bézier curve, where the circled points are derived using the program

```
n = length(x);
x1 = [x, x(3 : n) + .61,  x(3:n) + 2*.61,  x(3:n) + 3*.61];
y1 = [y, y(3 : n),  y(3 : n),  y(3 : n)];
B = bezier2d(x1, y1);
hold on;  plot(x1, y1, 'o - - r');
for i = 1 : (length(x1)/2 - 1),
  ezplot(B(1, i), B(2, i), [i - 1, i]);
end;                                                        % Figure 9.4(b)
```

4. Plot closed Bézier curves of two types:

(a) with the loop having a singular point, $k \in \mathbb{N}$.

```
k = 3;                                                      % enter k
i = 1 : 2*k;   x = sin(2*pi*i/(2*k - 1));    y = cos(2*pi*i/(2*k - 1));
B = bezier2d(x, y);                                         % Figure 9.3(d)
```

(b) C^1-continuous closed curve, $k \in \mathbb{N}$.

```
k = 3;                                                      % enter k
i = 1 : 2*k + 2;      x = sin(2*pi*i/(2*k));   y = cos(2*pi*i/(2*k));
x(1) = (x(2) + x(2*k + 1))/2;   y(1) = (y(2) + y(2*k + 1))/2;
x(2*k + 2) = x(1);   y(2*k + 2) = y(1);
B = bezier2d(x, y);                                         % Figure 9.3(e).
```

9.2.3 *Rational Bézier curves*

We can extend our field of view if we do not restrict ourself to polynomials.

The *rational Bézier curve* associated to control points $P_1, \ldots, P_{n+1}$ of $\mathbb{R}^N$ and corresponding *weights* $w_1, w_2, \ldots, w_{n+1}$ is given by

$$B(t) = \frac{\sum_{i=0}^{n} w_{i+1} B_{n,i}(t) P_{i+1}}{\sum_{i=0}^{n} w_{i+1} B_{n,i}(t)} = \sum_{i=0}^{n} \frac{w_{i+1} B_{n,i}(t)}{\sum_{j=0}^{n} w_{j+1} B_{n,j}(t)} P_{i+1}.$$

The rational Bézier curve is *projective invariant*, i.e.,

$$f\left(\sum_{i=0}^{n} \frac{w_{i+1} B_{n,i}(t)}{\sum_{j=0}^{n} w_{j+1} B_{n,j}(t)} P_{i+1} \right) = \sum_{i=0}^{n} \frac{w_{i+1} B_{n,i}(t)}{\sum_{j=0}^{n} w_{j+1} B_{n,j}(t)} f(P_{i+1})$$

for any projective transformation f.

We extend the program for Bézier curves (change its last four lines) to the case of a rational curve ($w_4 = w_6 = 5$, and other weights are equal to 1, Figure 9.2(b)). The result of f = **rbezier**(x, y, w) is the parametric polynomial curve $[f_1(t), f_2(t)]$. We derive and plot the plane curve

```
x = 2 : 8;    y = [32, 34, 36, 34, 39, 40, 37];    w = [1, 1, 5, 1, 5, 1, 1];
r = rbezier(x, y, w)                                          % Answer:
```

$$r = \left[\frac{3(2t-1)(20t^4 - 40t^3 + 20t^2 + 1)}{120t^6 - 360t^5 + 420t^4 - 240t^3 + 60t^2 + 1} + 5, \ \frac{4570t^6 - 13542t^5 + 15525t^4 - 8720t^3 + 2160t^2 + 12t + 32}{120t^6 - 360t^5 + 420t^4 - 240t^3 + 60t^2 + 1} \right]$$

```
hold on;    ezplot(r(1), r(2), [0, 1]);
plot(x, y, 'o - - r');                                        % Figure 9.2(b).
```

If all w_i are the same, then we obtain the usual Bézier curve.

One application of rational Bézier curves is the possibility of plotting all conic sections (circle, ellipse, hyperbola, parabola) without using trigonometric and hyperbolic functions.

Example. Given the three points $P_1 = (1,0)$, $P_2 = (1,1)$, $P_3 = (0,1)$, the corresponding Bézier curve is quadratic (see Example, p. 360) and is therefore a parabola $\mathbf{r}(t) = (1-t)^2 P_1 + 2t(1-t)P_2 + t^2 P_3 = [1 - t^2, 2t(1-t)]$.

The quadratic rational Bézier curve with the same control points and weights $w_1 = 1$, $w_2 = 1$, $w_3 = 2$ is $\mathbf{r}(t) = \left[\frac{1-t^2}{1+t^2}, \frac{2t}{1+t^2} \right]$ ($0 \le t \le 1$), which is the familiar parameterization of the unit quarter circle. In general, quadratic rational Bézier curves are conics:

parabola when $w_2^2 - w_1 w_3 = 0$;

ellipse $\mathbf{r}(t) = \left[a\frac{1-t^2}{1+t^2}, b\frac{2t}{1+t^2} \right]$ when $w_2^2 - w_1 w_3 < 0$;

hyperbola $\mathbf{r}(t) = \left[a\frac{1+t^2}{1-t^2}, b\frac{2t}{1-t^2} \right]$ when $w_2^2 - w_1 w_3 > 0$.

A rational Bézier curve with $m = 3$ is an *elementary* one. The gluing of many of these segments leads to a *piecewise* (long) *rational Bézier curve*.

A *piecewise rational Bézier curve* defined by $P_0, \ldots, P_{3m}$ ($m \ge 1$) is represented as the union of elementary rational Bézier curves $\gamma = \gamma^{(1)} \cup \cdots \cup \gamma^{(m)}$ ($0 \le t \le m-1$); the section $\gamma^{(i)}$ corresponds to the points $P_{3i-3}, P_{3i-2}, P_{3i-1}, P_{3i}$ for $i-1 \le t \le i$ and $i \le m$. By a special choice of weights on each segment, one obtains a C^2-continuous piecewise rational Bézier curve; see exercises. This property allows one to insert an elementary rational Bézier curve in the gap between any two just plotted C^2-continuous piecewise curves in such a way that the resulting curve is C^2-continuous.

9.3   Hermite Interpolation Curves

9.3.1   *Cubic Hermite curves*

Hermite cubic interpolation is based on two points P_1 and P_2 and two tangent vectors Q_1 and Q_2. It computes a curve segment that starts at P_1, moving with velocity Q_1, and ends at P_2, moving with velocity Q_2.

Given points P_1 and P_2 and nonzero vectors Q_1 and Q_2, the *elementary cubic Hermite curve* is defined by (5.16),

$$\mathbf{r}(t) = (1 - 3t^2 + 2t^3)P_1 + t^2(3 - 2t)P_2 + t(t-1)^2 Q_1 + t^2(t-1)Q_2, \qquad (9.7)$$

where $0 \le t \le 1$. The matrix presentation of (9.7) is $\mathbf{r}(t) = \mathbf{G}MT(t)$ $(0 \le t \le 1)$, where $\mathbf{r}(t) = [x(t);\ y(t);\ z(t)]$ and $T(t) = [t^3;\ t^2;\ t;\ 1]$ are column vectors, $\mathbf{G} = (P_1;\ P_2;\ Q_1;\ Q_2)$ is the *geometric matrix,* and

$$M = \begin{pmatrix} 2 & -3 & 0 & 1 \\ -2 & 3 & 0 & 0 \\ 1 & -2 & 1 & 0 \\ 1 & -1 & 0 & 0 \end{pmatrix} \qquad (9.8)$$

is the *basic matrix* of an elementary cubic Hermite curve.

Examples.

1. We call **hermite_my** to calculate the curve

```
xp = [.056, .287];  yp = [.820, .202];
xq = [2 -2];  yq = [1 -1];                                    % enter data
B = hermite_my(xp, yp, xq, yq);
```

and then plot the curve and the data:

```
hold on;    plot(xp, yp, 'o - -r');
arrow([xp(1), yp(1)],  [xq(1), yq(1)]/10, 'g');
arrow([xp(2), yp(2)],  [xq(2), yq(2)]/10, 'g');               % M-file arrow.m
ezplot(B(1), B(2),  [0, 1]);
```

2. Varying the magnitudes of both tangent vectors has an important geometric interpretation: it changes the *tension of the curve segment.* The matrix presentation of the

Hermite segment with the tension parameter $s > 0$ is $\mathbf{r}(t) = \mathbf{G}MT(t)$ $(0 \leq t \leq 1)$, where $\mathbf{G} = (P_1;\ P_2;\ sQ_1;\ sQ_2)$. The reader can play with the above program, replacing the data line by

```
s = 1.5;   xq = [2 -2]*s;  yq = [1 -1]*s;                    % enter data
```

3. Hermite interpolation provides a simple way to construct approximate circular arcs. Assume that an arc spanning an angle 2θ is needed and we place its endpoints P_1 and P_2 at locations $(\cos\theta, \sin\theta)$ and $(\cos\theta, \sin\theta)$, respectively. Since a circle is always perpendicular to its radius, we select as our start/end tangents two vectors $Q_1 = a(\sin\theta, \cos\theta)$ and $Q_2 = a(-\sin\theta, \cos\theta)$ that are perpendicular to P_1 and P_2. To determine a we require that the curve $\mathbf{r}(t)$ pass through the circular arc at its center, i.e., $\mathbf{r}(.5) = (1,0)$. This produces the equation

$$(1,0) = (\cos\theta + (a/4)\sin\theta,\ 0)$$

whose solution is $a = 4(1 - \cos\theta)/\sin\theta$. The plane curve can now be written in the matrix form $\mathbf{r}(t) = \mathbf{G}MT(t)$ $(0 \leq t \leq 1)$, where

$$G = \begin{pmatrix} \cos\theta & \cos\theta & 4(1-\cos\theta) & -4(1-\cos\theta) \\ \sin\theta & \sin\theta & 4(1-\cos\theta)/\tan\theta & 4(1-\cos\theta)/\tan\theta \end{pmatrix}.$$

This curve provides an excellent approximation to a circular arc, even for angles θ as large as 90°. You can verify this using the program below:

```
theta = 89/180*pi;                                          % enter data
xp = [1, 1]*cos(theta);  yp = [1, -1]*4*(1-cos(theta));
xq = [1 -1]*4*(1-cos(theta));  yq = [1, 1]*4*(1-cos(theta))/tan(theta);
B = hermite_my(xp, yp, xq, yq);       % approximation to a circular arc
t = linspace(-pi, pi, 200);    plot(cos(t), sin(t), '-r');
arrow([xp(1), yp(1)], [xq(1), yq(1)]/3, 'g');
arrow([xp(2), yp(2)], [xq(2), yq(2)]/3, 'g');               % M-file arrow.m
hold on;    ezplot(B(1), B(2), [0, 1]);
```

9.3.2 Piecewise Hermite curves

Using elementary cubic Hermite curves, the piecewise curve is determined by the points $P_1, \ldots, P_n$ $(n \geq 2)$ and nonzero vectors $Q_1, \ldots, Q_n$. Each quadruple P_i, Q_i, P_{i+1}, Q_{i+1} determines an elementary Hermite curve in the interval $t_{i-1} \leq t \leq t_i$. However, we obtain only a C^1-continuous piecewise curve. A C^2-continuous piecewise curve is also obtained in this way.

Example. We call **hermite2d** to derive a piecewise plane curve:

```
syms t;
xp = [.056, .287, .655, .716, .228, .269, .666, .929];
yp = [.820, .202, .202, .521, .521, .820, .820, .227];
xq = [.25, .25, -.25, -.25, .25, .25, -.25, -.25];
yq = [-.25, .25, -.25, .25, -.25, .25, -.25, .25];      % enter data
B = hermite2d(xp, yp, xq, yq);
```

Then we plot the curve and the data (points and arrows), Figure 9.5(a).

```
hold on;    plot(x, y, 'o - -r');
for i = 1 : length(x),   ezplot(B(1, i), B(2, i), [i - 1, i]);   end
arrow([xp(i), yp(i)], [xq(i), yq(i)], 'g');
```

The first two lines of this typical program will be referred to as **splot2**(x, y, B).

A C^2-continuous piecewise Hermite curve contains all control vertices, and at the endpoints has tangent vectors Q_1 and Q_n. Its curvature (at the vertices $P_2, \ldots, P_{n-1}$) is continuous.

Namely, a *piecewise (cubic) Hermite curve*, determined by the points $\mathbf{P} = P_1, \ldots, P_n$ ($n \geq 2$) and a pair of nonzero velocity vectors Q_1 and Q_n at the endpoints P_1 and P_n, is defined as a C^2-continuous curve that can be represented as a union of elementary Hermite curves $\gamma = \gamma^{(1)} \cup \cdots \cup \gamma^{(n-1)}$; the segment $\gamma^{(i)}$ corresponds to the points P_i, Q_i, P_{i+1}, Q_{i+1} for $i - 1 \leq t \leq i$, and the vectors $Q_2, \ldots, Q_{n-1}$ are derived as follows.

Changing a point in the array $\mathbf{P}$, adding a point, or changing one tangent vector at the endpoints leads to changing the whole Hermite curve. Let

$$\mathbf{r}_i(t) = (1 - 3t^2 + 2t^3)P_i + t^2(3 - 2t)P_{i+1} + t(1 - 2t + t^2)Q_i - t^2(1 - t)Q_{i+1}.$$

The end conditions, continuity, and C^1-smoothness are satisfied:

$$\mathbf{r}_1(0) = P_1, \quad \mathbf{r}_{n-1}(1) = P_n, \quad \mathbf{r}_1'(0) = Q_1, \quad \mathbf{r}_{n-1}'(1) = Q_n,$$
$$\mathbf{r}_{i-1}(1) = \mathbf{r}_i(0) = P_i, \quad \mathbf{r}_{i-1}'(1) = \mathbf{r}_i'(0) = Q_i \quad (2 \leq i \leq n-1).$$

The continuity of the second derivative $\mathbf{r}_{i-1}''(1) = \mathbf{r}_i''(0)$ ($2 \leq i \leq n - 1$) gives $n - 2$ vector equations

$$6P_{i-1} - 6P_i + 2Q_{i-1} + 4Q_i = -6P_i + 6P_{i+1} - 4Q_i - 2Q_{i+1}$$

or, equivalently,

$$Q_{i-1} + 4Q_i + Q_{i+1} = -3P_{i-1} + 3P_{i+1} \quad (2 \leq i \leq n-1). \tag{9.9}$$

Hence, the vectors $Q_2, \ldots, Q_{n-1}$ are derived from the matrix equation

$$
\begin{pmatrix} 1 & 4 & 1 & & \\ & 1 & 4 & 1 & \\ & & \cdots & & \\ & & 1 & 4 & 1 \end{pmatrix} \begin{pmatrix} Q_1 \\ Q_2 \\ \cdots \\ Q_n \end{pmatrix} = \begin{pmatrix} -3 & 0 & 3 & & \\ & -3 & 0 & 3 & \\ & & \cdots & & \\ & & -3 & 0 & 3 \end{pmatrix} \begin{pmatrix} P_1 \\ P_2 \\ \cdots \\ P_n \end{pmatrix} \qquad (9.10)
$$

involving two $(n-2) \times n$ matrices. The linear system (9.10) is solved in the procedure **hermite_cubic2d**.

Example. We call **hermite_cubic2d** to derive a piecewise plane curve:

```
xp = [.056, .287, .655, .716, .228, .269, .666, .929];
yp = [.820, .202, .202, .521, .521, .820, .820, .227];
xq = [.1, -.1];    yq = [.1, -.1];                              % can be modified
B = hermite_cubic2d(xp, yp, xq, yq);
```

Then we plot the curve with the data (points and arrows), using **splot2**(x, y, B), Figure 9.5(b), and additional lines.

```
arrow([xp(1), yp(1)], [xq(1), yq(1)], 'g');
arrow([xp(end), yp(end)], [xq(2), yq(2)], 'g');
```

The reader can play with a tension parameter $s > 0$ of Hermite curve, replacing one line in the above program:

```
s = 1.5;   xq = [.1, -.1]*s;   yq = [.1, -.1]*s;               % enter s
```

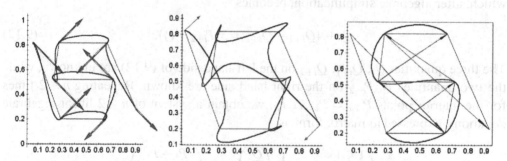

Fig. 9.5 Interpolated and piecewise Hermite curves. Catmull–Rom curve.

Practical curve methods that create a spline curve out of individual Hermite segments can add a tension parameter to the spline, thereby making the method more interactive (see the cardinal splines method).

9.3.3 *The cubic spline interpolation*

Based on a prescribed form of the common segment $S_i(x)$, we derived a spline interpolation of functions (see Section 1.6), used in the construction of spline curves.

Now we consider an alternative approach to applying the Hermite interpolant to deriving the cubic spline curve. The idea is to divide the set of n points into $n-1$ overlapping pairs of two points each, and to fit a Hermite segment (9.7), specified by two points and two tangents, to each pair.

We start with three adjacent points, P_i, P_{i+1}, P_{i+2}, of which P_{i+1} must be an interior point. The velocity vectors at these points are Q_i, Q_{i+1}, Q_{i+2}. The Hermite segments from P_i to P_{i+1} and from P_{i+1} to P_{i+2} are, respectively,

$$\mathbf{r}_i(t) = n_1(t)P_i + n_2(t)P_{i+1} + n_3(t)Q_i + n_4(t)Q_{i+1},$$
$$\mathbf{r}_{i+1}(t) = n_1(t)P_{i+1} + n_2(t)P_{i+2} + n_3(t)Q_{i+1} + n_4(t)Q_{i+2}, \qquad (9.11)$$

where $n_1 = 1 - 3t^2 + 2t^3$, $n_2 = 3t^2 - 2t^3$, $n_3 = t - 2t^2 + t^3$, and $n_4 = t^3 - t^2$.

Next, we require that the second derivatives of the two segments (linear functions of t) be equal at the interior points: $\mathbf{r}_i''(1) = \mathbf{r}_{i+1}''(0)$. Using $n_1'' = 12t - 6$, $n_2'' = 6 - 12t$, $n_3'' = 6t - 4$, $n_4'' = 6t - 2$, we obtain

$$[6(2t-1)P_i - 6(2t-1)P_{i+1} + 2(3t-2)Q_i + 2(3t-1)Q_{i+1}]_{t=1}$$
$$= [6(2t-1)P_{i+1} - 6(2t-1)P_{i+2} + 2(3t-2)Q_{i+1} + 2(3t-1)Q_{i+2}]_{t=0},$$

which, after algebraic simplification, becomes

$$Q_i + 4Q_{i+1} + Q_{i+2} = 3(P_{i+2} - P_i). \qquad (9.12)$$

The three quantities Q_i, Q_{i+1}, Q_{i+2} on the left-hand side of (9.12) are unknown, while the two quantities P_i, P_{i+2} on the right hand side are known. Repeating $n-2$ times for the interior points $P_{i+1} = P_2, \ldots, P_n$, we obtain a system of $n-2$ linear algebraic equations, expressed in matrix form as

$$\begin{pmatrix} 1 & 4 & 1 & & \\ & 1 & 4 & 1 & \\ & & \ldots & & \\ & & & 1 & 4 & 1 \end{pmatrix} \begin{pmatrix} Q_1 \\ Q_2 \\ \ldots \\ Q_n \end{pmatrix} = 3 \begin{pmatrix} P_3 - P_1 \\ P_4 - P_2 \\ \ldots \\ P_n - P_{n-2} \end{pmatrix} \qquad (9.13)$$

involving two $(n-2) \times n$ matrices. A practical approach to the solution is to let the reader specify the values of the two extreme tangents Q_1 and Q_n.

Since a cubic spline is based on Hermite segments, its *tension* can also be controlled in the same way. So a user-friendly algorithm inputs a parameter $T \in [0, 1]$ and multiplies each tangent vector by $s = \alpha(1 - T)$ for some predetermined α.

It turns out that splines of degrees higher than 3 are useful only for special applications because they are more computation-intensive and tend to have many undesirable inflection points (i.e., they tend to wiggle excessively). Splines of degree 1 are, of course, just segments connected to form a polygon, but quadratic splines can be useful in some applications.

As with the cubic spline, there are more unknowns than equations, and the standard technique is to ask the reader to provide a value for one of the unknown tangent vectors, normally Q_1.

9.3.4 Application: Cardinal spline curves

The cardinal spline is another example of how Hermite interpolation is applied to construct a spline curve.

Given points P_1, P_2, P_3, and P_4, the *elementary cardinal spline curve* $\mathbf{r}(t)$ ($0 \le t \le 1$) is defined by the property that the velocity at the endpoint $\mathbf{r}(0) = P_2$ is the vector $s \cdot \overrightarrow{P_1 P_3}$; the velocity at the endpoint $\mathbf{r}(1) = P_3$ is the vector $s \cdot \overrightarrow{P_2 P_4}$, where s is a real number (Figure 9.5(c)). The *tension parameter* is defined as $T = 1 - 2s$. (The *Catmull–Rom spline* is defined as the curve having tension zero in this case, i.e., $s = 1/2$).

To deduce the vector equation of the *elementary cardinal spline curve*, we substitute in (9.7) $Q_1 = s(P_3 - P_1)$, $Q_2 = s(P_4 - P_2)$ and obtain

$$\mathbf{r}(t) = (2t^2 - t - t^3)sP_1 + [1 + (2-s)t^3 + (s-3)t^2]P_2$$
$$+ [(s-2)t^3 + (3-2s)t^2 + ts]P_3 + (t^3 - t^2)sP_4. \quad (9.14)$$

For $s = 1/2$ we obtain $\mathbf{r}(t) = n_1(t)P_1 + n_2(t)P_2 + n_3(t)P_3 + n_4(t)P_4$, where the blending functions are

$$n_1 = -\frac{t}{2}(1-t)^2,\ n_2 = \frac{1}{2}(2 - 5t^2 + 3t^3),\ n_3 = \frac{t}{2}(1 + 4t - 3t^2),\ n_4 = -\frac{t^2}{2}(1-t). \quad (9.15)$$

The matrix presentation of (9.15) is easily calculated by applying Hermite interpolation (9.8), starting from $\mathbf{P}_h = (P_2\ P_3\ s(P_3 - P_1)\ s(P_4 - P_2))$ and the basic *geometric matrix* (9.8) of the Hermite spline. Hence, in the particular case of $s = 1/2$ we obtain the matrix equations $\mathbf{r}(t) = \mathbf{P}MT(t)$ ($0 \le t \le 1$), where $\mathbf{r}(t) = [x(t);\ y(t);\ z(t)]$ and $T(t) = [t^3;\ t^2;\ t;\ 1]$ are column vectors, $\mathbf{P} = (P_1\ P_2\ P_3\ P_4)$ is the basic *geometric matrix*, and

$$M = \frac{1}{2} \begin{pmatrix} -1 & 2 & -1 & 0 \\ 3 & -5 & 0 & 2 \\ -3 & 4 & 1 & 0 \\ 1 & -1 & 0 & 0 \end{pmatrix} \quad (9.16)$$

is the *basic matrix* (also termed the *parabolic blending matrix*).

A *piecewise cardinal spline curve* determined by the array $\mathbf{P} = P_1, \ldots, P_n$ ($n \geq 2$) is a curve $\gamma(t)$ ($0 \leq t \leq n-1$) represented as a union of elementary cardinal curves $\gamma^{(1)} \cup \cdots \cup \gamma^{(n-1)}$; the segment $\gamma^{(i)}(t)$ ($i-1 \leq t \leq i$) corresponds to the four points P_i, $P_{i+1}, P_{i+2}, P_{i+3}$, and $1 \leq i \leq n-1$.

The curve obtained is C^1-continuous, and it interpolates the points $\mathbf{r}(0) = P_2, \ldots,$ $\mathbf{r}(n-1) = P_{n-1}$. The tangent vectors at these points are: $\mathbf{r}'(i) = s(P_{i+2} - P_i)$ for $0 \leq i \leq n-1$ and some $s > 0$. Changing one point in $\mathbf{P}$ or adding one point changes only part of the cardinal curve: we need to recalculate the equations of the *four segments* $\gamma^{(i-1)}$, $\gamma^{(i)}, \gamma^{(i+1)}, \gamma^{(i+2)}$.

Examples.

1. Use **card_my** to derive the elementary curve.

```
s = 1/2;   x = 2 : 5;   y = [33.5, 34.8, 35, 33.5];          % enter data
B = card_my(x, y, s);
```

Then plot the curve and the data:

```
hold on;  plot(x, y, 'r');
arrow([x(2), y(2)], [(x(3) - x(1))/2, (y(3) - y(1))/2]/3, 'g');
arrow([x(3), y(3)], [(x(4) - x(2))/2, (y(4) - y(2))/2]/3, 'g');
ezplot(B(1), B(2),  [0, 1]);
```

2. Call **card2d** to derive a plane curve.

```
s = 1/2;   x = [.056, .287, .655, .716, .228, .269, .666, .929];
y = [.820, .202, .202, .521, .521, .820, .820, .227];          % enter data
B = card2d(x, y, s);
```

Plot the curve and play with the parameter s:

```
hold on;  axis equal;  plot(x, y, '--r');
for i = 1 :  length(x) - 3;
  arrow([x(i + 1), y(i + 1)], [(x(i + 2) - x(i))/2, (y(i + 2) - y(i))/2], 'g');
  arrow([x(i + 2), y(i + 2)], [(x(i + 3) - x(i + 1))/2, (y(i + 3) - y(i + 1))/2], 'g');
  ezplot(B(1, i), B(2, i),  [i - 1, i]);
end                                                          % Figure 9.5(b)
```

9.4 β-Spline Curves

Ideally, we would like a measure of continuity that is parameterization-independent. This can be done by using parameterizations to describe the point set comprising the piecewise curve, and then using an arc-length parameterization to determine smoothness.

9.4.1 *Geometrical continuity of curves*

An elementary curve γ: $\mathbf{r}(t)$ is called *geometrically C^k-continuous* if the natural pa-parameterization is regular C^k-continuous.

It is not necessary to check the natural parameterization for continuity. Any regular C^k-equivalent parameterization will do. Consider more carefully the cases $k = 1, 2$ in $\mathbb{R}^3$.

Definition 9.1. A curve γ: $\mathbf{r}(t)$ in $\mathbb{R}^3$ is called *geometrically*

(a) C^1-*continuous* if its tangent line (i.e., the unit vector $\tau = \mathbf{r}'/|\mathbf{r}'|$) changes continuously;

(b) C^2-*continuous*, if its main normal vector and the vector of curvature

$$\mathbf{k} = \frac{(\mathbf{r}' \times \mathbf{r}'') \times \mathbf{r}'}{|\mathbf{r}'|^4}$$

(derived by the formula of the *double product*) change continuously.

1. A C^2-continuous curve is characterized by the geometrically visible property that *the osculating circle continuously varies in t.*

2. C^2-continuity of a curve is a stronger condition than C^2-*regularity*, meaning C^1-continuity together with the continuity of $\mathbf{r}''(t)$.

Example. The piecewise plane curves, Figure 9.6, illustrate the fact that the continuity of the vectors $\mathbf{r}'$ and $\mathbf{r}''$ (depending on the choice of parameterization of the curve) does not imply the continuity of the vectors τ and $\mathbf{k}$ (which do not depend on the choice of parameterization of the curve), and hence does not always lead to a curve with nice geometry. Each curve in this example is composed of two segments: $\mathbf{r}_1(t)$ and $\mathbf{r}_2(t)$ ($0 \leq t \leq 1$), and the *condition of continuity* $\mathbf{r}_1(1) = \mathbf{r}_2(0)$ holds.

(a) $\mathbf{r}_1(t) = [\sin(\pi t^2/2), \cos(\pi t^2/2)]$, $\mathbf{r}_2(t) = [\cos(\pi t^2/2), -\sin(\pi t^2/2)]$ are two arcs of a circle; the curve is C^2-continuous.

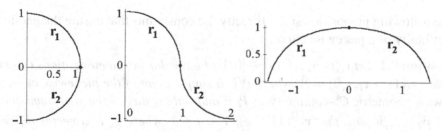

Fig. 9.6 Two-component curves.

(b) $\mathbf{r}_1(t) = [\cos(\frac{\pi}{2}(1-t)^3), \sin(\frac{\pi}{2}(1-t)^3)]$, $\mathbf{r}_2(t) = [2 - \cos(\frac{\pi}{2}t^3), -\sin(\frac{\pi}{2}t^3)]$ are arcs of two circles, The curve is C^1-continuous but is not C^2-continuous.

(c) $\mathbf{r}_1(t) = [t^2 + \frac{1}{2}(t-3), -t^2 + 2t]$, $\mathbf{r}_2(t) = [-t^2 + \frac{5}{2}t, -t^2 + 1]$. The second derivatives at the node are not equal, but the curve is C^2-continuous.

Each curve in this example (composed of two segments) is *continuous*: $\mathbf{r}_1(1) = \mathbf{r}_2(0)$. We will find the characteristics of the curve at its node, and then plot it.

(a) $\mathbf{r}_1(t) = [\sin(\pi t^2/2), \cos(\pi t^2/2)]$, $\mathbf{r}_2(t) = [\cos(\pi t^2/2), -\sin(\pi t^2/2)]$ are two arcs of a circle (Figure 9.6(a)). The derivatives at $(1,0)$ differ:

$$\mathbf{r}_1'(1) = [0, -\pi], \qquad\qquad \mathbf{r}_2'(0) = [0, 0],$$
$$\mathbf{r}_1''(1) = [-\pi^2, -\pi], \qquad\qquad \mathbf{r}_2''(0) = [0, -\pi].$$

However, the curvature vector $\mathbf{k}$ is continuous, because the whole curve coincides with the half-circle. The curve is C^2-continuous.

```
r1 = [sin(pi/2*t^2), cos(pi/2*t^2)];  t1 = 0;  t2 = 1;
r2 = [cos(pi/2*t^2), - sin(pi/2*t^2)];  t3 = 0;  t4 = 1;
```

(b) $\mathbf{r}_1(t) = [\cos(\frac{\pi}{2}(1-t)^3), \sin(\frac{\pi}{2}(1-t)^3)]$, $\mathbf{r}_2(t) = [2 - \cos(\frac{\pi}{2}t^3), -\sin(\frac{\pi}{2}t^3)]$ are arcs of two circles (Figure 9.6(b)). Although both derivatives at the node $(1,0)$ are equal, $\mathbf{r}_1'(1) = [0,0] = \mathbf{r}_2'(0)$, $\mathbf{r}_1''(1) = [0,0] = \mathbf{r}_2''(0)$, i.e., the vectors $\mathbf{r}'$ and $\mathbf{r}''$ are continuous, the main normal vectors $\mathbf{n}_1(1)$ and $\mathbf{n}_2(0)$ at the common point of the two half-circles are oppositely directed. The curve is C^1-continuous but is not C^2-continuous.

```
r1 = [cos(pi/2*(1 - t)^3), sin(pi/2*(1 - t)^3)];  t1 = 0;  t2 = 1;
r2 = [2-cos(pi/2*t^3), -sin(pi/2*t^3)];  t3 = 0;  t4 = 1;
```

(c) $\mathbf{r}_1(t) = [t^2 + \frac{1}{2}(t-3), -t^2 + 2t]$, $\mathbf{r}_2(t) = [-t^2 + \frac{5}{2}t, -t^2 + 1]$ (Figure 9.6(c)). The tangent lines at the node $(1,0)$ coincide, i.e., the first derivatives are equal and nonzero: $\mathbf{r}_1'(1) = [5/2, 0] = \mathbf{r}_2'(0)$. The second derivatives at the node are not equal, $\mathbf{r}_1''(1) = [2, -2]$, $\mathbf{r}_2''(0) = [-2, -2]$, but the curve is C^2-continuous because its curvature is continuous: $k_1(1) = k_2(0) = 8/25$.

```
r1 = [t^2+t/2-3/2, -t^2+2*t];  t1 = 0;  t2 = 1;
r2 = [-t^2+5*t/2, -t^2+1];  t3 = 0;  t4 = 1;
```

The following proposition states directly the constraints that ensure the geometrical C^k-continuity of a piecewise curve.

Proposition 9.1. *Let $\mathbf{r}_i(t)$, $\mathbf{r}_{i+1}(t)$ $(t \in [0,1])$ be regular parameterizations (segments) such that $\mathbf{r}_i(1) = \mathbf{r}_{i+1}(0) = \hat{P}_i$, where $\hat{P}_i$ is a simple point of the piecewise curve. They meet with geometric C^k-continuity at $\hat{P}_i$ if and only if there exist real numbers $\beta_1 > 0$, and $\beta_2, \ldots, \beta_k$ such that $\mathbf{r}_i^{(j)}(1) = g_j$, $1 \leq j \leq k$, where g_j is a vector differential expression which depends on $\mathbf{r}_i(t)$ and $\beta_1, \ldots, \beta_k$, and is computed using the following rules:*

1. *Expand $\frac{d^j \mathbf{r}_i}{du^j}(t(u))$ using the chain rule, treating $t: [\bar{u}, 1] \to [0, 1]$ as a regular C^k change of parameter. The expression should involve only derivatives of $\mathbf{r}_i^{(j)}(1)$ and derivatives of $t(u)$ with respect to u.*

2. *Evaluate at $\hat{P}_i$ and make substitutions $\beta_j = \frac{d^j t}{du^j}(1)$, $1 \le j \le k$.*

The process generates the following set of constraints:

$$\mathbf{r}'_{i+1}(0) = \beta_1 \mathbf{r}'_i(1),$$

$$\mathbf{r}''_{i+1}(0) = \beta_1^2 \mathbf{r}''_i(1) + \beta_2 \mathbf{r}'_i(1),$$

$$\mathbf{r}'''_{i+1}(0) = \beta_1^3 \mathbf{r}'''_i(1) + 2\beta_1 \beta_2 \mathbf{r}''_i(1) + \beta_3 \mathbf{r}'_i(1),$$

$$\mathbf{r}^{(4)}_{i+1}(0) = \beta_1^4 \mathbf{r}_i^{(4)}(1) + 6\beta_1^2 \beta_2 \mathbf{r}'''_i(1) + (4\beta_1 \beta_3 + 3\beta_1^2)\mathbf{r}''_i(1) + \beta_4 \mathbf{r}'_i(1),$$

$$\cdots$$

$$\mathbf{r}^{(k)}_{i+1}(0) = \beta_1^k \mathbf{r}_i^{(k)}(1) + \cdots + \beta_k \mathbf{r}'_i(1). \tag{9.17}$$

The idea is to construct splines that satisfy C^k constraints with matching parametric derivatives. The resulting spline will have the quantities $\beta_1, \ldots, \beta_k$ as parameters. Changing one of the β's will, in general, change the shape of the piecewise curve, but always in such a way that geometric smoothness is maintained; thus we refer to the β's as *shape parameters*.

Note that the shape parameter β_j is introduced in the constraint relating the jth derivatives in the parameterizations. For example, β_1, in the first equation of (9.17), controls the difference between the first parametric derivatives, but in such a way that the piecewise curve is geometrically smooth. If $\beta_1 = 1$ and $\beta_2 = 0$, the first two constraints of (9.17) reduce to parametric C^2-continuity. In general, if $\beta_1 = 1$ and $\beta_2 = \cdots = \beta_k = 0$, then C^k-continuity reduces to C^k, showing that geometric continuity is a strict generalization of parametric continuity. Using the β-constraints instead of requiring continuous parametric derivatives introduces n degrees of freedom (i.e., shape parameters). The shape parameters can be made available to a designer as a convenient method for changing the shape of the curve without altering the control polygon.

It is important to realize that the shape parameters are *local* at a joint. Thus, for a spline of n segments ($n-1$ joints) generated by C^k parameterizations, a total of $(n-1)k$ shape parameters are introduced. In what follows, for simplicity, we associate the same values of the k shape parameters with each of the joints, thereby making the assignment of shape parameters *global* to the piecewise curve.

Example. A *quadratic β-spline* is defined as a weighted sum of arbitrary control vertices $\mathbf{P} = P_1, \ldots, P_n$; the segments are

$$\mathbf{r}_i(t) = \sum_{j=1}^{3} b_j(t) P_{i+j-1} = b_1(t) P_i + b_2(t) P_{i+1} + b_3(t) P_{i+2} \quad (1 \le i \le n-1). \tag{9.18}$$

We will determine the functions $b_j(t)$ in such a way that $\mathbf{r}_i(t)$ and $\mathbf{r}_{i+1}(t)$ will meet with geometric C^1-continuity. The quadratic β-spline basis functions are of the form $b_j(t) = c_{j,1} + c_{j,2}t + c_{j,3}t^2$. The continuity condition $\mathbf{r}_{i+1}(0) = \mathbf{r}_i(1)$ gives us four equations:

$$0 = c_{1,1} + c_{1,2} + c_{1,3}, \; c_{1,1} = c_{2,1} + c_{2,2} + c_{2,3}, \; c_{2,1} = c_{3,1} + c_{3,2} + c_{3,3}, \; c_{3,1} = 0. \quad (9.19)$$

We will use the first derivative constraint of (9.17); for simplicity, assume a global assignment of the shape parameter β_1. We obtain four equations:

$$0 = b_1'(1), \quad b_1'(0) = \beta_1 b_2'(1), \quad b_2'(0) = \beta_1 b_3'(1), \quad b_3'(0) = 0,$$

or, using $b_i'(t) = c_{i,2} + 2c_{i,3}t$,

$$0 = c_{1,2} + 2c_{1,3}, \; c_{1,2} = \beta_1(c_{2,2} + 2c_{2,3}), \; c_{2,2} = \beta_1(c_{3,2} + 2c_{3,3}), \; c_{3,2} = 0. \quad (9.20)$$

The system (9.19)–(9.20) of 8 linear equations contains 9 coefficients $c_{i,j}$. The additional constraint is a normalization, which is chosen to be

$$b_1(0) + b_2(0) + b_3(0) = 1 \quad \Rightarrow \quad c_{1,1} + c_{2,1} + c_{3,1} = 1. \quad (9.21)$$

The solution can be shown to be

$$\begin{aligned}
&c_{1,1} = \beta_1/(\beta_1 + 1), &&c_{1,2} = -2\beta_1/(\beta_1 + 1), &&c_{1,3} = \beta_1/(\beta_1 + 1), \\
&c_{2,1} = 1/(\beta_1 + 1), &&c_{2,2} = 2\beta_1/(\beta_1 + 1), &&c_{2,3} = -1, &&(9.22) \\
&c_{3,1} = c_{3,2} = 0, &&c_{3,3} = 1/(\beta_1 + 1).
\end{aligned}$$

Hence,

$$b_1(t) = \beta_1 \frac{(t-1)^2}{\beta_1 + 1}, \qquad b_2(t) = \frac{2\beta_1 t + 1}{\beta_1 + 1} - t^2, \qquad b_3(t) = \frac{t^2}{\beta_1 + 1}. \quad (9.23)$$

This spline is of type C^1. When $\beta_1 = 1$, it reduces to the uniform quadratic B-spline (see Example 1, p. 384). A quadratic β-spline is called an approximating technique because the curve is not guaranteed to interpolate through the control vertices.

Given points P_1, P_2, P_3, an (elementary) quadratic β-spline curve is defined by the equation (9.18), where the three functional coefficients are given by (9.23). We plot the quadratic β-spline using the M-file **beta_s21d**.

```
syms t;    beta = 1;
x = [.056, .287, .655, .716, .228, .269, .666, .929];
y = [.820, .202, .202, .521, .521, .820, .820, .227];
B = beta_s21d(x, y, beta);                    % the quadratic β-spline
hold on;    plot(x, y, 'r');
for i = 1 : (length(x) - 2);
```

```
ezplot(B(1, i), B(2, i), [i - 1, i]);
end;
```

Compare it with a C^1 interpolating spline (cubic Catmull–Rom spline).

Example. Consider a *quadratic β-spline* in $\mathbb{R}^2$ (Example, p. 375), without a global assumption on β_1. Here β_{1i} $(1 \leq i \leq n)$ corresponds to the joint of the segments $\mathbf{r}_i(t)$ and $\mathbf{r}_{i+1}(t)$. Denote for simplicity $\beta_i = \beta_{1i}$. Given β_1, we wish to find the remaining β_i $(2 \leq i \leq n)$, subject to the condition that the curvature of $\mathbf{r}(t)$ is continuous at the joints.

For the first step we write down and solve the linear system for geometrical C^1-continuity of the curve. The basis functions depend on i and are of the form $b_{j,i}(t) = c^i_{j,1} + c^i_{j,2}t + c^i_{j,3}t^2$. The continuity condition $\mathbf{r}_{i+1}(0) = \mathbf{r}_i(1)$ gives us four equations (see Example, p. 375):

$$0 = b_{1,i}(1), \qquad\qquad b_{1,i+1}(0) = b_{2,i}(1),$$
$$b_{2,i+1}(0) = b_{3,i}(1), \qquad\qquad b_{3,i+1}(0) = 0,$$

or

$$0 = c^i_{1,1} + c^i_{1,2} + c^i_{1,3}, \qquad\qquad c^{i+1}_{1,1} = c^i_{2,1} + c^i_{2,2} + c^i_{2,3},$$
$$c^{i+1}_{2,1} = c^i_{3,1} + c^i_{3,2} + c^i_{3,3}, \qquad\qquad c^{i+1}_{3,1} = 0. \tag{9.24}$$

We will use the first derivative constraint of (9.17) and obtain four equations:

$$0 = b'_{1,i}(1), \qquad\qquad b'_{1,i+1}(0) = \beta_i b'_{2,i}(1),$$
$$b'_{2,i+1}(0) = \beta_i b'_{3,i}(1), \qquad\qquad b'_{3,i+1}(0) = 0,$$

or, using $b'_{j,i}(t) = c_{j,2} + 2c_{j,3}t$,

$$0 = c^i_{1,2} + 2c^i_{1,3}, \qquad\qquad c^{i+1}_{1,2} = \beta_i(c^i_{2,2} + 2c^i_{2,3}),$$
$$c^{i+1}_{2,2} = \beta_i(c^i_{3,2} + 2c^i_{3,3}), \qquad\qquad c^{i+1}_{3,2} = 0. \tag{9.25}$$

The additional constraint is a normalization, which is chosen to be

$$b_{1,i+1}(0) + b_{2,i+1}(0) + b_{3,i+1}(0) = 1 \quad \Rightarrow \quad c^{i+1}_{1,1} + c^{i+1}_{2,1} + c^{i+1}_{3,1} = 1. \tag{9.26}$$

From (9.24)(a,d) and (9.25)(a,d) it follows that

$$c^i_{1,1} = c^i_{1,3}, \quad c^i_{1,2} = -2c^i_{1,3} \quad \Rightarrow \quad b_{1,i}(t) = c^i_{1,3}(1-t)^2,$$
$$c^{i+1}_{3,1} = c^{i+1}_{3,2} = 0 \quad \Rightarrow \quad b_{3,i+1}(t) = c^{i+1}_{3,3}t^2.$$

From (9.24)(c), (9.25)(c) and (9.26) it follows that

$$c^{i+1}_{2,1} = 1 - c^{i+1}_{1,3}, \qquad c^{i+1}_{2,2} = 2\beta_i(1 - c^{i+1}_{1,3}), \qquad c^i_{3,3} = 1 - c^{i+1}_{1,3}. \tag{9.27}$$

From (9.24)(b) and (9.25)(b) (after excluding c_{23}^i) follows that

$$c_{1,3}^{i+1}(1+1/\beta_i) = c_{2,1}^i + (1/2)\, c_{2,2}^i. \tag{9.28}$$

Substituting (9.27)(b,c) into (9.28), we obtain the recursive formula

$$c_{1,3}^{i+1} = \frac{\beta_i}{1+\beta_i}(1+\beta_{i-1})(1-c_{1,3}^i). \tag{9.29}$$

We do not use the solution of (9.29). From the above, it follows that

$$c_{2,3}^{i+1} = -\frac{\beta_{i-1}(\beta_i+2)+1}{1+\beta_i}(1-c_{1,3}^i). \tag{9.30}$$

Hence, the functional coefficients of a quadratic β-spline curve are

$$b_{1,i}(t) = c_{1,3}^i (1-t)^2,$$

$$b_{3,i}(t) = (1-c_{1,3}^{i+1}) t^2,$$

$$b_{2,i}(t) = (1-c_{1,3}^i)(1+2\beta_{i-1}t) - \frac{\beta_{i-2}(\beta_{i-1}+2)+1}{1+\beta_{i-1}}(1-c_{1,3}^{i-1}) t^2.$$

For the second step, one may find shape parameters such that the curvature is continuous at the joints.

9.4.2   Cubic β-spline curve

Consider a *cubic β-spline* defined as a weighted sum of arbitrary control vertices $\mathbf{P} = P_1,\dots,P_n$; the segments are $\mathbf{r}_i(t) = \sum_{j=1}^4 b_j(t) P_{i+j-1}$ $(1 \le i \le n-2)$. We will determine functions $b_j(t)$ such that $\mathbf{r}_i(t)$ and $\mathbf{r}_{i+1}(t)$ will meet with C^2-continuity. The vectors τ (tangent) and $\mathbf{k}$ (curvature) are the main geometrical invariants using in the cubic β-spline curve construction.

Here $\beta_1 > 0$ and $\beta_2 \ge 0$ are called the *parameters of the shape* (β_1 is the *parameter of slant* or *displacement*; β_2 is the *parameter of tension*). For simplicity, we assume a global assignment of the shape parameters β_1, β_2.

Proposition 9.2. *Given points P_1, P_2, P_3, and P_4, the (elementary) cubic β-spline curve is defined by the equation*

$$\mathbf{r}(t) = b_1(t) P_1 + b_2(t) P_2 + b_3(t) P_3 + b_4(t) P_4 \qquad (0 \le t \le 1), \tag{9.31}$$

where the four functional coefficients are defined by the formulae

$$b_1(t) = \beta_1^3(1-t)^3/\delta_3,$$

$$b_2(t) = \frac{1}{\delta_3}[2\beta_1^3 t(t^2 - 3t + 3) + 2\beta_1^2(t^3 - 3t^2 + 2)$$
$$+ 2\beta_1(t^3 - 3t + 2) + \beta_2(2t^3 - 3t^2 + 1)],$$

$$b_3(t) = \frac{1}{\delta_3}[2\beta_1^2 t^2(3-t) + 2\beta_1 t(3-t^2) + \beta_2 t^2(3-2t) + 2(1-t^3)],$$

$$b_4(t) = 2t^3/\delta_3,$$

and $\delta_3 = 2\beta_1^3 + 4\beta_1^2 + 4\beta_1 + \beta_2 + 2.$

Proof. The proof is similar to the case of the quadratic β-spline in Example, p. 377. □

A (piecewise) *cubic β-spline curve* defined by the array $\mathbf{P} = P_1, \dots, P_n$ ($n \geq 4$) is a curve $\tilde{\gamma}(t)$ ($0 \leq t \leq n-3$) which is the union of $n-3$ elementary β-spline curves $\tilde{\gamma}^{(1)} \cup \dots \cup \tilde{\gamma}^{(n-3)}$; the segment $\tilde{\gamma}^{(i)}(t)$ ($i-1 \leq t \leq i$) corresponds to the elementary β-spline determined by the points $P_i, P_{i+1}, P_{i+2}, P_{i+3}$.

The ends of the curve lie in the triangles P_1, P_2, P_3 and P_{n-2}, P_{n-1}, P_n. In order that the β-spline smoothly ends at P_1 and P_n, with tangency of the segments $[P_1, P_2]$ and $[P_{n-1}, P_n]$, one should complete the resulting set of segments (vector-valued functions $\mathbf{r}_i(t)$) with four additional ones (two at each side), for example, by the method of *multiple points*. The extended set of $n+4$ control vertices becomes $\bar{\mathbf{P}} = P_1, P_1, \mathbf{P}, P_n, P_n$. Namely, we complete a curve with four segments $\gamma^{(1)}, \gamma^{(2)}$ and $\gamma^{(n)}, \gamma^{(n+1)}$, setting $\gamma^{(i+2)} = \tilde{\gamma}^{(i)}$ for $1 \leq i \leq n-3$, and obtain a smooth B-spline curve $\gamma = \gamma^{(1)} \cup \dots \cup \gamma^{(n+1)}$ ($0 \leq t \leq n+1$) with endpoints P_1 and P_n and segments $\gamma^{(1)}, \gamma^{(n+1)}$ tangent to the line segments $[P_1, P_2]$ and $[P_{n-1}, P_n]$.

For $\beta_1, \beta_2 \neq 0$ the piecewise curve is C^2-*continuous*. This is the main advantage of these curves over B-spline curves. Parameters of the form β_1 and β_2 can differ on different segments of the piecewise curve.

Changing one point P_{i_0} in $\mathbf{P}$ changes only part of the spline-curve: we recalculate the *four segments* $\gamma^{(i_0-1)}, \gamma^{(i_0)}, \gamma^{(i_0+1)}, \gamma^{(i_0+2)}$ (see exercises).

If we do not require nonnegativity of the parameters β_1 and β_2, then some *self-intersections and oscillations* of the curve can appear: see Figure 9.7.

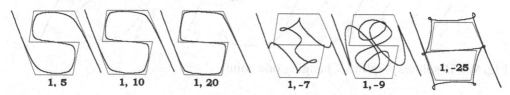

| 1, 5 | 1, 10 | 1, 20 | 1, -7 | 1, -9 | 1, -25 |

Fig. 9.7 Influence of β_1 and β_2 on the shape of β-spline.

Examples.

1. We call **beta_s2d** to derive the test curve in $\mathbb{R}^2$.

```
beta = [1, 7];    x = [.056, .287, .655, .716, .228, .269, .666, .929];
y = [.820, .202, .202, .521, .521, .820, .820, .227];              % enter data
B = beta_s2d(x, y, beta);
```

Then plot the plane curve, using the statement **splot2**(x, y, B) (Figure 9.8(b)).

2. We use **beta_s2d** and **betaspl2d_change** to recalculate four components of the β-spline

```
x = [.056, .287, .655, .716, .228, .269, .666, .929];
y = [.820, .202, .202, .521, .521, .820, .820, .227];
beta = [1, 7];                                            % enter data
B1 = beta_s2d(x, y, beta);
i0 = length(x) - 2;                           % enter data i0 ∈ [2, length(x) − 1]
delta = [rand - .5,  rand - .5]/2;                        % or enter data
xi = x;  xi(i0) = xi(i0) + delta(1);
yi = y;  yi(i0) = yi(i0) + delta(2);
B2 = bspline2d_change(x, y, beta, B1, i0, delta);
```

Then plot the old curve, its new part and the control polygon.

```
hold on;    plot(x, y, '–r');
plot(xi(i0 - 1 : i0 + 1),  yi(i0 - 1 : i0 + 1),  '–b');
for i = 1 :
 length(x),    ezplot(B1(1, i), B1(2, i),  [i - 1, i]);
end
for i = i0-1 : i0+2;
 h = ezplot(B2(1, i), B2(2, i),  [i - 1, i]);  set(h, 'Color', 'g', 'LineWidth', 2);
end                                                       % Figure 9.8(c)
```

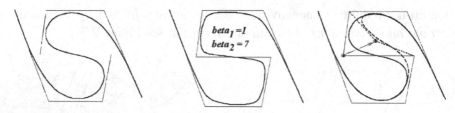

Fig. 9.8 B-spline. β-spline. Changing one point.

Given points P_i and weights w_i, define homogeneous control vertices $\bar{P}_i$ by

$$\bar{P}_i = \begin{cases} (w_i x_i, w_i y_i, w_i z_i, w_i) & \text{if } w_i \neq 0, \\ (x_i, y_i, z_i, 0) & \text{if } w_i = 0. \end{cases} \qquad (9.32)$$

The (*elementary*) *rational cubic β-spline curve* is defined by the equation $\mathbf{r}(t) = \left(\sum_{i=1}^{4} w_i b_i(t) P_i\right) / \left(\sum_{i=1}^{4} w_i b_i(t)\right)$. In homogeneous coordinates it is given by

$$\bar{\mathbf{r}}(t) = \sum_{i=1}^{4} \bar{b}_i(t)\, \bar{P}_i, \quad t \in [0,1],$$

where $\bar{b}_i(t) = b_i(t) / \left(\sum_{j=1}^{4} w_j b_j(t)\right)$, which takes the form of a polynomial elementary β-spline curve, but with homogeneous control vertices (9.32).

The *piecewise rational* (*cubic*) *β-spline curve* is defined similarly to the piecewise cubic β-spline curve.

Example. We call **rbetas_my** (see Section A.4) to compute elementary rational cubic β-spline curve.

```
x = 2 : 5;  y = [32, 38, 40, 33];
beta = [1, 2];  w = [1 2 3 1];                    % enter data
w1 = w/norm(w);   B = rbetas_my(x, y, w1, beta);
```

Then we plot it.

```
hold on;  plot(x, y, '--r');  ezplot(B(1), B(2), [0, 1]);
```

9.4.3 β-Spline curves in space

One can use cubic β-splines in $\mathbb{R}^3$. We develop similar programs (as in the 2-D case) for deriving and plotting in $\mathbb{R}^3$.

Example. We call **beta_s3d** to derive the curve

```
beta = [1, 7];
x = [.056, .287, .655, .716, .228, .269, .666, .929];
y = [.820, .202, .202, .521, .521, .820, .820, .227];
z = [1/4, 1/4, 0, 0, 0, 0, -1/4, -1/4];            % enter data
B = beta_s3d(x, y, z,  beta);
```

Then we plot the space curve, Figure 9.9(b), using **splot3**(x, y, z, B).

The *torsion* (an important invariant of space curves) may be discontinuous along cubic β-splines in $\mathbb{R}^3$. To improve the situation we develop *quartic β-splines*. The β-splines of order $d \geq 3$ may be useful for modeling of curves in $\mathbb{R}^{d-1}$. (For the case $d = 5$ see the exercises.)

Example. As our next example of the use of the C^k constraints, we will construct the geometric continuous *quartic* β-spline. The segment is represented by $\mathbf{r}_i(t) = \sum_{j=1}^{5} b_j(t) P_{i+j-1}$, where $b_i(t) = (1/\delta_4) \sum_{j=0}^{4} c_{i,j} t^j$. As for quadratic or cubic β-splines, we use the first three derivative constraints of (9.17), and obtain the linear system of $25 = 5 \times 5$ equations with 25 unknown coefficients $c_{i,j}$. The determinant of the system is

$$\delta_4 = (1 - \beta_1^2)\beta_3 + 2\beta_1\beta_2^2$$
$$+ 6(\beta_1 + 1)[\beta_2(\beta_1 + 1)^2 - \beta_1\beta_2 + (\beta_1 + 1)(\beta_1^2 + 1)(\beta_1^2 + \beta_1 + 1)].$$

For simplicity, we assume a global assignment of the shape parameters β_1, β_2, β_3. Given points P_1, P_2, P_3, P_4, and P_5, the (elementary) quartic β-spline curve is defined by the equation

$$\mathbf{r}(t) = b_1(t) P_1 + b_2(t) P_2 + b_3(t) P_3 + b_4(t) P_4 + b_5(t) P_5 \quad (0 \le t \le 1), \tag{9.33}$$

where the five functional coefficients are defined by the formulae

$$b_1(t) = -6\beta_1^6 (1 - t)^4 / \delta_4,$$
$$b_2(t) = [\beta_1(18\beta_1^4 + 30\beta_1^3 + 18\beta_1^2 + 6\beta_1^2\beta_2 + 12\beta_1\beta_2 - \beta_3\beta_1 + 2\beta_2^2) - 24\beta_1^2(\beta_1^4$$
$$- 2\beta_1^2 - 2\beta_1 - \beta_2)t + 6\beta_1(6\beta_1^5 + 4\beta_1^2\beta_2 - \beta_3\beta_1 + 2\beta_2^2 + 6\beta_1^4 - 6\beta_1^2)t^2$$
$$- 8\beta_1(3\beta_1^5 + 3\beta_1^2\beta_2 - \beta_3\beta_1 + 2\beta_2^2 + 3\beta_1\beta_2 + 3\beta_1^3 + 3\beta_1^4)t^3$$
$$+ 3\beta_1(2\beta_1^5 + 2\beta_1^4 + 2\beta_1^3 + 2\beta_1^2 + 2\beta_1^2\beta_2 + 4\beta_1\beta_2 - \beta_3\beta_1 + 2\beta_2^2)t^4] \, / \, \delta_4,$$
$$b_3(t) = [-6\beta_2 - \beta_3 - 18\beta_1^3 - 30\beta_1^2 - 12\beta_1\beta_2 - 18\beta_1 - 24\beta_1(2\beta_1^3$$
$$+ 2\beta_1^2 + \beta_1\beta_2 - 1)t + 6(6\beta_1^3 - 6\beta_1^5 - 2\beta_1\beta_2^2 + \beta_3\beta_1^2 + 2\beta_2 - 4\beta_1^3\beta_2 + 6\beta_1^2)t^2$$
$$+ 4(6\beta_1\beta_2 + 6\beta_1^3 + 6\beta_1^5 + 6\beta_1^2\beta_2 + 4\beta_1\beta_2^2 - 2\beta_3\beta_1^2 + 6\beta_1^3\beta_2 + 4\beta_1^4 + 4\beta_3)t^3$$
$$+ 3(\beta_3\beta_1^2 - 2\beta_1 - 4\beta_1\beta_2 - 4\beta_1^3 - 2\beta_1^5 - 4\beta_1^2\beta_2 - 2\beta_1\beta_2^2 - 2\beta_2 - 2\beta_1^3\beta_2$$
$$- 2\beta_1^4 - \beta_3 - 2\beta_1^2)t^4] \, / \, \delta_4,$$
$$b_4(t) = [-6 - 24\beta_1 t - 12(\beta_2 + 3\beta_1^2)t^2 - 4(\beta_3 + 6\beta_1\beta_2 + 6\beta_1^3)t^3$$
$$+ 3(\beta_3 + 2\beta_2 + 6 + 2\beta_1^3 + 2\beta_1^2 + 4\beta_1\beta_2 + 2\beta_1)t^4] \, / \, \delta_4,$$
$$b_5(t) = -6t^4 / \delta_4.$$

The matrix presentation of (9.33) is $\mathbf{r}(t) = \mathbf{P} M T(t)$ $(0 \le t \le 1)$, where $\mathbf{r}(t) = [x(t); y(t); z(t)]$ and $T(t) = [t^4; t^3; t^2; t; 1]$ are column vectors, $\mathbf{P} = (P_1; P_2; P_3; P_4; P_5)$ is the *geometric matrix* (see B-spline case), and M (too long to place here) is the *basic matrix* of the β-spline curve.

We apply **beta_s4d** to derive the space curve:

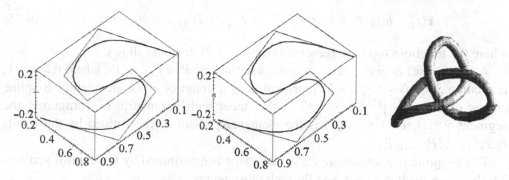

Fig. 9.9 B-spline and β-spline in space. Knot (trefoil).

```
beta = [1, 2, 7];
x = [.056, .287, .655, .716, .228, .269, .666, .929];
y = [.820, .202, .202, .521, .521, .820, .820, .227];
z = [1/4, 1/4, 0, 0, 0, 0, -1/4, -1/4];                              % enter data
B = beta_s4d(x, y, z, beta);
```

Plot the space curve using the typical program (see **splot3**(x, y, z, B)):

```
hold on;    plot3(x, y, z, 'r', 'LineWidth', 2);
for i = 1 : (length(x) + 2);
 ti = linspace(i - 1, i, 20);
 S = [subs(B(1, i), t, ti) + ti - ti; subs(B(2, i), t, ti) + ti - ti; subs(B(3, i), t, ti) + ti - ti];
 plot3(S(1, :), S(2, :), S(3, :), 'LineWidth', 2);
end;                                                                  % see Figure 9.9(b)
```

9.4.4 B-spline curves

We briefly survey the *cubic B-spline* ("B" stands for *basic*) that coincides with the case $\beta_1 = 1$, $\beta_2 = 0$ of the cubic β-spline. (We will not discuss the fundamental recursive definition or certain practical aspects such as the relation with least-squares fitting).

Let $\mathbf{r}_i(t)$, $\mathbf{r}_{i+1}(t)$ ($t \in [0, 1]$) be regular parametric segments such that $\mathbf{r}_i(1) = \mathbf{r}_{i+1}(0) = \hat{P}_i$, where $\hat{P}_i$ is a simple point of the piecewise curve. They meet with *parametric C^k-continuity* at $\hat{P}_i$ if and only if

$$\mathbf{r}_{i+1}^{(j)}(0) = \mathbf{r}_i^{(j)}(1), \qquad 1 \le j \le k. \tag{9.34}$$

Given points $P_1, P_2, \ldots, P_{k+1}$, the (*elementary*) *B-spline curve of order k* is defined by the equation

$$\mathbf{r}(t) = b_1(t)P_1 + b_2(t)P_2 + \cdots + b_{k+1}(t)P_{k+1} \qquad (0 \le t \le 1), \qquad (9.35)$$

where the functions $b_i(t)$ are presented (for $n = 2, 3$) in what follows.

A (piecewise) *B-spline curve of order k* defined by $\mathbf{P} = P_1, \ldots, P_n$, where $n \ge k+1$, is a curve $\tilde{\gamma}(t)$ $(0 \le t \le n-k)$ represented as a union of $n-k$ elementary B-spline curves of order k, $\tilde{\gamma}^{(1)} \cup \cdots \cup \tilde{\gamma}^{(n-k)}$ that meet with parametric C^k-continuity; the segment $\tilde{\gamma}^{(i)}(t)$ $(i-1 \le t \le i)$ is the elementary B-spline determined by the points $P_i, P_{i+1}, \ldots, P_{i+k-1}, P_{i+k}$.

The B-spline is a *smoothing* curve. Its shape is determined by the control vertices, but the curve itself does not pass through those points.

A (piecewise) *cubic B-spline* is defined as a weighted sum of arbitrary control vertices $\mathbf{P} = P_1, \ldots, P_n$ $(n \ge 4)$; the segments are

$$\mathbf{r}_i(t) = b_1(t)P_i + b_2(t)P_{i+1} + b_3(t)P_{i+2} + b_4(t)P_{i+3} \qquad (i \le n-3, \ 0 \le t \le 1).$$

The basis functions are

$$b_1(t) = \frac{(1-t)^3}{6}, \qquad\qquad\qquad b_2(t) = \frac{3t^3 - 6t^2 + 4}{6},$$

$$b_3(t) = \frac{-3t^3 + 3t^2 + 3t + 1}{6}, \qquad\qquad b_4(t) = \frac{t^3}{6}. \qquad (9.36)$$

It is the union of $n-3$ elementary cubic B-spline curves $\tilde{\gamma}^{(1)} \cup \cdots \cup \tilde{\gamma}^{(n-3)}$. The segment $\tilde{\gamma}^{(i)}(t)$ $(i-1 \le t \le i)$ corresponds to the elementary B-spline determined by the points $P_i, P_{i+1}, P_{i+2}, P_{i+3}$. In order that the B-spline ends smoothly at P_1 and P_n, with tangency of the segments $[P_1, P_2]$ and $[P_{n-1}, P_n]$, we complete the set of segments by the method of multiple (or phantom) points. The extended set $\bar{\mathbf{P}}$ of $n+4$ control vertices becomes $P_1, P_1, \mathbf{P}, P_n, P_n$. The resulting curve is C^2-regular, but only C^1-continuous.

Examples.

1. A *quadratic B-spline* is defined as a weighted sum of arbitrary control vertices $\mathbf{P} = P_1, \ldots, P_n$; the segments are

$$\mathbf{r}_i(t) = b_1(t)P_i + b_2(t)P_{i+1} + b_3(t)P_{i+2} \qquad (1 \le i \le n-2, \ 0 \le t \le 1),$$

where $b_1(t) = (t-1)^2/2$, $b_2(t) = -t^2 + t + 1/2$, and $b_3(t) = t^2/2$.

We determine $b_j(t)$ such that $\mathbf{r}_i(t)$ and $\mathbf{r}_{i+1}(t)$ meet with parametric C^1-continuity. (See Example, p. 375 with $\beta_1 = 1$, $\beta_2 = 0$.)

2. We derive and plot the cubic B-spline segment.

```
x = 2 : 5;   y = [32, 38, 40, 33];                    % enter data
B = betas_my(x, y, [1 0])                             % i.e., β₁ = 1, β₂ = 0.
hold on; plot(x, y, 'r'); ezplot(B(1), B(2), [0, 1]);
```

3. One may show that the degree-d (uniform) B-spline segment is given by $\mathbf{r}(t) = \mathbf{P}MT(t)$ $(0 \le t \le 1)$, where $T(t) = [t^d; \ldots; t^2; t; 1]$, $\mathbf{P} = (P_1 \; P_2 \; \cdots \; P_{d+1})$, and the elements m_{ij} of the basic matrix M are

$$m_{ij} = (n!)^{-1} \binom{n}{j} \sum_{k=i}^{n} (n-k)^j (-1)^{k-i} \binom{n+1}{k-j}.$$

For example,

$$M_4 = \frac{1}{4!} \begin{pmatrix} 1 & -4 & 6 & -4 & 1 \\ -4 & 12 & -6 & -12 & 11 \\ 6 & -12 & -6 & 12 & 11 \\ -4 & 4 & 6 & 4 & 1 \\ 1 & 0 & 0 & 0 & 0 \end{pmatrix}, \quad M_5 = \frac{1}{5!} \begin{pmatrix} -1 & 5 & -10 & 10 & -5 & 1 \\ 5 & -20 & 20 & 20 & -50 & 26 \\ -10 & 30 & 0 & -60 & 0 & 66 \\ 10 & -20 & -20 & 20 & 50 & 26 \\ -5 & 5 & 10 & 10 & 5 & 1 \\ 1 & 0 & 0 & 0 & 0 & 0 \end{pmatrix}.$$

4. We call **beta_s2d** to derive the plane cubic B-spline.

```
x = [.056, .287, .655, .716, .228, .269, .666, .929];
y = [.820, .202, .202, .521, .521, .820, .820, .227];        % enter data
B = beta_s2d(x, y, [1 0]);
```

Then we plot the curve and the data using **splot2(x, y, B)**.

5. One may experiment with the B-spline as a function of its knots, using the spline toolbox **bspligui**. The MATLAB® command **bspline** can also be used.

6. We show how to use B-splines to construct an *interpolating* cubic spline curve that passes through a set of $n+1$ given data points $P_1, \ldots, P_{n+1}$. The curve consists of n segments, and the idea is to use the P_i points to calculate a new set of points $\bar{P}_i$, and then use the new points as the control vertices of a cubic uniform B-spline curve.

Using $\bar{P}_i$ as our control vertices, (9.36) shows that the general segment $\mathbf{r}_i(t)$ terminates at $\mathbf{r}_i(1) = (\bar{P}_{i-2} + 4\bar{P}_{i-1} + \bar{P}_i)/6$. We require that the segment end at point P_{i-1}, which yields the equation:

$$P_{i-1} = (\bar{P}_{i-2} + 4\bar{P}_{i-1} + \bar{P}_i)/6 \tag{9.37}$$

When this equation is repeated for $0 \le i \le n$, we get a system of $n+1$ equations with the $\bar{P}_i$s as the unknowns. However, there are $n+3$ unknowns (P_{-1} through P_{n+1}), so we need two more equations. Consider the tangent vectors T_1, T_n of the interpolating curve at its two ends. They are given that by $T_1 = (P_1 - P_{-1})/2$ and $T_n = (P_{n+1} - P_{n-1})/2$. After these two relations are included, the resulting system of $n+3$ equations is

$$\frac{1}{6}\begin{pmatrix} -3 & 0 & 3 & & & \\ 1 & 4 & 1 & & & \\ & & \cdots & & & \\ & & & 1 & 4 & 1 \\ & & & -3 & 0 & 3 \end{pmatrix}\begin{pmatrix} \bar{P}_1 \\ \bar{P}_2 \\ \cdots \\ \bar{P}_{n-1} \\ \bar{P}_n \end{pmatrix} = \begin{pmatrix} T_1 \\ P_1 \\ \cdots \\ P_n \\ T_n \end{pmatrix}, \tag{9.38}$$

where the first and last equations are modified. Since the matrix is columnwise diagonally dominant, it is invertible. Thus the system of equations has a unique solution. The same construction can be developed for β-splines.

Let $P_1 = (1/6, 5/6)$, $P_2 = (1,1)$, $P_3 = (11/6, 5/6)$, and the two extreme tangents $T_1 = (1/2, 1/2)$ and $T_2 = (1/2, -1/2)$. Set up the linear system

$$\frac{1}{6}\begin{pmatrix} -3 & 0 & 3 & 0 & 0 \\ 1 & 4 & 1 & 0 & 0 \\ 0 & 1 & 4 & 1 & 0 \\ 0 & 0 & 1 & 4 & 1 \\ 0 & 0 & -3 & 0 & 3 \end{pmatrix}\begin{pmatrix} \bar{P}_1 \\ \bar{P}_2 \\ \bar{P}_3 \\ \bar{P}_4 \\ \bar{P}_5 \end{pmatrix} = \begin{pmatrix} T_1 \\ P_1 \\ P_2 \\ P_3 \\ T_2 \end{pmatrix}. \tag{9.39}$$

The solution is $\bar{P}_1 = (0,0)$, $\bar{P}_2 = (0,1)$, $\bar{P}_3 = (1,1)$, $\bar{P}_4 = (2,1)$, and $\bar{P}_5 = (2,0)$.

Given points P_1, P_2, P_3, P_4 and weights w_1, w_2, w_3, w_4, define homogeneous control vertices $\bar{P}_i$ by (9.32). The (*elementary*) *rational B-spline curve* is defined by the equation

$$\mathbf{r}(t) = \sum_{i=1}^{4} w_i n_i(t) P_i \Big/ \sum_{i=1}^{4} w_i n_i(t) = \sum_{i=1}^{4} \bar{N}_{4,i}(t) \bar{P}_i, \qquad t \in [0,1],$$

where $\bar{N}_{4,i}(t) = n_i(t) \big/ \sum_{j=1}^{4} w_j n_j(t)$, which takes the form of a polynomial elementary B-spline curve, but with homogeneous control vertices (modify Example, p. 381, for $\beta_1 = 1$, $\beta_2 = 0$).

The *nonuniform rational cubic B-spline* (NURBS for short) determined by points $P_1, \ldots, P_n$ and the weights $w_1, \ldots, w_n$ can take many shapes. Because of this, NURBS is today the de facto standard for curve design.

Remark 9.1. The knot vector approach to the B-spline curve of order k assumes that the curve is a weighted sum, $\mathbf{r}(t) = \sum_{i=0}^{n} N_{k+1,i}(t) P_i$ of the control vertices with the weight (or blending) functions $N_{k+1,i}(t)$ to be determined. The method is similar to that used in deriving Bézier curves (Section 9.2). The "gluing principle" is that each weight function (a piecewise kth-order polynomial) should have a bell shape, with maximum at its control vertex and support over the interval $[1 + k, i + k]$. To derive such a function, $N_{k+1,i}(t)$, we represent it as the union of $k + 1$ parts,

$$\bar{n}_1(t) = b_1(t - i + k), \quad \bar{n}_2(t) = b_2(t - i + k - 1), \quad \ldots, \quad \bar{n}_{k+1}(t) = b_{k+1}(t - i),$$

each defined over one unit of t. The following considerations are employed to set up equations to calculate the $\bar{n}_i(t)$ functions:

1. *they should be barycentric;*
2. *they should provide C^{k-1}-continuity at the k points where they meet;*
3. *$\bar{n}_1(t)$ and its first $k-1$ derivatives should be zero at the point $\bar{n}_1(0)$;*
4. *$\bar{n}_{k+1}(t)$ and its first $k-1$ derivatives should be zero at the point $\bar{n}_{k+1}(1)$.*

The support of $N_{k+1,i}(t)$ is the interval $[i-k, i+1]$. The weights all look the same and are shifted with respect to each other by using different ranges of t. The use of a knot vector is one reason why the B-spline curve is more general than the Bézier curve. The knots can be used as parameters and can be varied to obtain the desired shape of the curve.

9.5   The Cartesian Product Surface Patch

The concept of blending (introduced in Section 9.2) is important and useful in many curve and surface algorithms. The technique of blending quantities P_{ij} into a surface by means of weights taken from two curves is called the *Cartesian product*, although the term *tensor product* is also sometimes used. This section shows how blending can be used in surface design. Now examine the function

$$\mathbf{r}(u,v) = \sum_{i=1}^{n} \sum_{j=1}^{m} f_i(u) g_j(v) P_{ij} = \sum_{i=1}^{n} \sum_{j=1}^{m} h_{ij}(u,v) P_{ij}, \qquad (9.40)$$

where $h_{ij}(u,v) = f_i(u) g_j(v)$. The function (9.40) describes a surface; for any value of the pair (u,v), it computes a weighted sum of the points P_{ij}.

If $\sum_{i=1}^{n} f_i(u) = \sum_{j=1}^{m} g_j(v) = 1$, the surface $\mathbf{r}(u,v)$ is independent of the particular coordinate axes used, because

$$\sum_{i=1}^{n} \sum_{j=1}^{m} h_{ij}(u,v) = \sum_{i=1}^{n} f_i(u) \cdot \sum_{j=1}^{m} g_j(v) = 1.$$

The substitutions $u = 0, 1$ and $v = 0, 1$ reduce the expression to the four corner points. The four curves $\mathbf{r}(u,0)$, $\mathbf{r}(u,1)$, $\mathbf{r}(0,v)$, and $\mathbf{r}(1,v)$ are the *boundary curves* of the surface. Since there are four such curves, our surface is a *patch* that has a (roughly) rectangular shape. Given u_0, v_0, we call $\mathbf{r}(u,v_0)$ and $\mathbf{r}(u_0,v)$ the *u-curve* and *v-curve*, respectively. Two more special curves, the *surface diagonals*, are $\mathbf{r}(u,1-u)$ and $\mathbf{r}(u,u)$.

The particular case of (9.40) is $n = m$ and $f_i(t) = g_i(t)$, i.e.,

$$\mathbf{r}(u,v) = (f_1(u),\dots,f_n(u)) \begin{pmatrix} P_{11} & P_{12} & \dots & P_{1n} \\ \vdots & \vdots & \vdots & \vdots \\ P_{n1} & P_{n2} & \dots & P_{nn} \end{pmatrix} \begin{pmatrix} f_1(v) \\ \vdots \\ f_n(v) \end{pmatrix}. \tag{9.41}$$

If $f_i(t)$ are *polynomials of degree n*, then the last formula can be rewritten in the form

$$\mathbf{r}(u,v) = (u^{n-1},\dots,u,1)\, M^* \mathbf{P}_n M\, (v^{n-1},\dots,v,1)', \tag{9.42}$$

where $\mathbf{P}_n = \{P_{ij}\}_{i,j=1}^n$, and the *basic matrix* M can be found from

$$(f_1(t),\dots,f_n(t))' = M\,(t^n,\dots,t,1)'. \tag{9.43}$$

The *rational surface patch* corresponding to (9.40) and nonnegative weights $\{w_{ij}\}$ with positive sum is given by

$$\mathbf{r}(u,v) = \frac{\sum_{i,j=1}^n w_{ij} f_i(u) f_j(v) P_{ij}}{\sum_{i,j=1}^n w_{ij} f_i(u) f_j(v)}. \tag{9.44}$$

The following examples illustrate the importance of the Cartesian product.

Example. The parameterization of the *line segment* from P_1 to P_2 is

$$\mathbf{r}(t) = (1-t)P_1 + tP_2 = (B_{10}(t), B_{11}(t)) \begin{pmatrix} P_1 \\ P_2 \end{pmatrix}, \tag{9.45}$$

where B_{1i} are the Bernstein polynomials of degree 1. The Cartesian product of (9.45) with itself is the *bilinear surface* patch, see (9.40) for $m = n = 2$,

$$\begin{aligned} \mathbf{r}(u,v) &= (1-u,u)\,\mathbf{P}_2\,(1-v,v) \\ &= (1-u)(1-v)P_{11} + (1-u)vP_{12} + u(1-v)P_{21} + uvP_{22}. \end{aligned} \tag{9.46}$$

```
syms u v P11 P12 P21 P22;
P = [P11 P12; P21 P22];      r = [1 - u  u] * P * [1 - v; v]
```

A simple particular case of a bilinear surface (different from a plane) is defined by points $P_{11} = (0,0,1)$, $P_{21} = (1,0,0)$, $P_{12} = (1,1,1)$, $P_{22} = (0,1,0) \in \mathbb{R}^3$:

$$\mathbf{r}(u,v) = [u+v-2uv, \; v, \; 1-u].$$

```
rr = subs(r, {P11, P12, P21, P22}, {[0 0 1], [1 1 1], [1 0 0], [0 1 0]})
ezsurf(rr(1), rr(2), rr(3), [0, 1, 0, 1], 12)      % bilinear surface, Figure 9.10(a).
```

The tangent vectors and normal vector of the surface are easily calculated:

$$\partial_u \mathbf{r}(u,v) = [1 - 2v,\ 0,\ -1], \qquad\qquad \partial_v \mathbf{r}(u,v) = [1 - 2u,\ 1,\ 0],$$
$$\mathbf{n}(u,v) = \partial_u \mathbf{r} \times \partial_v \mathbf{r} = [1,\ -1 + 2u,\ 1 - 2v].$$

```
ru = diff(rr, u),   rv = diff(rr, v),   n = cross(ru, rv)
```

The vector $\partial_u \mathbf{r}(u,v)$ lies in the xz-plane, and $\partial_v \mathbf{r}(u,v)$ lies in the xy-plane. The *twist vector of a bilinear surface* is $P_{11} - P_{12} - P_{21} + P_{22}$. (The *twist vector* of a regular surface $\mathbf{r}(u,v)$ is the mixed derivative $\partial^2_{uv}\mathbf{r}(u,v)$.) It measures the deviation of P_{22} from the tangent plane to the surface at P_{11}.

```
ruv = diff(rr, u, v)                                      % obtain twist.
```

In what follows we compute and visualize surface patches. We base our investigations on the procedure **surf_patch**(P, M) (Section A.5), where P is an array of given points, and M is the basic matrix.

Examples.

1. The parametric cubic polynomial passing through four given points is $\mathbf{r}(t) = GMT(t)$; see (5.26). The Cartesian product of the curve with itself is a *bicubic surface patch* $\mathbf{r}(u,v)$ $(0 \le u, v \le 1)$ (see (9.42)),

$$\mathbf{r}(u,v) = (u^3, u^2, u, 1) M^* \mathbf{P}_4 M (v^3, v^2, v, 1)', \tag{9.47}$$

where $u, v \in \{0, 1/3, 2/3, 1\}$, that passes through 16 points.

Given 16 points, we evaluate and plot the surface patch:

```
M = [-9/2 9 -11/2 1;  27/2 -45/2 9 0;  -27/2 18 -9/2 0;  9/2 -9/2 1 0];
P = [ [0 0 0]; [1 0 0]; [2 0 0]; [3 0 0]; [0 1 0]; [1 1 1]; [2 1 -1/2]; [3 1 0];
[0 2 -1/2]; [1 2 0]; [2 2 -1/2]; [3 2 0]; [0 3 0]; [1 3 0]; [2 3 0]; [3 3 0] ];
r = surf_patch(P, M)                          % bicubic surface patch
ezsurf(r(1), r(2), r(3),  [0, 1, 0, 1],  30);           % Figure 9.10(c).
```

The surface and the 16 points on it may be exhibited using the program **view_patch**.

2. We use the Cartesian product for plotting Bézier patches.

(a) The cubic Bézier curve determined by four given points is $\mathbf{r}(t) = GMT(t)$; see (9.5). Applying the principle of the Cartesian product, we multiply this curve by itself in order to obtain a *Bézier bicubic surface* patch $\mathbf{r}(u,v)$ $(0 \le u, v \le 1)$, determined by 16 given points. Once the points are given, we evaluate and plot the surface patch replacing M and P in the program of Example 1, p. 389:

```
M = [-1 3 -3 1;  3 -6 3 0;  -3 3 0 0;  1 0 0 0];              % enter data: M and P.
P = [ [0 0 0]; [1 0 -1/2]; [2 0 -1/2]; [3 0 0]; [0 1 0]; [1 1 3/2]; [2 1 1]; [3 1 0];
[0 2 0]; [1 2 1]; [2 2 3/2]; [3 2 0]; [0 3 0]; [1 3 1]; [2 3 1]; [3 3 0] ];
r = surf_patch(P, M)                          % bicubic Bézier patch
ezsurf(r(1), r(2), r(3),  [0, 1, 0, 1],  30);           % Figure 9.11(b).
```

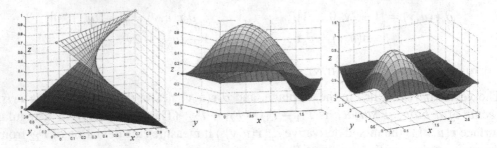

Fig. 9.10 Bilinear, biquadratic and bicubic Lagrange patches.

The surface and the 16 points may be shown using the program **view_patch**.

(b) The initial matrix representation of the nth-order Bézier curve determined by $n+1$ given points is (9.1). For various depictions of its, we use this matrix representation to obtain the *nth-order Bézier patch* $\mathbf{r}(u,v)$ ($0 \leq u,v \leq 1$), determined by $(n+1)^2$ given points,

$$\mathbf{r}(u,v) = [B_{n,n}(u), \ldots, B_{n,1}(u), B_{n,0}(u)]\mathbf{P}_{n+1}[B_{n,n}(v), \ldots, B_{n,1}(v), B_{n,0}(v)]'.$$

```
N = 5;                                                    % N = n + 1
P = [[0 0 0]; [1 0 -1/2]; [2 0 -1/2]; [3 0 0]; [4 0 0]; [0 1 0]; [1 1 2]; [2 1 1];
[3 1 0]; [4 1 0]; [0 2 0]; [1 2 1]; [2 2 3]; [3 2 0]; [4 2 0]; [0 3 0]; [1 3 1]; [2 3 1];
[3 3 0]; [4 3 0]; [0 4 0]; [1 4 1/2]; [2 4 1/2]; [3 4 0]; [4 4 0]];   % enter N² points
syms t u v;
j = 0 : (N - 1);
B = gamma(N)./gamma(j+1)./gamma(N-j).*t.^j.*(1-t).^(N-j-1);
X = P(:, 1);  Y = P(:, 2);  Z = P(:, 3);
rx = subs(B, t, u)*reshape(X, N, N)*subs(B, t, v)';
ry = subs(B, t, u)*reshape(Y, N, N)*subs(B, t, v)';
rz = subs(B, t, u)*reshape(Z, N, N)*subs(B, t, v)';
r = [rx, ry, rz];                                 % biquartic Bézier patch
ezsurf(r(1), r(2), r(3),  [0, 1, 0, 1], 20)        % Figure 9.11(c).
```

The surface and the $(n+1)^2$ points may be depicted using **view_patch**.

3. The following three examples illustrate the use of the Cartesian product for plotting Hermite patches.

(a) The quadratic Hermite curve is $\mathbf{r}(t) = \mathbf{G}MT(t)$, where (9.61) presents M. Thus, the *biquadratic Hermite patch*

$$\mathbf{r}(u,v) = (u^2, u, 1)\, M^*\, \hat{\mathbf{P}}_3\, M\, (v^2, v, 1)', \quad \text{where} \quad \hat{\mathbf{P}}_3 = \begin{pmatrix} P_{11} & P_{12} & Q_{11} \\ P_{21} & P_{22} & Q_{12} \\ Q_{21} & Q_{22} & T_{11} \end{pmatrix},$$

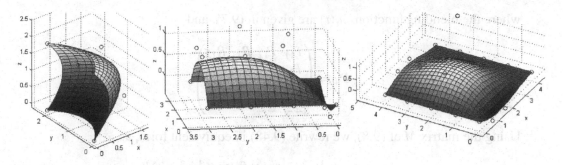

Fig. 9.11 Bézier patches: biquadratic, bicubic and biquartic.

is defined by the following nine quantities (Figure 9.12(a)):

- the four corner points P_{11}, P_{12}, P_{21}, and P_{22};
- the two tangents Q_{11} and Q_{12} in the u direction at P_{11} and P_{12};
- the two tangents Q_{21} and Q_{22} in the v direction at P_{11} and P_{21};
- the second derivative T_{11} (i.e. *twist*) at P_{11}.

Several methods exist to estimate the twist vectors of biquadratic and bicubic surface patches (we omit the details).

```
M = [ -1 0 1;  1 0 0;  -1 1 0 ]                    % enter data: M and PQ.
PQ = [ [0 0 0]; [1 0 0]; [1 1/2 1]; [0 1 0]; [1 1 0]; [2 0 0]; [1 1 0]; [1 0 1]; [1 0 0] ];
rr = surf_patch(PQ, M)
```

The surface, 4 points, and 5 arrows (Figure 9.13(a)) are shown by the program below.

```
P11 = [0 0 0];    P12 = [1 0 0];    Q21 = [1 1/2 1];
P21 = [0 1 0];    P22 = [1 1 0];    Q22 = [2 0 0];
Q11 = [1 1 0];    Q12 = [1 0 1];    T11 = [1 0 0];
PP = [P11; P12; P21; P22];
hold on;   view(3);   grid on;
plot3(PQ(:, 1), PQ(:, 2), PQ(:, 3), 'o');
arrow3(P11, Q21/3);    arrow3(P21, Q22);                    % M-file arrow3.m
arrow3(P11, Q11/2);   arrow3(P12, Q12/3);
arrow3(P11, T11/2, 'r');
ezsurf(rr(1), rr(2), rr(3),  [0, 1, 0, 1],  20)       % biquadratic Hermite patch.
```

(b) A *Ferguson surface patch* is an extension of the Hermite cubic curve segment. The patch is specified by its four corner points P_{ij} and by two tangent vectors Q_{ij} and R_{ij} in the u and v directions at each point, for a total of 12 three-dimensional quantities (Figure 9.12(b)). Here $Q_{ij} = \partial_u \mathbf{r}(P_{ij})$ and $R_{ij} = \partial_v \mathbf{r}(P_{ij})$. The patch has the following matrix representation:

$$\mathbf{r}(u,v) = (n_1(t), n_2(t), n_3(t), n_4(t)) \, \hat{\mathbf{P}}_4 \, (n_1(t), n_2(t), n_3(t), n_4(t))', \qquad (9.48)$$

where the blending functions $n_i(t)$ are given in (9.7), and

$$\hat{\mathbf{P}}_4 = \begin{pmatrix} P_{11} & P_{12} & R_{11} & R_{12} \\ P_{21} & P_{22} & R_{21} & R_{22} \\ Q_{11} & Q_{12} & 0 & 0 \\ Q_{21} & Q_{22} & 0 & 0 \end{pmatrix}.$$

Using the matrix M of (9.8), we rewrite (9.48) in equivalent form, as

$$\mathbf{r}(u,v) = (u^3, u^2, u, 1)\, M^* \,\hat{\mathbf{P}}_4 M\, (v^3, v^2, v, 1)'. \qquad (9.49)$$

Notice that the Ferguson surface is very similar to the bicubic Hermite patch, but is less flexible because it has zeros instead of the more general twist vectors. The Ferguson surface patch is easy to connect smoothly with other patches of the same type.

```
M = [2 -3 0 1; -2 3 0 0; 1 -2 1 0; 1 -1 0 0]              % enter data: M and PQ.
PQ = [ [0 0 0]; [1 0 0]; [2 0 -1/2]; [3 0 0]; [0 1 0]; [1 1 0]; [2 1 1]; [3 1 0];
     [0 2 0]; [1 2 1]; [0 0 0]; [0 0 0]; [0 3 0]; [1 3 1]; [0 0 0]; [0 0 0] ];
r = surf_patch(PQ, M)
ezsurf(r(1), r(2), r(3),  [0, 1, 0, 1], 20)                 % Ferguson surface patch.
```

The surface, 4 points, and 8 tangent vectors may be depicted using the program below.

```
P11 = [0 0 0];  P12 = [1 0 0];  P21 = [0 1 0];  P22 = [1 1 0];
R11 = [2 0 -1/2];  R12 = [3 0 0];  R21 = [2 1 1];  R22 = [3 1 0];
Q11 = [0 2 0];  Q12 = [1 2 1];  Q21 = [0 3 0];  Q22 = [1 3 1];
PP = [P11; P12; P21; P22];
hold on;   view(3);   grid on;
plot3(PQ(:,1),  PQ(:,2), PQ(:,3),  'o');
arrow3(P11, Q11/5);    arrow3(P11, R11/5);
arrow3(P12, Q12/5);    arrow3(P12, R12/5);
arrow3(P21, Q21/5);    arrow3(P21, R21/5);
arrow3(P22, Q22/5);    arrow3(P22, R22/5);            % see M-file arrow3.m
ezsurf(r(1), r(2), r(3),  [0, 1, 0, 1], 20)      % Ferguson patch, Figure 9.13(b).
```

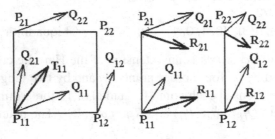

Fig. 9.12 Hermite type patches schemes: biquadratic, Ferguson.

(c) The cardinal spline (or the Catmull–Rom curve) can easily be extended to a surface that is fully defined by a rectangular grid of data points. A single *bicubic Catmull–Rom patch* is specified by 16 points, and is anchored at the four middle points. It is given in the form of (9.47) as

$$\mathbf{r}(u,v) = (u^3, u^2, u, 1)\, M^*\, \mathbf{P}_4 M\, (v^3, v^2, v, 1)', \tag{9.50}$$

where M is the parabolic blending matrix of (9.8). Tension can be added to a Catmull–Rom surface patch in the same way that it is added to a Catmull–Rom curve in the construction of a cardinal spline.

```
M = [-1 2 -1 0; 3 -5 0 2; -3 4 1 0; 1 -1 0 0]/2;        % enter data: M and P.
P = [ [0 0 0]; [1 0 0]; [2 0 0]; [3 0 0]; [0 1 0]; [.5 .5 1]; [2.5 .5 0]; [3 1 0];
[0 2 0]; [.5 2.5 0]; [2.5 2.5 1]; [3 2 0]; [0 3 0]; [1 3 0]; [2 3 0]; [3 3 0] ];
r = surf_patch(P, M)                                    % see Figure 9.13(c)
ezsurf(r(1), r(2), r(3),  [0, 1, 0, 1],  20)            % bicubic patch.
```

Plot the surface and the 16 points by means of the program **view_patch**.

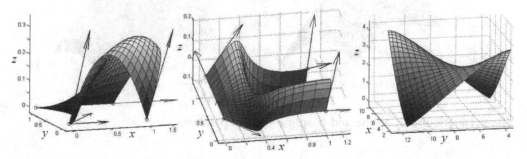

Fig. 9.13 Hermite type patches: biquadratic, Ferguson, cardinal.

4. We illustrate the use of the Cartesian product for the derivation and plotting of bicubic spline patches.

   (a) The cubic B-spline curve determined by four given points is $\mathbf{r}(t) = GMT(t)$; see (9.65). We calculate the *bicubic B-spline patch*.

```
M = [-1 3 -3 1; 3 -6 0 4; -3 3 3 1; 1 0 0 0]/6;         % enter data: M and P.
P = [ [0 0 0]; [1 0 -1/2]; [2 0 -1/2]; [3 0 0]; [0 1 0]; [1 1 3/2]; [2 1 1];
[3 1 0]; [0 2 0]; [1 2 1]; [2 2 3/2]; [3 2 0]; [0 3 0]; [1 3 1]; [2 3 1]; [3 3 0] ];
r = surf_patch(P, M)                                    % bicubic B-spline patch
ezsurf(r(1), r(2), r(3),  [0, 1, 0, 1],  20)            % Figure 9.14(a).
```

   (b) The cubic β-spline curve determined by four given points is $\mathbf{r}(t) = GMT(t)$; see (9.64). We compute and plot the *bicubic β-spline patch*.

```
b1 = 2;  b2 = 1/2;                        % enter data: β₁ and β₂. Then M and P.
xi = b1^2 + b1 + b2;
d3 = b2 + 2*b1^3 + 4*b1^2 + 4*b1 + 2;
M = [-2*b1^3  6*b1^3  -6*b1^3  2*b1^3;
 2*(b1^3+xi)  -3*(2*b1^3+2*b1^2+b2)  6*(b1^3-b1)  4*(b1^2+b1)+b2;
 -2*(xi+1)  3*(2*b1^2+b2)  6*b1 2;  2  0  0  0]/d3;
P = [[0 0 0]; [1 0 -1/2]; [2 0 -1/2]; [3 0 0]; [0 1 0]; [1 1 3/2]; [2 1 1];
 [3 1 0]; [0 2 0]; [1 2 1]; [2 2 3/2]; [3 2 0]; [0 3 0]; [1 3 1]; [2 3 1]; [3 3 0]];
r = surf_patch(P, M)                      % bicubic β-spline patch
ezsurf(r(1), r(2), r(3), [0, 1, 0, 1], 20)              % Figure 9.14(b).
```

The surfaces and 16 points may be shown by the program **view_patch**.

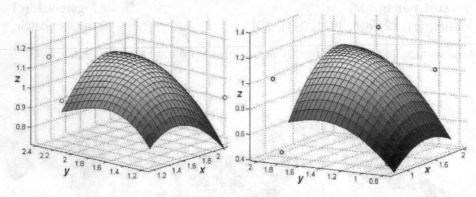

Fig. 9.14 Bicubic B-spline and β-spline surface patches.

9.6   Bicubic Spline Surfaces

Like a spline curve, a spline surface is specified by a set of *control vertices*.

A continuous surface **S** is called a *bicubic (B- or β-) spline surface* defined by a set of $(m+1)(n+1)$ vertices, called the *control graph*,

$$\mathbf{P} = \{P_{ij},\ i = l,\ldots,m+1,\ j = 1,\ldots,n+1\},\quad m \geq 4,\ n \geq 4$$

(Figure 9.15). The surface is a union $\mathbf{S} = S^{(1,1)} \cup \cdots \cup S^{(m-2,n-2)}$ of $(m-2)(n-2)$ elementary bicubic (B- or β-spline) patches $\{S^{(i,j)}\}_{j=1,\ldots,n-2}^{i=1,\ldots,m-2}$, where the (i,j)th fragment, $S^{(i,j)}$, is specified by the parametric equations

$$\mathbf{r}^{(i,j)}(u,v) = \sum_{k,l=1}^{4} n_k(u)\,n_l(v)P_{k-1+i,\,l-1+j}, \qquad 0 \le u,v \le 1. \qquad (9.51)$$

Equivalently, (9.51) is represented in matrix form as

$$\mathbf{r}^{(i,j)} = (n_1(u),\ldots,n_4(u)) \begin{pmatrix} P_{i,j} & P_{i,j+1} & P_{i,j+2} & P_{i,j+3} \\ P_{i+1,j} & P_{i+1,j+1} & P_{i+1,j+2} & P_{i+1,j+3} \\ P_{i+2,j} & P_{i+2,j+1} & P_{i+2,j+2} & P_{i+2,j+3} \\ P_{i+3,j} & P_{i+3,j+1} & P_{i+3,j+2} & P_{i+3,j+3} \end{pmatrix} \begin{pmatrix} n_1(v) \\ n_2(v) \\ n_3(v) \\ n_4(v) \end{pmatrix}$$

$$= (u^3, u^2, u, 1)\,M^* \mathbf{P}^{(i,j)} M\,(v^3, v^2, v, 1)^*, \qquad (9.52)$$

where M is the basic matrix (9.65) or (9.64) of the bicubic B- or β-spline surface (in the β-spline case, for simplicity assume a global assignment of the shape parameters β_1, β_2).

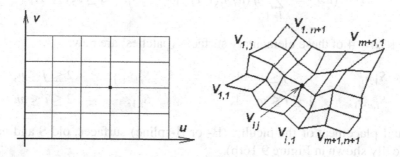

Fig. 9.15 Control graph for B- and β-spline surfaces.

The mutual placement of the vertices in the array $\mathbf{P}$ may be arbitrary (in particular, some of them may coincide). Some simple cases of vertex coincidence arise in relation to boundary or corner multiple vertices. As a rule, the vertices of the generated array do not belong to a bicubic (B- or β-spline) surface. It is natural to complete the control graph by a choice of auxiliary vertices such that the boundary curves of a new piecewise surface are close to or coincide with the boundary and corner vertices. It is common to use *multiple vertices* and *phantom vertices* for this purpose.

9.6.1 *Double vertices*

Additional surface patches are defined around the periphery of the interior ones naturally defined by the control vertices, by repeating boundary vertices in the spline surface formulation. The interior patches are then surrounded by additional patches. Define the $(m+3) \times (n+3)$ array $\mathbf{P}^* = \{P_{ij}^*\}_{j=1,\ldots,n+3}^{i=1,\ldots,m+3}$, setting

$$
\begin{aligned}
&P_{i,j}^* = P_{i-1,j-1}, \quad 2 \leq i \leq m+2, \quad 2 \leq j \leq n+2, \\
&P_{1,j}^* = P_{1,j-1}, \qquad P_{m+3,j}^* = P_{m+1,j}, \quad 2 \leq j \leq m+2, \\
&P_{i,1}^* = P_{i-1,1}, \qquad P_{i,m+3}^* = P_{i,n+1}, \quad 2 \leq i \leq m+2, \\
&P_{1,1}^* = P_{1,1}, \qquad P_{1,n+3}^* = P_{1,n+1}, \quad P_{m+3,1}^* = P_{m+1,1}, \quad P_{m+3,n+3}^* = P_{m+1,n+1}.
\end{aligned}
\tag{9.53}
$$

The $2m+2n+8$ new vertices "border" the given array $\mathbf{P}$ and, combined with the array, form a new array $\mathbf{P}^*$ consisting of $(m+3)(n+3)$ vertices. Note that the arrays $\mathbf{P}^*$ and $\mathbf{P}$ coincide as sets of points (geometrically we have no new vertices). The bicubic (B- or β-spline) surface $\mathbf{S}^*$ defined by the array $\mathbf{P}^*$ consists of mn fragments $\{S_{ij}\}_{j=1,\ldots,n}^{i=1,\ldots,m}$ with radius vectors calculated by the formulae

$$
\mathbf{r}^{(i,j)}(u,v) = \sum_{k,l=1}^{4} n_k(u)\, n_l(v)\, P_{k-1+i,l-1+j}^*, \quad 0 \leq u,v \leq 1.
\tag{9.54}
$$

Only $2m+2n-4$ of these elementary surfaces (patches) are new:

$$
\begin{aligned}
&S_{1,j}^*, && 1 \leq j \leq n-1, && S_{m-1,j}^*, && 2 \leq j \leq n, \\
&S_{i,n-1}^*, && 1 \leq i \leq m-1, && S_{i,1}^*, && 2 \leq i \leq m.
\end{aligned}
$$

The mutual placement of the bicubic (B- or β-spline) surfaces, old $\mathbf{S}$ and new $\mathbf{S}^*$, is schematically shown in Figure 9.16(b).

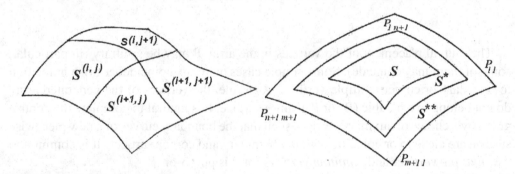

Fig. 9.16 Piecewise surfaces.

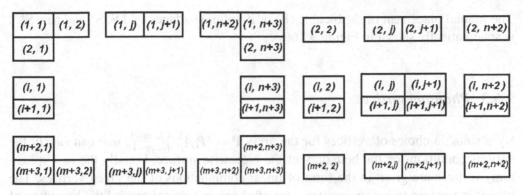

Fig. 9.17 Double vertices condition: additional and interior patches.

9.6.2 *Triple vertices*

This technique extends the double vertices technique by defining another set of additional surface patches around the periphery of those defined by $\mathbf{P}^*$. Define the third array $\mathbf{P}^{**} = \{P_{ij}^{**}\}_{j=1,\ldots,n+3}^{i=1,\ldots,m+3}$ by the same algorithm as that for the second array (see (9.53)), setting:

$$P_{i,j}^{**} = P_{i-1,j-1}^*, \quad 2 \leq i \leq m+4, \quad 2 \leq j \leq n+4,$$

$$P_{1,j}^{**} = P_{1,j-1}^*, \qquad P_{m+5,j}^{**} = P_{m+3,j}^*, \quad 2 \leq j \leq m+4,$$

$$P_{i,1}^{**} = P_{i-1,1}^*, \qquad P_{i,m+5}^{**} = P_{i,n+3}^*, \quad 2 \leq i \leq m+4,$$

$$P_{1,1}^{**} = P_{1,1}^*, \qquad P_{1,n+5}^{**} = P_{1,n+3}^*, \qquad P_{m+5,1}^{**} = P_{m+3,1}^*, \qquad P_{m+5,n+5}^{**} = P_{m+3,n+3}^*.$$

The $2m+2n+16$ new vertices "border" the array $\mathbf{P}^*$ and, combined with the array, form a new array $\mathbf{P}^{**}$ consisting of $(m+5)(n+5)$ vertices. Note that the arrays $\mathbf{P}^{**}$ and $\mathbf{P}$ coincide as sets of points (geometrically we have no new vertices). The bicubic (B- or β-spline) surface $\mathbf{S}^{**}$ defined by the array $\mathbf{P}^{**}$ consists of $(m+2)(n+2)$ fragments $\{S_{ij}\}_{j=1,\ldots,n+2}^{i=1,\ldots,m+2}$ with radius vectors calculated by the formulae

$$\mathbf{r}^{(i,j)}(u,v) = \sum_{k,l=1}^{4} n_k(u)\, n_l(v) P_{k-1+i,\, l-1+j}^*, \quad 0 \leq u,v \leq 1. \tag{9.55}$$

Only $2m+2n+4$ of these elementary surfaces (patches) are new:

$$S_{1,j}^{**}, \qquad 1 \leq j \leq n+1, \qquad\qquad S_{m+1,j}^{**}, \qquad 2 \leq j \leq n+2,$$

$$S_{i,n+1}^{**}, \qquad 1 \leq i \leq m+1, \qquad\qquad S_{i,1}^{**}, \qquad 2 \leq i \leq m+2.$$

The mutual placement of the bicubic (B- or β-spline) surfaces (old $\mathbf{S}$, $\mathbf{S}^*$ and new $\mathbf{S}^{**}$) is schematically shown in Figure 9.16(b).

9.6.3 Phantom vertices

By a suitable choice of vertices for the array $\mathbf{P} = \{P_{ij}\}_{j=1,\ldots,n+1}^{i=1,\ldots,m+1}$, one can satisfy the various conditions on the boundary of the piecewise surface. Usually the new (phantom) vertices are given in the form of linear combinations of the given vertices. The phantom vertices technique creates a set of phantom vertices around the boundary of the original control graph which defines additional surface patches. These patches are of normal size, and the complete surface consists of mn patches.

Consider a bicubic β-spline surface $\mathbf{S}^*$ generated by the array $\mathbf{P}^* = \{P_{ij}^*\}_{j=1,\ldots,n+3}^{i=1,\ldots,m+3}$, whose $2m + 2n + 8$ vertices are defined by means of the following formulae:

$$P_{i,j}^* = P_{i-1,j-1}, \quad 2 \le i \le m+2, \quad 2 \le j \le n+2,$$
$$P_{1,j}^* = A_1(P_{1,j-1} - P_{2,j-1}) + P_{1,j-1},$$
$$P_{m+3,j}^* = A_1(P_{m+1,j-1} - P_{m,j-1}) + P_{m+1,j-1}, \quad 2 \le j \le n+2,$$
$$P_{i,1}^* = A_2(P_{i-1,1} - P_{i-1,2}) + P_{i-1,1},$$
$$P_{i,n+3}^* = A_2(P_{i-1,n+1} - P_{i-1,n}) + P_{i-1,n+1}, \quad 2 \le i \le m+2,$$
$$P_{1,1}^* = A_1(P_{1,2}^* - P_{1,3}^*) + P_{1,2}^*,$$
$$P_{1,n+3}^* = A_2(P_{1,n+2}^* - P_{1,n+1}^*) + P_{1,n+2}^*,$$
$$P_{m+3,1}^* = A_2(P_{m+1,2}^* - P_{m+1,3}^*) + P_{m+1,2}^*,$$
$$P_{m+3,n+3}^* = A_1(P_{m+1,n+2}^* - P_{m+1,n+1}^*) + P_{m+1,n+2}^*, \tag{9.56}$$

where $A_1 = \frac{2\beta_1^2 + \beta_2}{2\beta_1^3}$, $A_2 = \frac{1}{2}(2\beta_1 + \beta_2)$. Note that a solution (9.56) defines each of the four corner phantom vertices in terms of other phantom vertices and expresses them explicitly in terms of the original control vertices. For $\beta_1 = 1$, $\beta_2 = 0$ these formulae reduce to the B-spline phantom vertices:

$$P_{i,j}^* = P_{i-1,j-1}, \quad 2 \le i \le m+2, \quad 2 \le j \le n+2,$$
$$P_{1,j}^* = 2P_{1,j-1} - P_{2,j-1}, \qquad P_{m+3,j}^* = 2P_{m+1,j-1} - P_{m,j-1}, \quad 2 \le j \le n+2,$$
$$P_{i,1}^* = 2P_{i-1,1} - P_{i-1,2}, \qquad P_{i,n+3}^* = 2P_{i-1,n+1} - P_{i-1,n}, \quad 2 \le i \le m+2,$$
$$P_{1,1}^* = 2P_{1,2}^* - P_{1,3}^*, \qquad P_{1,n+3}^* = 2P_{1,n+2}^* - P_{1,n+1}^*,$$
$$P_{m+3,1}^* = 2P_{m+1,2}^* - P_{m+1,3}^*, \qquad P_{m+3,n+3}^* = 2P_{m+1,n+2}^* - P_{m+1,n+1}^*. \tag{9.57}$$

The bicubic (B- or β-spline) surface $\mathbf{S}^*$ defined by $\mathbf{P}^*$ has the following property: *the curvature of the common curve of the neighboring patches at a point lying on the boundary curve of the surface is zero.*

In the case of phantom vertices, the twist vectors at the corner points of the surface $\mathbf{S}^*$ are equal to the twist vectors of the corresponding bilinear surfaces constructed on the edges adjacent to the vertices.

9.6.4 *Examples*

We now compute and visualize spline surfaces. We base our investigations on the procedure **surf_spline**(X, Y, Z, 'method') (see Section A.5), where X, Y, Z are coordinate arrays of given points, and 'method' is either 'BB' or 'BT'.

The reader can alternatively fit surfaces to data and view plots with the *Surface Fitting Tool* interface.

Examples.

1. Approximate the *graph of a function* $z = x^2 + y^2$ on the rectangle $\Pi = \{-1 \leq x \leq 2, \; |y| \leq 3\}$. First, prepare the data.

```
a = -1;   b = 2;   c = -3;   d = 3;   n = 6;   m = 5;              % enter data
[uu, vv] = meshgrid(0 : 1/(n - 1) : 1, 0 : 1/(m - 1) : 1);
X = a + uu.*(b - a);
Y = c + vv.*(d - c);
Z = (a + uu.*(b - a)).^2 + (c + vv.*(d - c)).^2                    % enter function
```

Then call **surf_spline** using either of two lines:

```
S = surf_spline(X, Y, Z, 'BB');                                   % B-spline,
S = surf_spline(X, Y, Z, 'BT');                                   % β-spline.
```

Finally, plot the piecewise surface (see Figure 9.18) with a typical program, called **view_surf** in what follows.

```
syms u v;
[ui, vi] = meshgrid(0 : .2 : 1);
hold on;    view(3);    grid on;    plot3(X, Y, Z, 'o--');        % control graph
for i = 1 : (n + 1)*(m + 1);
 SX = double(subs(S(i, 1), {u v}, { ui vi }));
 SY = double(subs(S(i, 2), {u v}, { ui vi }));
 SZ = double(subs(S(i, 3), {u v}, { ui vi }));
 c = i/(n + 1)/(m + 1)*ones(size(SX, 1));                         % coloring
 surf(SX, SY, SZ,  c);                                            % ith patch of a piecewise surface
end;
```

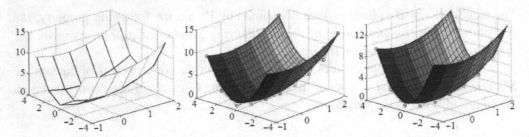

Fig. 9.18 (a) Control graph. (b,c) B-spline and β-spline surfaces, $\beta = [1,7]$.

2. The "Figure Eight" shape of the Klein bottle is obtained by rotating a figure eight about an axis while placing a twist in it, and is given by equations (8.8). The illustrative Figure 9.19(a) is obtained by the program below

```
syms u v;
ezsurf((R + cos(u/2)*sin(v) - sin(u/2)*sin(2*v))*cos(u), (R + cos(u/2)*sin(v) -
sin(u/2)*sin(2*v))*sin(u), sin(u/2)*sin(v) + cos(u/2)*sin(2*v), [-pi, pi*3/4, -pi, pi]);
```

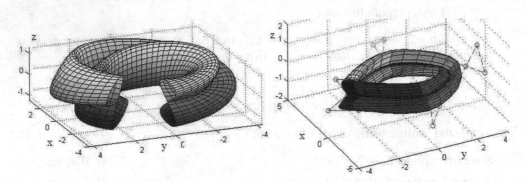

Fig. 9.19 Immersed Klein bottle and its β-spline approximation, $\beta = [1,1]$.

9.7 Exercises

Section 9.2

1. Given three points P_1, P_2, P_3, calculate the parabola that goes from P_1 to P_3 and whose start and end tangent vectors point in directions $P_2 - P_1$ and $P_3 - P_2$, respectively.

2. Represent the Bézier curve recursively by $\mathbf{r}_n^{(n)}(t)$, where

$$\mathbf{r}_i^{(0)}(t) = P_i, \quad \mathbf{r}_i^{(j)}(t) = (1-t)\,\mathbf{r}_{i-1}^{(j-1)}(t) + t\,\mathbf{r}_i^{(j-1)}(t) \quad \text{for } j > 0. \tag{9.58}$$

3. Represent the Bézier curve also in a very compact form as

$$\mathbf{r}(t) = (1 - t + tE)^n P_1, \tag{9.59}$$

where E is the *shift operator* defined by $EP_i = P_{i+1}$ (i.e., applying E to P_i produces P_{i+1}). To prove (9.59), show

$$\mathbf{r}(t) = \sum_{i=0}^{n} B_{n,i}(t)\, P_{i+1} = \sum_{i=0}^{n} \binom{n}{i} t^i (1-t)^{n-i} E^i P_1$$

$$= \sum_{i=0}^{n} \binom{n}{i} (tE)^i (1-t)^{n-i} P_1 = ((1-t) + tE)^n P_1$$

where the last step follows from the binomial theorem.

4. Show that (for any n) the result of the procedure $\mathbf{r} = \mathbf{bezier}(x, y)$ (Section A.2) is the polynomial curve $[f_1(t), \ldots, f_N(t)] \subset \mathbb{R}^N$. Try $N = 2, 3$.

5. Show that a Bézier curve is invariant under affine transformations.

6. Show that the elementary Bézier curve is equivalently defined by (9.4).

7. Verify that a piecewise Bézier curve is C^1-*continuous* if the triples of vertices $P_{3i-1}, P_{3i}, P_{3i+1}$ are collinear (lie on the same line).

8. For C^2-*continuity* of a Bézier curve in $\mathbb{R}^3$ it is necessary that all five points P_{3i-2}, $P_{3i-1}, P_{3i}, P_{3i+1}, P_{3i+2}$ are coplanar (lie in the same plane).

9. Plot 3 elementary Bézier curves using the procedure $\mathbf{bezier2d}(x, y)$.

10. Find conditions for a closed piecewise Bézier curve to be continuous.

Answer: If $P_0 = P_{3m}$ and the triples $P_{3m-1}P_{3m} = P_0P_1$ are collinear, a closed piece-wise Bézier curve is C^1-continuous.

11. Let $P_i = (x_i, y_i, z_i)$. Define homogeneous control points $\bar{P}_i$ by

$$\bar{P}_i = \begin{cases} (w_i x_i, w_i y_i, w_i z_i, w_i) & \text{if } w_i \neq 0, \\ (x_i, y_i, z_i, 0) & \text{if } w_i = 0. \end{cases}$$

Show that in homogeneous coordinates the rational Bézier curve is given by $\bar{\mathbf{r}}(t) = \sum_{i=0}^{n} \bar{B}_{n,i}(t) \bar{P}_{i+1}$, $t \in [0, 1]$ where $\bar{B}_{n,i}(t) = B_{n,i}(t) \Big/ \sum_{j=0}^{n} w_{j+1} B_{n,j}(t)$. This is a poly-nomial Bézier curve, but with homogeneous control points.

12. Given the array P_0, P_1, P_2, P_3, and nonnegative real numbers k_0 and k_3, find weights w_0, w_1, w_2, w_3 such that the values of the curvature of the *rational elementary Bézier curve* at the boundary control points coincide with the given real number (k_0 at the point P_0 and k_3 at P_3).

Hint. $w_0 = 1$, $w_1 = \frac{4}{3}\left(\frac{c_0{}^2 c_3}{k_0{}^2 k_3}\right)^{1/3}$, $w_2 = \frac{4}{3}\left(\frac{c_3{}^2 c_0}{k_3{}^2 k_0}\right)^{1/3}$, $w_3 = 1$, $c_0 = \frac{|(P_1 - P_0) \times (P_2 - P_0)|}{2|P_1 - P_0|^3}$,

$c_3 = \frac{|(P_2 - P_1) \times (P_2 - P_0)|}{2|P_3 - P_2|^3}$.

13. Develop an M-file **rbezier2d**(x, y) for deriving piecewise rational Bézier curves in analogy to **bezier2d**(x, y).

Section 9.3   1. Plot the four *Hermite blending functions* of (9.7):

```
t = linspace(0, 1, 50);
n1 = 1 - 3*t.^2 + 2*t.^3;   n2 = 1 - n1;
n3 = t - 2*t.^2 + t.^3;   n4 = t.^2.*(t - 1);
plot(t, [n1; n2; n3; n4])
```

Show that the weights $n_1(t)$ and $n_2(t) = 1 - n_1(t)$ assigned to the points P_1 and P_2 are monotonic for $t \in [0,1]$ functions. The function $n_3(t)$ is a bit trickier. It starts at zero, has a maximum at $t = 1/3$, then drops slowly back to zero (verify!). The function $n_4(t)$ behaves similarly.

2. Given the endpoints $P_1 = (0,0)$ and $P_2 = (1,0)$ and the velocity vectors $Q_1 = \alpha(\cos\theta, \sin\theta)$ and $Q_1 = \alpha(\cos\theta, \sin\theta)$, calculate the value of θ for which the Hermite segment from P_1 to P_2 has a cusp.

3. Given endpoints P_1 and P_2, unit tangent vectors T_1 and T_2, and a third point P_3, find scale factors α and β such that the Hermite segment $\mathbf{r}(t)$ determined by P_1, P_2 and velocities αT_1, βT_2, respectively, will pass through P_3. Also find t_0 for which $\mathbf{r}(t_0) = P_3$.

4. Hermite interpolation can be applied to compute (approximate) conic sections. Given three points P_1, P_2, and P_3 and a scalar a, we take the 4-tuple P_1, P_2, $a(P_3 - P_1)$, $a(P_2 - P_3)$ as our two points and two extreme tangent vectors, and compute the Hermite segment that approximates a conic section. Show that we obtain an ellipse when $0 \le a < 0.5$, a parabola when $a = 0.5$, and a hyperbola when $0.5 < a \le 1$.

5. A variation of the Hermite method is the case where two points and just one tangent are given. Each segment $\mathbf{r}_i(t)$ is therefore a quadratic polynomial defined by its two endpoints P_1 and P_2 and by its beginning tangent vector Q_1. Let $\mathbf{r}(t) = \mathbf{a}t^2 + \mathbf{b}t + \mathbf{c}$ be a quadratic polynomial satisfying the conditions $\mathbf{r}(0) = P_1$, $\mathbf{r}(1) = P_2$, $\mathbf{r}'(0) = Q_1$. Show that $\mathbf{a} = P_2 - P_1 - Q_1$, $\mathbf{b} = Q_1$, $\mathbf{c} = P_1$, and hence

$$\mathbf{r}(t) = (1 - t^2)P_1 + t^2 P_2 + (t - t^2)Q_1 \qquad (0 \le t \le 1). \tag{9.60}$$

6. What are the conditions for closed C^1-continuous and C^2-continuous piecewise Hermite curves?

7. Show that the matrix presentation of (9.60) is $\mathbf{r}(t) = \mathbf{G}\mathbf{M}T(t)$ $(0 \le t \le 1)$, where $T(t) = [t^2; t; 1]$ is a column vector, $\mathbf{G} = (P_1\ P_2\ Q_1)$ is the *geometric matrix* and

$$M = \begin{pmatrix} -1 & 0 & 1 \\ 1 & 0 & 0 \\ -1 & 1 & 0 \end{pmatrix} \tag{9.61}$$

is the *basic matrix* of a quadratic Hermite curve.

8. Show that the Bézier curve determined by four points P_1, P_2, P_3, and P_4 is a Hermite curve (having identical shape) determined by the points P_1, P_4, $3(P_2 - P_1)$, and $3(P_4 - P_3)$. Hence, one can add a tension parameter to a (cubic) Bézier curve by the rule P_1, P_4, $3s(P_2 - P_1)$, $3s(P_4 - P_3)$. Find the parameterization of this curve and its matrix representation.

9. The quadratic spline is based on the Hermite segment of (9.60).

(a) Compute the quadratic spline curve. Show that the neighboring segments

$$\mathbf{r}_i(t) = (1 - t^2)P_i + t^2 P_{i+1} + (t - t^2)Q_i,$$
$$\mathbf{r}_{i+1}(t) = (1 - t^2)P_{i+1} + t^2 P_{i+2} + (t - t^2)Q_{i+1} \tag{9.62}$$

have the velocity vectors $2(P_{i+1} - P_i) - Q_i$ and Q_{i+1} at the node P_{i+1}. Requiring C^1-continuity, deduce the system of $n - 1$ linear equations

$$\begin{pmatrix} 1 & 1 & & \\ & 1 & 1 & \\ & & \cdots & \\ & & & 1 & 1 \end{pmatrix} \begin{pmatrix} Q_1 \\ Q_2 \\ \cdots \\ Q_n \end{pmatrix} = 2 \begin{pmatrix} P_2 - P_1 \\ P_3 - P_2 \\ \cdots \\ P_n - P_{n-1} \end{pmatrix}. \tag{9.63}$$

(b) Create an M-file to calculate the quadratic spline, and then plot it.

10. Find explicitly the matrix presentation of the elementary cardinal spline curve (generalization of (9.16)).

Section 9.4   1. Show that the matrix presentation of (9.18) is $\mathbf{r}(t) = \mathbf{P}MT(t)$ $(0 \le t \le 1)$, where $\mathbf{r}(t) = [x(t); y(t); z(t)]$ and $T(t) = [t^2; t; 1]$ are column vectors, $\mathbf{P} = (P_1 \ P_2 \ P_3)$, and

$$M = \frac{1}{\beta_1 + 1} \begin{pmatrix} \beta_1 & -2\beta_1 & \beta_1 \\ -\beta_1 - 1 & 2\beta_1 & 1 \\ \beta_1 & 0 & 0 \end{pmatrix}.$$

2. Find the shape parameters $\{\beta_{1i}\}$ such that the curvature of a quadratic β-spline (see Example, p. 377) is continuous at the joints.

3. Verify particular cases of cubic β-splines related to the cubic B-spline:
(a) *the equality $\beta_1 = 1$ is equivalent to the continuity of $\mathbf{r}'$*;
(b) *the system $\beta_1 = 1$, $\beta_2 = 0$ is equivalent to the continuity of $\mathbf{r}'$ and $\mathbf{r}''$*.

4. Verify that:
(a) For $\beta_1 = 1$ and $\beta_2 = 0$ one obtains an elementary cubic B-spline.

(b) The $b_i(t)$ are nonnegative for $0 \leq t \leq 1$, and their sum is equal to 1.

(c) The elementary β-spline lies in the convex hull of its control vertices.

(d) The β-spline curve is invariant under affine transformations.

5. Show that the matrix presentation of (9.31) is $\mathbf{r}(t) = \mathbf{P}MT(t)$ $(0 \leq t \leq 1)$, where $\mathbf{r}(t) = [x(t);\ y(t);\ z(t)]$ and $T(t) = [t^3;\ t^2;\ t;\ 1]$ are column vectors, $\mathbf{P} = (P_1,\ P_2,\ P_3,\ P_4)$ is the *geometric matrix*, and

$$M = \frac{1}{\delta_3} \begin{pmatrix} -2\beta_1^3 & 6\beta_1^3 & -6\beta_1^3 & 2\beta_1^3 \\ 2(\beta_1^3 + \xi) & -3(2\beta_1^3 + 2\beta_1^2 + \beta_2) & 6(\beta_1^3 - \beta_1) & 4(\beta_1^2 + \beta_1) + \beta_2 \\ -2(\xi + 1) & 3(2\beta_1^2 + \beta_2) & 6\beta_1 & 2 \\ 2 & 0 & 0 & 0 \end{pmatrix} \quad (9.64)$$

(where $\xi = \beta_1^2 + \beta_1 + \beta_2$) is the *basic matrix* of the β-spline curve.

6. Calculate and plot the self-repeating plane β-spline by the same method used for the Bézier curves, only replacing one line by:

    B = **beta_s2d**(x1, y1, beta);   % Figure 9.4

7. Play with **beta_s2d** using the data in (a) and (b) below.

(a) Plot your hand (see [Moler] for realistic input using the mouse):

    x = [6.5 6 5.8 4.5 3 3.5 4.5 5.7 7.5 7 6.8 7.2 8 8.8 9.8 10 10 10.5 11.8 12
    11.8 12.8 13.2 14 14.8 14.7 14.1 13.9 14.2 16 17 18.4 19.1 19.3 17.8 15.9
    14.6 14.6 6.5];
    y = [1 5 8.5 11 14.2 15.2 15 13 10.5 14.8 18.8 19.5 19 15.7 12 16 20 21
    20.2 16.3 11.8 15.8 19 19.8 18.6 15 11.5 9 7.8 8 9.5 10 9.6 8.5 6.4 4 1 1 1];

In order to use polar coordinates, the data must be centered so that it lies on a curve that is *starlike* with respect to the origin (that is, every ray emanating from the origin meets the data only once). This means that you must be able to find values x_0 and y_0 so that the MATLAB statements

    xp = x - x0; yp= y - y0;
    theta = atan2(yp, xp);   r = sqrt(xp.^2 + yp.^2)
    plot(theta, r)

produce a set of points that can be interpolated with a single-valued function, $r = r(\theta)$. For the data obtained by sampling the outline of your hand, the point (x_0, y_0) is located near the base of your palm. Furthermore, in order to use **interp1**, it is also necessary to order the data so that theta is monotonically increasing.

    ti = linspace(theta(1), theta(n - 2), 102);
    B = **interp1**(theta(1 : n - 2), r(1 : n - 2), ti, 'cubic');
    hold on;   polar(theta, r);   polar(ti, B, 'r');

(b) Data for a car profile; Figure 9.20.

x = [226 348 356 400 410 452 470 452 438 496 416 412 430 428 394
344 232 164 58 38 36 98 100 148 158 228 226];
y = [21 17 5 4 18 24 55 62 82 89 83 79 79 63 68 94 87 62 53 34 21 16 2 1
13 21 21];

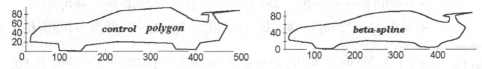

Fig. 9.20 Modeling of a car profile: $\beta_1 = 1, \beta_2 = 9$.

8. What are the conditions for a closed (piecewise) cubic β-spline curve?
Hint. Similar to those for a cubic B-spline curve; see Section 9.4.4.

9. Check that a cubic β-spline curve lies in the union of the convex hulls of four consecutive control vertices P_{i-1}, P_i, P_{i+1}, P_{i+2}.

10. Do the following experiments with test data given after the procedure **beta_2d(x, y, beta)**:

(a) Change $\beta_1 > 0$ for fixed $\beta_2 \geq 0$ (for $\beta_1 = 0$ we get the polygon).

(b) Change $\beta_2 > 0$ for fixed $\beta_1 > 0$ (for $\beta_2 \to \infty$ each node of the spline runs to the corresponding control vertex; Figure 9.7(a–c)).

(c) For variety, consider $\beta_2 < 0$ for fixed $\beta_1 > 0$ (the convexity condition does not hold; Figure 9.7(d–f)).

Conclude that if we do not require nonnegativity of the parameters β_1 and β_2, then some *self-intersections and oscillations* of the curve can appear.

11*. Develop M-files for deriving rational cubic β-splines similar to **beta_s2d(x, y, z, beta)** and **beta_s3d(x, y, z, beta)**. Use **rbetas_my** for elementary rational curves. Study the properties of these curves.

12*. Using MATLAB, construct the *quintic β-spline* in $\mathbb{R}^4$. Verify that for $\beta_1 = 1$, $\beta_2 = \beta_3 = \beta_4 = 0$ it coincides with *quintic B-spline*. Plot (for various β's) projections of the curve onto coordinate 2-planes and coordinate 3-spaces.

Hint. The segment is

$$\mathbf{r}_i(t) = \sum_{j=1}^{6} b_j(t) P_{i+j-1}, \qquad \text{where} \qquad b_i(t) = (1/\delta_5) \sum_{j=0}^{5} c_{i,j} t^j.$$

Similarly to quadratic, cubic, or quartic β-splines, use the first four derivative constraints of (9.17), and obtain a linear system of $36 = 6 \times 6$ equations with 36 unknown coefficients $c_{i,j}$. Assume a global assignment of the shape parameters β_1, β_2, β_3, β_4. The determinant is

$$\delta_5 = 72 + 288\beta_1 + 72\beta_1\beta_2^2 + 3\beta_4 + 96\beta_1\beta_3 + 1440\beta_1^4 + 36\beta_2\beta_1\beta_3 + 1188\beta_1^3$$
$$+ 108\beta_2 - 30\beta_1^4\beta_2\beta_3 - 9\beta_4\beta_1^2 + 1764\beta_1^4\beta_2 + 432\beta_1^2\beta_2^2 + 72\beta_1^10$$
$$+ 1023\beta_1^5\beta_2 + 432\beta_1^3\beta_2^2 - 108\beta_1^4\beta_3 - 9\beta_4\beta_1^3 + 4\beta_1\beta_3^2 + 24\beta_3 + 336\beta_1\beta_2$$
$$+ 1764\beta_1^3\beta_2 + 111\beta_3\beta_1^2 + 1080\beta_1^7 + 336\beta_1^6\beta_2 - 96\beta_1^5\beta_3 + 288\beta_1^4\beta_2^2$$
$$+ 1059\beta_1^2\beta_2 + 1368\beta_1^6 + 288\beta_1^9 + 54\beta_1^2\beta_2\beta_3 - 2\beta_4\beta_1\beta_2 + 48\beta_1^5\beta_2^2$$
$$+ 36\beta_1^3\beta_2^3 + 720\beta_1^2 + 1476\beta_1^5 + 648\beta_1^8 + 3\beta_1^5\beta_4 - 24\beta_1^6\beta_3 + 108\beta_1^7\beta_2.$$

Section 9.4.4 1. Verify that the matrix form of the quadratic B-spline is $\mathbf{r}(t) = \mathbf{P}MT(t)$ ($0 \le t \le 1$), where $\mathbf{r}(t) = [x(t); \ y(t); \ z(t)]$ and $T(t) = [t^2; \ t; \ 1]$ are column vectors, $\mathbf{P} = (P_1 \ P_2 \ P_3)$, and

$$M = \frac{1}{2}\begin{pmatrix} 1 & -2 & 1 \\ 1 & 2 & -2 \\ 0 & 0 & 1 \end{pmatrix}.$$

2. Show that, given the 4 control vertices $P_1 = (-1,0)$, $P_2 = (-1,2)$, $P_3 = (1,2)$, and $P_4 = (1,0)$, the first quadratic spline segment is

$$\mathbf{r}_1(t) = \frac{1}{2}(P_1, P_2, P_3)\begin{pmatrix} 1 & -2 & 1 \\ -2 & 2 & 1 \\ 1 & 0 & 0 \end{pmatrix}\begin{pmatrix} t^2 \\ t \\ 1 \end{pmatrix}$$
$$= \frac{1}{2}(t-1)^2(-1,0) + (-t^2 + t + 1/2)(-1,2) + t^2/2(1,2) = (t^2 - 1, 1 + 2t - t^2).$$

Deduce from this that the first segment starts with velocity $\mathbf{r}_1(0) = (0,2)$ and ends with velocity $\mathbf{r}_1(1) = (2,0)$. Calculate the second segment, its velocity, and the joint point $\mathbf{r}_2(1)$.

3. Show that the functional weight multipliers of the cubic B-spline

$$n_1(t) = \frac{(1-t)^3}{6}, \quad n_2(t) = \frac{3t^3 - 6t^2 + 4}{6}, \quad n_3(t) = \frac{-3t^3 + 3t^2 + 3t + 1}{6}, \quad n_4(t) = \frac{t^3}{6}$$

(see (9.36)) are nonnegative for $0 \le t \le 1$ and their sum is equal to 1.

```
t = linspace(0, 1, 50);
n1 = (1 - t).^3/6;   n2 = (3*t.^3 - 6*t.^2 + 4)/6;
n3 = (-3*t.^3 + 3*t.^2 + 3*t + 1)/6;   n4 = t.^3/6;
plot(t, [n1; n2; n3; n4])
```

4. Verify that the matrix presentation of (9.36) is $\mathbf{r}(t) = \mathbf{P}MT(t)$ ($0 \le t \le 1$), where $\mathbf{r}(t) = [x(t); \ y(t); \ z(t)]$ and $T(t) = [t^3; \ t^2; \ t; \ 1]$ are column vectors, $\mathbf{P} = (P_1 \ P_2 \ P_3 \ P_4)$ is the *geometric matrix*, and

$$M = \frac{1}{6}\begin{pmatrix} -1 & 3 & -3 & 1 \\ 3 & -6 & 0 & 4 \\ -3 & 3 & 3 & 1 \\ 1 & 0 & 0 & 0 \end{pmatrix} \qquad (9.65)$$

is the *basic matrix* of an elementary B-spline curve.

5. Given a cubic B-spline segment $\mathbf{r}(t)$ based on the four control vertices P_1, P_2, P_3, and P_4, find four control vertices Q_1, Q_2, Q_3, and Q_4 such that the Bézier curve $\tilde{\mathbf{r}}(t)$ they define will have the same shape as $\mathbf{r}(t)$.

Hint. This can be done by equating the matrices of (9.65) that define $\mathbf{r}(t)$ and those of (5.26) that define $\tilde{\mathbf{r}}(t)$, i.e., $\mathbf{PM} = \mathbf{Q}\tilde{\mathbf{M}}$. The answer is $Q_1 = (P_1 + 4P_2 + P_3)/6$, $Q_2 = (4P_2 + 2P_3)/6$, $Q_3 = (2P_2 + 4P_3)/6$, $Q_4 = (P_2 + 4P_3 + P_4)/6$.

6. Compute $M \cdot n!$ for $n = 3, 4, 5, \ldots$. See Example 3, p. 385.

```
n = 2;                                        % enter data
for i = 0 : n;   for j = 0 : n;
 Q = 0;
 for k = j : n;
  Q = Q + (n-k).^i*(-1).^(k-j).*gamma(n+2)./gamma(k-j+1)./gamma(n-k+j+2);
 end;
 M(j + 1, i + 1) = gamma(n + 1)./gamma(i + 1)./gamma(n - i + 1).*Q;
end;   end
M                                             % matrix M multiplied by n!
```

7. Call **beta_s3d** to calculate the test piecewise curve in $\mathbb{R}^3$.

```
x = [.056, .287, .655, .716, .228, .269, .666, .929];
y = [.820, .202, .202, .521, .521, .820, .820, .227];
z = [1/4, 1/4, 0, 0, 0, 0, -1/4, -1/4];            % enter data
B = beta_s3d(x, y, z,  [1 0]);
```

Then plot the curve in space (again one may use the + ti - ti trick to avoid a one-element zero array), Figure 9.9(a).

```
syms t real;
hold on;    plot3(x, y, z, 'o - - r');
for i = 1 : (length(x) + 1);
 ti = linspace(i - 1, i, 20);
 S = [subs(B(1, i), t, ti) + ti - ti; subs(B(2, i), t, ti) + ti - ti; subs(B(3, i), t, ti) + ti - ti];
 plot3(S(1, :), S(2, :), S(3, :), 'LineWidth', 2);
end
```

This typical program will be referred to as **splot3**(x, y, z, B).

8. As a simple application, we plot the knot *trefolium* using coordinates of vertices of the triangle $A = (-2,0), B = (0,4), C = (2,0)$ and midpoints of its edges. The arrays x, y and z below contain the data.

x = [2, 0, -1, 0, 1, 0, -2, -1, 1, 2, 2]; y = [0, 0, 2, 4, 2, 0, 0, 2, 2, 0, 0];
z = [0, 1, 2, 2, 0, 2, 2, 1, 2, 1, 0];   % enter data
B = **beta_s3d**(x, y, z, [1 0]);
splot3(x, y, z, B)   % Figure 9.9(c)

9. Calculate and plot the self-repeating plane B-spline by the same method used for the Bézier curve (Figure 9.4), only replacing one line by the line below.

B = **beta_s2d**(x1, y1, [1 0]);

10. Show that the B-spline curve is invariant under affine transformations.

11. What are the conditions for a *closed* (piecewise) cubic B-spline curve?

Hint. For closing a C^2-regular B-spline curve, one should complete the array $P_1, \ldots, P_n$ with three points $P_{n+1} = P_1$, $P_{n+2} = P_2$, $P_{n+3} = P_3$.

12. Check that a B-spline curve lies in the union of the convex hulls of the four points $\{P_{i-1}, P_i, P_{i+1}, P_{i+2}\}$ for $i > 1$.

Section 9.5 1. Verify that (9.40) can also be written in the matrix form

$$\mathbf{r}(u,v) = (f_1(u), \ldots f_n(u)) \begin{pmatrix} P_{11} & P_{12} & \ldots & P_{1m} \\ \vdots & \vdots & & \vdots \\ P_{n1} & P_{n2} & \ldots & P_{nm} \end{pmatrix} \begin{pmatrix} g_1(v) \\ \vdots \\ g_m(v) \end{pmatrix}. \tag{9.66}$$

2. Show that if $f_i(t)$ are *polynomials of degree n*, then (9.41) can be rewritten in the form (9.42), where $\mathbf{P}_n = \{P_{ij}\}_{i,j=1}^n$, and the *basic matrix M* can be found from (9.43).

3. The quadratic B-spline and β-spline curves determined by three given points are studied in Section 9.4. Use the Cartesian product for plotting *biquadratic B-spline* and β-*spline surface patches*.

4. Represent the parametric quadratic polynomial that passes through three given points in the form $\mathbf{r}(t) = \mathbf{G}MT(t)$; see (5.14). Apply the principle of the Cartesian product to multiply this curve by itself. Obtain a *biquadratic surface* patch $\mathbf{r}(u,v)$ ($0 \leq u, v \leq 1$; see (9.42) for $n = 3$)

$$\mathbf{r}(u,v) = (u^2, u, 1) M^* \mathbf{P}_3 M (v^2, v, 1)' \tag{9.67}$$

that passes through 9 given points $\{P_{ij}\}_{i,j=1}^3$, corresponding to $u, v \in \{0, 1/2, 1\}$. They should be roughly equally spaced over the surface.

Evaluate and plot the surface patch given by the 9 points:

M = [2 -3 1; -4 4 0; 2 -1 0];
P = [[0 0 0]; [1 0 0]; [2 0 0]; [0 1 0]; [1 1 1]; [2 1 -1/2]; [0 2 0]; [1 2 0]; [2 2 0]];
r = **surf_patch**(P, M)   % biquadratic surface patch
ezsurf(r(1), r(2), r(3), [0, 1, 0, 1], 20);   % Figure 9.10(b)

Plot the surface and the 9 points using **view_patch**; see Section A.5.

5. The quadratic Bézier curve determined by three given points is $\mathbf{r}(t) = \mathbf{G}MT(t)$; see (9.2). Apply the principle of the Cartesian product to multiply this curve by itself in order to obtain a *Bézier biquadratic surface* patch $\mathbf{r}(u,v) = (u^2, u, 1) M \mathbf{P}_3 M (v^2, v, 1)'$ $(0 \leq u, v \leq 1)$, determined by 9 given points. Once the points are given, evaluate and plot the surface patch:

```
M = [1 -2 1; -2 2 0; 1 0 0];                        % symmetric matrix
P = [[0 0 0]; [1 0 1]; [0 0 2]; [1 1 0]; [4 1 1]; [1 1 2]; [0 2 0]; [1 2 1]; [0 2 2]];
r = surf_patch(P, M)                                % biquadratic Bézier patch
ezsurf(r(1), r(2), r(3),  [0, 1, 0, 1],  20);       % Figure 9.11(a).
```

Plot the surface and the 9 points using the program **view_patch**.

6. Verify that the rational patch (9.44) in homogeneous coordinates is

$$\bar{\mathbf{r}}(u,v) = \sum_{i,j=1}^{n} \bar{h}_{ij}(u,v)\,\bar{P}_{ij}, \qquad \bar{h}_{ij}(u,v) = \frac{f_i(u)f_j(v)}{\sum_{i,j=1}^{n} w_{ij}\,f_i(u)f_j(v)}, \qquad u, v \in [0,1],$$

and plot this patch using MATLAB.

> **Section 9.6**

1. Show that the twist vector at the corner point of the B-spline patch $S^{(1,1)}$ is equal to $P_{11} - P_{12} + P_{22} - P_{21}$.

2. Show that the twist vectors at the corner vertices of four corner patches $S^{(1,1)}$, $S^{(1,n-1)}$, $S^{(m-1,1)}$, $S^{(m-1,n-1)}$ of the β-spline surface $\mathbf{S}^*$ are

$$\mathbf{r}_{uv}^{(1,1)}(0,0) = (36/\delta^2)(P_{1,1} - P_{1,2} + P_{2,2} - P_{2,1}),$$

$$\mathbf{r}_{uv}^{(1,n)}(0,1) = (36/\delta^2)(P_{1,n} - P_{1,n+1} + P_{2,n+1} - P_{2,n}),$$

$$\mathbf{r}_{uv}^{(m,n)}(1,1) = (36/\delta^2)(P_{m,n} - P_{m,n+1} + P_{m+1,n+1} - P_{m+1,n}),$$

$$\mathbf{r}_{uv}^{(m,1)}(1,0) = (36/\delta^2)(P_{m,1} - P_{m,2} + P_{m+1,2} - P_{m+1,1}), \qquad (9.68)$$

where $\delta = \beta_2 + 2\beta_1^3 + 4\beta_1^2 + 4\beta_1 + 2$; see Proposition 9.2. These are equal (up to a coefficient) to the twist vectors of the bilinear surfaces defined by the quadruples of vertices

$$P_{1,1}, P_{1,2}, P_{2,2}, P_{2,1}; \qquad\qquad P_{1,n}, P_{1,n+1}, P_{2,n+1}, P_{2,n};$$

$$P_{m,n}, P_{m,n+1}, P_{m+1,n+1}, P_{m+1,n}; \qquad\qquad P_{m,1}, P_{m,2}, P_{m+1,2}, P_{m+1,1}.$$

For $\delta = 12$, (9.68) reduce to the twist vectors at B-spline patches.

3. Show that:

(a) The corner patches, $S^{(1,1)}$, $S^{(1,n+1)}$, $S^{(m+1,1)}$, $S^{(m+1,n+1)}$ of the bicubic (B- or β-spline) surface $\mathbf{S}^{**}$ are *bilinear surfaces*. The twist vectors at these β-spline patches are, respectively,

$$\mathbf{r}_{uv}^{(1,1)}(u,v) = (36/\delta)\,u^2 v^2 (P_{1,1} - P_{1,2} + P_{2,2} - P_{2,1}),$$

$$\mathbf{r}_{uv}^{(1,n)}(u,v) = (36/\delta)\,u^2 v^2 (P_{1,n} - P_{1,n+1} + P_{2,n+1} - P_{2,n}),$$

$$\mathbf{r}_{uv}^{(m,n)}(u,v) = (36/\delta)\,u^2 v^2 (P_{m,n} - P_{m,n+1} + P_{m+1,n+1} - P_{m+1,n}),$$

$$\mathbf{r}_{uv}^{(m,1)}(u,v) = (36/\delta)\,u^2 v^2 (P_{m,1} - P_{m,2} + P_{m+1,2} - P_{m+1,1}).$$

For $\delta = 12$ these formulae reduce to the twist vectors at B-spline patches.

Hint. Show, for instance, that

$$\mathbf{r}^{(1,1)} = (1 - \tilde{u})(1 - \tilde{v})P_{1,1} + \tilde{u}(1 - \tilde{v})P_{2,1} + (1 - \tilde{u})\tilde{v}P_{1,2} + \tilde{u}\tilde{v}P_{2,2},$$

where the new parameters are defined by $\tilde{u} = n_4(u)$ and $\tilde{v} = n_4(v)$.

(b) All four corner vertices lie at the corners of the bicubic B- or β-spline surface $\mathbf{S}^{**}$ constructed using the triple vertices:

$$\mathbf{r}^{(1,1)}(0,0) = P_{1,1}, \qquad\qquad \mathbf{r}^{(1,n+2)}(0,1) = P_{1,n+1},$$

$$\mathbf{r}^{(m+2,1)}(1,0) = P_{m+1,1}, \qquad\qquad \mathbf{r}^{(m+2,n+2)}(1,1) = P_{m+1,n+1}.$$

4. Approximate the following *parametric surfaces* and plot the results.

(a) Prepare the data for the *single-sheeted hyperboloid of revolution*, and then call **surf_spline**.

```
n = 4;   m = 5;                                  % enter n > 2, m > 3
[uu, vv] = meshgrid(-1 : 2/(n - 1) : 1, 0 : 2*pi/(m - 1) : 2*pi);
X = cosh(uu).*cos(vv);
Y = cosh(uu).*sin(vv);
Z = sinh(uu);
S = surf_spline(X, Y, Z, 'BT');                  % β-spline
```

Plot the piecewise surface, Figure 9.21(b), using the program **view_surf**.

(b) Prepare the data for the *torus of revolution*, and then call **surf_spline**:

```
R = 1;   n = 5;   m = 5;                % enter R ∈ (0,3/2), n > 3, m > 3
[uu, vv] = meshgrid(0:2*pi/(n-1):2*pi+2*pi/(n-1), 0:2*pi/(m-1):2*pi+2*pi/(m-1));
X = (3 + R*cos(uu)).*cos(vv);
Y = (3 + R*cos(uu)).*sin(vv);
Z = R*sin(uu);
S = surf_spline(X, Y, Z, 'BT');                  % β-spline
```

Plot the piecewise surface, Figure 9.21(a), using the program **view_surf**.

(c) Prepare the data for the immersed *Klein bottle*, and then call **surf_spline**:

```
n = 5;   m = 6;
[uu, vv] = meshgrid(-pi : 2*pi/(n - 1) : pi, -pi : 2*pi/(m - 1) : pi);
X = (R + cos(uu/2).*sin(vv) - sin(uu/2).*sin(2*vv)).*cos(uu);
```

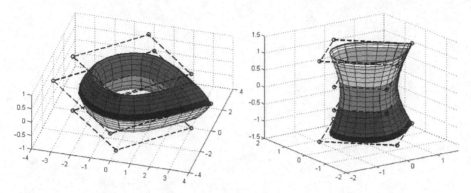

Fig. 9.21 Spline surfaces for torus and single-sheeted hyperboloid, $\beta = [1,1]$.

```
Y = (R + cos(uu/2).*sin(vv) - sin(uu/2).*sin(2*vv)).*sin(uu);
Z = sin(uu/2).*sin(vv) + cos(uu/2).*sin(2*vv);
S = surf_spline(X, Y, Z, 'BT');                          % β-spline
```

Plot the surface, Figure 9.19(b), by a typical program **view_surf**.

5. Prepare the control graph and plot the following spline surfaces: sphere or ellipsoid, Möbius band, helicoid, and monkey saddle.

6. Show that β- (B-) spline surfaces are affine invariant, i.e.,

$$f\left(\sum_{i,j=0}^{n} N_{n,i}(u)N_{n,j}(v)P_{i+1,j+1}\right) = \sum_{i=0}^{n} N_{n,i}(u)N_{n,j}(v)f(P_{i+1,j+1})$$

for any affine transformation f.

Appendix

M-Files

This chapter contains our procedures in the form of M-files.

A.1 Polynomial Interpolation

```
function v = polyinterp(x, y, u)              % several Lagrange polynomials
   n = length(x);   d = size(y, 1);   v = zeros(length(u), d)';
   for k = 1 : n;
   w = ones(size(u));
   for j = [1 : k-1  k + 1 : n];
   w = (u - x(j)). / (x(k) - x(j)).*w;
   end
   if d == 1;   v = v + w*y(k);
   else  v = v + w*y(:, k);
   end;   end
   v = simplify(v);
end
```

```
function  H = hermite_interp2d(x, y, z)              % Hermite polynomial
   syms t real;
   n = length(x) - 1;   H = zeros(1, 2*(n + 1));
   for i = 1: (n + 1);
   yi = 1 - sign(abs([1 : n + 1] - i));
   L = polyfit(x, yi, n);
   Lx = polyder(L);
   H = H + conv(y(i)*([0 1] - 2*polyval(Lx, x(i))*[1-x(i)]) + z(i)*[1-x(i)], conv(L, L));
   end;
end
```

```
function  S = cubic_spline(x, y, m_in, m_fin)                    % cubic spline
    syms t real;
    n = length(x) - 1;
    h = x(2 : n+1) - x(1 : n);    d = (y(2 : n+1) - y(1 : n))./h;
    B = [[zeros(n - 2, 1),  diag(h(2 : end - 1))]; zeros(1, n - 1)];
    C = B';
    A = 2*diag(h(1 : n - 1) + h(2 : n)) + B + C;
    u = 6*(d(2 : n) - d(1 : n-1));
    u(1) = u(1) - h(1)*m_in;    u(end) = u(end) - h(end)*m_fin;
    m = A \ u';
    m = [m_in  m'  m_fin];
    for k = 1 : n;
     S(k) = m(k)/(6*h(k))*(x(k+1) - t)^3 + m(k+1)/(6*h(k))*(t - x(k))^3 +
     (y(k)/h(k) - m(k)*h(k)/6)*(x(k+1) - t) + (y(k+1)/h(k) - m(k+1)*h(k)/6)*(t - x(k));
    end
end
```

A.2 Bézier Curves

```
function bez = bezier(x, y)                                      % Bézier curve
    syms t real;
    n = length(x) - 1;   j = 0 : n;
    B = gamma(n + 1)./gamma(j + 1)./gamma(n - j + 1).*t.^j.*(1 - t).^(n - j);
    bez = simplify([x;  y]*B');
end
```

```
function rbez = rbezier(x, y, w)                         % rational Bézier curve
    syms t real;
    n = length(x) - 1;   j = 0 : n;
    RB = gamma(n+1)./gamma(j+1)./gamma(n - j + 1).*t.^j.*(1-t).^(n-j).*w(j+1);
    rbez = simplify([x; y]*RB' / sum(RB));
end
```

```
function  bez2 = bezier2d(x, y)          % piecewise Bézier curve, calls bezier.m
    syms t real;
    m = length(x)/2;   p = [x', y'];
    p(1, :) = 2*p(1, :) - p(2, :);     p(2*m, :) = 2*p(2*m, :) - p(2*m-1, :);
    bez2 = [ ];
    for i = 1 : m - 1;
     pi = [(p(2*i-1, :) + p(2*i, :))/2; p(2*i, :); p(2*i+1, :); (p(2*i+1, :) + p(2*i+2, :))/2];
     yi = bezier(pi(:, 1)', pi(:, 2)');
     bez2 = [bez2,  simplify(subs(yi,  t,  t - i + 1))];
    end;
end
```

A.3 Hermite Curves

```
function  hspl = hermite_my(xp, yp, xq, yq)          % elementary cubic curve
    syms t real;
    n1 = 1 - 3*t.^2 + 2*t.^3;
    n2 = 3*t.^2 - 2*t.^3;
    n3 = t - 2*t.^2 + t.^3;
    n4 = -t.^2.*(1 - t);
    hspl = [xp; yp]*[n1 n2]' + [xq; yq]*[n3 n4]';
end
```

```
function  hspl2 = hermite2d(xp, yp, xq, yq)          % C^1-continuous curve
    syms t real;
    hspl2 = [  ];
    for i = 1 : length(xp) - 1;
     yi = hermite_my(xp(i : i + 1), yp(i : i + 1), xq(i : i + 1), yq(i : i + 1));
     hspl2 = [hspl2, simplify(subs(yi, t, t - i + 1))];
    end
end
```

```
function  hcub2 = hermite_cubic2d(xp, yp, xq, yq)    % C^2-continuous curve
    syms t real;
    n = length(xp);
    A1 = eye(n) + 4*diag(ones(1, n - 1), 1) + diag(ones(1, n - 2), 2);
    A2 = -3*eye(n) + 3*diag(ones(1, n - 2), 2);
    Qsol = A2(1 : n - 2, :)*[xp; yp]'\A1(1 : n - 2, :)
    Xq = [xq(1) Qsol(1, :) xq(2)];   Yq = [yq(1) Qsol(2, :) yq(2)];
    hcub2 = [  ];
    for i = 1 : n-1;
     yi = hermite_my(xp(i : i + 1), yp(i : i + 1), Xq(i : i + 1), Yq(i : i + 1));
     hcub2 = [hcub2, simplify(subs(yi, t, t - i + 1))];
    end
end
```

```
function  cardspl = card_my(x, y, s)        % elementary cardinal spline curve
    syms t real;
    n1 = (2*t.^2 - t - t.^3)*s;
    n2 = 1 + (2 - s)*t.^3 + (s - 3)*t.^2;
    n3 = (s - 2)*t.^3 + v(3 - 2*s)*t.^2 + t*s;
    n4 = (t.^3 - t.^2)*s;
    cardspl = simplify([x; y]*[n1 n2 n3 n4]');
end
```

```
function  cardspl2 = card2d(x, y, s)            % piecewise cardinal spline
    syms t real;
    cardspl2 = [  ];
```

```
    for i = 1 : length(x) - 3;
      yi = card_my(x(i : i + 3),  y(i : i + 3),  s);
      cardspl2 = [cardspl2,  simplify(subs(yi,  t,  t - i + 1))];
    end
  end
```

A.4 β-Spline Curves

```
function betas1 = betas1_my(x, y, beta)                    % elementary
  syms t real;                                  % quadratic β-spline in ℝ²
  n1 = beta*(t - 1).^2;
  n2 = -(beta + 1)*t.^2 + 2*beta*t + 1;
  n3 = beta*t.^2;
  betas1 = ([x; y]*[n1, n2, n3]'/(beta + 1));
end
```

```
function beta_s1 = beta_s1d(x, y, beta)         % quadratic β-spline in ℝ²
  syms t real;
  n = length(x);
  p = [x;  y];
  beta_s1 = [ ];
  for i = 1 : n - 2;
    yi = betas1_my(p(1, i : i + 2),  p(2, i : i + 2),  beta);
    beta_s1 = [beta_s1,  simplify(subs(yi, t,  t - i + 1))];
  end
end
```

```
function  betas = betas_my(x, y, beta)       % elementary cubic β-spline in ℝ²
  syms t real;
  b1 = beta(1);    b2 = beta(2);
  n1 = 2*b1^3*(1-t).^3;
  n2 = 2*b1^3*t.*(t.^2 - 3*t + 3) + 2*b1^2*(t.^3 - 3*t.^2 + 2)
       +2*b1*(t.^3 - 3*t + 2) + b2*(2*t.^3 - 3*t.^2 + 1);
  n3 = 2*b1^2*t.^2*(-t+3) + 2*b1*t*(-t.^2+3) + b2*t.^2*(-2*t+3) + 2*(-t.^3+1);
  n4 = 2*t.^3;
  d = 2*b1^3 + 4*b1^2 + 4*b1 + b2 + 2;
  betas = ([x;  y]*[n1, n2, n3, n4]'/d);
end
```

```
function  beta_s2 = beta_s2d(x, y, beta)       % piecewise cubic β-spline in ℝ²
  syms t real;
  n = length(x);
  p = [x(1)  x(1)  x  x(n)  x(n);  y(1)  y(1)  y  y(n)  y(n)];
  beta_s2 = [ ];
```

```
    for i = 1 : n;
      yi = betas_my(p(1, i : i + 3), p(2, i : i + 3), beta);
      beta_s2 = [beta_s2,  simplify(subs(yi, t, t - i + 1))];
    end
end
```

To change one vertex we create an M-file function (a similar program was developed for B-splines) that calls **betas_my**:

```
function  beta2c = betaspl2d_change(x, y, beta, B, i0, delta)
    syms t real;
    n = length(x);
    p = [x(1) x(1) x x(n) x(n); y(1) y(1) y y(n) y(n)];
    p(:, i0 + 2) = p(:, i0 + 2) + delta';
    beta2c = B;
    for i = i0 - 1 : i0 + v2;
    yi = betas_my(p(1, i : i + 3), p(2, i : i + 3), beta);
     beta2c(:, i) = simplify(subs(yi, t, t - i + 1));
    end
end
```

```
function rbetas = rbetas_my(x, y, w, beta)              % elementary rational
                                                        % cubic β-spline in ℝ²
    syms t real;  b1 = beta(1);  b2 = beta(2);
    n1 = 2*b1^3*(1 - t).^3;
    n2 = 2*b1^3*t.*(t.^2 - 3*t + 3) + 2*b1^2*(t.^3 - 3*t.^2 + 2)
    + 2*b1*(t.^3 - 3*t + 2) + b2*(2*t.^3 - 3*t.^2 + 1);
    n3 = 2*b1^2*t.^2*(3 - t) + 2*b1*t*(3 - t.^2) + b2*t.^2*(3 - 2*t) + 2*(1 - t.^3);
    n4 = 2*t.^3;
    W = w*[n1, n2, n3, n4]';
    rbetas = [x; y]*[w(1)*n1, w(2)*n2, w(3)*n3, w(4)*n4]'/W;
end
```

```
function betas3 = betas3_my(x, y, z, beta)  % elementary cubic β-spline in ℝ³
    syms t real;
    b1 = beta(1);    b2 = beta(2);
    n1 = 2*b1^3*(1 - t).^3;
    n2 = 2*b1^3*t.*(t.^2-3*t+3) + 2*b1^2*(t.^3 - 3*t.^2+2) + 2*b1*(t.^3-3*t+2)
        + b2*(2*t.^3 - 3*t.^2+1);
    n3 = 2*b1^2*t.^2*(3-t) + 2*b1*t*(3-t.^2) + b2*t.^2*(3-2*t) + 2*(1-t.^3);
    n4 = 2*t.^3;
    d = b2 + 2*b1^3 + 4*b1^2 + 4*b1 + 2;
    betas3 = simplify(([x; y; z]*[n1, n2, n3, n4]'/d));
end
```

```
function  betas3 = beta_s3d(x, y, z, beta)     % piecewise cubic β-spline in ℝ³
  syms t real;
  n = length(x);
  p = [x(1) x(1) x x(n) x(n);  y(1) y(1) y y(n) y(n);  z(1) z(1) z z(n) z(n)];
  betas3 = [ ];
  for i = 1 : n + 1;
   yi = betas3_my(p(1, i : i + 3), p(2, i : i + 3), p(3, i : i + 3), beta);
   betas3 = [betas3, simplify(subs(yi, t, t - i + 1))];
  end
end
```

```
function betas4 = betas4_my(x, y, z, beta)        % elementary quartic β-spline
  syms t real;
  b1 = beta(1);  b2 = beta(2);  b3 = beta(3);
  d = (1-b1^2)*b3+6*b1*b2-6*b1^2*b2+2*b1*b2^2+6*b2*(1+b1)^3
  +6*(1+b1)^2*(1+b1^2)*(1+b1+b1^2);
n1 = -6*b1^6+24*b1^6*t-36*b1^6*t.^2+24*b1^6*t.^3-6*b1^6*t.^4;
n2 = -(18*b1^2+30*b1^3+12*b1*b2+2*b2^2-b1*b3+6*b1^2*b2
  +18*b1^4)*b1 -24*b1^2*(-2*b1-2*b1^2-b2+b1^4)*t
  +6*b1*(6*b1^5-6*b1^2+2*b2^2-b1*b3+4*b1^2*b2+6*b1^4)*t.^2
  -8*b1*(3*b1^5+2*b2^2-b1*b3+3*b1^2*b2+3*b1^3+3*b1*b2+3*b1^4)*t.^3
  +3*b1*(2*b1^5+2*b1^2+2*b1^3+4*b1*b2+2*b2^2-b1*b3
  +2*b1^2*b2+2*b1^4)*t.^4;
n3 = -18*b1^3-30*b1^2-18*b1-12*b1*b2-6*b2-b3-24*b1*(2*b1^2+2*b1^3
  +b1*b2-1)*t +(36*b1^2+36*b1^3+12*b2-12*b1*b2^2+6*b1^2*b3
  -24*b1^3*b2-36*b1^5)*t.^2
  +(24*b1^3+24*b1^4+4*b3+24*b1*b2 + 24*b1^2*b2+16*b1*b2^2
  -8*b1^2*b3+24*b1^3*b2+24*b1^5)*t.^3
  +(-12*b1^3-6*b1^4-12*b1^2*b2-6*b1*b2^2+3*b1^2*b3-6*b1^3*b2
  -6*b1^5-6*b1^2-6*b1-12*b1*b2-6*b2-3*b3)*t.^4;
n4 = -6-24*b1*t-(12*b2+36*b1^2)*t.^2-(24*b1^3+24*b1*b2+4*b3)*t.^3
  +(6*b1^3+6*b1^2+12*b1*b2+6*b1+6+3*b3+6*b2)*t.^4;
n5 = -6*t.^4;
  betas4 = simplify(([x; y; z]*[n1, n2, n3, n4, n5]'/ d));
end
```

```
function beta4 = beta_s4d(x, y, z, beta)     % piecewise quartic β-spline in ℝ³
  syms t real;
  n = length(x);
  p = [x(1) x(1) x(1) x x(n) x(n) x(n); y(1) y(1) y(1) y y(n) y(n) y(n);
    z(1) z(1) z(1) z z(n) z(n) z(n)];
  beta4 = [ ];
  for i = 1 : n + 2;
   yi = betas4_my(p(1, i : i + 4), p(2, i : i + 4), p(3, i : i + 4), beta);
   beta4 = [beta4, simplify(subs(yi, t, t - i + 1))];
  end;
end
```

A.5 Spline Surfaces

```
function surf = surf_patch(P, M)                              % for n = 3, 4, 5
   n = length(M);   if length(P)  = n^2;  disp('verify dimension of P ');
   else
      syms u v real;
      X = P(:, 1);   Y = P(:, 2);   Z = P(:, 3);
      if n == 3;         U = [u^2 u 1];   V = [v^2; v; 1];
      elseif n == 4;   U = [u^3 u^2 u 1];   V = [v^3; v^2; v; 1];
      elseif n == 5;   U = [u^4 u^3 u^2 u 1];   V = [v^4; v^3; v^2; v; 1];
      end;
      XX = reshape(X, n, n);   YY = reshape(Y, n, n);   ZZ = reshape(Z, n, n);
      rx = U*M'*XX*M*V;   ry = U*M'*YY*M*V;   rz = U*M'*ZZ*M*V;
      surf = [rx, ry, rz];
   end
end
```

```
function surfs = surf_spline(X, Y, Z,  method)        % 'method' can be BB, BT
   syms u v real;                                            % cubic surface spline
   [m  n] = size(X);
   X2 = cat(2, X(:, 1), X(:, 1), X, X(:, n), X(:, n));
   XX = cat(1, X2(1, :), X2(1, :), X2, X2(m, :), X2(m, :));
   Y2 = cat(2, Y(:, 1), Y(:, 1), Y, Y(:, n), Y(:, n));
   YY = cat(1, Y2(1, :), Y2(1, :), Y2, Y2(m, :), Y2(m, :));
   Z2 = cat(2, Z(:, 1), Z(:, 1), Z, Z(:, n), Z(:, n));
   ZZ = cat(1, Z2(1, :), Z2(1, :), Z2, Z2(m, :), Z2(m, :));              % building P**
   if method(2) == 'B';
      M = [-1 3 -3 1;  3 -6 0 4;  -3 3 3 1;  1 0 0 0]/6;
   elseif method(2) == 'T';
      beta1 = 1;    beta2 = 7;                                % define β₁, β₂
      xi = beta1^2 + beta1 + beta2;
      delta3 = beta2 + 2*beta1^3 + 4*beta1^2 + 4*beta1 + 2;
      M = [-2*beta1^3 6*beta1^3 -6*beta1^3 2*beta1^3;
      2*(beta1^3 + xi)  -3*(2*beta1^3+2*beta1^2 + beta2)  6*(beta1^3 - beta1)
      4*(beta1^2 + beta1) + beta2;  -2*(xi + 1)  3*(2*beta1^2 + beta2)  6*beta1  2;
      2 0 0 0] / delta3;
   else
      error([method, ' is an invalid method.']);
   end;
   U = [u^3  u^2  u  1];    V = [v^3; v^2; v; 1];
   surfs = [ ];
for j = 1 : n + 1;    for i = 1 : m + 1;
   Xij = XX(i : i + 3, j : j + 3);  Yij = YY(i : i + 3, j : j + 3);  Zij = ZZ(i : i + 3, j : j + 3);
   rx = U*M'*Xij*M*V;   ry = U*M'*Yij*M*V;   rz = U*M'*Zij*M*V;
   surfs = [surfs;  [rx, ry, rz] ];
```

```
    end;    end
end
```

hold on; view(3); grid on;   % the program **view_patch**
plot3(P(:, 1), P(:, 2), P(:, 3), 'o');   % *P* – control vertices
ezsurf(r(1), r(2), r(3), [0, 1, 0, 1], 20)   % *r* – parameterization of a surface

A.6 Semi-Regular Polyhedra

1) **p33334**

```
c = 1; v = 0.7044022*c; u = 0.4524646*c;
V = [-c+u c -c+v; -c+v c c-u; c-u c c-v; c-v c -c+u; c c-u -c+v; c -c+v -c+u; c -c+u
c-v; c c-v c-u; c-v c-u c; c-u -c+v c; -c+v -c+u c; -c+u c-v c; -c+v c-u -c; -c+u
-c+v -c; c-v -c+u -c; c-u c-v -c; -c c-v -c+u; -c c-u c-v; -c -c+v c-u; -c -c+u -c+v;
c-u -c -c+v; -c+v -c -c+u; -c+u -c c-v; c-v -c c-u];
F4 = [ ];  j = 1 : 4; for i = 1 : 6;  F4 = [F4; 4*(i-1)+j]; end;
F3 = [4 5 3; 3 9 2; 2 18 1; 4 1 13; 8 3 5; 7 10 8; 6 21 7; 5 16 6; 9 8 10; 12 2 9;
11 19 12; 11 10 24; 13 17 14; 13 16 4; 16 15 6; 15 14 22; 24 23 11; 23 22 20;
21 15 22; 21 24 7; 19 18 12; 17 1 18; 17 20 14; 20 19 23; 9 3 8; 10 7 24; 11
23 19; 18 2 12; 4 16 5; 13 1 17; 22 14 20; 6 15 21];
hold on;  axis equal;  view(3)
patch('Vertices', V, 'Faces', F4, 'FaceColor', [0 1 1])
patch('Vertices', V, 'Faces', F3, 'FaceColor', [0 .5 1])
```

2) **p3434**

```
c = 1;
V = [c c 0; c 0 -c; 0 c -c; -c 0 -c; -c c 0; -c 0 c; -c -c 0; 0 c c; 0 -c c; c -c 0; c 0 c;
0 -c -c];
F3 = [1 2 3; 5 3 4; 7 4 12; 10 12 2; 8 11 1; 8 5 6; 9 6 7; 11 9 10];
F4 = [1 8 5 3; 3 4 12 2; 12 7 9 10; 9 6 8 11; 2 10 11 1; 5 6 7 4];
hold on;  axis equal;  view(3)
patch('Vertices', V, 'Faces', F4, 'FaceColor', [0 1 1])
patch('Vertices', V, 'Faces', F3, 'FaceColor', [0 .5 1])
```

3) **p3444_1**

```
c = 1; d = c*sqrt(2)/(2+sqrt(2));
V = [d d c; -d d c; -d -d c; d -d c; c d d; c d -d; c -d -d; c -d d; -d c d; -d c -d; d c
-d; d c d; d -c d; d -c -d; -d -c -d; -d -c d; -c -d d; -c -d -d; -c d -d; -c d d; d d -c;
-d d -c; -d -d -c; d -d -c];
F4 = [ ];  j = 1 : 4; for i = 1 : 6;  F4 = [F4; 4*(i-1)+j]; end;
F4b = [1 4 8 5; 4 13 16 3; 3 17 20 2; 2 9 12 1; 12 5 6 11; 8 13 14 7; 16 17 18
15; 20 9 10 19; 6 7 24 21; 14 15 23 24; 18 23 22 19; 10 11 21 22];
```

F3 = [1 12 5; 4 8 13; 3 16 17; 2 20 9; 6 11 21; 7 24 14; 15 23 18; 19 22 10];
hold on; axis equal; view(3)
patch('Vertices', V, 'Faces', F4, 'FaceColor', [0 1 1])
patch('Vertices', V, 'Faces', F4b, 'FaceColor', [0 .5 1])
patch('Vertices', V, 'Faces', F3, 'FaceColor', [0 .2 .5])

4) **p3535**

c = 1; p = c*(sqrt(5)-1)/2; v = p/2;
V = [c/2 c+v c/2+v; 0 c+p 0; c/2 c+v -c/2-v; c+v c/2+v -c/2; c+v c/2+v c/2; c/2+v
c/2 c+v; c/2+v -c/2 c+v; c+v -c/2-v c/2; c+p 0 0; c/2+v -c/2 -c-v; c/2 -c-v -c/2-v;
0 -c-p 0; c/2 -c-v c/2+v; -c/2 -c-v c/2+v; -c/2-v -c/2 c+v; 0 0 c+p; -c/2-v c/2 c+v;
-c/2 c+v c/2+v; -c-v c/2+v c/2; -c-v c/2+v -c/2; -c/2 c+v -c/2-v; -c/2-v c/2 -c-v; 0
0 -c-p; c/2+v c/2 -c-v; -c/2-v -c/2 -c-v; -c/2 -c-v -c/2-v; -c-v -c/2-v -c/2; -c-p 0 0;
-c-v -c/2-v c/2; c+v -c/2-v -c/2];
F3 = [1 2 18; 2 3 21; 1 6 5; 17 18 19; 7 6 16; 16 17 15; 7 13 8; 14 15 29; 4 24
3; 20 21 22; 14 12 13; 26 11 12; 5 9 4; 9 8 30; 19 20 28; 28 27 29; 23 24 10;
22 23 25; 25 26 27; 30 11 10];
F5 = [6 1 18 17 16; 7 16 15 14 13; 3 24 23 22 21; 23 10 11 26 25; 3 2 1 5 4; 2
21 20 19 18; 27 26 12 14 29; 12 11 30 8 13; 8 9 5 6 7; 4 9 30 10 24; 20 22 25
27 28; 19 28 29 15 17];
hold on; axis equal; view(3)
patch('Vertices', V, 'Faces', F5, 'FaceColor', [0 .5 1])
patch('Vertices', V, 'Faces', F3, 'FaceColor', [0 .2 .5])

5) **p366**

c = 1; d = c/3;
V = [d c -d; c d -d; d d -c; -d d c; -d c d; -c d d; c -d d; d -d c; d -c d; -d -d -c;
-c -d -d; -d -c -d];
F3 = [1 2 3; 4 5 6; 7 8 9; 10 11 12];
F6 = [2 7 8 4 5 1; 2 7 9 12 10 3; 8 9 12 11 6 4; 10 11 6 5 1 3];
hold on; axis equal; view(3)
patch('Vertices', V, 'Faces', F6, 'FaceColor', [0 .5 1])
patch('Vertices', V, 'Faces', F3, 'FaceColor', [0 .2 .5])

6) **p388**

c = 1; d = (2*c)/(2 + sqrt(2));
V = [c c c-d; c c d-c; -c c c-d; -c c d-c; c -c c-d; c -c d-c; -c -c c-d; -c -c d-c;
c c-d c; c d-c c; c c-d -c; c d-c -c; -c d-c -c; -c c-d -c; -c c-d c; -c d-c c;
c-d c -c; d-c c -c; c-d c c; d-c c c; c-d -c c; d-c -c c; c-d -c -c; d-c -c -c];
F3 = [17 2 11; 9 19 1; 12 23 6; 5 10 21; 8 13 24; 7 22 15; 4 14 18; 3 20 16];
F8 = [13 8 7 15 16 3 4 14; 23 6 5 21 22 7 8 24; 12 6 5 10 9 1 2 11;
17 2 1 19 20 3 4 18; 11 12 23 24 13 14 18 17; 10 9 19 20 16 15 22 21];
hold on; axis equal; view(3)
patch('Vertices', V, 'Faces', F8, 'FaceColor', [0 .5 1])
patch('Vertices', V, 'Faces', F3, 'FaceColor', [0 .2 .5])

7) **p3_10_10**

```
c = 1;  p = (sqrt(5)-1)*c/2;  m = c*(sqrt(5)-1)/(2+2*sin(54*pi/180));
k = c*(sqrt(5)-1)-2*m;  d = p*(c*(sqrt(5)-1)-m)/(c*(sqrt(5)-1));
t = m*p/(c*(sqrt(5)-1));  r = m*cos(72*pi/180);
a(1, :) = [c-k/2; c-r; c+t]; a(2, :) = [k/2; p+r; c+d]; a(3, :) = [0; p-m; c+p];
a(4, :) = [0; -p+m; c+p]; a(5, :) = [k/2; -p-r; c+d]; a(6, :) = [c-k/2; r-c; c+t];
a(7, :) = [c+t; -c+k/2; c-r]; a(8, :) = [c+d; -k/2; p+r]; a(9, :) = [c+d; k/2; p+r];
a(10, :) = [c+t; c-k/2; c-r]; a(11, :) = [c-r; c+t; c-k/2]; a(12, :) = [p+r; c+d; k/2];
a(13, :) = [p+r; c+d; -k/2]; a(14, :) = [c-r; c+t; -c+k/2]; a(15, :) = [c+t; c-k/2; r-c];
a(16, :) = [c+d; k/2; -p-r]; a(17, :) = [c+p; 0; -p+m]; a(18, :) = [c+p; 0; p-m];
a(19, :) = [c+d; -k/2; -p-r]; a(20, :) = [c+t; -c+k/2; r-c]; a(21, :) = [c-r; -c-t; -c+k/2];
a(22, :) = [p+r; -c-d; -k/2]; a(23, :) = [p+r; -c-d; k/2]; a(24, :) =[ c-r; -c-t; c-k/2];
a(25, :) = [p-m; -c-p; 0]; a(26, :) = [-p+m; -c-p; 0]; a(27, :) = [-p-r; -c-d; -k/2];
a(28,:) = [-c+r; -c-t; -c+k/2]; a(29,:) = [-c+k/2; r-c; -c-t]; a(30,:) = [-k/2; -p-r; -c-d];
a(31, :) = [k/2; -p-r; -c-d]; a(32, :) = [c-k/2; r-c; -c-t]; a(33, :) = [c-k/2; c-r; -c-t];
a(34, :) = [k/2; p+r; -c-d]; a(35, :) = [-k/2; p+r; -c-d]; a(36, :) = [-c+k/2; c-r; -c-t];
a(37, :) = [-c+r; c+t; -c+k/2]; a(38, :) = [-p-r; c+d; -k/2]; a(39, :) = [-p+m; c+p; 0];
a(40, :) = [p-m; c+p; 0]; a(41, :) = [0; p-m; -c-p]; a(42, :) = [0; -p+m; -c-p];
a(43, :) = [-p-r; c+d; k/2]; a(44, :) = [-c+r; c+t; c-k/2]; a(45, :) = [-c-t; c-k/2; c-r];
a(46, :) = [-c-d; k/2; p+r]; a(47, :) = [-c-d; -k/2; p+r]; a(48, :) = [-c-t; -c+k/2; c-r];
a(49, :) = [-c+k/2; r-c; c+t]; a(50, :) = [-k/2; -p-r; c+d]; a(51, :) = [-k/2; p+r; c+d];
a(52, :) = [-c-p; 0; p-m]; a(53, :) = [-c-p; 0; -p+m]; a(54, :) = [-c-d; -k/2; -p-r];
a(55, :) = [-c-t; -c+k/2; r-c]; a(56, :) = [-p-r; -c-d; k/2]; a(57, :) =[-c+r; -c-t; c-k/2];
a(58, :) = [-c+k/2; c-r; c+t]; a(59, :) = [-c-t; c-k/2; r-c]; a(60, :) = [-c-d; k/2; -p-r];
V = [ ];  for i = 1 : 60;  V = [V; a(i, :)];  end;
F10 = [1 2 3 4 5 6 7 8 9 10; 3 4 50 49 48 47 46 45 58 51; 15 16 19 20 32 31
42 41 34 33; 41 42 30 29 55 54 60 59 36 35; 7 8 18 17 19 20 21 22 23 24; 9
18 17 16 15 14 13 12 11 10; 59 60 53 52 46 45 44 43 38 37; 53 54 55 28 27
56 57 48 47 52; 14 33 34 35 36 37 38 39 40 13; 11 12 40 39 43 44 58 51 2 1;
29 30 31 32 21 22 25 26 27 28; 25 23 24 6 5 50 49 57 56 26];
F3 = [1 10 11; 2 51 3; 58 44 45; 46 52 47; 48 57 49; 4 50 5; 6 24 7; 8 18 9; 12
13 40; 43 39 38; 27 26 56; 22 23 25; 19 16 17; 15 33 14; 34 41 35; 36 59 37;
60 54 53; 55 29 28; 31 30 42; 21 32 20];
hold on;  axis equal;  view(3)
patch('Vertices', V, 'Faces', F3, 'FaceColor', [0 1 1])
patch('Vertices', V, 'Faces', F10, 'FaceColor', [0 .5 1])
```

8) **p466**

```
c = 1;  t = c/3;  u = c-t;
V = [0 u t; -t u 0; 0 u -t; t u 0; u 0 t; u t 0; u 0 -t; u -t 0; 0 -u t; t -u 0; 0 -u -t; -t -u 0;
-u 0 t; -u -t 0; -u 0 -t; -u t 0; t 0 u; 0 t u; -t 0 u; 0 -t u; t 0 -u; 0 t -u; -t 0 -u; 0 -t -u];
F4 = [ ];  j = 1 : 4;  for i = 1 : 6;  F4 = [F4; 4*(i - 1) + j];  end;
F6 = [5 17 18 1 4 6; 10 9 20 17 5 8; 11 24 23 15 14 12; 23 15 16 2 3 22; 1 2
16 13 19 18; 9 20 19 13 14 12; 7 21 24 11 10 8; 6 7 21 22 3 4];
hold on;  axis equal;  view(3)
```

patch('Vertices', V, 'Faces', F4, 'FaceColor', [0 1 1])
patch('Vertices', V, 'Faces', F6, 'FaceColor', [0 .5 1])

9) p566

c = 1; t = c/3; p = c*(sqrt(5)-1)/2; d = p/3; v = (c+2*p)/3; m = (2*c+p)/3;
V = [d 0 c; 2*d t m; v d c-t; v -d c-t; 2*d -t m; d 0 -c; 2*d t -m; v d t-c; v -d t-c;
2*d -t -m; -d 0 c; -2*d t m; -v d c-t; -v -d c-t; -2*d -t m; -d 0 -c; -2*d t -m; -v d t-c;
-v -d t-c; -2*d -t -m; -t m 2*d; -d c-t v; d c-t v; t m 2*d; 0 c d; -t m -2*d; -d c-t -v;
d c-t -v; t m -2*d; 0 c -d; -t -m 2*d; -d t-c v; d t-c v; t -m 2*d; 0 -c d; -t -m -2*d; -d
t-c -v; d t-c -v; t -m -2*d; 0 -c -d; c-t v d; m 2*d t; c d 0; m 2*d -t; c-t v -d; t-c v d;
-m 2*d t; -c d 0; -m 2*d -t; t-c v -d; c-t -v d; m -2*d t; c -d 0; m -2*d -t; c-t -v -d;
t-c -v d; -m -2*d t; -c -d 0; -m -2*d -t; t-c -v -d];
F5 = [ ]; j = 1 : 5; for i = 1 : 12; F5 = [F5; 5*(i-1)+j]; end;
F6 = [11 12 22 23 2 1; 5 33 32 15 11 1; 23 24 41 42 3 2; 42 43 53 52 4 3; 52
51 34 33 5 4; 51 55 39 40 35 34; 60 56 31 35 40 36; 56 57 14 15 32 31; 14 57
58 48 47 13; 12 13 47 46 21 22; 21 46 50 26 30 25; 24 25 30 29 45 41; 28 7 8
44 45 29; 44 8 9 54 53 43; 9 10 38 39 55 54; 10 6 16 20 37 38; 36 37 20 19
59 60; 58 59 19 18 49 48; 50 49 18 17 27 26; 28 27 17 16 6 7];
hold on; axis equal; view(3)
patch('Vertices', V, 'Faces', F5, 'FaceColor', [0 1 1]);
patch('Vertices', V, 'Faces', F6, 'FaceColor', [0 .5 1]);

10) p3454

c = 1; p = c*(sqrt(5)-1)/2;
a(1, :) = [c/3,c/3,c+2*p/3]; a(2, :) = [(p+c)/3,(2*c+p)/3,2*(c+p)/3];
a(3, :) = [0,c+p/3,(2*c+p)/3]; a(4, :) = [-(p+c)/3,(c*2+p)/3,(2*p+2*c)/3];
a(5, :) = [-c/3,c/3,c+2*p/3]; a(6, :) = [-(2*c+p)/3,0,c+p/3];
a(7, :) = [-(2*c+2*p)/3,(p+c)/3,(2*c+p)/3]; a(8, :) = [-c-2*p/3,c/3,c/3];
a(9, :) = [-c-2*p/3,-c/3,c/3]; a(10, :) = [-(2*c+2*p)/3,-(c+p)/3,(2*c+p)/3];
a(11, :) = [-(2*c+p)/3,(2*c+2*p)/3,(c+p)/3]; a(12, :) = [-c/3,c+2*p/3,c/3];
a(13, :) = [-c/3,c+2*p/3,-c/3]; a(14, :) = [-(2*c+p)/3,(2*c+2*p)/3,-(c+p)/3];
a(15, :) = [-c-p/3,(2*c+p)/3,0]; a(16, :) = [(2*c+p)/3,(2*c+2*p)/3,(c+p)/3];
a(17, :) = [c+p/3,(2*c+p)/3,0]; a(18, :) = [(2*c+p)/3,(2*c+2*p)/3,-(c+p)/3];
a(19, :) = [c/3,c+2*p/3,-c/3]; a(20, :) = [c/3,c+2*p/3,c/3];
a(21, :) = [-(c+p)/3,(2*c+p)/3,-(2*c+2*p)/3]; a(22, :) = [0,c+p/3,-(2*c+p)/3];
a(23, :) = [(c+p)/3,(2*c+p)/3,-(2*c+2*p)/3]; a(24, :) = [c/3,c/3,-c-2*p/3];
a(25, :) = [-c/3,c/3,-c-2*p/3]; a(26, :) = [-(2*c+p)/3,0,-c-p/3];
a(27, :) = [-(2*c+2*p)/3,-(c+p)/3,-(2*c+p)/3]; a(28, :) = [-c-2*p/3,-c/3,-c/3];
a(29, :) = [-c-2*p/3,c/3,-c/3]; a(30, :) = [-(2*c+2*p)/3,(c+p)/3,-(2*c+p)/3];
a(31, :) = [-c/3,-c/3,c+2*p/3]; a(32, :) = [-(c+p)/3,-(2*c+p)/3,(2*c+2*p)/3];
a(33, :) = [0,-c-p/3,(2*c+p)/3]; a(34, :) = [(c+p)/3,-(2*c+p)/3,(2*c+2*p)/3];
a(35, :) = [c/3,-c/3,c+2*p/3]; a(36, :) = [(2*c+p)/3,0,c+p/3];
a(37, :) = [(2*c+2*p)/3,-(c+p)/3,(2*c+p)/3]; a(38, :) = [c+2*p/3,-c/3,c/3];
a(39, :) = [c+2*p/3,c/3,c/3]; a(40, :) = [(2*c+2*p)/3,(c+p)/3,(2*c+p)/3];
a(41, :) = [(2*c+p)/3,-(2*c+2*p)/3,(c+p)/3]; a(42, :) = [c/3,-c-2*p/3,c/3];
a(43, :) = [c/3,-c-2*p/3,-c/3]; a(44, :) = [(2*c+p)/3,-(2*c+2*p)/3,-(c+p)/3];

```
a(45, :) = [c+p/3,-(2*c+p)/3,0]; a(46, :) = [c+2*p/3,-c/3,-c/3];
a(47, :) = [(2*c+2*p)/3,-(c+p)/3,-(2*c+p)/3]; a(48, :) = [(2*c+p)/3,0,-c-p/3];
a(49, :) = [(2*c+2*p)/3,(c+p)/3,-(2*c+p)/3]; a(50, :) = [c+2*p/3,c/3,-c/3];
a(51, :) = [-(c+p)/3,-(2*c+p)/3,-(2*c+2*p)/3]; a(52, :) = [-c/3,-c/3,-c-2*p/3];
a(53, :) = [c/3,-c/3,-c-2*p/3]; a(54, :) = [(c+p)/3,-(2*c+p)/3,-(2*c+2*p)/3];
a(55, :) = [0,-c-p/3,-(2*c+p)/3]; a(56, :) = [-c/3,-c-2*p/3,-c/3];
a(57, :) = [-c/3,-c-2*p/3,c/3]; a(58, :) = [-(2*c+p)/3,-(2*c+2*p)/3,(c+p)/3];
a(59, :) = [-c-p/3,-(2*c+p)/3,0]; a(60, :) = [-(2*c+p)/3,-(2*c+2*p)/3,-(p+c)/3];
V = [ ];  for i = 1 : 60; V = [V; a(i, :)]; end;
F3 = [4 11 7; 3 20 12; 19 22 13; 14 21 30; 8 15 29; 34 41 37; 33 57 42; 38 45
46; 44 54 47; 43 56 55; 1 35 36; 16 2 40; 50 17 39; 49 23 18; 24 48 53; 52 26
25; 51 60 27; 59 9 28; 31 5 6; 58 32 10; 30 11 10];
F4 = [2 16 20 3; 3 12 11 4; 4 7 6 5; 7 11 15 8; 15 14 30 29; 14 13 22 21; 13 12
20 19; 19 18 23 22; 21 25 26 30; 29 28 9 8; 35 34 37 36; 34 33 42 41; 33 32
58 57; 57 56 43 42; 43 55 54 44; 44 47 46 45; 45 38 37 41; 38 46 50 39; 47 54
53 48; 55 56 60 51; 1 36 40 2; 16 40 39 17; 17 50 49 18; 23 49 48 24; 24 53
52 25; 26 52 51 27; 27 60 59 28; 9 59 58 10; 10 32 31 6; 5 31 35 1];
F5 = [1 2 3 4 5; 6 7 8 9 10; 11 12 13 14 15; 16 17 18 19 20; 21 22 23 24 25;
26 27 28 29 30; 31 32 33 34 35; 36 37 38 39 40; 41 42 43 44 45; 46 47 48 49
50; 51 52 53 54 55; 56 57 58 59 60];
hold on;  axis equal;  view(3);  grid
patch('Vertices', V, 'Faces', F3, 'FaceColor', [0 1 1])
patch('Vertices', V, 'Faces', F4, 'FaceColor', [0 .2 1])
patch('Vertices', V, 'Faces', F5, 'FaceColor', [0 1 .3])
```

11) **p468**

```
c = 1;  d = 2*c*(3*sqrt(2)+2)/7;  m = d*(3*sqrt(2)-4)/4;
l = m+d*(2-sqrt(2))/4;  t = d*(2-sqrt(2))/4;
V = [c-l -t c; t -c+l c; -t -c+l c; -c+l -t c; -c+l t c; -t c-l c; t c-l c; c-l t c; c t c-l; c -t
c-l; c -c+l t; c -c+l -t; c -t -c+l; c t -c+l; c c-l -t; c c-l t; c-l c t; c-l c -t; t c -c+l; -t c
-c+l; -t c-l -c; t c-l -c; c-l t -c; -c+l t -c; -c+l c -t; -c c-l -t; -c t -c+l; -c -t -c+l; -c+l -t
-c; -t c c-l; t c c-l; -c+l c t; -c c-l t; -c t c-l; -c -t c-l; -c -c+l t; -c+l -c t; -c+l -c -t; -c
-c+l -t; -t -c -c+l; -t -c+l -c; t -c+l -c; t -c -c+l; c-l -c -t; c-l -t -c; t -c c-l; -t -c c-l; c-l
-c t];
F8 = [1 2 3 4 5 6 7 8; 17 18 19 20 25 32 30 31; 22 23 45 42 41 29 24 21; 37
38 40 43 44 48 46 47; 9 10 11 12 13 14 15 16; 26 27 28 39 36 35 34 33];
F6 = [5 6 30 32 33 34; 3 4 35 36 37 47; 1 2 46 48 11 10; 7 8 9 16 17 31; 20 25
26 27 24 21; 28 39 38 40 41 29; 14 23 22 19 18 15; 12 13 45 42 43 44];
F4 = [6 7 31 30; 1 10 9 8; 2 3 47 46; 4 5 34 35; 32 25 26 33; 36 37 38 39; 11
12 44 48; 15 16 17 18; 19 20 21 22; 27 24 29 28; 40 41 42 43; 13 14 23 45];
hold on;  axis equal;  view(3)
patch('Vertices', V, 'Faces', F4, 'FaceColor', [0 1 1])
patch('Vertices', V, 'Faces', F6, 'FaceColor', [0 .2 1])
patch('Vertices', V, 'Faces', F8, 'FaceColor', [0 .5 .5])
```

12) **p4610**

c = 1; p = (sqrt(5)-1)*c/2; m = c*(sqrt(5)-1)/(2+2*sin(54*pi/180));
k = c*(sqrt(5)-1)-2*m; d = (2*p-m)/2; t = m/2; r = m*cos(72*pi/180);
tau = (sqrt(5)-1)/2; l = tau/(sqrt(3+tau)*(1+tau)+2*tau);
a(1, :) = [c-k/2; c-r+2*l*(r-c); c+t];
a(2, :) = [k/2+l*(c+t-k/2); p+r+l*(k/2-c-p-r); c+d-l*(r+d)];
a(3, :) = [l*(c+d); p-m+l*(m-p-k/2); c+p+l*(r-c)];
a(4, :) = [l*(c+d); -p+m-l*(m-p-k/2); c+p+l*(r-c)];
a(5, :) = [k/2+l*(c+t-k/2); -p-r-l*(k/2-c-p-r); c+d-l*(r+d)];
a(6, :) = [c-k/2; -c+r-2*l*(r-c); c+t];
a(7, :) = [c+t+l*(k/2-c-t); k/2-c+l*(p+r-k/2+c); c-r+l*(d+r)];
a(8, :) = [c+d-l*(c+d); -k/2+l*(p-m+k/2); p+r+l*(c-r)];
a(9, :) = [c+d-l*(c+d); k/2-l*(p-m+k/2); p+r+l*(c-r)];
a(10, :) = [c+t+l*(k/2-c-t); -k/2+c-l*(p+r-k/2+c); c-r+l*(d+r)];
a(11, :) = [c-r+l*(d+r); c+t+l*(k/2-c-t); -k/2+c-l*(p+r-k/2+c)];
a(12, :) = [c+t; c-k/2; c-r+2*l*(r-c)];
a(13, :) = [c+d-l*(r+d); k/2+l*(c+t-k/2); p+r+l*(k/2-c-p-r)];
a(14, :) = [c+p+l*(r-c); l*(c+d); p-m+l*(m-p-k/2)];
a(15, :) = [c+p+l*(r-c); l*(c+d); -p+m-l*(m-p-k/2)];
a(16, :) = [c+d-l*(r+d); k/2+l*(c+t-k/2); -p-r-l*(k/2-c-p-r)];
a(17, :) = [c+t; c-k/2; -c+r-2*l*(r-c)];
a(18, :) = [c-r+l*(d+r); c+t+l*(k/2-c-t); k/2-c+l*(p+r-k/2+c)];
a(19, :) = [p+r+l*(c-r); c+d-l*(c+d); -k/2+l*(p-m+k/2)];
a(20, :) = [p+r+l*(c-r); c+d-l*(c+d); k/2-l*(p-m+k/2)];
a(21, :) = [p+r+l*(k/2-c-p-r); c+d-l*(r+d); k/2+l*(c+t-k/2)];
a(22, :) = [p-m+l*(m-p-k/2); c+p+l*(r-c); l*(c+d)];
a(23, :) = [-p+m-l*(m-p-k/2); c+p+l*(r-c); l*(c+d)];
a(24, :) = [-p-r-l*(k/2-c-p-r); c+d-l*(r+d); k/2+l*(c+t-k/2)];
a(25, :) = [-c+r-2*l*(r-c); c+t; c-k/2];
a(26, :) = [k/2-c+l*(p+r-k/2+c); c-r+l*(d+r); c+t+l*(k/2-c-t)];
a(27, :) = [-k/2+l*(p-m+k/2); p+r+l*(c-r); c+d-l*(c+d)];
a(28, :) = [k/2-l*(p-m+k/2); p+r+l*(c-r); c+d-l*(c+d)];
a(29, :) = [-k/2+c-l*(p+r-k/2+c); c-r+l*(d+r); c+t+l*(k/2-c-t)];
a(30, :) = [c-r+2*l*(r-c); c+t; c-k/2];
a(31, :) = [-k/2+l*(p-m+k/2); p+r+l*(c-r); -c-d+l*(c+d)];
a(32, :) = [k/2-c+l*(p+r-k/2+c); c-r+l*(d+r); -c-t-l*(k/2-c-t)];
a(33, :) = [-c+r-2*l*(r-c); c+t; -c+k/2];
a(34, :) = [-p-r-l*(k/2-c-p-r); c+d-l*(r+d); -k/2-l*(c+t-k/2)];
a(35, :) = [-p+m-l*(m-p-k/2); c+p+l*(r-c); -l*(c+d)];
a(36, :) = [p-m+l*(m-p-k/2); c+p+l*(r-c); -l*(c+d)];
a(37, :) = [p+r+l*(k/2-c-p-r); c+d-l*(r+d); -k/2-l*(c+t-k/2)];
a(38, :) = [c-r+2*l*(r-c); c+t; -c+k/2];
a(39, :) = [-k/2+c-l*(p+r-k/2+c); c-r+l*(d+r); -c-t-l*(k/2-c-t)];
a(40, :) = [k/2-l*(p-m+k/2); p+r+l*(c-r); -c-d+l*(c+d)];
a(41, :) = [l*(c+d); p-m+l*(m-p-k/2); -c-p-l*(r-c)];
a(42, :) = [k/2+l*(c+t-k/2); p+r+l*(k/2-c-p-r); -c-d+l*(r+d)];
a(43, :) = [c-k/2; c-r+2*l*(r-c); -c-t];

```
a(44, :) = [c+t+l*(k/2-c-t); -k/2+c-l*(p+r-k/2+c); -c+r-l*(d+r)];
a(45, :) = [c+d-l*(c+d); k/2-l*(p-m+k/2); -p-r-l*(c-r)];
a(46, :) = [c+d-l*(c+d); -k/2+l*(p-m+k/2); -p-r-l*(c-r)];
a(47, :) = [c+t+l*(k/2-c-t); k/2-c+l*(p+r-k/2+c); -c+r-l*(d+r)];
a(48, :) = [c-k/2; -c+r-2*l*(r-c); -c-t];
a(49, :) = [k/2+l*(c+t-k/2); -p-r-l*(k/2-c-p-r); -c-d+l*(r+d)];
a(50, :) = [l*(c+d); -p+m-l*(m-p-k/2); -c-p-l*(r-c)];
a(51, :) = [c-r+l*(d+r); -c-t-l*(k/2-c-t); k/2-c+l*(p+r-k/2+c)];
a(52, :) = [c+t; -c+k/2; -c+r-2*l*(r-c)];
a(53, :) = [c+d-l*(r+d); -k/2-l*(c+t-k/2); -p-r-l*(k/2-c-p-r)];
a(54, :) = [c+p+l*(r-c); -l*(c+d); -p+m-l*(m-p-k/2)];
a(55, :) = [c+p+l*(r-c); -l*(c+d); p-m+l*(m-p-k/2)];
a(56, :) = [c+d-l*(r+d); -k/2-l*(c+t-k/2); p+r+l*(k/2-c-p-r)];
a(57, :) = [c+t; -c+k/2; c-r+2*l*(r-c)];
a(58, :) = [c-r+l*(d+r); -c-t-l*(k/2-c-t); -k/2+c-l*(p+r-k/2+c)];
a(59, :) = [p+r+l*(c-r); -c-d+l*(c+d); k/2-l*(p-m+k/2)];
a(60, :) = [p+r+l*(c-r); -c-d+l*(c+d); -k/2+l*(p-m+k/2)];
a(61, :) = [-p+m-l*(m-p-k/2); -c-p-l*(r-c); -l*(c+d)];
a(62, :) = [-p-r-l*(k/2-c-p-r); -c-d+l*(r+d); -k/2-l*(c+t-k/2)];
a(63, :) = [-c+r-2*l*(r-c); -c-t; -c+k/2];
a(64, :) = [k/2-c+l*(p+r-k/2+c); -c+r-l*(d+r); -c-t-l*(k/2-c-t)];
a(65, :) = [-k/2+l*(p-m+k/2); -p-r-l*(c-r) ; -c-d+l*(c+d)];
a(66, :) = [k/2-l*(p-m+k/2); -p-r-l*(c-r); -c-d+l*(c+d)];
a(67, :) = [-k/2+c-l*(p+r-k/2+c); -c+r-l*(d+r); -c-t-l*(k/2-c-t)];
a(68, :) = [c-r+2*l*(r-c); -c-t; -c+k/2];
a(69, :) = [p+r+l*(k/2-c-p-r); -c-d+l*(r+d); -k/2-l*(c+t-k/2)];
a(70, :) = [p-m+l*(m-p-k/2); -c-p-l*(r-c); -l*(c+d)];
a(71, :) = [p-m+l*(m-p-k/2); -c-p-l*(r-c); l*(c+d)];
a(72, :) = [p+r+l*(k/2-c-p-r); -c-d+l*(r+d); k/2+l*(c+t-k/2)];
a(73, :) = [c-r+2*l*(r-c); -c-t; c-k/2];
a(74, :) = [-k/2+c-l*(p+r-k/2+c); -c+r-l*(d+r); c+t+l*(k/2-c-t)];
a(75, :) = [k/2-l*(p-m+k/2); -p-r-l*(c-r); c+d-l*(c+d)];
a(76, :) = [-k/2+l*(p-m+k/2); -p-r-l*(c-r); c+d-l*(c+d)];
a(77, :) = [k/2-c+l*(p+r-k/2+c); -c+r-l*(d+r); c+t+l*(k/2-c-t)];
a(78, :) = [-c+r-2*l*(r-c); -c-t; c-k/2];
a(79, :) = [-p-r-l*(k/2-c-p-r); -c-d+l*(r+d); k/2+l*(c+t-k/2)];
a(80, :) = [-p+m-l*(m-p-k/2); -c-p-l*(r-c); l*(c+d)];
a(81, :) = [-c+r-l*(d+r); -c-t-l*(k/2-c-t); k/2-c+l*(p+r-k/2+c)];
a(82, :) = [-p-r-l*(c-r); -c-d+l*(c+d); -k/2+l*(p-m+k/2)];
a(83, :) = [-p-r-l*(c-r); -c-d+l*(c+d); k/2-l*(p-m+k/2)];
a(84, :) = [-c+r-l*(d+r); -c-t-l*(k/2-c-t); -k/2+c-l*(p+r-k/2+c)];
a(85, :) = [-c-t; -c+k/2; c-r+2*l*(r-c)];
a(86, :) = [-c-d+l*(r+d); -k/2-l*(c+t-k/2); p+r+l*(k/2-c-p-r)];
a(87, :) = [-c-p-l*(r-c); -l*(c+d); p-m+l*(m-p-k/2)];
a(88, :) = [-c-p-l*(r-c); -l*(c+d); -p+m-l*(m-p-k/2)];
a(89, :) = [-c-d+l*(r+d); -k/2-l*(c+t-k/2); -p-r-l*(k/2-c-p-r)];
```

a(90, :) = [-c-t; -c+k/2; -c+r-2*l*(r-c)];
a(91, :) = [-c-t-l*(k/2-c-t); k/2-c+l*(p+r-k/2+c); -c+r-l*(d+r)];
a(92, :) = [-c-d+l*(c+d); -k/2+l*(p-m+k/2); -p-r-l*(c-r)];
a(93, :) = [-c-d+l*(c+d); k/2-l*(p-m+k/2); -p-r-l*(c-r)];
a(94, :) = [-c-t-l*(k/2-c-t); -k/2+c-l*(p+r-k/2+c); -c+r-l*(d+r)];
a(95, :) = [-c+k/2; c-r+2*l*(r-c); -c-t];
a(96, :) = [-k/2-l*(c+t-k/2); p+r+l*(k/2-c-p-r); -c-d+l*(r+d)];
a(97, :) = [-l*(c+d); p-m+l*(m-p-k/2); -c-p-l*(r-c)];
a(98, :) = [-l*(c+d); -p+m-l*(m-p-k/2); -c-p-l*(r-c)];
a(99, :) = [-k/2-l*(c+t-k/2); -p-r-l*(k/2-c-p-r); -c-d+l*(r+d)];
a(100, :) = [-c+k/2; -c+r-2*l*(r-c); -c-t];
a(101, :) = [-c-t; c-k/2; c-r+2*l*(r-c)];
a(102, :) = [-c+r-l*(d+r); c+t+l*(k/2-c-t); -k/2+c-l*(p+r-k/2+c)];
a(103, :) = [-p-r-l*(c-r); c+d-l*(c+d); k/2-l*(p-m+k/2)];
a(104, :) = [-p-r-l*(c-r); c+d-l*(c+d); -k/2+l*(p-m+k/2)];
a(105, :) = [-c+r-l*(d+r); c+t+l*(k/2-c-t); k/2-c+l*(p+r-k/2+c)];
a(106, :) = [-c-t; c-k/2; -c+r-2*l*(r-c)];
a(107, :) = [-c-d+l*(r+d); k/2+l*(c+t-k/2); -p-r-l*(k/2-c-p-r)];
a(108, :) = [-c-p-l*(r-c); l*(c+d); -p+m-l*(m-p-k/2)];
a(109, :) = [-c-p-l*(r-c); l*(c+d); p-m+l*(m-p-k/2)];
a(110, :) = [-c-d+l*(r+d); k/2+l*(c+t-k/2); p+r+l*(k/2-c-p-r)];
a(111, :) = [-c-d+l*(c+d); k/2-l*(p-m+k/2); p+r+l*(c-r)];
a(112, :) = [-c-d+l*(c+d); -k/2+l*(p-m+k/2); p+r+l*(c-r)];
a(113, :) = [-c-t-l*(k/2-c-t); k/2-c+l*(p+r-k/2+c); c-r+l*(d+r)];
a(114, :) = [-c+k/2; -c+r-2*l*(r-c); c+t];
a(115, :) = [-k/2-l*(c+t-k/2); -p-r-l*(k/2-c-p-r); c+d-l*(r+d)];
a(116, :) = [-l*(c+d); -p+m-l*(m-p-k/2); c+p+l*(r-c)];
a(117, :) = [-l*(c+d); p-m+l*(m-p-k/2); c+p+l*(r-c)];
a(118, :) = [-k/2-l*(c+t-k/2); p+r+l*(k/2-c-p-r); c+d-l*(r+d)];
a(119, :) = [-c+k/2; c-r+2*l*(r-c); c+t];
a(120, :) = [-c-t-l*(k/2-c-t); -k/2+c-l*(p+r-k/2+c); c-r+l*(d+r)];
V = [ ]; for i = 1 : 120; V = [V; a(i, :)]; end;
F10 = [ ];   j = 1 : 10; for i = 1 : 12; F10 = [F10; 10*(i-1)+j]; end;

F4 = [1 29 28 2; 3 117 116 4; 5 75 74 6; 7 57 56 8; 9 13 12 10; 14 55 54 15; 16 45 44 17; 18 38 37 19; 11 20 21 30; 22 36 35 23; 24 103 102 25; 26 119 118 27; 58 73 72 59; 51 60 69 68; 52 47 46 53; 48 67 66 49; 41 50 98 97; 42 40 39 43; 31 96 95 32; 33 105 104 34; 61 70 71 80; 62 82 81 63; 64 100 99 65; 76 115 114 77; 78 84 83 79; 85 113 112 86; 87 109 108 88; 90 89 92 91; 93 107 106 94; 101 110 111 120];

F6 = [1 10 12 11 30 29; 8 56 55 14 13 9; 6 74 73 58 57 7; 4 116 115 76 75 5; 2 28 27 118 117 3; 15 54 53 46 45 16; 17 44 43 39 38 18; 19 37 36 22 21 20; 23 35 34 104 103 24; 25 102 101 120 119 26; 32 95 94 106 105 33; 31 40 42 41 97 96; 47 52 51 68 67 48; 49 66 65 99 98 50; 59 72 71 70 69 60; 61 80 79 83 82 62; 63 81 90 91 100 64; 77 114 113 85 84 78; 86 112 111 110 109 87; 88 108 107 93 92 89];
hold on; axis equal; view(3)

```
patch('Vertices', V, 'Faces', F4, 'FaceColor', [0 1 1])
patch('Vertices', V, 'Faces', F6, 'FaceColor', [0 .2 1])
patch('Vertices', V, 'Faces', F10, 'FaceColor', [0 .5 .5])
```

13) **p3444_2** (Ashkinuze)

```
c = 1;  d = c*sqrt(2)/(2+sqrt(2));
a(1, :) = [d; d; c]; a(2, :) = [-d; d; c]; a(3, :) = [-d; -d; c]; a(4, :) = [d; -d; c];
a(5, :) = [c; d; d]; a(6, :) = [c; d; -d]; a(7, :) = [c; -d; -d]; a(8, :) = [c; -d; d];
a(9, :) = [-d; c; d]; a(10, :) = [-d; c; -d]; a(11, :) = [d; c; -d]; a(12, :) = [d; c; d];
a(13, :) = [d; -c; d]; a(14, :) = [d; -c; -d]; a(15, :) = [-d; -c; -d]; a(16, :) = [-d; -c; d];
a(17, :) = [-c; -d; d]; a(18, :) = [-c; -d; -d]; a(19, :) = [-c; d; -d]; a(20, :) = [-c; d; d];
a(21, :) = [d*sqrt(2); 0; -c]; a(22, :) = [0; d*sqrt(2); -c];
a(23, :) = [-d*sqrt(2); 0; -c]; a(24, :) = [0; -d*sqrt(2); -c];
V = [ ];  for i = 1 : 24;  V = [V; a(i)];  end;
F4 = [ ]; j = 1 : 4;  for i = 1 : 6;  F4 = [F4; 4*(i-1)+j];  end;
F4b = [1 4 8 5; 4 13 16 3; 3 17 20 2; 2 9 12 1; 12 5 6 11; 8 13 14 7; 16 17 18
15; 20 9 10 19; 11 22 21 6; 7 21 24 14; 15 24 23 18; 10 19 23 22];
F3 = [1 12 5; 4 8 13; 3 16 17; 2 20 9; 6 21 7; 14 24 15; 18 23 19; 10 22 11];
hold on;  axis equal;  view(3)
patch('Vertices', V, 'Faces', F4, 'FaceColor', [0 1 1])
patch('Vertices', V, 'Faces', F4b, 'FaceColor', [0 .2 1])
patch('Vertices', V, 'Faces', F3, 'FaceColor', [0 .5 .5])
```

A.7 Hyperbolic Geometry

```
function f = segment(A, B)                           % M-file segment.m
   hold on;
   if A(1) == B(1)
   f = plot([A(1), B(1)], [A(2), B(2)], 'LineWidth',2);
   plot([A(1), B(1)], [0, 2*max(A(2), B(2))], 'r:');
   else  syms x t;
                %% x0 = solve((B(1)-x)^2 + B(2)^2 - (A(1)-x)^2 - A(2)^2);
   x0 = (B(1)^2 + B(2)^2 - A(1)^2 - A(1)^2)/(B(1) - A(1))/2;
   R = sqrt((B(1) - x0)^2 + B(2)^2);
   t1 = acos((B(1) - x0)/R);    t2 = acos((A(1) - x0)/R);
   X = x0 + R*cos(t);  Y = R*sin(t);
   t = linspace(t1, t2, 101);
   f = plot(x0 + R*cos(t), R*sin(t), 'LineWidth',3);
   tt = linspace(0,pi,101);  plot(x0 + R*cos(tt), R*sin(tt), 'r:');
end
```

```
function f = distance(A, B)                          % M-file distance.m
   if A(1) == B(1)
```

```
      f = abs(log(B(2)/A(2)))
      else syms x;
      x0 = solve((B(1) - x)^2 + B(2)^2 - (A(1) - x)^2 - A(2)^2);
      R = sqrt((B(1) - x0)^2 + B(2)^2);
      t1 = acos((B(1) - x0)/R);    t2 = acos((A(1) - x0)/R);
      f = abs(log( tan(t2/2)/tan(t1/2) ));
      end
end
```

```
function f = triangle(A, B, C)                       % M-file triangle.m
    hold on;  axis equal;  title('Triangle');
    segment(A, B);  segment(B, C);  segment(A, C);
    text(A(1), A(2), 'A');  text(B(1), B(2), 'B');  text(C(1), C(2), 'C');
end
```

```
function f = perimeter(A, B, C)                      % M-file perimeter.m
    distance(A, B) + distance(C, B) + distance(A, C);
end
```

```
function f = angle(B, A, C)                          % M-file angle.m
    a = distance(C, B);  b = distance(A, C);  c = distance(A, B);
    f = acos((cosh(c)*cosh(b) - cosh(a))/(sinh(c)*sinh(b)))
end
```

```
function f = sum_angles(A, B, C)                % M-file sum of angles.m
    f = angle(A, B, C) + angle(C, A, B) + angle(B, C, A);
end
```

```
function f = lambert(a1, a2)                         % M-file lambert.m
    if abs(sinh(a1)*sinh(a2)) >= 1;   disp ('No solutions')
    else
    hold on;  title('Quadrangle of Lambert');  axis equal;
    A = [0, 1]; B = [0, exp(a2)];
    t1 = 2*atan(exp(-a1));  C = [cos(t1), sin(t1)];
    R1 = tan(t1); x1 = 1/cos(t1);
    F(1) = (exp(2*a2) - R1^2 + x1^2)/(2*x1); F(2) = sqrt(exp(2*a2) - F(1)^2);
    phi = angle(B, [F(1), F(2)], C);
    f = segment(A, B);  segment(A, C);  segment(B, F);  segment(C, F);
    text(A(1), A(2), 'A');  text(B(1), B(2), 'B');
    text(C(1), C(2), 'C');  text(F(1), F(2), 'F');
    end
end
```

```
function f = deduce_angle( )                     % M-file deduce angle.m
    T = pi/6;   f = segment([0, 1], [cos(T), sin(T)]);  axis equal;
    segment([cos(T), sin(T)], [cos(T), sin(T) + 1]);  segment([0, 0], [0, 2]);
```

```
    plot([0, cos(T)], [0, sin(T)], 'r:');  title('The angle of parallelism (proof)');
    text(cos(T) - .1, sin(T) + .3, 't');  text(0.4, .1, 't');  text(.3, 1.1, 'd');
end
```

```
function f = parallel(A, P, Q)                        % M-file parallel.m
    if P(1) == Q(1)
    B1 = [P(1), 0];  B2 = [A(1), A(2) + 1];
    else  syms x;
    x1 = eval(solve((P(1) - x)^2+P(2)^2 - (Q(1) - x)^2-Q(2)^2));
    R1 = sqrt((P(1) - x1)^2 + P(2)^2);
    B1 = [x1 - R1, 0]; B2 = [x1 + R1, 0];
    phi = angle(B1, A, B2)/2
    f = segment(A, B1);  segment(A, B2);  segment(P, Q);
    title('The angle of parallelism');
    text(A(1), A(2)+.2, 'A');  text(P(1)-.2, P(2), 'P');  text(Q(1), Q(2)+.2, 'Q');
    end;
end
```

```
function f = perpendicular(A, P, Q)                   % M-file perpendicular.m
    if P(1) == Q(1);
     B = [P(1), sqrt((A(1) - P(1))^2 + A(2)^2)];
    else  syms x X xx;
    x1 = eval(solve((P(1) - x)^2 + P(2)^2 - (Q(1) - x)^2 - Q(2)^2));
    R1 = sqrt((P(1) - x1)^2 + P(2)^2);
    if x1 == A(1);
     B = [x1, R1];
     else
     x2 = eval(solve((A(1) - X)^2 + A(2)^2 - (X - x1)^2 + R1^2));
     R2 = sqrt((A(1) - x2)^2 + A(2)^2);
     B(1) = eval(solve(R1^2 - (x1 - xx)^2 - R2^2 + (x2 - xx)^2));
     B(2) = sqrt(R1^2 - (x1 - B(1))^2); B = [B(1), B(2)];
     end;
    end;
    d = distance(A, B), title('The distance from a point to a line');
    hold on;  f = segment(P, Q);  segment(A, B);
    text(A(1), A(2), 'A');  text(B(1), B(2), 'B');
    text(P(1), P(2), 'P');  text(Q(1), Q(2), 'Q');
end
```

```
function f = biorthogonal(A, B, C, D)                 % M-file biorthogonal.m
    syms x xx;
    if A(1) ~= B(1)
    x1 = eval(solve((A(1)-x)^2 + A(2)^2 - (B(1)-x)^2 - B(2)^2));
    R1 = sqrt((A(1) - x1)^2 + A(2)^2);
    end;
    if C(1) ~= D(1)
```

```
    x2 = eval(solve((C(1)-x)^2 + C(2)^2 - (D(1)-x)^2 - D(2)^2));
    R2 = sqrt((C(1) - x2)^2 + C(2)^2);
  end;
  if A(1) == B(1)
    x3 = A(1);  R3 = sqrt((x3 - x2)^2 - R2^2);
    K(1) = x3;  K(2) = R3;
    H(1) = -(- x2^2 + x3^2 - R3^2 + R2^2)/(x2 - x3)/2;
    H(2) = sqrt(R3^2 - (H(1) - x3)^2);
  elseif C(1) == D(1);
    x3 = C(1);  R3 = sqrt((x3 - x1)^2 - R1^2);
    H(1) = x3; H(2) = R3;
    K(1) = -(- x1^2 + x3^2 - R3^2 + R1^2)/(x1 - x3)/2;
    K(2) = sqrt(R3^2 - (K(1) - x3)^2)
  else if R1 == R2
    x3 = (x1 + x2)/2;
  else    x0 = eval(solve((x1 - xx)/R1 - (x2 - xx)/R2));
    xE = x1 - R1^2/(x1 - x0); xF = x2 - R2^2/(x2 - x0);
    x3 = (xE + xF)/2;
  end;
    R3 = sqrt((x3 - x1)^2 - R1^2);
    K(1) = -(- x1^2 + x3^2 - R3^2 + R1^2)/(x1 - x3)/2;
    H(1) = -(- x2^2 + x3^2 - R3^2 + R2^2)/(x2 - x3)/2;
    K(2) = sqrt(R3^2 - (K(1) - x3)^2);
    H(2) = sqrt(R3^2 - (H(1) - x3)^2);
  end;
  H = [H(1), H(2)];  K = [K(1), K(2)];
  d = distance(H, K)
  f = segment(A, B); segment(C, D); segment(H, K);
  title('The distance between two super-parallel lines')
  text(A(1), A(2), 'A'); text(B(1), B(2), 'B'); text(C(1), C(2), 'C');
  text(D(1), D(2), 'D'); text(H(1), H(2), 'H'); text(K(1), K(2), 'K');
end
```

```
function f = transversal(A, B, P, Q, theta)          % M-file transversal.m
  tt = cos(theta);  syms x xx x33 R33;
  if A(1) ~= B(1)
    x1 = solve((A(1) - x)^2 + A(2^2 - (B(1) - x)^2 - B(2)^2));
    R1 = sqrt((A(1) - x1)^2 + A(2)^2);
  end;
  if P(1) ~= Q(1)
    x2 = solve((P(1) - x)^2 + P(2)^2 - (Q(1) - x)^2 - Q(2)^2);
    R2 = sqrt((P(1) - x2)^2 + P(2)^2);
  end;
  for ii = 1 : 2;   for jj = 1 : 2
    if A(1) == B(1)
      [RR3 xx3] = solve((A(1) - x33) / R33 - (-1)^ii*tt,
```

```
      (R2^2 + R33^2 - (x2 - x33)^2)/(2*R2*R33) - (-1)^jj*tt, x33, R33);
    if eval(RR3(1)) < 0
     R3 = eval(RR3(2));  x3 = eval(xx3(2));
    else  R3 = eval(RR3(1));  x3 = eval(xx3(1));
    end
    H1 = A(1);  H2 = sqrt(R3^2 - (H1 - x3)^2);
    K1 = solve(R3^2 - (x3 - x)^2 - R2^2 + (x2 - x)^2);
    K2 = sqrt(R2^2 - (x2 - K1)^2);
  elseif  P(1) == Q(1)
    [RR3 xx3] = solve((P(1) - x33)/R33 - (-1)^ii*tt,
    (R1^2 + R33^2 - (x1 - x33)^2)/(2*R1*R33) - (-1)^jj*tt, x33, R33);
    if eval(RR3(1)) < 0
     R3 = eval(RR3(2));  x3 = eval(xx3(2));
    else  R3 = eval(RR3(1));  x3 = eval(xx3(1));
    end
    K1 = P(1);  K2 = sqrt(R3^2 - (K1 - x3)^2);
    H1 = eval(solve(R3^2 - (x3 - x)^2 - R1^2 + (x1 - x)^2));
    H2 = eval(sqrt(R1^2 - (x1 - H1)^2));
  else
    [RR3 xx3] = solve((R1^2 + R33^2 - (x1 - x33)^2)/(2*R1*R33) - (-1)^ii*tt,
    (R2^2 + R33^2-(x2 - x33)^2)/(2*R2*R33) - (-1)^jj*tt, x33, R33);
    if eval(RR3(1)) < 0
     R3 = eval(RR3(2));  x3 = eval(xx3(2));
    else
     R3 = eval(RR3(1));  x3 = eval(xx3(1));
    end
    H1 = eval(solve(R3^2 - (x3 - x)^2 - R1^2 + (x1 - x)^2));
    H2 = eval(sqrt(R1^2 - (x1 - H1)^2));
    K1 = eval(solve(R3^2 - (x3 - xx)^2 - R2^2 + (x2 - xx)^2, xx));
    K2 = eval(sqrt(R2^2 - (x2 - K1)^2));
  end;
  K(:, :, ii + 2*(jj - 1)) = [K1, K2];  H(:, :, ii + 2*(jj - 1)) = [H1, H2];
  d(ii + 2*(jj - 1)) = distance(H(:, :, ii + 2*(jj - 1)), K(:, :, ii + 2*(jj - 1)));
  end;  end
  disp('The four distances are');    d
  f = segment(A, B);  segment(P, Q);
  for ii = 1 : 4;    segment(H(:, :, ii), K(:, :, ii));    end
  title('The four transversal between two lines');
  text(A(1), A(2), 'A');  text(B(1), B(2), 'B');
  text(P(1), P(2), 'P');  text(Q(1), Q(2), 'Q');
end
```

```
function f = circle(A, R)                              % M-file circle.m
  yE = A(2)*cosh(R);  R1 = A(2)*sinh(R);
  t = linspace(-pi, pi, 101);
  hold on;  axis equal;
```

```
    f = segment([A(1) - A(2), 0], [A(1) + A(2), 0]);
    plot(A(1) + R1*cos(t), yE + R1*sin(t), 'r', 'LineWidth', 2);
    plot(A(1), A(2), 'o');  text(A(1), A(2), 'A');
end
```

```
function f = equidistant(A, B, d, n)                    % M-file equidistant.m
    syms x t;  t = linspace(0, pi, 101);
    hold on;  axis equal;  title('Equidistant (of a line)');
    text(A(1), A(2), 'A');  text(B(1), B(2), 'B');
    if A(1) == B(1)
    x1 = A(1);  y1 = max(A(2), B(2)) + 1;
    y2 = 2*atan(exp(-d));
    tt = linspace(x1, x1 + y1, 101);
    f = plot(x1, t, 'LineWidth', 2);  plot(tt, tan(y2)*(tt - x1), 'LineWidth', 2);
    for ii = 1 : n - 1;
      plot(x1 + (y1*ii/n)*cos(t), (y1*ii/n)*sin(t), 'g:');
    end
    else
    x1 = solve((B(1) - x)^2 + B(2)^2 - (A(1) - x)^2 - A(2)^2);
    R = eval(sqrt((B(1) - x1)^2 + B(2)^2));
    y1 = R*exp(d);  y2 = (y1^2 - R^2)/(2*y1);  R2 = y1 - y2;
    for ii = 1 : n-1
     ti = pi*ii/(2*n);  xi(ii) = eval(x1 + R*cos(ti) + R*sin(ti)*tan(ti));
     Ri(ii) = eval(sqrt((x1 + R*cos(ti) - xi(ii))^2 + (R*sin(ti))^2));
    end;
    Y3 = asin(y2/R2);  tt = linspace(-Y3, pi + Y3, 101);
    f = plot(eval(x1 + R*cos(t)), R*sin(t),'LineWidth',2);
    plot(eval(x1) + R2*cos(tt), y2 + R2*sin(tt), 'LineWidth',2);
    for ii = 1 : n - 1
     plot(xi(ii) + Ri(ii)*cos(t), Ri(ii)*sin(t), 'r');
     plot(eval(2*x1) - xi(ii) + Ri(ii)*cos(t), Ri(ii)*sin(t), 'r:');
    end;
end
```

```
function f = horocycle(A, B, n)                         % M-file horocycle.m
    syms x y t tt ttt;
    hold on;  axis equal;
    if A(1) == B(1)
    x1 = A(1);  R = max(A(2), B(2)) + 1 R2 = A(2)/2;
    else
    x0 = solve((B(1) - x)^2 + B(2)^2 - (A(1) - x)^2 - A(2)^2);
    R = eval(sqrt((B(1) - x0)^2 + B(2)^2));  x1 = eval(x0 + R);
    R2 = eval(solve(y^2 - (A(1) - x1)^2 - (A(2) - y)^2));
    end;
    t = linspace(0, pi, 101);
    tt = linspace(0, R, 101);
```

```
    ttt = linspace(-pi, pi, 101);
    f = plot(x1 + R2*cos(ttt), R2*(1 + sin(ttt)), 'LineWidth', 2);
    if A(1) == B(1);
     plot(x1, tt, 'g -','LineWidth', 2);
    else
     plot(eval(x0) + R*cos(t), R*sin(t), 'LineWidth', 2);
    end
    for ii = 1 : n-1
     plot(x1 + (R*ii/n) + (R*ii/n)*cos(t), (R*ii/n)*sin(t));
     plot(x1 - (R*ii/n) + (R*ii/n)*cos(t), (R*ii/n)*sin(t));
    end;
    title('The horocycle');  text(A(1), A(2), 'A');  text(B(1), B(2), 'B');
end
```

```
function f = fifth(A, B, a, n)        % Program fifth.m (case of vertical base line)
    tt = linspace(0, pi, 101);  hold on;
    if A(1) ~= B(1);  disp('A1 is not equal B1')
    else
     for ii = 1 : n
      [t, y] = ode45(@ f1, [0 4/(ii + 1)], [0, ii]);                    % see subprogram
      p(:, ii) = y(:, 1);  q(:, ii) = y(:, 2);
     end
     f = plot(p + A(1), q, 'r', 'LineWidth', 2);  title('The fifth line');
     for ii = 1 : n;
      plot(A(1) + ii*cot(a) + ii*cos(tt)/sin(a), ii*sin(tt)/sin(a), ':');
     end;
    end
    segment(A, B);  axis equal;
    text(A(1), A(2), 'A');     text(B(1), B(2), 'B');
```

```
.....................................................................
function dy1 = f1(t, y);                           % Subprogram: function f₁
    global a;   T = (-cos(a)*y(1)+sqrt(y(1)^2+y(2)^2*sin(a)^2))/sin(a);
    dy1 = zeros(2, 1);  dy1 = [y(1) - T*cot(a); y(2)];
```

```
function f = H3segment(A, B)
    D = [A; B];
    if A(1) == B(1) & A(2) == B(2)
    plot3([A(1), B(1)],  [A(2), B(2)],  [0, 2*max(A(3), B(3))], 'r:');
    hold on;  view(3);  grid;  axis equal;
    plot3([A(1), B(1)],  [A(2), B(2)],  [A(3), B(3)], 'LineWidth', 2);
    plot3(D(:, 1), D(:, 2), D(:, 3), 'o')
    else
    x0 = 1/2*(B(2)^2*B(1)+B(2)^2*A(1)-2*B(2)*A(1)*A(2)-2*A(2)*B(1)*B(2)
    +A(1)^3-B(3)^2*A(1)-A(1)^2*B(1)-B(1)^2*A(1)+A(2)^2*B(1)+B(3)^2*B(1)
    +B(1)^3-A(3)^2*B(1)+A(2)^2*A(1)+A(3)^2*A(1))/(B(1)^2
    -2*B(1)*A(1)+A(1)^2+B(2)^2-2*B(2)*A(2)+A(2)^2);
```

```matlab
    y0 = 1/2*(-2*B(1)*A(1)*B(2)+A(2)*B(1)^2+A(1)^2*B(2)-2*A(1)*A(2)*B(1)
    +B(1)^2*B(2)+B(2)^3-B(2)^2*A(2)+B(3)^2*B(2)-B(3)^2*A(2)+A(1)^2*A(2)
    -A(2)^2*B(2)+A(2)^3-A(3)^2*B(2)+A(3)^2*A(2))/(B(1)^2-2*B(1)*A(1)
    +A(1)^2+B(2)^2-2*B(2)*A(2)+A(2)^2);

    R = sqrt((B(1) - x0)^2 + (B(2) - y0)^2 + B(3)^2);
    C = [x0, y0, 0];
    tb = R/sqrt((B(1) - A(1))^2 + (B(2) - A(2))^2);
    N = R*[0, 0, 1];  e = tb*[B(1) - A(1), B(2) - A(2) 0];
    if [B(1) - x0, B(2) - y0]*[B(1) - A(1), B(2) - A(2)]' > 0;
     t1 = asin(B(3)/R);
    else t1 = pi - asin(B(3)/R);
    end
    if [A(1) - x0, A(2) - y0]*[B(1) - A(1), B(2) - A(2)]' > 0;
     t2 = asin(A(3)/R);
    else t2 = pi - asin(A(3)/R);
    end
    syms t;
    Ct = C + e*cos(t) + N*sin(t);
    T = linspace(t1, t2, 21);  T0 = linspace(0, pi, 21);
    plot3(subs(Ct(1), t, T)+T-T, subs(Ct(2), t, T)+T-T, subs(Ct(3), t, T));
    hold on; view(3);  grid;  axis equal;
    plot3(subs(Ct(1), t, T0)+T-T, subs(Ct(2), t, T0)+T-T, subs(Ct(3), t, T0), 'r:');
    plot3(D(:, 1), D(:, 2), D(:, 3), 'o')
    end
end


function f = H3distance(A, B)
    if A(1) == B(1) & A(2) == B(2)
    f = abs(log(B(3)/A(3)));
    else
    syms x;
    x0 = 1/2*(B(2)^2*B(1)+B(2)^2*A(1)-2*B(2)*A(1)*A(2)-2*A(2)*B(1)*B(2)
    +A(1)^3-B(3)^2*A(1)-A(1)^2*B(1)-B(1)^2*A(1)+A(2)^2*B(1)+B(3)^2*B(1)
    +B(1)^3-A(3)^2*B(1)+A(2)^2*A(1)+A(3)^2*A(1))/(B(1)^2-2*B(1)*A(1)
    +A(1)^2+B(2)^2-2*B(2)*A(2)+A(2)^2);

    y0 = 1/2*(-2*B(1)*A(1)*B(2)+A(2)*B(1)^2+A(1)^2*B(2)-2*A(1)*A(2)*B(1)
    +B(1)^2*B(2)+B(2)^3-B(2)^2*A(2)+B(3)^2*B(2)-B(3)^2*A(2)
    +A(1)^2*A(2)-A(2)^2*B(2)+A(2)^3-A(3)^2*B(2)
    +A(3)^2*A(2))/(B(1)^2-2*B(1)*A(1)+A(1)^2+B(2)^2-2*B(2)*A(2)+A(2)^2);
    R = sqrt((B(1) - x0)^2 + (B(2) - y0)^2 + B(3)^2);
    t1 = asin(B(3)/R);  t2 = asin(A(3)/R);
    f = abs(log( tan(t2 / 2) / tan(t1 / 2) ));
    end
end
```

```
function f = H3triangle(A, B, C)
   hold on; title('Triangle'); axis equal;
   H3segment(A, B); H3segment(B, C); H3segment(A, C);
   text(A(1), A(2), A(3)+.3, 'A'); text(B(1), B(2), B(3)+.3, 'B');
   text(C(1), C(2), C(3)+.3, 'C'); D = [A; B; C]; plot3(D(:, 1), D(:, 2), D(:, 3), 'o')
end
```

```
function f = H3plane(A, B, C)
   syms u v;
   D = [A; B; C];
   DD = det([1 A(1) A(2); 1 B(1) B(2); 1 C(1) C(2)]);
   if DD == 0
    F0 = [A(1) + u*(C(1) - A(1)) + eps*u, A(2) + u*(C(2) - A(2)) + eps*v];
   else
   x0 = 1/2*(-B(1)^2*C(2)+B(1)^2*A(2)-B(3)^2*C(2)+A(3)^2*C(2)-B(2)*A(2)^2
   -A(2)*C(3)^2-A(2)*C(1)^2-B(2)*A(1)^2-A(2)*C(2)^2+B(2)^2*A(2)
   +B(2)*C(3)^2-B(2)^2*C(2)+A(2)^2*C(2)+B(3)^2*A(2)-B(2)*A(3)^2
   +B(2)*C(1)^2+B(2)*C(2)^2+A(1)^2*C(2))/(C(1)*B(2)-C(1)*A(2)-A(1)*B(2)
   -C(2)*B(1)+C(2)*A(1)+A(2)*B(1));

   y0 = -1/2*(-C(1)*B(1)^2-C(1)*B(2)^2-C(1)*B(3)^2+C(1)*A(1)^2+C(1)*A(2)^2
   +C(1)*A(3)^2+A(1)*B(1)^2+A(1)*B(2)^2+A(1)*B(3)^2+C(1)^2*B(1)
   -C(1)^2*A(1)+C(2)^2*B(1)-C(2)^2*A(1)+C(3)^2*B(1)-C(3)^2*A(1)
   -A(1)^2*B(1)-A(2)^2*B(1)-A(3)^2*B(1))/(C(1)*B(2)-C(1)*A(2)-A(1)*B(2)
   -C(2)*B(1)+C(2)*A(1)+A(2)*B(1));

   R = sqrt((B(1) - x0)^2 + (B(2) - y0)^2 + B(3)^2);
   end
   plot3(D(:, 1), D(:, 2), D(:, 3), 'o', 'LineWidth', 2)
   hold on; grid; axis equal;
   if DD == 0;
    ezmesh(F0(1), F0(2), v, [0, 1, 0, 2], 10);
   else
    ezmesh(x0+R*sin(u)*cos(v), y0+R*sin(u)*sin(v), R*cos(u), [0, pi/2, 0, 2*pi],
    20);
   end
end
```

```
function f = H3perp02(P, A, B, C)
   D = [P; A; B; C];
   if det([1 A(1) A(2); 1 B(1) B(2); 1 C(1) C(2)]) == 0
    A0 = [A(1) A(2)]; B0 = [B(1) B(2)]; C0 = [C(1) C(2)];
    P0 = [P(1) P(2)]; t0 = ((A0 - P0)*(C0 - A0)')/((C0 - A0)*(C0 - A0)');
    Q0 = A0 + t0*(C0 - A0); Q = [Q0'; sqrt(P(3)^2 + (Q0 - P0)*(Q0 - P0)')]';
   else
    x0 = 1/2*(-B(1)^2*C(2)+B(1)^2*A(2)-B(3)^2*C(2)+A(3)^2*C(2)
    -B(2)*A(2)^2-A(2)*C(3)^2-A(2)*C(1)^2-B(2)*A(1)^2-A(2)*C(2)^2
    +B(2)^2*A(2)+B(2)*C(3)^2-B(2)^2*C(2)+A(2)^2*C(2)+B(3)^2*A(2)
```

```
    -B(2)*A(3)^2+B(2)*C(1)^2+B(2)*C(2)^2+A(1)^2*C(2))/(C(1)*B(2)
    -C(1)*A(2)-A(1)*B(2)-C(2)*B(1)+C(2)*A(1)+A(2)*B(1));

    y0 = -1/2*(-C(1)*B(1)^2-C(1)*B(2)^2-C(1)*B(3)^2+C(1)*A(1)^2
    +C(1)*A(2)^2+C(1)*A(3)^2+A(1)*B(1)^2+A(1)*B(2)^2+A(1)*B(3)^2
    +C(1)^2*B(1)-C(1)^2*A(1)+C(2)^2*B(1)-C(2)^2*A(1)+C(3)^2*B(1)
    -C(3)^2*A(1)-A(1)^2*B(1)-A(2)^2*B(1)-A(3)^2*B(1))/(C(1)*B(2)
    -C(1)*A(2)-A(1)*B(2)-C(2)*B(1)+C(2)*A(1)+A(2)*B(1));

    R = sqrt((B(1) - x0)^2 + (B(2) - y0)^2 + B(3)^2);
    t1 = 1/2*(R^2+P(1)^2-2*P(1)*x0+x0^2+P(2)^2-2*P(2)*y0+y0^2
    +P(3)^2)/(P(1)^2-2*P(1)*x0+x0^2+P(2)^2-2*P(2)*y0+y0^2);
    C1 = [x0 + t1*(A(1) - x0), y0 + t1*(A(2) - y0) 0];
    tq = R^2/((A(1) - x0)^2 + (A(2) - y0)^2)/t1;
    h = sqrt(R^2 - tq^2*((A(1) - x0)^2 + (A(2) - y0)^2));
    Q = [tq*(A(1) - x0), tq*(A(2) - y0), h];
  end;
  d = H3distance(P, Q);
  hold on;  grid;  axis equal;  title('Distance');
  H3segment(P, Q);  H3plane(A, B, C);
  plot3(D(:, 1), D(:, 2), D(:, 3), 'o', 'LineWidth', 2)
  text(A(1), A(2), A(3), 'A');  text(B(1), B(2), B(3), 'B');  text(C(1), C(2), C(3), 'C');
  text(P(1), P(2), P(3), 'P');  text(Q(1), Q(2), Q(3), 'Q');
end
```

```
function f = H3parallel(A, P, Q)
  if P(1) == Q(1) & P(2) == Q(2)
  B1 = [P(1), P(2), 0];  B2 = [A(1), A(2), A(3) + 1];
  else
  x1 = 1/2*(Q(2)^2*Q(1)+Q(2)^2*P(1)-2*Q(2)*P(1)*P(2)-2*P(2)*Q(1)*Q(2)
  +P(1)^3 - Q(3)^2*P(1) - P(1)^2*Q(1) - Q(1)^2*P(1) + P(2)^2*Q(1)
  +Q(3)^2*Q(1) + Q(1)^3 - P(3)^2*Q(1) + P(2)^2*P(1) + P(3)^2*P(1))/
  (Q(1)^2 - 2*Q(1)*P(1) + P(1)^2 + Q(2)^2 - 2*Q(2)*P(2) + P(2)^2);

  y1 = 1/2*(-2*Q(1)*P(1)*Q(2)+P(2)*Q(1)^2+P(1)^2*Q(2)-2*P(1)*P(2)*Q(1)
  +Q(1)^2*Q(2) + Q(2)^3 - Q(2)^2*P(2) + Q(3)^2*Q(2) - Q(3)^2*P(2)
  +P(1)^2*P(2) - P(2)^2*Q(2) + P(2)^3 - P(3)^2*Q(2) + P(3)^2*P(2))/
  (Q(1)^2 - 2*Q(1)*P(1) + P(1)^2 + Q(2)^2 - 2*Q(2)*P(2) + P(2)^2);

  R = sqrt((Q(1) - x1)^2 + (Q(2) - y1)^2 + Q(3)^2);
  C = [x1, y1, 0];
  tb = R/sqrt((Q(1) - P(1))^2 + (Q(2) - P(2))^2);
  B2 = [x1, y1, 0] + tb*[Q(1) - P(1), Q(2) - P(2), 0];
  B1 = [x1, y1, 0] - tb*[Q(1) - P(1), Q(2) - P(2), 0];
  end;
  hold on;  axis equal;
  title('Angle of parallelism');
  H3segment(A, B1);  H3segment(A, B2);  H3segment(P, Q);
end
```

A.8 Investigation of a Surface in $\mathbb{R}^3$

The program will help you to compute the characteristics of a surface.

```
syms u v positive;  syms x y z real;
```
% **Input (two types of surfaces)**
% (a) *Graph* of a function $f(x,y)$
```
X = u;  Y = v;  Z = (u^2 - v^2)/5;              % define functions
u1 = -1;  u2 = 1;  v1 = -1;  v2 = 1;  P = [1, 0];    % define initial data
```
% (b) *Parametric surface* $\mathbf{r}(u,v)$
```
X = u*cos(v);  Y = u*sin(v);  Z = u;             % define functions
u1 = -pi;  u2 = pi;  v1 = 0;  v2 = 2*pi;  P = [2, 3];   % define initial data
```
% **Coordinate System**
```
r = [X, Y, Z];  r_p = subs(r, [u v], P);          % parametric equations and P
u_curve = subs(r, v, P(2));  v_curve = subs(r, u, P(1)); % coordinate curves
```
% **Tangent plane**
```
ru = diff(r, u); rv = diff(r, v);        % velocity vectors of u-curve and v-curve
n1 = cross(ru, rv);  n = simplify(n1/sqrt(n1*n1'));        % normal vector
ru_p = subs(ru, [u v], [P(1), P(2)]); rv_p = subs(rv, [u v], [P(1), P(2)]);
n_p = subs(n, [u v], [P(1), P(2)]);
TM = simplify(dot(n, [x,y,z] - r));        % 1st method: implicit equation
TpM = eval(r_p + ru_p*u + rv_p*v);         % 2nd method: parametric eq-s
```
% **First fundamental form and related problems**
```
E = simplify(dot(ru, ru));  F = simplify(dot(ru, rv));  G = simplify(dot(rv, rv));
E_p = subs(E, [u v], P);  F_p = subs(F, [u v], P);  G_p = subs(G, [u v], P);
                                            % coefficients E, F, G
g = simplify(E*G - F^2);  g_p = subs(g, [u v], P);        % determinant g
syms du dv t;
I_f = E*du^2 + 2*F*du*dv + G*dv^2;
I_p = E_p*du^2 + 2*F_p*du*dv + G_p*dv^2;      % first fundamental form
theta_uv = acos(F/sqrt(E*G));                % angle between u- and v- curves
theta_uvP = subs(theta_uv, [u v], P);
length_u = double(int(subs(sqrt(E), [u v], [t,P(2)]), t, u1, u2));    % lengths
                                            % of u- and v- curves
length_v = double(int(subs(sqrt(G), [u v], [P(1), t]), t, v1, v2));
U = t;  V = t;               % equations of a curve on a $M^2$ (insert data)!
t1 = 0;  t2 = pi;    ut = diff(U, t);  vt = diff(V, t);
c_gamma = subs(r, [u v], [U, V]);            % a curve and its length
l = double(int(sqrt(subs(E*ut^2 + F*ut*vt + G*vt^2, [u v], [U V])), t, t1, t2));
area_M = double(int(int(sqrt(g), u, 0, u2), v, 0, v2));    % surface area
```

% **Second fundamental form and related problems**

```
ruu = diff(ru, u);  ruv = diff(ru, v);  rvv = diff(rv, v);
ruu_p = subs(ruu, [u v], P);  ruv_p = subs(ruv, [u v], P);
rvv_p = subs(rvv, [u v], P);              % second derivatives for further use
L = simplify(dot(ruu, n));  M = simplify(dot(ruv, n));  N = simplify(dot(rvv, n));
L_p = subs(L, [u v], P);  M_p = subs(M, [u v], P);  N_p = subs(N, [u v], P);
                              % coefficients of the second fundamental form
II = L*du^2 + 2*M*du*dv + N*dv^2;
II_p = L_p*du^2 + 2*M_p*du*dv + N_p*dv^2;   % second fundamental form
GK = simplify((L*N - M^2)/g);
GK_p = subs(GK, [u v], P);                   % Gaussian curvature
H = simplify((L*G + N*E - 2*M*F)/g/2);
H_p = subs(H, [u v], P);                     % mean curvature
syms k;    ki = solve(k^2*g - k*(L*G + N*E - 2*M*F) + (L*N - M^2), k);
ki_p = subs(ki, [u v], P);                   % principal curvatures
kv = simplify(N/G);  kv_p = subs(kv, [u v], P);   % normal curvatures
ku = simplify(L/E);  ku_p = subs(ku, [u v], P);   % of coordinate lines
syms w1 w2;
eq1 = simplify(det([-w2^2, w1*w2, -w1^2; E, F, G; L, M, N]));
solve(eq1, w1);                              % principal directions

eq1_p = simplify(det([-w2^2, w1*w2, -w1^2; E_p, F_p, G_p; L_p, M_p, N_p]));
solve(eq1_p, w1);
solve(L_p*w1^2 + 2*M_p*w1*w2 + N_p*w2^2);    % asymptotic directions
```

% **Differential equations of geodesics**

```
Eu = diff(E, u);  Ev = diff(E, v);
Fu = diff(F, u);  Fv = diff(F, v);
Gu = diff(G, u);  Gv = diff(G, v);
eq1 = diff('u(t)', t, 2) + subs((G*Eu - F*(2*Fu - Ev))/(2*g)*diff('u(t)', t)^2
      + (G*Ev - F*Gu) / g*diff('u(t)', t)*diff('v(t)', t)
      + (G*(2*Fv - Gu) - F*Gv)/(2*g)*diff('v(t)', t)^2, {u v}, {'u(t)', 'v(t)'});
eq2 = diff('v(t)',t,2) + subs((E*(2*Fu - Ev) - F*Eu)/(2*g)*diff('u(t)', t)^2
      + (E*Gu - F*Ev) / g*diff('u(t)', t)*diff('v(t)', t)
      + (E*Gv - F*(2*Fv - Gu))/(2*g)*diff('u(t)', t)^2, {u v}, {'u(t)', 'v(t)'});
```

The function M-file displays a level surface $f(x, y, z) = c$.

```
function  out = impl(f, corners, c)                 % see impl.m in [Cooper]
    xmin = corners(1);  ymin = corners(3);  zmin = corners(5);
    xmax = corners(2);  ymax = corners(4);  zmax = corners(6);
    x = linspace(xmin, xmax ,21);  y = linspace(ymin, ymax, 21);
    z = linspace(zmin, zmax, 21);  [XX, YY, ZZ] = meshgrid(x, y, z);
    W = feval(f, XX, YY, ZZ);
    m = min(min(min(W)));  M = max(max(max(W)));
```

```
sprintf( 'The max over this domain is % 5.5f %', M)
sprintf( 'The min over this domain is % 5.5f %', m)
if c < m | c > M
 sprintf('In this domain, the function does not take on the value % 5.5f %', c)
else
[X, Y] = meshgrid(x, y);
dz = (zmax - zmin)/40;
for zz = zmin : dz : zmax
 Z = feval(f, X, Y, zz);
 con = contours(X, Y, Z, [c,c]);
 nn = size(con, 2);
 if nn > 0
  j = 1;        while j < nn
   npairs = con(2, j);
   xdata = con(1, j + 1: j + npairs);
   ydata = con(2, j + 1: j + npairs);
   plot3(xdata, ydata, zz + 0*xdata)
   j = j + npairs + 1;  hold on
  end;
 end;
 end;
end
axis(corners);
xlabel('x');  ylabel('y');  zlabel('z');  hold off
```

References

1. Anderson, J.W., *Hyperbolic Geometry*. Springer, SUMS (2005)
2. Blyth, T.S., and Robertson, E.F., *Basic Linear Algebra*. Springer, SUMS (2007)
3. Blyth, T.S., and Robertson, E.F., *Further Linear Algebra*. Springer, SUMS (2002)
4. Crossley, M.D., *Essential Topology*. Springer, SUMS (2007)
5. Dineen, S., *Multivariate Calculus and Geometry*. Springer, SUMS (2001)
6. Fenn, R., *Geometry*. Springer, SUMS (2006)
7. Hirst, K.E., *Calculus of One Variable*. Springer, SUMS (2006)
8. Howie, J.M., *Complex Analysis*. Springer, SUMS (2003)
9. Johnson, D.L., *Symmetries*. Springer, SUMS (2002)
10. Marsh, D., *Applied Geometry for Computer Graphics & CAD*. Springer, SUMS (2005)
11. Matthews, P.C., *Vector Calculus*. Springer, SUMS (2000)
12. Pressley, A.N., *Elementary Differential Geometry*. Springer, SUMS (2001)
13. Searcóid, M., *Metric Spaces*. Springer, SUMS (2007)
14. Smith, G., *Introductory Mathematics: Algebra and Analysis*. Springer, SUMS (2000)
15. Wallace, D., *Groups, Rings and Fields*. Springer, SUMS (2001)

Geometry & modeling books

16. Armstrong, M.A., *Groups and Symmetry*. Springer (1988)
17. Baragar, A., *A Survey of Classical and Modern Geometries with Computer Activities*. Prentice Hall (2001)
18. Beardon, A.F., *Algebra and Geometry*. Cambridge (2005)
19. Berger, M., *Geometry I, II*. Springer-Universitext, Paris (1992, 1996)

20. Brannan, D.A., Esplen M.F., and Gray J.J., *Geometry*. Cambridge University Press (2007)

21. Bursky, R., *Computer Graphics and Geometric Modeling Using Beta-Splines*. Springer (1987)

22. Casse, R., *Projective Geometry: An Introduction*. Oxford University Press (2006)

23. Cromwell, P.R., *Polyhedra*. Cambridge University Press (1997)

24. Duzhin, S.V., and Tchebotarevsky, B.D., *Transformation Groups for Beginners*. AMS (2004)

25. Kühnel, W., *Differential Geometry: Curves – Surfaces – Manifolds*. AMS (2006)

26. Farin, G., *Curves and Surfaces for Computer-Aided Geometric Design: A Practical Guide*. Academic Press (1988)

27. Farin, G., *Practical Linear Algebra*. A K Peters, Ltd. (2005)

28. Gallier, J., *Geometric Methods and Applications for Computer Science and Engineering*. Springer (2001)

29. Gibson, C.G., *Elementary Euclidean Geometry. An Introduction*. Cambridge University Press (2004)

30. Gray, A., *Modern Differential Geometry of Curves and Surfaces with Mathematica*. Springer (1998)

31. Gutenmacher, V., and Vasilyev, N., *Lines and Curves: A Practical Geometry Handbook*. Birkhäuser (2004)

32. Hanson, A.J., *Visualizing Quaternions*. Elsevier (2006)

33. Martin, G.E., *Transformation Geometry: An Introduction to Symmetry*. Springer (1997)

34. Montiel, S., and Ros, A., *Curves and Surfaces*. AMS (2005)

35. Needham, T., *Visual Complex Analysis*. Clarendon Press, Oxford (2000)

36. Pedow, D., *Geometry and the Liberal Arts*. Penguin Books (1976)

37. Prasolov, V.V., and Tikhomirov, V.M., *Geometry*. AMS (2001)

38. Reid, M., and Szendrői, B., *Geometry and Topology*. Cambridge University Press (2005)

39. Rovenski, V., *Geometry of Curves and Surfaces with Maple*. Birkhäuser, Boston (2000)

40. Rutter, J.W., *Geometry of Curves*. Chapman & Hall/CRC (2000)

41. Salomon, D., *Curves and Surfaces for Computer Graphics*. Springer (2006)

42. Salomon, D., *Transformations and Projections in Computer Graphics*. Springer (2006)

43. Shikin, E.V., and Plis, A.I., *Handbook on Splines for the User*. CRC Press (1995)

44. Shores, T.S., *Applied Linear Algebra and Matrix Analysis*. Springer, Undergraduate Texts in Math. (2007)

45. Stillwell, J., *The Four Pillars of Geometry.* Springer, Undergraduate Texts in Math. (2005)

46. Tapp, K., *Matrix Groups for Undergraduates.* AMS (2005)

47. Toponogov, V.A., *Differential Geometry of Curves and Surfaces: A Concise Guide.* Birkhäuser (2006)

48. Toth, G., *Glimpses of Algebra and Geometry.* Springer (2002)

49. Wunsch, A.D., *Complex Variables with Applications.* Addison Wesley (2005)

50. Yamaguchi, F., *Curves and Surfaces in Computer-Aided Geometric Design.* Springer-Verlag (1988)

MATLAB books

51. Cooper, J., *A MATLAB Companion for Multivariable Calculus.* Academic Press (2001)

52. Chen, K., Giblin, P., and Irving, A., *Mathematical Explorations with MATLAB.* Cambridge University Press (1999)

53. Danaila I. et al., *An Introduction to Scientific Computing: Twelve Computational Projects Solved with MATLAB.* Springer (2006)

54. Devis, T.A., and Sigmon, K., *MATLAB Primer.* Chapman & Hall/CRC (2005)

55. Gander, W., and Hřebíček, J., *Solving Problems in Scientific Computing Using MAPLE and MATLAB.* Springer-Verlag (1997)

56. Hahn, B.D., and Valentine, D.T., *Essential MATLAB for Engineers and Scientists.* Elsevier (2007)

57. Higham, D.J., and Higham, N.J., *MATLAB Guide.* SIAM (2005)

58. Hunt, B.R., Lipsman, R.L., and Rosenberg, J.M., *A Guide to MATLAB for Beginners and Experienced Users.* Cambridge Univ. Press (2006)

59. Kalechman, M., *Practical MATLAB Applications for Engineers.* CRC Press (2009)

60. Kiusalaas, J., *Numerical Methods in Engineering with MATLAB.* Cambridge Univ. Press (2005)

61. McMahon, D., *MATLAB Demystified.* The McGraw-Hill Companies (2007)

62. Klima, R.E., Sigmon, N., and Stitzinger, E., *Applications of Abstract Algebra with MAPLE and MATLAB.* Boca Raton: Chapman & Hall (2007)

63. Lynch, S., *Dynamical Systems with Applications Using MATLAB.* Birkhäuser (2004)

64. Marchand, P., and Holland, O.T., *Graphics and GUIs with MATLAB.* Chapman & Hall/CRC (2003)

65. Manassah, J.T., *Elementary Mathematical and Computational Tools for Electrical and Computer Engineers Using MATLAB.* CRC Press (2001)

66. Mathews, J.H., and Fink, K.D., *Numerical Methods Using* MATLAB. Prentice Hall (1999)

67. Moler, C., *Numerical Computing with* MATLAB. SIAM (2004)

68. Otto, S.R., and Denier, J.P., *An Introduction to Programming and Numerical Methods in* MATLAB. Springer (2005)

69. Quarteroni, A., and Saleri, F., *Scientific Computing with MATLAB and Octave.* Springer (2006)

70. Sayood, K., *Learning Programming Using* MATLAB. Morgan & Claypool Publ. (2007)

71. Yang, W.-Y., *Applied Numerical Methods Using* MATLAB. Wiley & Sons, Inc., 2005

Commentary

The reader can find additional reading for each chapter
in the following Bibliography:

Chapter 1: 6, 7, 14, 15, 35, 39, 51–71.
Chapter 2: 2, 3, 6, 8–11, 13, 15, 16, 18–20, 23, 24, 27, 29,
                31–33, 35, 37, 38, 45, 46, 48, 49.
Chapter 3: 6, 17, 19, 20, 22, 24, 28, 33, 37, 38, 42, 44–46, 48.
Chapter 4: 1, 8, 17–20, 32, 35, 45, 49.
Chapter 5: 7, 30, 39, 40, 51, 58.
Chapter 6: 4, 7, 11, 12, 25, 28, 30, 31, 34, 36, 40, 47.
Chapter 7: 4, 5, 11, 12, 25, 28, 30, 34, 39, 47.
Chapter 8: 5, 30, 39, 51.
Chapter 9: 5, 10, 21, 26, 27, 39, 41, 43, 50, 52.

Index

binormal, 335
boundary curves, 387
canal, 340
Cartesian product, 387
closed, 289
compact, 289
complete, 289
conchoidal, 346
conic, 332, 353
conoid, 332
coordinate curve, 387
developable, 332, 333
elementary, 286
envelope, 337
generalized sphere, 168
hodograph, 287
level, 290
normal, 335
of revolution, 323
one-sided, 288
osculating paraboloid, 304
parallel (equidistant), 343
patch, 387
pedal, 345
podoid, 345
quadric, 154
rectifying, 354
regular, 289
Riemann sphere, 100
ruled, 331
screw, 327
self-intersecting, 286
simple, 286
smooth, 289
transcendental, 329
translation, 341
tube (over a curve), 340
twist vector, 389
twisted, 343
surface name
astroidal sphere, 351
Bohemian dome, 342
catenoid, 311, 326
circular cylinder, 286, 325
cone, 286, 325
conoid of Wallis, 333
cyclide of Dupin, 350
elliptical cone, 353
Enneper, 312
helicoid, 81, 312, 327, 332
hyperbolic paraboloid, 353

hyperboloid, 325, 353
Klein bottle, 286, 355, 410
Möbius band, 288, 327, 355
monkey saddle, 351
paraboloid, 305
pinched torus, 286, 328
Plücker's conoid, 333
pseudosphere, 326
revolution of the astroid, 350
sphere, 286, 325
swallow-tail, 329, 332
torus, 81, 286, 326, 410
twisted sphere, 327
Whitney umbrella, 301, 351
wings, 319, 351

torsion of a space curve, 278
transformation
affine, 135
cissoidal, 39, 346
compression, 136
dilation, 181
expansion, 136
glide, 88
half-turn, 90
hyperbolic translation, 179
inversion, 159, 161
inversive, 163
isometry, 170
linear, 124
Lorentz, 137
Möbius, 163, 177
parabolic translation, 179
perspective, 148
Poincaré extension, 180
projection, 127
projective, 147, 148, 153
reflection, 88, 98, 153, 159, 161
rigid motion (isometry), 87, 98
rotation, 88
scaling, 136, 138
screw motion, 90
shear, 136
stereographic projection, 99
translation, 87, 88, 126
transposition, 121
triangle, 70, 98
inequality, 10, 170

vector, 16
binormal, 283

Printed in the United States
By Bookmasters

Printed in the United States
By Bookmasters